AF859290

COLLOIDAL HYDRODYNAMICS

COLLOID SCIENCE

Editors
R. H. Ottewill and R. L. Rowell

In recent years colloid science has developed rapidly and now frequently involves both sophisticated mathematical theories and advanced experimental techniques. However, many of the applications in this field require simple ideas and simple measurements. The breadth and the interdisciplinary nature of the subject have made it virtually impossible for a single individual to distill the subject for all to understand. The need for understanding suggests that the approach to an interdisciplinary subject should be through the perspectives gained by individuals.

The series consists of separate monographs, each written by a single author or by collaborating authors. It is the aim that each book will be written at a research level but will be readable by the average graduate student in chemistry, chemical engineering or physics. Theory, experiment and methodology, where necessary, are arranged to stress clarity so that the reader may gain in understanding, insight and predictive capability. It is hoped that this approach will also make the volumes useful for non-specialists and advanced undergraduates.

The author's role is regarded as paramount, and emphasis is placed on individual interpretation rather than on collecting together specialist articles.

The editors simply regard themselves as initiators and catalysts.

1. J. Mahanty and B. W. Ninham: Dispersion Forces.
2. R. J. Hunter: Zeta Potential in Colloid Science.
3. D. H. Napper: Polymeric Stabilization of Colloidal Dispersions.
4. T. G. M. van de Ven: Colloidal Hydrodynamics.

COLLOIDAL HYDRODYNAMICS

THEO G. M. VAN DE VEN

Pulp and Paper Research Institute of Canada
and
Department of Chemistry, McGill University,
Montreal, Quebec,
Canada

541.345
V217
ALVM-1

PPG INDUSTRIES, INC.
TECHNICAL INFORMATION CENTER
POB 1009/4325 ROSANNA DRIVE
ALLISON PARK, PA 15101 USA

4-12-91 USB 77⁵⁰

ACADEMIC PRESS
Harcourt Brace Jovanovich, Publishers
London San Diego New York Berkeley
Boston Sydney Tokyo Toronto

ACADEMIC PRESS LIMITED
24–28 Oval Road
London NW1 7DX

United States Edition published by
ACADEMIC PRESS INC.
San Diego, CA 92101

Copyright © 1989 by
ACADEMIC PRESS LIMITED

All Rights Reserved
No part of this book may be reproduced in any form,
by photostat, microfilm or any other means,
without written permission from the publishers

British Library Cataloguing in Publication Data

van de Ven, T. G. M.
Colloidal hydrodynamics.
1. Colloids. Chemical reactions & physical
processes. Kinetics
I. Title II. Series
541.3′451

ISBN 0–12–710770–3

Typeset by Colset Private Limited, Singapore
Printed in Great Britain by
T. J. Press (Padstow) Ltd, Padstow, Cornwall.

à Francine

Preface

Everywhere we look we find examples of fluids containing colloidal particles, e.g. in our bodies, our food, our environment. Many industrial processes deal with colloidal dispersions, and they are encountered in agricultural, pharmaceutical and medical applications. In most of these applications the colloidal particles are subjected to a hydrodynamic flow field of one kind or another when they are transported, stirred, shaken and so on. It is no wonder that there exists a great interest in the flow behaviour of colloidal dispersions, as can be witnessed by the tremendous growth in the number of articles published in the scientific literature in this field. Surprisingly, however, very few textbooks or monographs are available to guide the student or researcher through the labyrinth of the scientific literature.

It is hoped that this book will fill this gap. It is intended for graduate students in chemistry or chemical engineering who are interested in the hydrodynamic aspects of colloidal phenomena, as well as a reference book for researchers in the field.

Colloidal hydrodynamics can be considered a sub-field of physicochemical hydrodynamics, just as colloid science is considered a sub-field of physical chemistry. Besides being of interest to people working in colloid science, it is hoped that this book will also appeal to hydrodynamicists. Many fundamental problems in colloidal hydrodynamics are as yet unsolved and could be a challenge for people with a background in hydrodynamics or applied mathematics.

The mathematics required to follow the text is minimal. However, some knowledge of differential equations and some notion about tensor analysis would certainly benefit the reader. In many instances mathematical details about obtaining solutions from differential equations are omitted. Often the governing equations are stated and the solutions given without detail. This book contains no problems for students, but serious students could attempt to verify solutions when derivations are absent.

I would like to express here my thanks to the various persons who over the years have stimulated my interest in colloid science and hydrodynamics. I look back with great pleasure to my graduate years at the van 't Hoff Laboratory in Utrecht, where Professors Theo Overbeek and Agienus Vrij introduced me to colloid science and the intricacies of light scattering. The person who, undoubtedly, had the most influence on my career was my Ph.D. supervisor at McGill University, the late Professor Stanley Mason. He gave me confidence in my abilities, showed by his own example that studying science is great fun, and gave me uncountable words of advice and wisdom. Much of his research is summarized in this book and it is unfortunate that he is no longer with us. I would have loved to hear his comments! Another colleague at McGill who has influenced me greatly is Professor Raymond G. Cox who, with his marvellous combination of profound physical insight and mathematical skills, showed me the importance of mathematical rigor. During innumerable tea-breaks in our laboratory he has kept me mathematically "honest". I would like to thank Professors Bob Hunter, who introduced me to rheology, Harry Goldsmith, who aroused my interest in biological systems, Howard Brenner, who in frequent visits to Montreal strongly influenced my thinking, and Ron Ottewill, in whose laboratory most of the second half of this book was written while I was on a six-month sabbatical leave. I would also like to thank all my students and post-doc's for their stimulating discussions, and especially my Polish connection, some members of which were co-authors of a review article on particle deposition (Z. Adamczyk, T. Dabros, J. Czarnecki and T. G. M. van de Ven, *Adv. Colloid Interface Sci.* **19** (1983) 183), which forms the basis of Chapter 6 of this book, and Dr. Richard Janse van Rensburg, who critically read the manuscript, suggested various improvements and, hopefully removed most of my "Dutchisms". Finally, I would like to thank Audrey McGuinness for her patience and the care with which she typed the manuscript, and the Pulp and Paper Research Institute of Canada, whose enlightened policy it is to support people like me to do fundamental research of interest to the paper industry. Needless to say the field of colloidal hydrodynamics is extremely relevant to papermaking, since paper is made from pulp suspensions usually containing a large amount of colloidal material.

On a more personal note, remembering how my late father (who, in his peculiar, intuitive way, introduced me to probability and combinatorial theory as applied to the game of bridge, highly relevant to the topic of this book) enchanted us with his large repertoire of funny limericks, I would like to make use of this occasion to dedicate a limerick to him:

Confronted with fluid dynamics,
A student in colloids, he panics!
To help him in stopping
His latex from dropping
Is *Colloidal Hydrodynamics*

T. G. M. van de Ven

Montreal, July, 1988

Contents

Chapter 1

Basic Concepts of Fluid Dynamics and Colloidal Forces

1.1 Introduction

Traditionally, most studies in colloid science have dealt with the stability of colloidal dispersions under quiescent conditions in the absence of external forces or fields, when the driving force for bringing particles together is translational Brownian motion. All small colloidal particles are subject to Brownian motion. This is probably the most characteristic feature of colloids. Throughout this book we define colloidal particles as being either solid or liquid, whose size is within (or close to) the range 5 nm to 5 μm. Particles smaller than 5 nm are considered to be of molecular size, and larger ones do not exhibit a sufficient amount of Brownian motion to remain suspended.

Although the stability of quiescent colloidal dispersions has taught us a lot about the forces involved in interactions among colloidal particles, these studies are of limited use in many practical applications of colloidal dispersions. During various processes, transportation and end uses, these dispersions are stirred, shaken, made to flow through pipes and vessels, subjected to vibrations, etc., and, therefore, the stability of colloidal dispersions subjected to various hydrodynamic flow fields is of tremendous practical interest. On the other hand, the study of colloidal dispersions under well-defined hydrodynamic conditions can also increase our theoretical understanding of the fundamental forces and phenomena which occur in such systems.

In studying the dynamics of colloidal systems it is obviously necessary to bridge the gap between the fields of hydrodynamics and colloid science. In both fields excellent textbooks exist, but in this book we shall concentrate on the interplay between the two. However, in this first chapter the essentials of both fields are summarized in order to make the book comprehensible to readers with various backgrounds.

In the following chapters the dynamics of colloidal systems are treated systematically, starting with isolated particles in quiescent media and subsequently increasing the complexity of these systems by introducing external fields and particle interactions.

Hydrodynamic or other forces acting on colloidal particles affect their motion and transport and determine to a large extent the spatial distribution of particles in a fluid medium. We will first discuss the basic equations that describe these distributions since they clearly show the relative importance of colloidal and hydrodynamic parameters.

1.2 The number and orientation distribution of colloidal particles

Most of the properties of a dispersion can be evaluated if the position, orientation, size, shape and velocity of each particle in suspension are known. This information is usually not available and we have to be content with less. However, it is sufficient to know the position and orientation distributions, which tell us how many particles are in a given region of space and how many are pointing in a given direction, respectively. These distributions are governed by the convective-diffusion equations which will be described in more detail in Chapter 2. These equations can be written as

$$\frac{\partial n}{\partial t} = \nabla \cdot (\boldsymbol{D}_{\mathrm{t}} \cdot \nabla n) - \nabla \cdot (\boldsymbol{u} n) \tag{1.1}$$

$$\frac{\partial p}{\partial t} = \nabla \cdot (\boldsymbol{D}_{\mathrm{r}} \cdot \nabla p) - \nabla \cdot (\omega p) \tag{1.2}$$

These equations are, in essence, conservation equations which state that the increase or decrease in the number of particles having positions or orientations in a given volume element of physical or orientational space (term on left-hand side of eqs (1.1) and (1.2)) is equal to the difference between the amount that diffuses in and out of that element (first term on right-hand side) and the amount that flows in and out of it (second term on right-hand side). Equation (1.1) is the translational convective-diffusion equation and eq (1.2) is the rotary convective-diffusion equation. In eq (1.1), n is the number density distribution of particles, which is a function of position $\boldsymbol{r}$ and time t, $\boldsymbol{D}_{\mathrm{t}}$ is the translational-diffusion tensor and $\boldsymbol{u}$ the translational velocity of the particle due to flow and/or other external or colloidal forces. Analogously, in eq (1.2) p is the orientation distribution function, $\boldsymbol{D}_{\mathrm{r}}$ the rotary diffusion tensor and $\boldsymbol{\omega}$ the angular velocity of the particle. The symbol ∇ is the common differential operator, for which expressions in various coordinate systems can be found in Happel and Brenner (1983).

Equations (1.1) and (1.2) are only valid for particles with such a high degree of symmetry that any coupling between rotary and translational motion is absent (Brenner, 1974). This is the case for most particles considered in this book. Equations (1.1) and (1.2) can be applied to: (i) a system of many particles, in which case n and p are the number of particles in a given position and pointing in a given direction, respectively; or (ii) a single particle, in which case n and p denote the probability that the particle is at a given position or points in a given direction, respectively. Particle interactions can be included in eqs (1.1) and (1.2) by changing the values of the coefficients in the tensors $\boldsymbol{D}_\mathrm{t}$ and $\boldsymbol{D}_\mathrm{r}$ and in the vectors $\boldsymbol{u}$ and $\boldsymbol{\omega}$.

When trying to solve eqs (1.1) and (1.2), it is necessary to have information about $\boldsymbol{D}_\mathrm{t}$, $\boldsymbol{D}_\mathrm{r}$, $\boldsymbol{u}$ and $\boldsymbol{\omega}$. Evaluating these tensors and vectors requires, in general, a knowledge of both hydrodynamics and colloid science. Apart from the viscosity of the medium, the components of $\boldsymbol{D}_\mathrm{t}$ and $\boldsymbol{D}_\mathrm{r}$ depend, in general, on the size and shape of the suspended particles, as well as on the location and orientation of the particles with respect to neighbouring particles and bounding walls. For neutral particles, the diffusion tensors are related to the resistance tensors $\boldsymbol{f}$ by the generalized Stokes–Einstein equations:

$$\boldsymbol{D}_\mathrm{t}:\boldsymbol{f}_\mathrm{t} = kT \tag{1.3}$$

and

$$\boldsymbol{D}_\mathrm{r}:\boldsymbol{f}_\mathrm{r} = kT \tag{1.4}$$

Here k is the Boltzmann constant and T the absolute temperature. (The double dot product between two tensors results in a scalar, i.e. $\boldsymbol{D}_\mathrm{t} = \boldsymbol{f}_\mathrm{t}^{-1}kT$, etc.) The translational resistance tensor $\boldsymbol{f}_\mathrm{t}$ can be found from the hydrodynamic forces $\boldsymbol{F}$ exerted on a particle kept fixed in a uniform fluid flow of velocity $\boldsymbol{v}$ (or equivalently from the drag force experienced by a particle moving through the liquid with the same velocity):

$$\boldsymbol{F} = -\boldsymbol{f}_\mathrm{t}\cdot\boldsymbol{v} \tag{1.5}$$

Similarly, the rotary friction tensor $\boldsymbol{f}_\mathrm{r}$ can be found from the torque $\boldsymbol{L}$ experienced by a particle rotating with angular velocity $\boldsymbol{\omega}$:

$$\boldsymbol{L} = -\boldsymbol{f}_\mathrm{t}\cdot\boldsymbol{\omega} \tag{1.6}$$

In addition, the coefficients of $\boldsymbol{D}_\mathrm{t}$ and $\boldsymbol{D}_\mathrm{r}$ also depend on colloidal parameters. For colloidal particles with electrical charges on their surfaces (which is usually the case, especially in polar media), the coefficients depend on the particle surface charge density and on the nature and concentration of the ions in solution. For colloidal particles with adsorbed macromolecules, the components depend on the size and shape of these macromolecules and on their configuration on the surface.

The velocity of a particle is usually the result of many components. For a particle subjected to hydrodynamic forces only, its translational velocity is not always equal to that of the fluid. For a particle with sufficient symmetry, such as spheres, disks, rods, ellipsoids, ribbon- or brick-shaped particles, etc., in the absence of inertia effects, the centre of symmetry follows the motion of the fluid, except when it is affected by neighbouring particles and walls. Hence for most rigid particles far from walls and other particles:

$$\boldsymbol{u}_{\mathrm{hydr}} = \boldsymbol{v} \tag{1.7}$$

where $\boldsymbol{u}_{\mathrm{hydr}}$ denotes the velocity of the particle and $\boldsymbol{v}$ that of the fluid. For deformable particles, such as liquid drops, a sideways motion is superimposed upon the fluid motion. More generally, in the neighbourhood of other particles and walls:

$$\boldsymbol{u}_{\mathrm{hydr}} = \boldsymbol{g}_{\mathrm{t}} \cdot \boldsymbol{v} \tag{1.8}$$

where $\boldsymbol{g}_{\mathrm{t}}$ is a correction tensor which depends on the distance between the particle and the neighbouring particles and walls, and on orientation. The correction tensor $\boldsymbol{g}_{\mathrm{t}}$ also determines the corrections of $\boldsymbol{f}_{\mathrm{t}}$ for interactions with particles and walls.

The rotational motion $\boldsymbol{\omega}$ of a particle is determined by the hydrodynamic torque exerted by a flowing liquid. For a freely mobile particle, in the absence of inertia, the angular velocity $\boldsymbol{\omega}$ adjusts itself in such a way that the net torque on the particle is zero. Again corrections must be applied when the particle is near other particles or walls:

$$\boldsymbol{\omega}_{\mathrm{hydr}} = \boldsymbol{g}_{\mathrm{r}} \cdot \boldsymbol{\omega}_0 \tag{1.9}$$

where $\boldsymbol{g}_{\mathrm{r}}$ is a correction tensor and $\boldsymbol{\omega}_0$ the velocity in an unbounded infinitely dilute medium.

When a particle is subjected to additional forces besides hydrodynamic ones, then

$$\boldsymbol{u} = \boldsymbol{u}_{\mathrm{hydr}} + \boldsymbol{u}_{\mathrm{ext}} + \boldsymbol{u}_{\mathrm{coll}} \tag{1.10}$$

where $\boldsymbol{u}_{\mathrm{hydr}}$ is the velocity of the particle in the absence of external and colloidal forces. The velocities $\boldsymbol{u}_{\mathrm{ext}}$ and $\boldsymbol{u}_{\mathrm{coll}}$ can be evaluated from a knowledge of the external and colloidal forces acting on the particles:

$$\boldsymbol{F}_{\text{ext}} = \boldsymbol{f}_{\text{t}} \cdot \boldsymbol{u}_{\text{ext}} \tag{1.11}$$

and

$$\boldsymbol{F}_{\text{coll}} = \boldsymbol{f}_{\text{t}} \cdot \boldsymbol{u}_{\text{coll}} \tag{1.12}$$

The external force can be gravity, or can arise from applied electric, magnetic or other fields. The colloidal force $\boldsymbol{F}_{\text{coll}}$ usually consists of several contributions, such as van der Waals dispersion forces, electrostatic forces due to the overlap of ionic diffuse double layers surrounding electrically charged particles and, often, forces arising from adsorbed polymers. Some aspects of these forces are discussed in more detail in Section 1.4.

By analogy, the rotational motion of a particle depends on the hydrodynamic, external and colloidal torques exerted on it:

$$\boldsymbol{\omega} = \boldsymbol{\omega}_{\text{hydr}} + \boldsymbol{\omega}_{\text{ext}} + \boldsymbol{\omega}_{\text{coll}} \tag{1.13}$$

with $\boldsymbol{\omega}_{\text{ext}}$ and $\boldsymbol{\omega}_{\text{coll}}$ derived from the external and colloidal torques $\boldsymbol{L}_{\text{ext}}$ and $\boldsymbol{L}_{\text{coll}}$:

$$\boldsymbol{L}_{\text{ext}} = \boldsymbol{f}_{\text{r}} \cdot \boldsymbol{\omega}_{\text{ext}} \tag{1.14}$$

and

$$\boldsymbol{L}_{\text{coll}} = \boldsymbol{f} \cdot \boldsymbol{\omega}_{\text{coll}} \tag{1.15}$$

We have seen that in order to predict or describe the properties of colloidal dispersions subjected to flow and/or other fields, one must solve the translational and rotary convective-diffusion eqs (1.1) and (1.2). To solve these equations a knowledge of $\boldsymbol{D}_{\text{t}}$ and $\boldsymbol{D}_{\text{r}}$ (or $\boldsymbol{f}_{\text{t}}$ and $\boldsymbol{f}_{\text{r}}$) and $\boldsymbol{u}$ and $\boldsymbol{\omega}$ is required. All these tensors and vectors are determined by a variety of hydrodynamic and colloidal parameters. In this book we outline how the various parameters can be obtained and give examples of the properties of dispersions that can be predicted by solving the convective-diffusion equations. In the remaining part of this chapter we will deal with the underlying fluid dynamic and colloidal theories.

1.3 Basic equations of fluid dynamics

1.3.1 The continuity equation

In the above we introduced the translational convective-diffusion equation (eq (1.1)) to describe the concentration distribution of colloidal particles. In the form of a mass-balance equation it applies equally well to the concentration of fluid molecules. In the absence of diffusion, eq (1.1) reduces to

$$\frac{\partial \rho}{\partial t} + \nabla \cdot (\rho v) = 0 \tag{1.16}$$

where the concentration n has been replaced by the density ρ of the fluid. For incompressible fluids $\partial \rho / \partial t = 0$ and eq (1.16) reduces to

$$\nabla \cdot v = 0 \tag{1.17}$$

In Cartesian coordinates, eq (1.17) can be written as

$$\frac{\partial v_1}{\partial x_1} + \frac{\partial v_2}{\partial x_2} + \frac{\partial v_3}{\partial x_3} = 0 \tag{1.18}$$

where v_i is the component of the velocity in the x_i direction. Traditionally eq (1.16) or eq (1.17) is referred to as the *continuity equation.*

1.3.2 The Navier–Stokes equation

The equation of motion of a fluid is in essence a reformulation of Newton's second law which states that the force exerted on a body of mass m equals m times the acceleration $\boldsymbol{a}$. For a fluid volume V, the mass (or density) and acceleration can, in principle, vary from place to place. The volume V can be divided into infinitesimally small volumes $\mathrm{d}V$ over which m (or ρ) and $\boldsymbol{a}$ are all constant and subsequently integrated over the whole volume. Hence for a liquid

$$\boldsymbol{F} = m\,\boldsymbol{a} = \int_V \rho\,\boldsymbol{a}\,\mathrm{d}V \tag{1.19}$$

The force $\boldsymbol{F}$ exerted on a fluid volume V usually consists of two distinct contributions: (i) a body force which is exerted on the total volume, and (ii) a surface force which is exerted on the boundary A of volume V. Hence

$$\boldsymbol{F} = \boldsymbol{F}_{\text{body}} + \boldsymbol{F}_{\text{surf}} \tag{1.20}$$

Common body forces are gravity or external electric or magnetic fields. Defining $\boldsymbol{F}_{\text{ext}}$ as the body force per unit mass we can write

$$\boldsymbol{F}_{\text{body}} = \int_V \rho \boldsymbol{F}_{\text{ext}} \mathrm{d}V \tag{1.21}$$

The surface forces can be found by considering the stresses exerted on boundary A of the fluid:

$$\boldsymbol{F}_{\text{surf}} = \int_A \Pi \cdot \mathrm{d}\boldsymbol{A} \tag{1.22}$$

where Π is the *stress tensor* which specifies the local normal and tangential forces per unit area exerted on a fluid element, and $\mathrm{d}\boldsymbol{A}$ is a vector pointing perpendicularly to the surface, its magnitude determined by the surface area of the infinitesimal element $\mathrm{d}A$. The product of Π and $\mathrm{d}\boldsymbol{A}$ integrated over the total boundary area yields the total surface force exerted on V.

Using Green's theorem, eq (1.22) can be written as

$$\boldsymbol{F}_{\text{surf}} = \int_V \nabla \cdot \Pi \mathrm{d}V \tag{1.23}$$

By combining eqs (1.19) to (1.23), the equation of motion for a fluid can be written as

$$\rho \boldsymbol{a} = \nabla \cdot \Pi + \rho \boldsymbol{F}_{\text{ext}} \tag{1.24}$$

The acceleration $\boldsymbol{a}$ occurring in the above expression requires some comment. An acceleration is defined as a change in velocity with time. For a fluid there are two ways in which a flow field can be specified. One way is to specify the fluid velocity $\boldsymbol{v}$ at every point $\boldsymbol{r}$ and time t. Knowledge of $\boldsymbol{v}(\boldsymbol{r},t)$ provides a spatial distribution of the fluid velocity. Specifying $\boldsymbol{v}(\boldsymbol{r},t)$ is known as the *Eulerian* description.

Alternatively, the fluid can be divided into a large number of material elements and their positions specified at some initial time t_0. Fluid motion can then be specified by the location r_i of the centre of mass of each of these elements at subsequent times. This is called the *Lagrangian* description which provides the material velocity $\boldsymbol{U}[\boldsymbol{r}(\mathrm{t})]$. Both descriptions are equivalent. We will use both descriptions several times in subsequent chapters when we describe the motion of colloidal particles.

From the derivation of eq (1.24) it is clear that we are interested in the acceleration of a material fluid element. Hence

$$\boldsymbol{a} = \frac{\partial \boldsymbol{U}}{\partial t} = \frac{\partial \boldsymbol{v}}{\partial t} + \boldsymbol{v} \cdot \nabla \boldsymbol{v} \tag{1.25}$$

where the right-hand side expresses the acceleration in Eulerian velocities. The two terms on the right-hand side of eq (1.25) can be interpreted as the local rate of change in velocity ($\partial \boldsymbol{v}/\partial t$) due to changes at position $\boldsymbol{r}$ and a convective rate of change in velocity ($\boldsymbol{v} \cdot \nabla \boldsymbol{v}$) due to the transport of the element to a different position. The tensor $\nabla \boldsymbol{v}$ is called the gradient of the velocity field. In Cartesian coordinates it can be written as

$$\nabla \boldsymbol{v} = \begin{vmatrix} \dfrac{\partial v_1}{\partial x_1} & \dfrac{\partial v_2}{\partial x_1} & \dfrac{\partial v_3}{\partial x_1} \\[2ex] \dfrac{\partial v_1}{\partial x_2} & \dfrac{\partial v_2}{\partial x_2} & \dfrac{\partial v_3}{\partial x_2} \\[2ex] \dfrac{\partial v_1}{\partial x_3} & \dfrac{\partial v_2}{\partial x_3} & \dfrac{\partial v_3}{\partial x_3} \end{vmatrix} \tag{1.26}$$

The gradient tensor $\nabla \boldsymbol{v}$ can be split into symmetric and antisymmetic parts:

$$\nabla \boldsymbol{v} = \tfrac{1}{2}(\nabla \boldsymbol{v} + \nabla \boldsymbol{v}^{\dagger}) + \tfrac{1}{2}(\nabla \boldsymbol{v} - \nabla \boldsymbol{v}^{\dagger}) \tag{1.27}$$

where $\nabla \boldsymbol{v}^{\dagger}$ is the transpose of $\nabla \boldsymbol{v}$. Equation (1.27) can be rewritten as

$$\nabla \boldsymbol{v} = \boldsymbol{S} + \Lambda \tag{1.28}$$

where $\boldsymbol{S}$ is the *rate-of-strain tensor* which describes fluid deformation, and Λ is the *vorticity tensor* which corresponds to a solid-body rotation.

To make use of eq (1.24), we still have to find an expression for the stress tensor $\boldsymbol{\Pi}$. In the absence of fluid flow, the stresses in a fluid are caused by the static fluid pressure which is isotropic, i.e.

$$\Pi = -p\boldsymbol{I} \tag{1.29}$$

where p is the hydrostatic pressure and $\boldsymbol{I}$ the idem or unit tensor. In matrix notation

$$\Pi = -\begin{pmatrix} p & 0 & 0 \\ 0 & p & 0 \\ 0 & 0 & p \end{pmatrix} = -p\begin{pmatrix} 1 & 0 & 0 \\ 0 & 1 & 0 \\ 0 & 0 & 1 \end{pmatrix}$$

When fluid flow occurs, the stress tensor usually becomes anisotropic and can be expressed as the sum of an isotropic pressure p (usually different from the static pressure) and a non-isotropic contribution. For incompressible Newtonian liquids this non-isotropic contribution is given by $2\mu\boldsymbol{S}$, where μ is the viscosity of the fluid. This proportionality between the stress and the rate of strain is a phenomenological relationship, which many fluids have been found to obey. Newtonian liquids are defined as liquids for which this relationship holds. Hence for Newtonian liquids (to which this book is restricted)

$$\Pi = -p\boldsymbol{I} + 2\mu\boldsymbol{S} \tag{1.30}$$

Taking the divergence (and using $\nabla v = 0$) gives:

$$\nabla \cdot \Pi = -\nabla p + \mu \nabla^2 v \tag{1.31}$$

Substituting eqs (1.25) and (1.31) into eq (1.24) gives the equation of motion:

$$\rho\left(\frac{\partial v}{\partial t} + v \cdot \nabla v\right) = -\nabla p + \mu \nabla^2 v + \rho \boldsymbol{F}_{\text{ext}} \tag{1.32}$$

Equation (1.32) is referred to as the Navier–Stokes equation which, together with the continuity equation (1.17), describes the fluid motion of incompressible Newtonian liquids. When temperature gradients are present, μ and ρ must be specified as functions of temperature. This introduces an additional term, proportional to $\nabla\mu$ in the Navier–Stokes equation.

In order to solve eqs (1.17) and (1.32) for the fluid velocity v, the initial and boundary conditions must be specified. For steady-state problems, for which the term $\partial v/\partial t$ has reached a zero value, only the boundary conditions need specification. For flow through channels or other vessels, one can specify the flux or volume flow rate through it, together with the condition of zero fluid velocity at the bounding walls. This boundary condition is called the no-slip (or stick) boundary condition, i.e. the velocity of fluid neighbouring a solid surface is the same as the velocity of that surface. For the flow around a colloidal particle it is usually assumed that far from the particle (at "infinity") the flow equals the undisturbed flow field, while at the particle's surface the no-slip boundary condition again applies. For flow around a fluid drop, at the drop surface, continuity of normal and tangential stresses over the liquid/fluid interface is required. For very small particles or ions, the no-slip boundary condition no longer applies and corrections for slip must be made.

1.3.3 Low Reynolds number fluid dynamics

The Navier–Stokes equation is a highly non-linear equation and complete solutions of it exist only for simple geometries. It is often possible to simplify the equation considerably. In many problems it is sufficient to consider steady fluid motion only, in which the term $\partial v/\partial t$ can be neglected. It is possible to write eq (1.32) in non-dimensional form by introducing a reference length scale ℓ and velocity v_0. (At the same time this introduces a reference time scale of ℓ/v_0.) Defining the dimensionless parameters

$$\tilde{x}_i = x_i/\ell\ ;\ \tilde{v} = v/v_0 \text{ and } \tilde{p} = \ell p/\mu v_0$$

and substituting them into eq (1.32), neglecting the non-steady term as well as external forces, gives

$$\mathrm{Re}\,\tilde{v}\cdot\tilde{\nabla}\tilde{v} = -\ \tilde{\nabla}\tilde{p} + \tilde{\nabla}^2\tilde{v} \tag{1.33}$$

where the tildes refer to non-dimensional parameters and operators. Re is the *Reynolds number*, defined as

$$\mathrm{Re} = \frac{\ell v_0 \rho}{\mu} \tag{1.34}$$

and can be interpreted as the ratio of inertial to viscous forces.

Two different flow regimes can be distinguished, depending on the magnitude of the value of Re. When $\mathrm{Re} \gg 1$, the two terms on the right-hand side of eq (1.33) can be neglected, resulting in the so-called potential flow, except for regions near solid surfaces where the viscous terms cannot be neglected. These regions are referred to as hydrodynamic boundary layers. For Reynolds numbers above a critical value the flow is turbulent and the flow must be described by statistical laws. The critical Reynolds number depends on the geometry of the flow. At low Reynolds number the flow is laminar.

When $\mathrm{Re} \ll 1$, the term $\mathrm{Re}\,\tilde{v}\cdot\tilde{\nabla}\tilde{v}$ in eq (1.33) can be neglected and the Navier–Stokes equation reduces to the Stokes or *creeping-flow* equation. In dimensional form eq (1.33) can be written as

$$\mu\nabla^2 v = \nabla p \tag{1.35}$$

or, in Cartesian coordinates,

$$\begin{aligned}
\mu\left(\frac{\partial^2 v_1}{\partial x_1^2}+\frac{\partial^2 v_1}{\partial x_2^2}+\frac{\partial^2 v_1}{\partial x_3^2}\right) &= \frac{\partial p}{\partial x_1}\\
\mu\left(\frac{\partial^2 v_2}{\partial x_1^2}+\frac{\partial^2 v_2}{\partial x_2^2}+\frac{\partial^2 v_2}{\partial x_3^2}\right) &= \frac{\partial p}{\partial x_2}\\
\mu\left(\frac{\partial^2 v_3}{\partial x_1^2}+\frac{\partial^2 v_3}{\partial x_2^2}+\frac{\partial^2 v_3}{\partial x_3^2}\right) &= \frac{\partial p}{\partial x_3}
\end{aligned} \tag{1.36}$$

The creeping-flow equation is particularly useful in describing the flow around colloidal particles because in this case the characteristic length ℓ occurring in the Reynolds number can be identified with the particle size. Even for a colloidal particle as large as 1 μm in water, assuming $\rho \simeq 10^3$ kg m^{-3}, the Reynolds number approaches unity at velocities of about 1 m s^{-1}. Hence for colloidal particles the condition Re $\ll 1$ is almost always satisfied. It must be pointed out that even when the Reynolds number based on the size of a colloidal particle is usually less than one, the Reynolds number based on some macroscopic dimension of the flow (e.g. the diameter of a tube) can be much larger than one. Even in turbulent flow, if the particle is smaller than the microscale of the turbulence, the local flow in the neighbourhood of a colloidal particle is laminar even though the "distant" velocity it experiences is a fluctuating function of time. This is usually the case.

Further simplifications of the creeping-flow equation can be made for flow geometries that are either two-dimensional or axisymmetric. In such cases one can introduce the concept of a stream function. For two-dimensional flow the velocity $\boldsymbol{v}$ can be split into two components v_1 and v_2 in the X_1 and X_2 directions, respectively ($v_3 = 0$). A stream function ψ is then defined by the equations

$$\frac{\partial \psi}{\partial x_2} = v_1\,;\quad \frac{\partial \psi}{\partial x_1} = -v_2 \tag{1.37}$$

The continuity equation for this case

$$\frac{\partial v_1}{\partial x_1}+\frac{\partial v_2}{\partial x_2}=0 \tag{1.38}$$

can be written as

$$\frac{\partial^2\psi}{\partial x_2\partial x_1} - \frac{\partial^2\psi}{\partial x_1\partial x_2} = 0 \tag{1.39}$$

and is automatically satisfied for all ψ.

The creeping-flow equation reduces to

$$\nabla^4\psi = 0 \tag{1.40}$$

where $\nabla^4\psi = \nabla^2(\nabla^2\psi)$, as can be seen by differentiating $\partial p/\partial x_1$ by x_2 and differentiating $\partial p/\partial x_2$ by x_1 and subtracting the two terms.

In spherical coordinates (r, ϕ), ψ can be defined as

$$v_r = -\frac{1}{r^2\sin\phi}\frac{\partial\psi}{\partial\phi}\ ;\ v_\phi = \frac{1}{r\sin\phi}\frac{\partial\psi}{\partial r} \tag{1.41}$$

and the creeping-flow equation reduces to

$$E^4\psi = 0 \tag{1.42}$$

where

$$E^2 = \frac{\partial}{\partial r^2} + \frac{\sin\phi}{r^2}\frac{\partial}{\partial\phi}\left(\frac{1}{\sin\phi}\frac{\partial}{\partial\phi}\right) \tag{1.43}$$

1.3.4 Simple flow geometries

1.3.4.1 Poiseuille flow

Let us consider the flow through a circular tube or a parallel-plate channel. To obtain the velocity profile in such a channel one must solve the Navier–Stokes equation together with the continuity equation, with the no-slip boundary condition on the channel wall. Consider first a parallel-plate channel (see Fig. 1.1). Choosing the origin on the channel axis of symmetry, it follows from the geometry that the velocities in the X_1 and X_3 directions (X_3 directed towards the viewer) equal zero. The only non-zero velocity is in the X_2 direction. Here we will restrict ourselves to creeping flow and hence the flow is given by

$$\nabla^4\psi = 0 \tag{1.44}$$

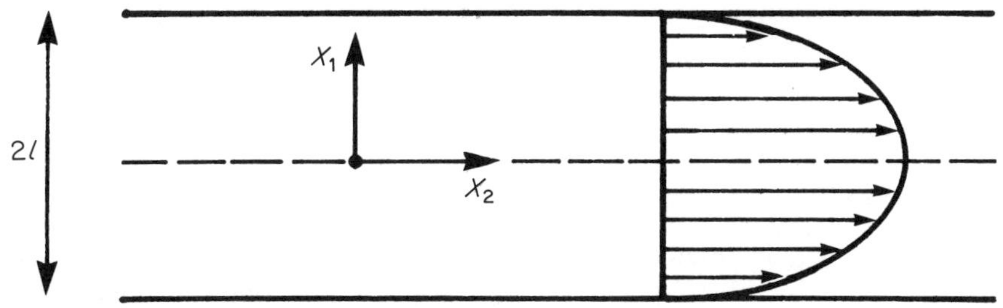

Fig. 1.1. Coordinate system for describing the flow profile in a flat channel of width 2ℓ.

which, as we have already seen, satisfies the continuity equation. Because $v_1(=\partial\psi/\partial x_2)=0$, it follows from the definition of ψ that ψ is a function of x_1 only. Therefore, eq (1.44) can be written as

$$\frac{d^4\psi}{dx_1^4}=0 \tag{1.45}$$

the solution of which is

$$\psi = Ax_1^3 + Bx_1^2 + Cx_1 + D \tag{1.46}$$

Equation (1.46) contains four constants of integration because eq (1.45) is a fourth-order differential equation. From eq (1.46) it follows that

$$v_2 = -\frac{\partial\psi}{\partial x_1} = -(3Ax_1^2 + 2Bx_1 + C) \tag{1.47}$$

The constants appearing in eq (1.47) can be determined from the no-slip boundary condition and from the volumetric flow rate through the channel. Let the distance between the origin and the channel walls be ℓ, then according to the no-slip boundary condition:

$$v_2(-\ell)=0\ ;\ v_2(\ell)=0\ ,$$

from which it follows that $B=0$ and $C=3A\ell^2$. Hence

$$v_2 = 3A\ell^2\left[1-\left(\frac{x_1}{\ell}\right)^2\right] \tag{1.48}$$

Equation (1.48) represents a parabola with the maximum velocity, v_{max}, occurring at $x_1 = 0$. Hence we can write

$$v_2 = v_{max}\left[1 - \left(\frac{x_1}{\ell}\right)^2\right] \tag{1.49}$$

The velocity v_{max} is related to the amount of fluid passing through the channel. A volume of $2\bar{v}\ell$ per unit length is passing through the channel, $\bar{v}$ being the average fluid velocity:

$$\bar{v} = \frac{1}{2\ell}\int_{-\ell}^{\ell} v_2 \mathrm{d}x_1 = \frac{2}{3} v_{max} \tag{1.50}$$

Similarly, the fluid velocity in a circular tube of radius R is given by

$$v = v_{max}\left[1 - \left(\frac{r}{R}\right)^2\right] \tag{1.51}$$

where r is the distance from the tube axis. For a circular tube

$$\pi R^2 \bar{v} = \int_0^R 2\pi r v \, \mathrm{d}r = \frac{1}{2}\pi R^2 v_{max}$$

or

$$\bar{v} = \frac{1}{2} v_{max} \tag{1.52}$$

The parabolic velocity profile in tubes and channels is usually called *Poiseuille flow*. As a matter of fact, eqs (1.49) and (1.51) are not only solutions of the creeping-flow equation, but also of the complete Navier–Stokes equation. These equations apply up to the critical Reynolds number at which the flow becomes turbulent ($\mathrm{Re}_{crit} \simeq 2000$).

Let us now consider the flow of a dilute suspension of colloidal particles through a channel. Neglecting gravity (or buoyancy), which causes the particles to sediment or to rise (depending on whether the density of the particles is larger or smaller than that of the medium), all spheres will follow the fluid streamlines (when $\mathrm{Re} \ll 1$) and have the same velocity as the fluid would have at the centre if the particles were absent, except for particles close to the wall which are somewhat slowed down. The centres of non-spherical particles with sufficient symmetry (triaxial particles such as ellipsoids and brick-shaped particles, and biaxial particles such as cylinders, rods and disks) also follow the fluid streamlines (except near walls). Particles with less symmetry either oscillate around a streamline with a constant period (axisymmetric particles without fore–aft symmetry, e.g. a doublet of unequal sized spheres), or drift sideways, as do liquid drops. In addition, inertia causes a drift across streamlines, but for colloidal particles the condition $Re \ll 1$ is almost always satisfied because of the small size of the particles.

Because the velocity of particles flowing through a channel depends on their position in the channel, not all particles have the same velocity and, as a result, collisions occur between particles, the faster-moving particles colliding with slower-moving particles closer to the wall. The number of collisions is determined by the relative velocities between particles which in turn are determined by the velocity gradient of the flow. In tube flow, from eq (1.51), the velocity gradient is given by

$$G = \left|\frac{\mathrm{d}v}{\mathrm{d}r}\right| = \frac{2v_{\max} r}{R^2} = \frac{\bar{v} r}{R^2} \tag{1.53}$$

The velocity gradient G is often referred to as the *shear rate*. It has the dimensions of inverse seconds (s^{-1}). In Poiseuille flow the shear rate is zero at the tube centre and maximum at the tube wall ($G_{\max} = \bar{v}/R$).

Two particles, at distances r and $r + \delta r$ from the axis, interacting during a collision will be subjected to shear rates of $\bar{v}r/R^2$ and $\bar{v}(r+\delta r)/R^2$, respectively. When two particles of radius a collide, $\delta r \lesssim 2a$ and, usually, $a \ll r$. Hence collisions can be considered to occur at constant G. Therefore, in order to understand collisions or interactions among colloidal particles in Poiseuille flow, it is sufficient to understand the interactions in flows with constant G. Such flows are called simple shear flows; they are of fundamental importance in colloidal hydrodynamics.

1.3.4.2 Simple shear flow

A simple shear flow can be generated by moving two parallel plates, separated by a fluid, in opposite directions (see Fig. 1.2). If one plate is moved with velocity u and the second with velocity $-u$, the velocity gradient is

$$G = \frac{\Delta u}{2\ell} = \frac{u}{\ell}$$

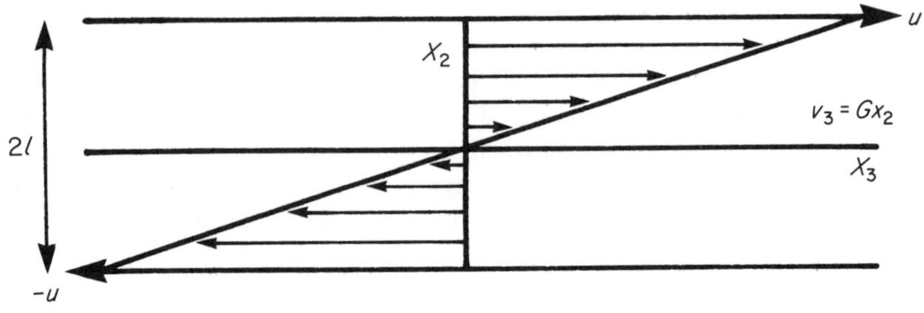

Fig. 1.2. Simple shear flow, created by moving two plates, a distance 2ℓ apart, in opposite directions with velocity u. At any distance x_2 between the plates, the velocity is given by $v_3 = Gx_2$. The X_1 axis (not shown) is directed towards the viewer.

where 2ℓ is the distance between the plates. Hence a simple shear flow can be represented by

$$v_1 = v_2 = 0\ ;\ v_3 = Gx_2 \tag{1.54}$$

It can readily be shown that this flow field satisfies the creeping-flow equation with the no-slip boundary condition.

Experimentally, a simple shear flow can be more easily created in the gap between two concentric cylinders. Rotating one or both cylinders creates a simple shear flow in the gap, provided the radius of the cylinders is much larger than the gap width (see Fig. 1.3a). For one cylinder rotating with angular velocity ω, the shear rate is given by $G \simeq \omega r/d$, where r is the cylinder radius and d the gap width ($d/r \ll 1$).

Another experimental technique for creating a simple shear flow is the cone and plate technique (see Fig. 1.3b). At a distance r from the centre and with either the cone or the plate rotating with angular velocity ω, the rate of shear is $G = \omega r/d$, where d is the gap width at distance r. For a cone and plate, $r/d = \cot\alpha$, where α is the cone angle, and G is therefore constant throughout the cell.

In simple shear, the velocity gradient tensor ∇v takes the simple form (cf. eq (1.54)):

$$\nabla v = G \begin{pmatrix} 0 & 0 & 0 \\ 0 & 0 & 1 \\ 0 & 0 & 0 \end{pmatrix}$$

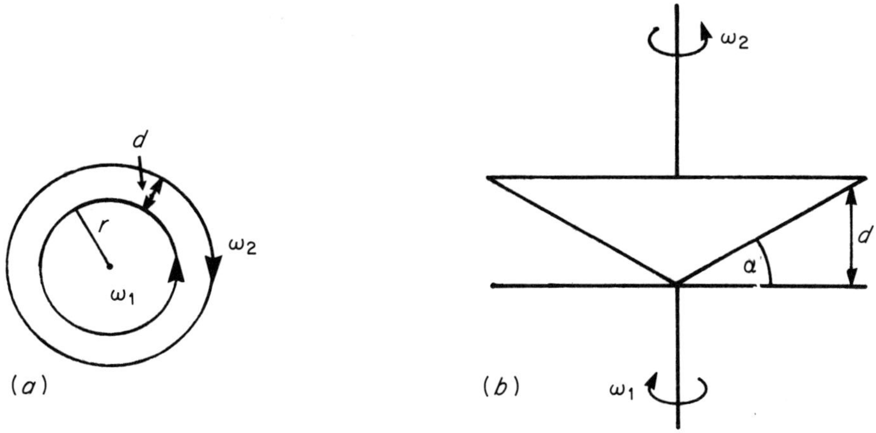

Fig. 1.3. Experimental devices for creating simple shear flow.
(*a*) Couette apparatus: $G \simeq (\omega_1 - \omega_2) r/d$ when $r \gg d$.
(*b*) Cone and plate apparatus: $G \simeq (\omega_1 - \omega_2)\cot\alpha$.

or

$$\nabla v = \tfrac{1}{2} G \begin{pmatrix} 0 & 0 & 0 \\ 0 & 0 & 1 \\ 0 & 1 & 0 \end{pmatrix} + \tfrac{1}{2} G \begin{pmatrix} 0 & 0 & 0 \\ 0 & 0 & 1 \\ 0 & -1 & 0 \end{pmatrix}$$

i.e. (cf. eqs (1.27) and (1.28))

$$\nabla v = \boldsymbol{S} + \Lambda$$

where $\boldsymbol{S}$ is the rate-of-strain tensor and Λ the vorticity tensor. The velocity fields given by $\boldsymbol{S}$ and Λ are shown schematically in Fig. 1.4. It can be seen that a simple shear flow can be considered as a superposition of a pure straining and a pure rotational flow. As a result, suspended particles are under tension when $\phi = 45°$ (see Fig. 1.4), under compression when $\phi = -45°$ and execute a rotation around the vorticity axis X_1. This axis is called the vorticity axis because the rotational motion is around this axis only. The components Λ_{ij} of Λ can be written as

$$\Lambda_{ij} = -\epsilon_{ijk}\omega_k$$

where ω_k are the components of the angular velocity vector ω, and ϵ_{ijk} is the alternating unit vector. From the definition of Λ it follows that

$$\omega = \tfrac{1}{2} \nabla x v \qquad (1.55)$$

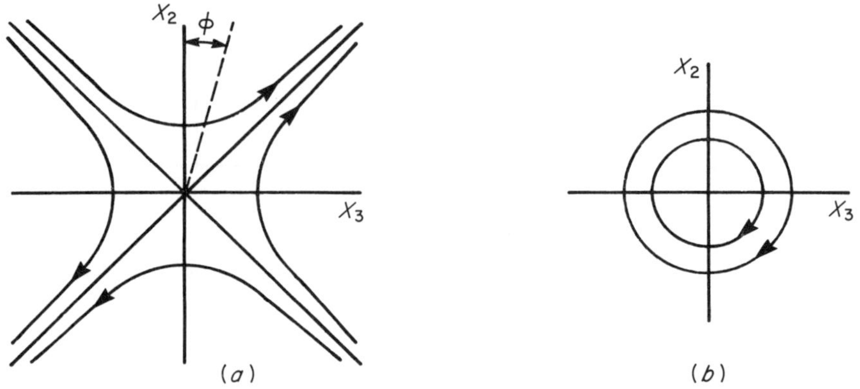

Fig. 1.4. The flow fields given by a pure straining motion (*a*) and a pure rotational motion (*b*)

and for simple shear $\omega = (\frac{1}{2}G, 0, 0)$. The vorticity can be interpreted as the angular velocity of a small spherical tracer particle suspended in the fluid. For simple shear, such a particle will rotate with angular velocity $\frac{1}{2}G$ about the vorticity axis X_1.

1.3.4.3 Flow around a translating sphere

A standard problem in hydrodynamics and colloid science is finding the velocity of a particle on which one or several forces are acting. Here we consider the simple example of a sphere which moves under the action of a constant external force (e.g. gravity) through a quiescent fluid (see Fig. 1.5).

The moving sphere creates a disturbance flow in its surrounding neighbourhood which becomes zero at large distances. We assume that after some time the sphere will have reached a steady-state velocity, u. In order to determine u, we must find the fluid flow field v around the sphere and calculate the stresses exerted on the sphere surface by the fluid. When $\mathrm{Re} \ll 1$, we must solve the creeping-flow equation (eq (1.42))

$$E^4\psi = 0 \tag{1.56}$$

with the appropriate boundary conditions.

A trial solution of eq (1.56), suggested by the geometry of the problem, is

$$\psi = f(r)\sin^2\phi \tag{1.57}$$

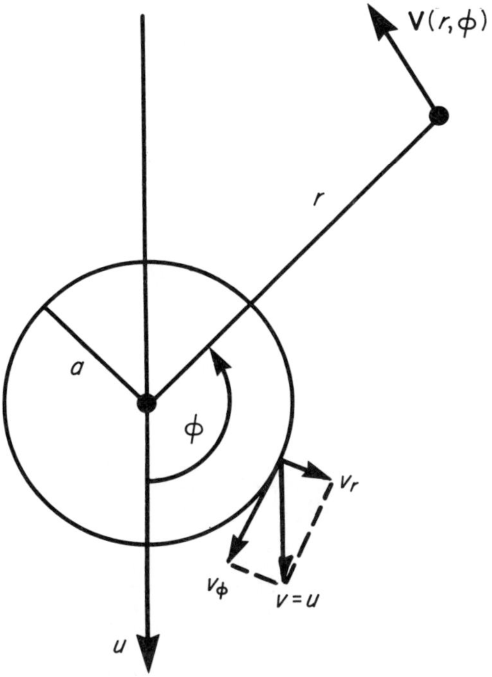

Fig. 1.5. The geometry of a translating sphere with velocity u. The fluid velocity v is a function of the radial distance r and the angle ϕ. At the surface the no-slip boundary condition requires that $v_r = u\cos\phi$ and $v_\phi = -u\sin\phi$.

Substituting this solution into eq (1.56) gives

$$E^2\psi = \sin^2\phi \left(f'' - \frac{2f}{r^2}\right) = g(r) \sin^2\phi \tag{1.58}$$

and

$$E^4\psi = \sin^2\phi \left(g'' - \frac{2g}{r^2}\right) = 0 \tag{1.59}$$

Hence

$$g'' - \frac{2g}{r^2} = 0 \tag{1.60}$$

which has the solution

$$g(r) = Ar^2 + \frac{B}{r} \tag{1.61}$$

Substitution into eq (1.58) gives

$$f'' - \frac{2f}{r^2} = Ar^2 + \frac{B}{r} \tag{1.62}$$

which has the solution

$$f(r) = \frac{A}{10} r^4 + Cr^2 - \frac{B}{2} r + \frac{D}{r} \tag{1.63}$$

From eqs (1.41), (1.57) and (1.63) it follows that the velocity field is given by

$$v_{\mathrm{r}} = -2\left(\frac{Ar^2}{10} + C - \frac{B}{2r} + \frac{D}{r^3}\right)\cos\phi \tag{1.64a}$$

$$v_{\phi} = \left(\frac{2Ar^2}{5} + 2C - \frac{B}{2r} - \frac{D}{r^3}\right)\sin\phi \tag{1.64b}$$

The constants A, B, C and D must be determined from the boundary conditions. Far from the sphere, the disturbance flow must reduce to zero, hence

$$v_{\mathrm{r}} = v_{\phi} = 0 \text{ at } r = \infty$$

from which it follows that $A = C = 0$. At the surface the no-slip boundary condition requires that (cf. Fig. 1.5)

$$v_{\mathrm{r}}(a,\phi) = u\cos\phi \tag{1.65a}$$

$$v_{\phi}(a,\phi) = -u\sin\phi \tag{1.65b}$$

from which it follows that

$$\frac{B}{a} - \frac{2D}{a^3} = u \text{ and } \frac{B}{2a} + \frac{D}{a^3} = u \tag{1.66a}$$

and thus

$$B = \frac{3}{2} ua \text{ and } D = \frac{1}{4} ua^3 \tag{1.66b}$$

Hence

$$v_r = \frac{1}{2} u \cos\phi \left(\frac{a}{r}\right)^2 \left(\frac{3r}{a} - \frac{a}{r}\right) \tag{1.67a}$$

and

$$v_\phi = -\frac{1}{4} u \sin\phi \left(\frac{a}{r}\right) \left[\left(\frac{a}{r}\right)^2 + 3\right] \tag{1.67b}$$

From the flow field given by eq (1.67), the stress tensor can be calculated from eq (1.30), from which the hydrodynamic force on the sphere can be found from eq (1.22). Performing these algebraic calculations gives

$$F = -6\pi\mu a u \tag{1.68}$$

This is the famous Stokes drag force on a sphere.

1.4 Nature of colloidal forces

1.4.1 Introduction

One of the basic problems in colloid science is to find the net force acting between two (or more) colloidal particles. The net force determines the stability of the colloidal system. If the net force is attractive, particles will come together and form clusters or aggregates, or, in the case of liquid particles, will coalesce, and the system will be unstable. If the net force is repulsive, the system is stable and all particles move around as individual entities. The problem is complicated by the fact that the net force between two particles is often attractive at some distances but repulsive at others. This results in energy barriers preventing the particles from reaching the most energetically favourable distance, which is usually close contact. Dispersions in which particles are prevented from making contact by energy barriers can, for practical purposes, be regarded as stable, although they are unstable in a thermodynamic sense. The height of the barrier can be regarded as a measure of stability: a low barrier results in rather unstable dispersions while a high barrier results in relatively stable ones. Besides dispersions that are thermodynamically unstable, others exist that are thermodynamically stable. Examples are association colloids (micelles, microemulsions, etc.) and, in some cases, particles with polymers anchored to their surface. In such systems entropic contributions to the free energy of interaction can cause thermodynamic stability. Thermodynamically stable and unstable dispersions are also called lyophilic and lyophobic dispersions, respectively.

Here we shall deal mainly with lyophobic colloids, with the exception of polymer-coated colloids.

There are many different forces that determine the stability of colloidal dispersions. The most important ones are van der Waals dispersion forces, electrostatic forces and forces due to adsorbed polymer interactions. Additional forces that are sometimes important are solvation and hydrophobic forces. In the following sections the essential nature of these forces is outlined. Several textbooks on colloid science are available for more details (see, for example, Hiemenz, 1977, 1986; Hunter 1987).

1.4.2 Van der Waals dispersion forces

Van der Waals dispersion forces are forces acting between neutral atoms or molecules. Three different types of this force can be distinguished, depending on the nature of the molecules. If both molecules have a permanent electric dipole moment, then there is attraction or repulsion between the molecules, depending on how the molecules are oriented. Parallel dipoles will repel each other, while antiparallel dipoles will attract one another. By averaging over all orientations, it is found that a net attraction exists between permanent-dipole molecules. This is due to the fact that some dipole orientations are more likely to occur than others because dipoles spend more time in them. In particular, antiparallel orientations are favoured over parallel ones. This attraction between permanent dipoles is called *Keesom attraction*, after Keesom who first calculated it. He found

$$V_{\text{att}} = -\frac{\mu^4}{(4\pi\epsilon)^2 kTr^6} \tag{1.69}$$

where μ is the dipole moment, ϵ the permittivity of the medium, kT the thermal energy, and r the distance between the molecules.

Debye calculated that a net attraction also exists between a permanent dipole and a non-polar molecule or atom. The permanent dipole induces a dipole in the non-polar molecule and the permanent and induced dipoles attract each other. This results in the *Debye attraction*:

$$V_{\text{att}} = -\frac{\alpha\mu^2}{(4\pi\epsilon)^2 r^6} \tag{1.70}$$

where α is the polarizability of the non-polar molecule.

The third type of van der Waals dispersion energy is between two neutral non-polar molecules or atoms and is called *London dispersion*. This force arises from the fact that at every moment in time an atom or molecule can be considered to have a dipole. The dipole, however, continuously fluctuates in direction and magnitude. Such an instantaneous dipole will induce a dipole in a neighbouring molecule, resulting in an attraction. A quantum-mechanical calculation of this interaction gives:

$$V_{\text{att}} = -\frac{3\alpha^2 h\nu_0}{64\pi^2\epsilon^2 r^6} \tag{1.71}$$

where $\frac{1}{2}h\nu_0$ is the ground-state energy of the atom or molecule, h is Planck's constant, and ν_0 is the frequency of vibration.

All three types of dispersion forces, or their combinations, can be written as

$$V_{\text{disp}} = -\frac{\lambda}{r^6} \tag{1.72}$$

In deriving eq (1.72) it is tacitly assumed that the electromagnetic interactions between atoms and molecules act instantaneously. In reality it takes a certain time for an electromagnetic signal to travel back and forth between two molecules. When an electromagnetic signal returns, it will find the original emitting molecule with a dipole moment that has changed direction and magnitude. Because of this effect, the dispersion force is weakened or retarded when the molecules are far apart. In such cases

$$V_{\text{disp}} \propto \frac{1}{r^7} \tag{1.73}$$

For two colloidal particles, each consisting of a large number of molecules, the dispersion energy can be obtained by adding the contributions of all the possible pairs of atoms. This can be achieved by integrating over the volumes of both particles, as shown schematically in Fig. 1.6, and hence the attractive energy between two colloidal particles is

$$V_{\text{disp}} = -\lambda n^2 \int_{V_1}\int_{V_2} \frac{\mathrm{d}V_1\,\mathrm{d}V_2}{r^6} \tag{1.74}$$

where n is the number of molecules per unit volume.

For two equal spheres eq (1.74) gives

$$V_{\text{disp}} = -\frac{A}{6}\left[\frac{2a^2}{h^2+4ah} + \frac{2a^2}{(h+2a)^2} + \ln\frac{h^2+4ah}{(h+2a)^2}\right] \tag{1.75}$$

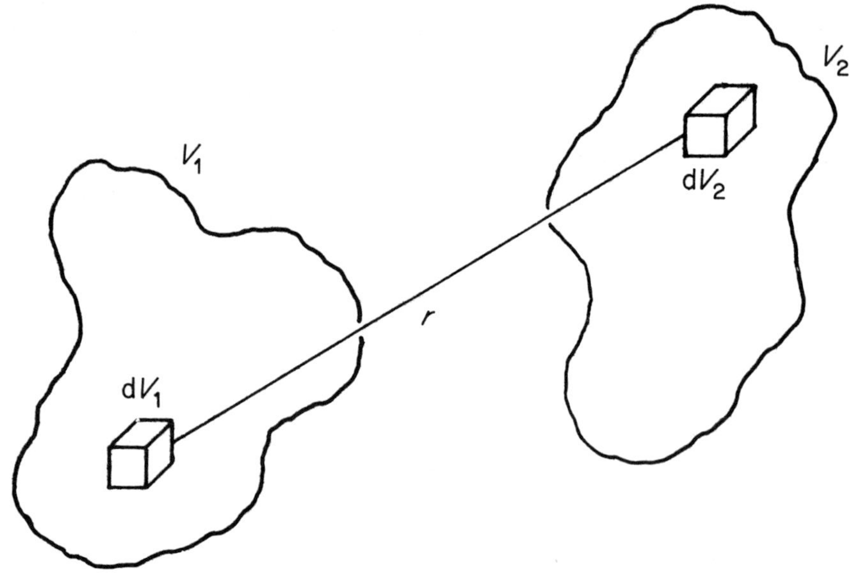

Fig. 1.6. The dispersion interaction between volume elements dV_1 and dV_2 equals $-n^2(\lambda/r^6)dV_1dV_2$. The total interaction can be found by integrating over volumes V_1 and V_2.

where h is the gap between the spheres and A is the *Hamaker constant*, $A = \pi^2\lambda n^2$. Typically, A is of the order of 10^{-19}–10^{-21}J. At large distances, eq (1.75) is inaccurate because of retardation effects. A useful approximation for the van der Waals force between two spherical particles of radii a_1 and a_2, that can be used for all distances, is (Schenkel and Kitchener, 1960):

$$F_{\text{disp}} = -\frac{\partial V_{\text{disp}}}{\partial h} = -\frac{Aa_1a_2}{6h^2(a_1+a_2)}f(p) \tag{1.76}$$

where $f(p) = \dfrac{1+3\cdot 54p}{1+1\cdot 77\,\text{p}}, \; p < 1$

and $f(p) = \dfrac{0.98}{p} - \dfrac{0\cdot 434}{p^2} + \dfrac{0\cdot 067}{p^3}, \; p > 1$

where $p = 2\pi h/\lambda_L$; λ_L is the London retardation wavelength, usually of the order of 100 nm.

The Hamaker constant, A, depends on both the properties of the particle and those of the medium. For example, for two particles of composition 1 suspended in medium 2, the effective Hamaker constant is (Visser, 1972):

$$A_{121} = A_{11} + A_{22} - A_{12} \simeq (A_{11} - A_{22})^{1/2} \tag{1.77}$$

It is possible to evaluate the van der Waals dispersion force for geometries other than spheres. For two parallel flat plates of thickness d, the energy per unit area is:

$$V_{\text{disp}} = -\left[\frac{A}{12\pi}\frac{1}{h^2} + \frac{1}{(h+2d)^2} - \frac{2}{(h+d)^2}\right] \tag{1.78}$$

or for distances of interest, $d \gg h$,

$$V_{\text{disp}} = -\frac{A}{12\pi h^2} \tag{1.79}$$

In summing up the dispersion contributions of all the molecules that constitute the colloidal particles, one tacitly assumes that these contributions are additive. In general this assumption is untrue. As dispersion forces result from charge fluctuations in individual molecules, it is clear that these fluctuations will be affected by the presence of neighbouring molecules. An alternative approach is to calculate the dispersion forces using the continuum approach of Lifshitz and co-workers (Lifshitz, 1956; Dzyaloshinskii *et al.*, 1959). In its simpliest form, Lifshitz and co-workers consider the interaction between two half spaces a distance h apart. The results of this theory are expressed in terms of the permeabilities ϵ_1, ϵ_2 and ϵ_3 of the left half space, the medium in between, and the right half space, respectively. These permeabilities are evaluated as a discrete set of frequencies ξ_n:

$$\xi_n = \frac{2\pi kT}{\hbar} n \qquad n = 0, 1, 2 \ldots . \tag{1.80}$$

where $\hbar = h/2\pi$.

When the permeabilities in the three phases do not differ much from one another, the energy of interaction per unit area between the plates can be written as (Parsegian, 1975)

$$V_{\text{disp}} = -\frac{kT}{8\pi h^2}\sum_{n=0}^{\infty}\left(\frac{\epsilon_3 - \epsilon_2}{\epsilon_3 + \epsilon_2}\right)\left(\frac{\epsilon_1 - \epsilon_2}{\epsilon_1 + \epsilon_2}\right)(1 + r_n)e^{-r_n} \tag{1.81}$$

where $r_n = 2h\epsilon_2^{1/2}\xi_n/c$, in which c is the speed of light. The parameter r_n is the ratio of the time it takes an electromagnetic signal to travel between the gap and the period of vibration. Comparing eq (1.81) with eq (1.79) yields an expression for the Hamaker constant in terms of the Lifshitz theory:

$$A = \frac{3kT}{2}\sum_{n=0}^{\infty}\left(\frac{\epsilon_1-\epsilon_2}{\epsilon_1+\epsilon_2}\right)\left(\frac{\epsilon_3-\epsilon_2}{\epsilon_3+\epsilon_2}\right)(1+r_n)e^{-r_n} \tag{1.82}$$

Notice that for equal plates, $\epsilon_1 = \epsilon_3$, A is positive and there is always attraction. For unequal plates, A can be positive or negative depending on the value of ϵ_2. Hence the van der Waals interaction between two particles of different composition suspended in a third phase can be either attractive or repulsive, depending on the type of material. The interaction between similar particles is always attractive.

It is of interest to note that eq (1.82) describes correctly the retardation effect. As $h \to \infty$:

$$\sum (1+r_n)e^{-r_n} \simeq \int_0^{\infty} (1+r_n)e^{-r_n}dn = (\epsilon_2^{1/2}c\xi_n h)^{-1} \tag{1.83}$$

i.e. at large h, V varies as h^{-3}, rather than as h^{-2}.

1.4.3 Electrostatic interactions between colloidal particles

1.4.3.1 Ion distribution around a colloidal particle

In most cases, but particularly in polar media, colloidal particles possess an electrostatic charge. Because a colloidal dispersion as a whole is electrically neutral, it follows that there must be an excess of ions of opposite charge in solution. These oppositely charged ions will be found preferentially in the neighbourhood of the suspended colloidal particles. They form what is called a diffuse (electrical) double layer around each colloidal particle. One can say that a "cloud" of ions is associated with each colloidal particle. Because of electrostatic interactions, this cloud contains an excess of counterions (of charge opposite to that on the colloidal particle) and a deficiency of co-ions (of the same charge). The presence of these clouds or double layers affects considerably the forces between colloidal particles.

To understand the interactions between charged particles, we must first discuss the properties of the double layer surrounding each particle. The

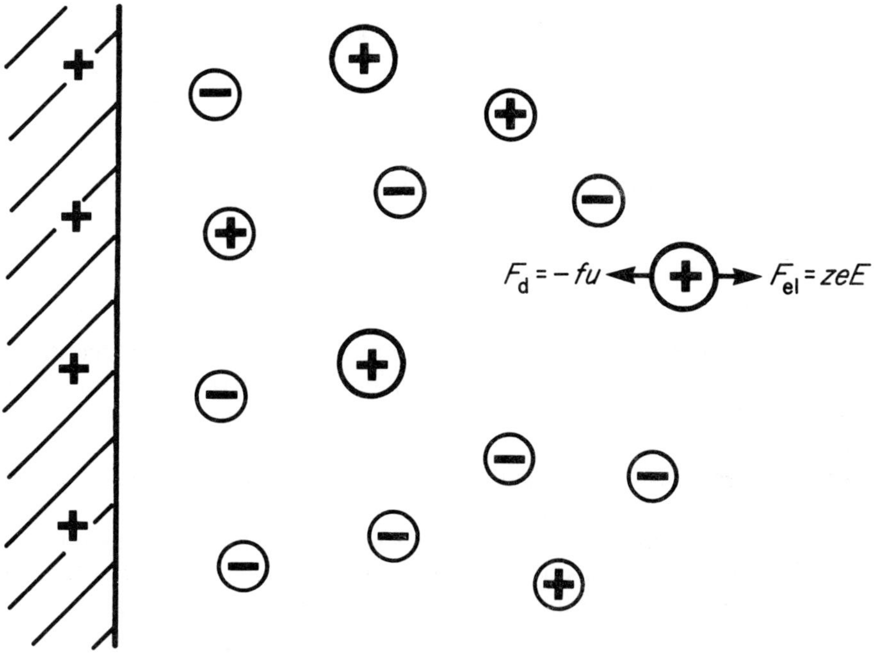

Fig. 1.7. Schematic diagram of a diffuse double layer near a flat boundary, together with the forces acting upon the positive and negative ions.

distribution of ions can readily be found from the steady-state convective-diffusion equation (see eq (1.1)):

$$\nabla \cdot (D_i \nabla n_i) - \nabla \cdot (u_i n_i) = 0 \tag{1.84}$$

where D_i is the diffusion coefficient of ion i, n_i its number concentration, and u_i its velocity due to the electric field acting on it (caused by the presence of a charged colloidal particle). This velocity is given by (cf. eq (1.11))

$$F_{e\ell} = z_i e E = f_i u_i = \frac{kT}{D_i} u_i \tag{1.85}$$

where we have used the Stokes–Einstein equation (eq (1.3)). Here z_i is the valency of ion i, e the (positive) unit electric charge (the charge of a proton), and E the field strength. Combining eqs (1.84) and (1.85), and using $E = -\nabla\psi$, gives:

$$\nabla \cdot \left(\nabla n_i + n_i \frac{z_i e}{kT} \nabla \psi \right) = 0 \tag{1.86}$$

It can be seen that n_i is independent of D_i. Equation (1.86) can readily be solved for the ion distribution near a flat plate and around a spherical particle. In each case

$$n_i = n_i^0 \mathrm{e}^{-z_i e \psi / kT} \tag{1.87}$$

The ion distribution near a flat plate is shown schematically in Fig. 1.7, together with the forces exerted on the ions. The potential, ψ, appearing in eq (1.87) is a function of x (for a flat plate) or of r (for a sphere). The exponential distribution given by eq (1.87) is known as a *Boltzmann distribution*.

It can be seen that for a negatively charged colloidal particle ($\psi < 0$), the exponent in eq (1.87) is positive for positive ions and negative for negative ones. Hence the concentration of positive ions is a maximum at the surface; the opposite is true for positively charged particles. From eq (1.87) it follows that in order to find the distribution of ions near a colloidal particle, the potential, ψ, as a function of distance must be found. According to standard electrostatic theory, for a diffuse charge of charge density ρ (charge per unit volume), the relation between the potential and the charge density is given by the *Poisson equation*:

$$\nabla^2 \psi = -\frac{\rho}{\epsilon} \tag{1.88}$$

where ϵ is the permittivity of the medium (product of dielectric constant and permittivity of free space). For ions in solution the charge density is given by

$$\rho = \sum_i n_i z_i e \tag{1.89}$$

Equations (1.87) to (1.89) determine the ion and potential distribution around a colloidal particle. Combining these equations results in the so-called *Poisson–Boltzmann equation*:

$$\nabla^2 \psi = -\frac{e}{\epsilon} \sum_i n_i^0 z_i e^{-z_i e \psi / kT} \tag{1.90}$$

The solution of the Poisson–Boltzmann equation determines ψ and hence n_i by means of eq (1.87). Equation (1.90) is restricted to relatively low electrolyte concentrations (ca. 1 mM), the main reason being that the ions are treated as point charges. The finite size of the particles prevents the build-up of large charge densities near their surface.

In general, the Poisson–Boltzmann equation must be solved numerically and many such solutions have been reported in the literature. However, a few limiting solutions provide valuable insight into the equation.

1.4.3.2 Solutions of the Poisson–Boltzmann equation

(a) Debye–Hückel approximation

For low potentials ($z_i e\psi \ll kT$), the exponential occurring in the Poison–Boltzmann equation can be expanded:

$$\exp\left(-\frac{z_i e\psi}{kT}\right) = 1 - \frac{z_i e\psi}{kT} + \ldots\ldots \tag{1.91}$$

Substituting this relation into eq (1.90) and making use of the electroneutrality condition in the bulk, i.e.

$$\sum_i n_i^0 z_i e = 0 \tag{1.92}$$

results in

$$\nabla^2\psi = \kappa^2\psi \tag{1.93}$$

where

$$\kappa^2 = \frac{e^2}{\epsilon kT}\sum_i n_i^0 z_i^2 \tag{1.94}$$

For a double layer adjacent to a flat plate, the solution of eq (1.93) is

$$\psi = \psi_0 \mathrm{e}^{-\kappa x} \tag{1.95}$$

while for the double layer around a spherical particle, the solution is

$$\psi = \psi_0 \frac{\mathrm{e}^{-\kappa(r-a)}}{r/a} \tag{1.96}$$

where ψ_0 is the potential at the surface. Thus for low potentials, the potential decays exponentially with distance from the surface. The characteristic length over which the potential is changing is κ^{-1}, which is called the Debye length or, more usually, the double-layer "thickness". This is an important parameter in determining double-layer interactions and, as can be seen from eq (1.94), it can be varied by changing the concentration and the valency of an electrolyte. The concentration of the electrolyte is related to n_i^0 by:

$$n_i^0 = 1000 N_A c_i^0 \tag{1.97}$$

where c_i^0 is the bulk concentration of ions of species i (in mol l^{-1}) and N_A is Avogadro's number (6.02×10^{23}). Typically, κ^{-1} varies from about 100 nm at 10^{-5} M to less than 1 nm at 0.1 M.

(b) Gouy–Chapman theory

Another useful solution of the Poisson–Boltzmann equation can be found for symmetrical electrolytes for which

$$z_+ = -z_- = z$$

and

$$n_+^0 = n_-^0 = n_0$$

In this case the Poisson–Boltzmann equation reduces to (for a flat plate)

$$\frac{d^2\psi}{dx^2} = \frac{2en_0 z}{\epsilon} \sinh\left(\frac{ze\psi}{kT}\right) \tag{1.98}$$

the solution of which is

$$\psi = \frac{2kT}{ze} \ln \frac{1+\gamma_0 e^{-\kappa x}}{1-\gamma_0 e^{-\kappa x}} \tag{1.99}$$

where $\gamma_0 = \tanh(ze\psi_0/kT)$.

The solution of the Poisson–Boltzmann equation allows the surface potential, ψ_0, to be related to the surface-charge density, σ. Because of electroneutrality, the excess charge in solution must balance the charges on the particle surface:

$$\sigma = -\int_0^\infty \rho \, dx \tag{1.100}$$

Using the Poisson equation (eq (1.88)):

$$\sigma = -\int_0^\infty \epsilon \frac{d^2\psi}{dx^2}\, dx = \epsilon \frac{d\psi}{dx}\bigg|_0^\infty = -\epsilon \left(\frac{d\psi}{dx}\right)_{x=0} \qquad (1.101)$$

Differentiation of eq (1.99) and substitution into eq (1.101) gives

$$\sigma = \frac{2\epsilon\kappa kT}{ze} \sinh\left(\frac{ze\psi_0}{kT}\right) \qquad (1.102)$$

It is of interest to note that for low potentials eq (1.99) reduces to eq (1.95) and eq (1.102) reduces to

$$\sigma = \epsilon\kappa\psi_0 \qquad (1.103)$$

This relation implies that at low potentials the double layer can be considered as a capacitor consisting of two plates a distance κ^{-1} apart. For spherical particles, similar reasoning leads to

$$\sigma = \epsilon \frac{1+\kappa a}{a} \psi_0 \qquad (1.104)$$

1.4.3.3 Double-layer overlap

When two colloidal particles, each surrounded by an electrical diffuse double layer, approach one another, their respective double layers begin to overlap. As a result, the ion and potential distribution around a particle in the neighbourhood of a second particle is no longer symmetrical. This causes asymmetrical stresses of electrical origin on the particle surface and, as a result, the particle experiences a force. In general, the force $\boldsymbol{F}$ on a particle can be found from (cf. eq (1.22))

$$\boldsymbol{F} = \int_A \boldsymbol{\Pi}_{el} \cdot dA \qquad (1.105)$$

where $\boldsymbol{\Pi}$ is the stress tensor which must be integrated over the total surface area, A, of the particle. The stress tensor is determined by the force balance (cf. eq (1.24)):

$$\nabla \cdot \boldsymbol{\Pi}_{el} - \rho \nabla \psi = 0 \qquad (1.106)$$

where $\rho \nabla\psi$ is the electrical body force, ρ being the charge density and ψ the potential. As in the case of a single particle, ψ is determined by the Poisson equation and the convective-diffusion equation for each of the ions:

$$\nabla^2\psi = -\frac{\rho}{\epsilon}\,;\, \rho = \sum_i n_i z_i e \qquad (1.107)$$

and

$$\nabla\cdot(\nabla n_i + \frac{n_i z_i e}{kT}\nabla\psi) = 0 \qquad (1.108)$$

Equations (1.107) and (1.108), together with the appropriate boundary conditions, completely determine the charge and potential distribution around a particle. The stress tensor can be found by solving eq (1.106) and, subsequently, the force can be calculated from eq (1.105).

The appropriate boundary conditions for eqs (1.107) and (1.108) need some comment. The boundary conditions for eq (1.108) are: (i) at large distances, n_i approaches the bulk concentration n_i^0; and (ii) at the particle surface the non-penetration boundary condition applies. For eq (1.107) it is required that $\psi \to 0$ as $r \to \infty$. The boundary condition at the surface of each particle depends on whether or not the approach of two particles occurs under conditions of thermodynamic equilibrium. If thermodynamic equilibrium prevails, the surface potential, ψ_0, is constant and this serves as the appropriate boundary condition. Under such conditions, the surface charge density, σ (as well as the total charge), varies upon approach. The reason for this is shown in Fig. 1.8. When thermodynamic equilibrium is not maintained during approach, the surface potential varies. In general, both σ and ψ can vary. When σ is constant during approach, the second boundary condition can be taken as the conservation of charge which for two spheres, with the origin defined as being midway between them, can be written as

$$\int_{-\infty}^{0}\int_{-\infty}^{\infty}\int_{-\infty}^{\infty} \rho\,\mathrm{d}x_1\,\mathrm{d}x_2\,\mathrm{d}x_3 = -4\pi a^2\sigma$$

The expected equipotential surfaces around two spheres or aligned cylinders when the surface potential, ψ_0, is constant are shown schematically in Fig. 1.9.

The calculation of the ion and potential distribution around two spheres is a difficult problem that can only be solved numerically. A simpler problem, that can be readily solved by numerical finite-difference methods, is to calcu-

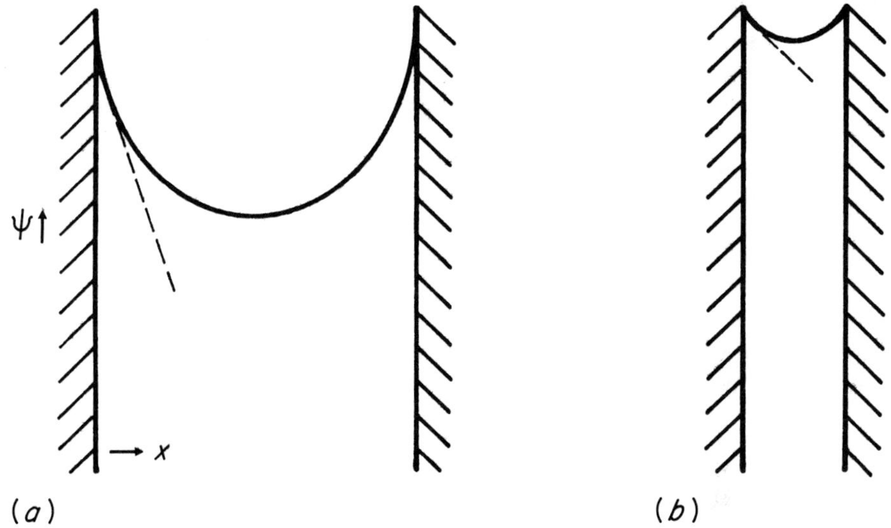

Fig. 1.8. Schematic potential profile between two flat plates (*a*) far apart and (*b*) close together. The slope at the surface (indicated by the dashed lines) decreases as the surfaces approach one another. Because $\sigma \propto -d\psi/dx$ (see eq (1.101)), it follows that the surface charge density decreases upon approach.

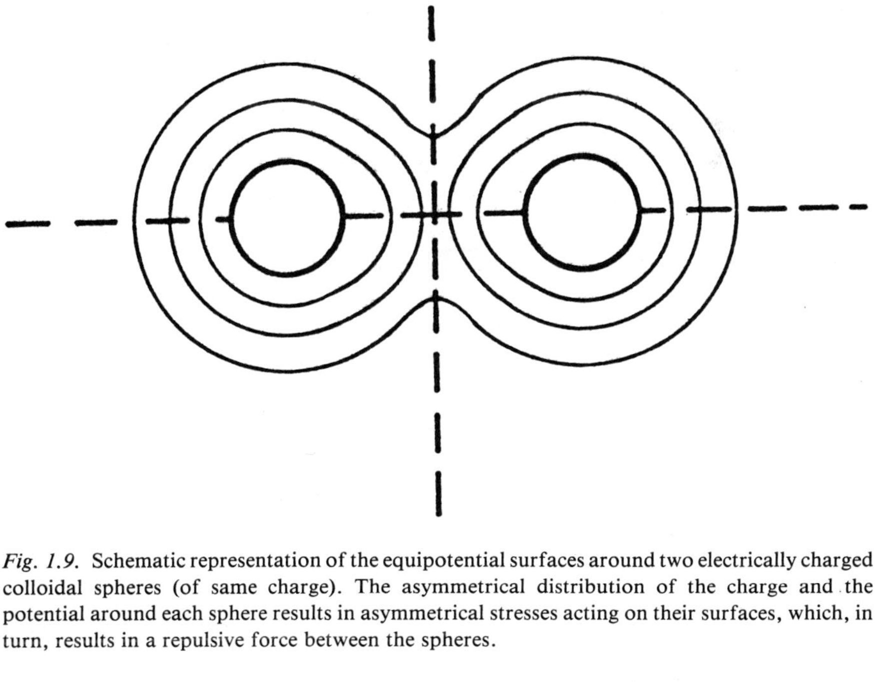

Fig. 1.9. Schematic representation of the equipotential surfaces around two electrically charged colloidal spheres (of same charge). The asymmetrical distribution of the charge and the potential around each sphere results in asymmetrical stresses acting on their surfaces, which, in turn, results in a repulsive force between the spheres.

late the force between two parallel cylinders. To the author's knowledge, this has not been attempted. Instead of solving the exact problem, the most common (but far less accurate) approach is to calculate the force between two plates and to find approximate expressions for other geometries based on the solution for two plates.

The potential distribution for two plates is shown schematically in Fig. 1.10. In this case eq (1.106) reduces to

$$\frac{\mathrm{d}p}{\mathrm{d}x}-\rho\frac{\mathrm{d}\psi}{\mathrm{d}x}=0 \tag{1.109}$$

Substituting for ρ using eqs (1.89) and (1.87) and only considering symmetrical electrolytes, gives

$$\mathrm{d}p=2zen_0\sinh\left(\frac{ze\psi}{kT}\right)\mathrm{d}\psi \tag{1.110}$$

The force, per unit area, that drives the plates apart is given by

$$F_\mathrm{R}=p_\mathrm{m}-p_\infty \tag{1.111}$$

where p_m is the pressure midway between the plates and p_∞ the pressure at $x=\infty$. Hence, combining eqs (1.110) and (1.111):

$$F_\mathrm{R}=\int_{p_\infty}^{p_m}\mathrm{d}p=2zen_0\int_0^{\psi_m}\sinh\left(\frac{ze\psi}{kT}\right)\mathrm{d}\psi \tag{1.112}$$

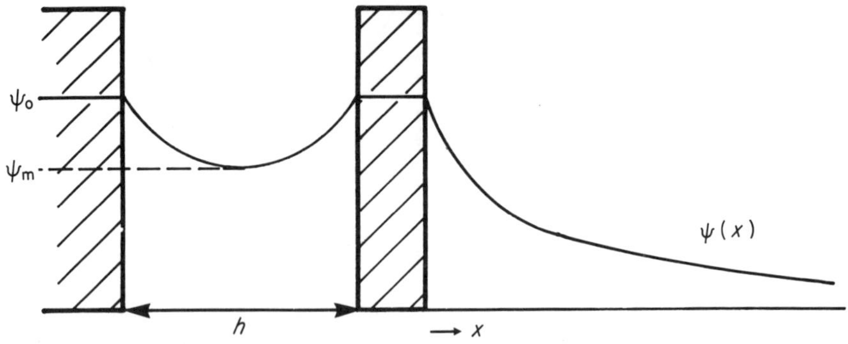

Fig. 1.10. Schematic representation of the double-layer overlap between two flat plates of surface potential ψ_0. ψ_m is the potential midway between the plates.

which, upon integration, results in

$$F_R = 2kTn_0\left[\cosh\left(\frac{ze\psi_m}{kT}\right) - 1\right] \tag{1.113}$$

a result first derived by Langmuir (1938) and independently by Derjaguin (1940). The usual assumption made in finding ψ_m (Verwey and Overbeek, 1948) is simply assuming that the potential midway between the plates is the sum of the potentials of both (undisturbed) double layers:

$$\psi_m = 2\psi\left(\frac{h}{2}\right) \tag{1.114}$$

This is justified for small potentials, which means that the approximation is only valid when the two plates are far apart. At contact, eq (1.114) is incorrect by a factor of two. When $ze\psi_m/kT \ll 1$, the cosh term in eq (1.113) can be expanded:

$$\cosh\left(\frac{ze\psi_m}{kT}\right) = 1 + \frac{1}{2}\left(\frac{ze\psi_m}{kT}\right)^2 + \ldots \tag{1.115}$$

Combining eqs (1.115), (1.114) and (1.113) and using eq (1.99), yields

$$F_R = 32\gamma_0^2\epsilon\kappa^2\frac{k^2T^2}{z^2e^2}e^{-\kappa h} \tag{1.116}$$

Equation (1.116) shows that the electrostatic repulsion between two charged plates falls off exponentially, the characteristic length being given by the Debye length, κ^{-1}.

An alternative way of looking at the repulsive force between two charged colloidal particles is to consider the osmotic pressure, π, in the gap, given by van 't Hoff's law

$$\pi = nkT \tag{1.117}$$

The osmotic pressure difference between the midpoint between the plates and the bulk produces a repulsive force between the plates:

$$F_R = \Delta\pi = (\Sigma n_i^m - \Sigma n_i^0)kT \tag{1.118}$$

Substituting the Boltzmann distribution into eq (1.118) and making the same assumptions as above again leads to eq (1.116).

1.4.3.4. Forces between particles with constant charge

When the surface charge density, σ, remains constant as two particles approach one another, the equations given above for the force between two plates must be modified. This has been done by Frens and Overbeek (1972), who modified the Gouy–Chapman theory. The calculation involves numerical procedures which, although easily performed, prevent a simple analytical insight. Here we will look at the case of low potentials, in which case the theory simplifies considerably. Applying the capacitor model (cf. eq (1.103)) of the double layer to two plates of constant charge results in

$$\sigma = \epsilon\kappa(\psi_\sigma - \psi_m) \tag{1.119}$$

where ψ_σ is the surface potential (now a function of the distance h between the plates) and ψ_m is the potential midway between the plates. Equation (1.119) applies to all plate separations, including infinity, and because σ is constant it follows that

$$\psi_\sigma - \psi_m = \psi_0 \tag{1.120}$$

where ψ_0 is the surface potential of an isolated sphere (i.e. $\psi_0 = \psi_\sigma(\infty)$). Making the same approximations as before we can write

$$\psi_m = 2\psi_\sigma e^{-\kappa h/2} \tag{1.121}$$

and

$$\psi_\sigma = \frac{\psi_0}{1 - 2e^{-\kappa h/2}} \tag{1.122}$$

As we have seen, the potential midway between the plates determines the force between the plates, hence

$$F_R = 2\epsilon\kappa^2\psi_0^2 \frac{e^{-\kappa h}}{1 - 4e^{-\kappa h/2}} \tag{1.123}$$

At short separations, eq (1.123) is no longer valid because as $h \to 0$, $\psi_m \to \psi_\sigma$ and eq (1.121) overestimates ψ_m by a factor of two. The same is true for approach at constant potential. To overcome this problem the approximate equation (Takamura *et al.*, 1981a)

$$F_R \simeq 2\epsilon\,\kappa^2\psi_0^2 \frac{e^{-\kappa h}}{1 \pm e^{-\kappa h}} \tag{1.124}$$

can be used, where the plus sign applies to approach at constant potential and the minus sign to approach at constant charge. At large separations the distinction between constant charge and constant potential becomes unimportant. Equations (1.123) and (1.124) show that the repulsive force at constant charge is larger than the force at constant potential. As $h \to 0$, $F_R \to \infty$ for constant charge (as a Coulomb repulsion between two charged particles), while $F \to$ constant for approach at constant ψ_0.

1.4.3.5 Forces between parallel cylinders and spheres

To evaluate the forces between two surfaces other than flat plates is more difficult. A useful approximation is to replace particles with curved boundaries by a series of flat strips, since for each of these the force per unit area can be found from eqs (1.116) or (1.124). This is known as the *Derjaguin approximation* (Derjaguin, 1934). Let us consider the repulsive force between two equally charged parallel cylinders (see Fig. 1.11) of radius a and gap separation h. From the geometry it can be seen that

$$\frac{h_z - h}{2} = a - (a^2 - z^2)^{1/2} \tag{1.125}$$

where h_z and z are defined in Fig. 1.11.

The force per unit area between two strips a distance h_z apart equals

$$F_z = B\mathrm{e}^{-\kappa h} \tag{1.126}$$

where

$$B = 32\gamma_0^2 \epsilon\kappa^2 \mathrm{k}^2 T^2/(z^2 e^2)$$

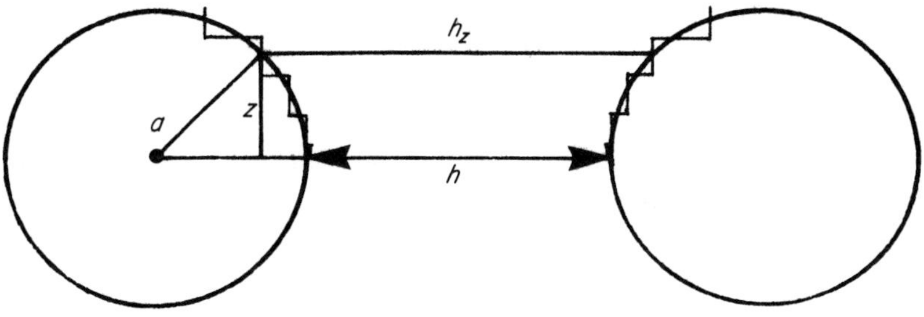

Fig. 1.11. Geometry of two parallel cylinders or spheres, approximated by a series of flat strips or rings.

or for low potentials

$$B = 2\epsilon\psi_0^2\kappa^2$$

Hence the force per unit length between two parallel cylinders is

$$F_R = 2B\int_0^a e^{-\kappa h_z}dz \qquad (1.127a)$$

or, using eq (1.125):

$$F_R = 2B\int_0^a e^{-\kappa h}e^{-2\kappa a + 2\kappa\sqrt{(a^2 - z^2)}}dz \qquad (1.127b)$$

When the cylinders are close together, the main contribution to the integral comes from strips near $z = 0$ and the integral can be approximated by expanding the square root in the exponential:

$$F_R \simeq 2Be^{-\kappa h}\int_0^\infty e^{-\kappa z^2/a}dz = B\left(\frac{\pi a}{\kappa}\right)^{1/2} e^{-\kappa h} \qquad (1.128)$$

From the force per unit length between two cylinders, the force between two spheres can be found. Consider again Fig. 1.11, but now let the two circles represent the cross-sections of two spheres. Then the force between two spheres can be approximated as consisting of the sum of the forces between a larger number of infinitesimally thin cylinders of radius $(a^2 - z^2)^{1/2}$:

$$F_R = 2B\left(\frac{\pi}{\kappa}\right)^{1/2}\int_0^a (a^2 - z^2)^{1/4}e^{-\kappa h_z}dz \qquad (1.129)$$

Evaluating this integral with the same assumptions as above yields, for the force between two spheres of radius a:

$$F_R = B\pi\left(\frac{a}{\kappa} - \frac{1}{8\kappa^2}\right)e^{-\kappa h} \qquad (1.130)$$

or when $\kappa a \gg 1/8$

$$F_R = \frac{B\pi a}{\kappa} e^{-\kappa h} \tag{1.131}$$

The results for cylinders (eq (1.128)) and spheres (eq (1.131)) are valid for relatively large particles. The condition $\kappa a \gg 1/8$ is usually satisfied for colloidal particles, except for very small particles at very low electrolyte concentrations, e.g. for a 10-nm particle at 10^{-5}M 1 – 1 electrolyte, $\kappa a \simeq 0 \cdot 1$.

From the potential around spheres, Ball *et al.* (1970) found that for large distances r ($r = 2a + h$),

$$F_R = \epsilon \psi_0^2 a^2 \frac{1 + \kappa r}{r^2} e^{-\kappa h} \tag{1.132}$$

The Derjaguin approximation for unequal-sized spheres of radii a_1 and a_2 and surface potentials ψ_1 and ψ_2, respectively, can also be used. For $\psi_1 = \psi_2 = 0$ (Hogg *et al.*, 1966), $F_R = -\partial V_R/dh$ with

$$V_R = \frac{4\pi\epsilon\epsilon_0 a_1 a_2 \psi_0^2}{(a_1 + a_2)} \ln\left(1 + e^{-2\kappa r/(a_1 + a_2)}\right) \tag{1.133}$$

1.4.4 Classical DLVO theory

Derjaguin and Landau (1941) and Verwey and Overbeek (1948) combined the van der Waals dispersion forces with the electrostatic double-layer forces to explain the stability of lyophobic colloids. Nowadays this theory is commonly referred to as the DLVO theory.

The total force exerted on a colloidal particle in the vicinity of a second particle is the sum of several contributions. In the DLVO theory it is customary to represent the interactions by the free energy of interaction, rather than by the forces exerted on the particles. The energy of interaction is related to the force by

$$F_{int} = -\frac{\partial V_{int}}{\partial r} \tag{1.134}$$

For the case of two spherical particles at short separations (cf. eqs (1.75) and (1.131)):

$$V_{int} = -\frac{Aa}{12h} + Ce^{-\kappa h} \tag{1.135}$$

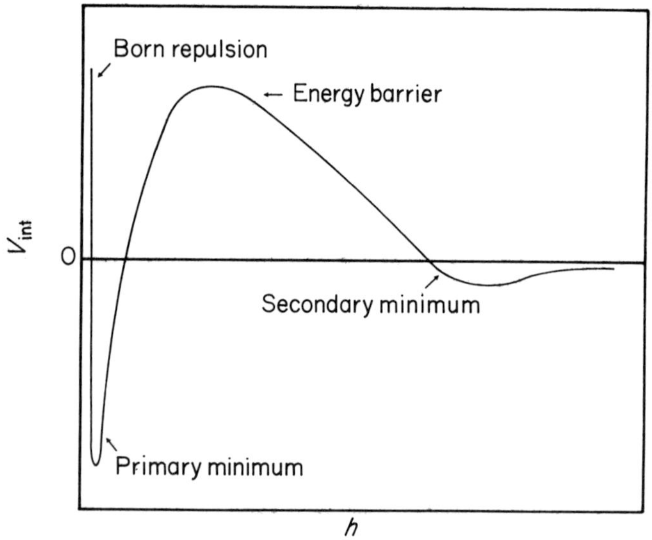

Fig. 1.12. Schematic representation of the energy of interaction between two colloidal particles as a function of the distance h between them.

where C is a constant. Equation (1.135) gives the interaction energy in its simplest form, but can be modified by taking more sophisticated expressions or numerical results for the dispersion and electrostatic interactions. The energy of interaction is shown schematically in Fig. 1.12 in which an additional repulsive force, the Born repulsion, has been introduced which operates at extremely short distances and prevents the particles from interpenetration.

As can be seen from Fig. 1.12, when the Born repulsion is neglected, the van der Waals force dominates at short and long distances and the energy of interaction is negative (attractive). At intermediate distances the forces are repulsive, provided the electrostatic repulsion is sufficiently large. The curve shows three positions of zero force, two corresponding to energy minima (referred to as the primary minimum and the secondary minimum, respectively) and one corresponding to an energy maximum. The height of the energy maximum determines the stability of the suspension. When the barrier is high ($V_{max} > \sim 10kT$), few particles can overcome it and no coagulation occurs (in the absence of external force fields). The height of the barrier can be varied by varying either the surface potential, ψ_0, or the ionic strength which affects the double-layer thickness, κ^{-1}. At low potentials or high ionic strength the barrier disappears. Figure 1.13 shows the calculated results for two spheres of radius 1 μm.

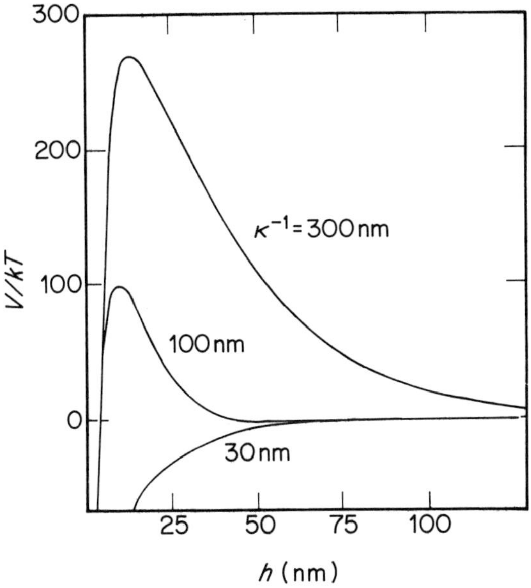

Fig. 1.13. Energy of interaction between two spheres of radius 1 μm at various ionic strengths: $A = 10^{-19}$J, $\psi_0 = 26$mV.

The DLVO theory can easily be extended by considering additional forces between particles.

1.4.5 Forces between colloidal particles in the presence of polymer

1.4.5.1 General remarks

Adding polymers to a colloidal dispersion can result in a variety of interesting phenomena. If the polymers adsorb onto the particles they can either stabilize or destabilize the dispersion. Stabilization occurs when the interaction between adsorbed polymers leads to steric or elastic repulsion during the approach of two particles. Depending on the dominant mechanism, this is called *steric* or *elastic stabilization*. When particles with adsorbed molecules are suspended in a poor solvent, attraction occurs as the polymers phase separate. When the particles are only partly covered with polymer, the adsorbed polymer can form a macromolecular bridge by adsorbing on two particles simultaneously. This results in *bridging flocculation*. When the polymer has no affinity for the colloidal particles but remains freely dissolved in the solvent then, depending on conditions, either stabilization or destabilization can take place. For two colloidal particles to approach each other in a dispersion containing free polymer, the polymer must be squeezed out of the gap between the particles. If this removal of the polymer requires a large force, *depletion stabilization* occurs. At small separations the gap contains no polymer and therefore, an osmotic-pressure difference exists between the liquid in the gap and the bulk solution. As a result the liquid wants to leave the gap, resulting in attraction between the particles. This can lead to *depletion flocculation*.

1.4.5.2 Particles with adsorbed polymer

The main difference between polymers adsorbed or anchored onto a colloidal particle and polymers free in solution is the change in their configurations. When in contact with a solid, the number of configurations available to a polymer chain is severely restricted. An important property of adsorbed polymers is their segment distribution, which describes the distribution of polymer segments at the surface. In general, one distinguishes between segments that are in tails (or dangling ends), loops or trains, as shown schematically in Fig. 1.14.

It is possible to make predictions of the segment-density profile (the number of polymer segments per unit volume as a function of distance from the surface) either by generating polymer configurations using Monte Carlo methods (Napper, 1983) or by considering various lattice models in which each site is either occupied by a solvent molecule or a polymer segment (Fleer and Scheutjens, 1982).

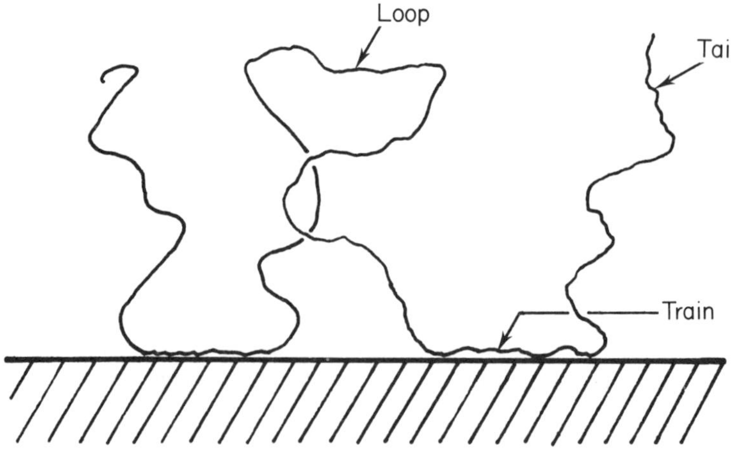

Fig. 1.14. Schematic diagram of a polymer configuration at a solid surface.

When two particles, each with adsorbed polymer, are brought into close proximity with each other, their respective segment-density profiles will change, resulting in changes in the entropy of the system. At the same time, enthalpic changes occur because of changes in the interaction of polymer segments with solvent molecules and among themselves.

The change in the Gibbs free energy, ΔG, of the system is given by

$$\Delta G = \Delta H - T\Delta S \tag{1.136}$$

where H is the enthalpy, T the absolute temperature, and S the entropy. Depending on the relative magnitudes of ΔH and $T\Delta S$, one can speak of enthalpic or entropic stability. In order to predict the forces between two colloidal particles in the presence of polymer, we must calculate the changes in the Gibbs free energy, from which the force can be readily determined.

$$F = -\nabla(\Delta G) \tag{1.137}$$

For two approaching particles three different regimes can be distinguished: (i) the non-interpenetration domain where the particles are so far apart that the respective segment distributions of the adsorbed polymers are not affected; (ii) the interpenetration domain in which the adsorbed polymer segments penetrate each other; and (iii) the interpenetration plus compression domain in which the particles are so close together that the adsorbed polymer layers are compressed, resulting in an elastic repulsion. The changes in free energy upon approach depend on the domain corresponding to a given separation. For the non-penetration domain $\Delta G = 0$ and the particles experience no force. The free energy changes during the approach of polymer-coated particles have been calculated to various degrees of approximation by a variety of authors. For a review the reader is referred to the excellent textbook by Napper (1983). Because of the relatively low volume fraction of polymer in the interpenetration domain, the dominant contribution to ΔG (by far) comes from the free energy of mixing between the adsorbed polymers in the gap between the particles and the solvent molecules. In the interpenetration plus compression domain, in addition to the mixing energy, the elastic energy also becomes important.

According to the Flory–Huggins theory (Huggins 1942a, b; Flory, 1969), the entropy of mixing of n_1 solvent molecules with n_2 polymer molecules is given by

$$\Delta S^{\mathrm{M}} = -k(n_1 \ln c_1 + n_2 \ln c_2) \tag{1.138}$$

where k is the Boltzmann constant and c_1 and c_2 are the volume fractions of solvent and polymer, respectively. The enthalpy of mixing is given by

$$\Delta H^{M} = n_1 c_2 \chi kT \tag{1.139}$$

where χ is an interaction parameter; χkT represents the excess free energy associated with the transfer of one solvent molecule from the pure polymer to a polymer solution of volume fraction c_2. Combining eqs (1.138) and (1.139) yields

$$\Delta G = kT(n_1 \ln c_1 + n_2 \ln c_2 + n_1 c_2 \chi) \tag{1.140}$$

Equation (1.140) applies to the mixing of free dissolved polymers. The second term on the right-hand side of eq (1.140) is related to the entropy of the polymer, which is usually much less than the entropy associated with the solvent molecules. When a polymer is attached to a colloidal particle, this term becomes negligible:

$$\Delta G^{M} = n_1 kT[\ln(1 - c_2) + \chi c_2] \tag{1.141}$$

Consider a volume element $\mathrm{d}V$ within the adsorbed polymer layer, with a number of solvent molecules $\mathrm{d}n_1$, each of volume V_1. Let the number of polymer segments per unit volume be ρ_2 and the volume of a segment V_s (i.e. $\rho_2 V_s = c_2$), then

$$V_1 \mathrm{d}n_1 = (1 - \rho_2 V_s)\mathrm{d}V \tag{1.142}$$

and the free energy of mixing in element $\mathrm{d}V$ equals

$$\mathrm{d}(\Delta G^{M}) = kT\left[-\frac{V_s}{V_1}(1-\chi)\rho_2 + \frac{V_s^2}{V_1}(\tfrac{1}{2} - \chi)\rho_2^2 + \ldots\right]\mathrm{d}V \tag{1.143}$$

where the logarithmic term in eq (1.141) has been expanded. If two particles, initially very (infinitely) far apart, approach one another to a distance h, then

$$\Delta G^{M} = \Delta G^{M}(h) - \Delta G^{M}(\infty) \tag{1.144}$$

Evaluating $\mathrm{d}(\Delta G^{M})$ at distance h and at infinity, integrating over the total polymer volume, and noting that $\int \rho_2 \mathrm{d}V$ is constant, gives (Napper, 1983):

$$\Delta G^{M} = kT\frac{V_s^2}{V_1}(\tfrac{1}{2} - \chi)P(h) \tag{1.145}$$

where

$$P(h) = \int_V \rho_2^2(h)\,\mathrm{d}V - \int_V \rho_2^2(\infty)\,\mathrm{d}V \tag{1.146}$$

Equation (1.145), together with eq (1.137), describes the force between two polymer-coated particles. If the segment-density distribution is known, $P(h)$ can be evaluated. For flat plates, eq (1.145) can be written as

$$\Delta G^{\mathrm{M}} = kT\frac{V_{\mathrm{s}}^2}{V_1}\left(\tfrac{1}{2} - \chi\right)\nu^2 i^2 P^*(h) \tag{1.147}$$

where ν is the number of polymer chains per unit area of surface, and i is the number of segments per molecule. $P^*(h)$ is a function depending on the normalized segment-density profile. For a constant segment density in the interpenetration domain

$$P^*(h) = \frac{2}{L}\left(1 - \frac{h}{2L}\right) \tag{1.148}$$

and in the interpenetration plus compression domain

$$P^*(h) = \frac{3}{h} - \frac{2}{L} \tag{1.149}$$

where L is the thickness of the steric layer. $P^*(h)$ for various other segment distributions is given by Napper (1983). As noted earlier, segment densities can be found from numerical Monte Carlo calculations or lattice models.

Besides the segment-density distribution, the parameter χ appearing in eq (1.147) also plays an important role. It is usually a strong function of temperature and depends also on the volume fraction, c_2. Both enthalpic and entropic effects are incorporated in χ:

$$\frac{1}{2} - \chi = \psi - \kappa = \psi\left(1 - \frac{\theta}{T}\right) \tag{1.150}$$

where the θ temperature is defined as $(\kappa/\psi)T$; ψ and κ can be interpreted as the enthalpy and entropy of dilution, respectively. At the θ temperature, $\chi = \frac{1}{2}$ and the free energy of mixing is zero (see eq (1.147)). In general, a polymer–solvent system can have up to three different temperatures at which $\chi = \frac{1}{2}$. It follows that a dispersion can sometimes be destabilized by cooling, sometimes by heating and sometimes by both. If $\chi < \frac{1}{2}$, the solvent is said to be a better than θ solvent. In such a good solvent, polymer interpenetration is disfavoured and the force between the particles is repulsive. If $\chi > \frac{1}{2}$, the solvent is said to be a worse than θ solvent; interpenetration is favoured and the force between two particles is attractive.

When particles come within the compression domain, an elastic contribution must be added to the free energy (Napper, 1983)

$$\Delta G^{el} = -2\nu kT \ln \frac{\Omega(h)}{\Omega(\infty)}, \qquad (1.151)$$

where $\Omega(h)$ is the configurational probability at distance h.

For spheres and particles of other shapes, the energy or the force of interaction can be found from the flat-plate result (eq (1.147)) using the Derjaguin approximation (see Section 1.4.3.5).

The energy of interaction for various solvents is shown schematically in Fig. 1.15; the elastic contribution to the interaction energy at short separations can be seen. The deep minimum in curve 3 is due to the mutual attraction of adsorbed polymer segments.

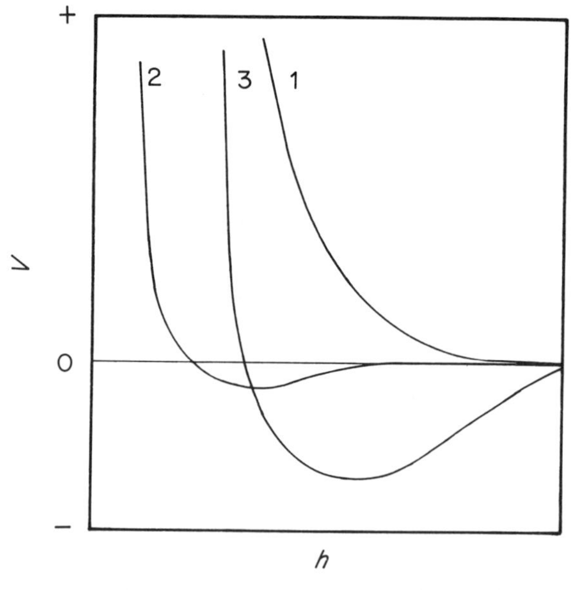

Fig. 1.15. The interaction energy between two polymer-coated particles in various solvents (schematic). Curve 1: a better than θ solvent with negligible van der Waals attraction. Curve 2: a better than θ solvent with appreciable van der Waals attraction. Curve 3: a worse than θ solvent (after Napper, 1983).

1.4.5.3 Free polymer in solution

To calculate the forces between two particles suspended in a polymer solution, in the absence of polymer adsorption, it is necessary to know the enthalpy and entropy changes which occur when the polymer is squeezed out of the gap (resulting in repulsion) and the osmotic contribution when the solvent wants to leave the gap spontaneously after the polymer is squeezed out (resulting in attraction). The changes in free energy can be calculated by generating the segment-density distributions between two plates, using either Monte Carlo or lattice models. For reasons of complexity, such calculations can only be carried out numerically. Figure 1.16 gives examples of interaction curves that clearly show the existence of a maximum which, if sufficiently high, is responsible for stability, and a minimum responsible for flocculation. The existence of a maximum has recently been questioned (Evans, 1988).

1.4.6 Solvation, hydration and hydrophobic forces

When two particles approach one another to within very small distances, the assumption that the suspending medium can be treated as a continuum

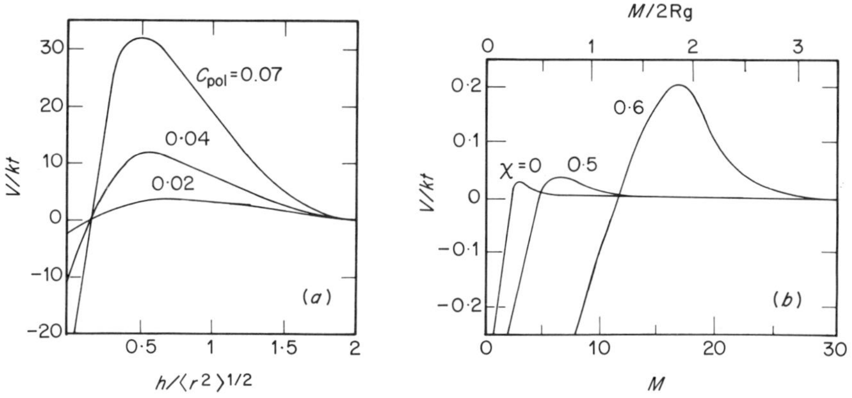

Fig. 1.16. Plots of the interaction energy between two particles suspended in a polymer solution, in the absence of polymer adsorption.
(*a*) Total energy of interaction between two spheres of radius 77.5 nm in a poly(oxyethylene) solution of molar mass 20 000, for polymer volume fractions of 0·07, 0·04 and 0·02, respectively (after Napper, 1983).
(*b*) Energy per lattice site between two plates in a polymer solution for various values of the interaction parameter χ: volume fraction 0·2; polymer of 100 segments; *M* is number of lattice layers (after Fleer and Scheutjens, 1982).

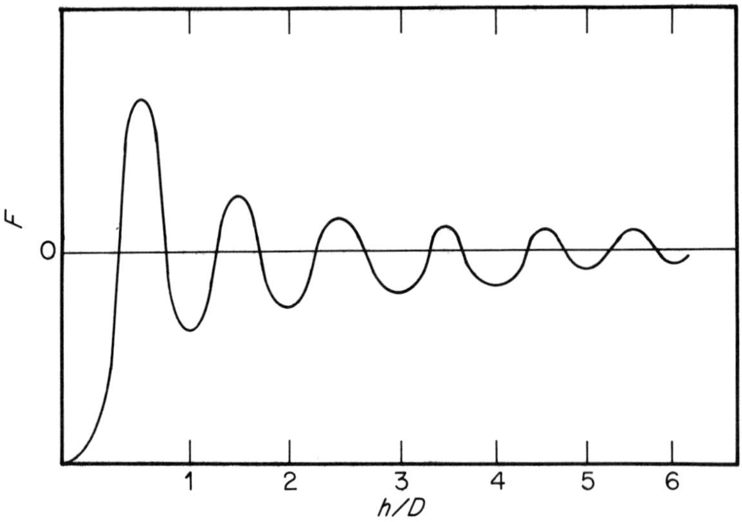

Fig. 1.17. Force per unit area between two plates as a function of distance, normalized by the diameter, D, of the solvent molecule (schematic).

becomes invalid. Instead the distribution of molecules between the two particles may exhibit spatial structure such as being ordered in layers. For instance, when the distance between two plates approaches the diameter of a solvent molecule, it will be difficult to bring the plates any closer together because the layer of molecules in between the plates must be squeezed out of the gap. If the distance between the plates is $1 \cdot 5$ times the molecular diameter, however, there is more empty space (free volume) available in the gap which gives rise to an attraction. It is clear that at this level the force per unit area must be an oscillatory function of separation, as shown schematically in Fig. 1.17.

These forces have been measured experimentally by Horn and Israelachvili (1981) and Christenson and Horn (1983) between two mica plates in octamethylcyclotetrasiloxane and in propylene carbonate, respectively. As yet, no quantitative theory for these forces is available.

In aqueous media the solvation forces are called hydration forces. In contrast to the forces in many organic solvents which are oscillatory, the hydration force in aqueous media is of short range and has an approximately exponential decay with a characteristic length of a few nm. It has been measured between glass fibres, silica surfaces and liquid bilayers (Israelachvili, 1985). An example is shown in Fig. 1.18. The repulsion between two hydrophilic surfaces is not fully understood. Ruckenstein (1986) has com-

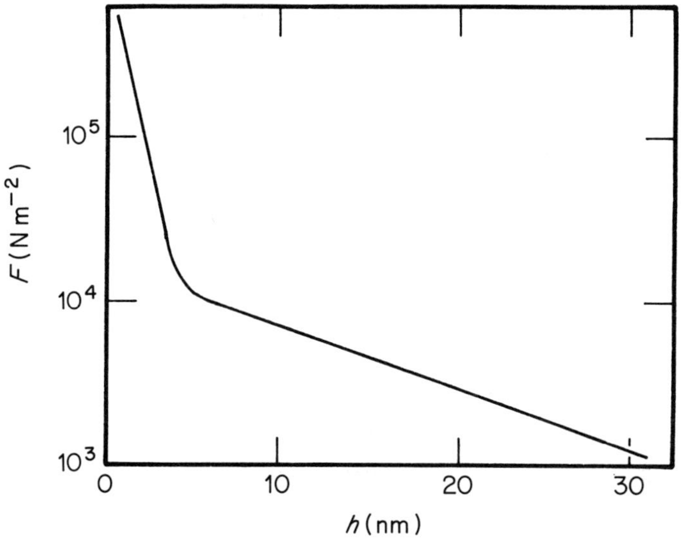

Fig. 1.18. Experimentally observed force per unit area between two silica surfaces in water. At distances of $h \geqslant 4$–5 nm, there is double-layer repulsion; at $h < 4$ nm, a repulsive hydration force is observed (after Peschel *et al.*, 1982).

pared the forces between two negatively charged plates on which positive ions were adsorbed with those between two bilayers with zwitterionic head groups. The repulsion could be a result of the degree of polarization of the water molecules in the gap and (in agreement with experiment) would depend on ionic strength. Belouschek *et al.* (1988) have ascribed the hydration force (in the absence of ions) to polarization of the hydrogen bonds between water molecules, but claimed that the structure of water between two hydrophilic plates is unaffected by this and is the same as bulk water.

Besides the short-range repulsive force that operates between two hydrophilic particles, there also exists a short-range attractive force between two hydrophobic surfaces which, for obvious reasons, is called a hydrophobic force. Little is known about this force and there is some confusion about it in the literature. According to the experiments of Pashley and Israelachvili (1981a,b), who measured this force between mica surfaces covered with HTAB monolayers in water, the hydrophobic attraction becomes stronger than the van der Waals attraction at distances below 6 nm.

It is not known whether or not any short-range force acts between a hydrophobic surface and a hydrophilic surface.

Chapter 2

Brownian Motion of Non-interacting Colloidal Particles

2.1 Brownian motion of neutral colloidal particles

2.1.1 Qualitative description of Brownian motion

Brownian motion can be defined as the perpetual irregular motion executed by small suspended particles due to fluctuations in the mean force exerted on them by the molecules of the suspending fluid. Under normal conditions a Brownian particle will suffer about 10^{21} collisions per second with fluid molecules and it is the combined impact of these collisions which causes the chaotic motion of the particle. This motion was first observed by the botanist Robert Brown (1828), who observed the motion of small pollen grains suspended in water. He was, however, unaware of its cause.

The most direct way to study Brownian motion is by observing, filming and analysing the motion under a microscope. Figure 2.1 shows four sequential frames taken of a suspension of latex spheres suspended in water. By carefully examining and comparing frames such as these, the displacement of each sphere during the time interval between frames can be measured accurately. Similarly, by photographing non-spherical particles the changes in particle orientation caused by rotary Brownian motion can be observed. An example is shown in Fig. 2.2 which shows the Brownian motion of tungsten oxide particles suspended in water.

An attempt to measure the distances the particles travel (or the angle through which they rotate), reveals that the distance (or angle) depends on the film speed used in recording the motion. For example, looking at the position of a particle in four sequential frames, it is found that the particle has travelled a certain distance, as shown schematically in Fig. 2.3a. Suppose the film speed is n frames per second, then the time interval between frames is n^{-1} seconds and the apparent velocity of the particle is

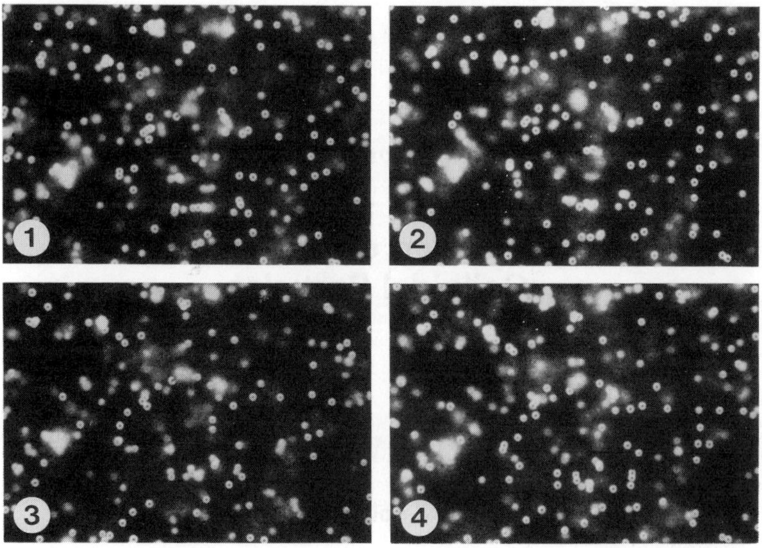

Fig. 2.1. Brownian motion of spherical latex particles suspended in water. Particle radius $a = 0.5\ \mu m$. The time interval between frames is 1 s.

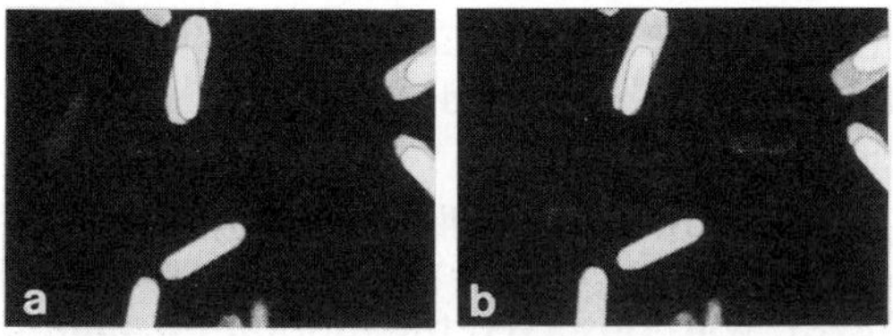

Fig. 2.2. Rotary Brownian motion of non-spherical tungsten oxide particles of length $\ell \simeq 5\ \mu m$ suspended in water. The time interval in between frames is 1 s.

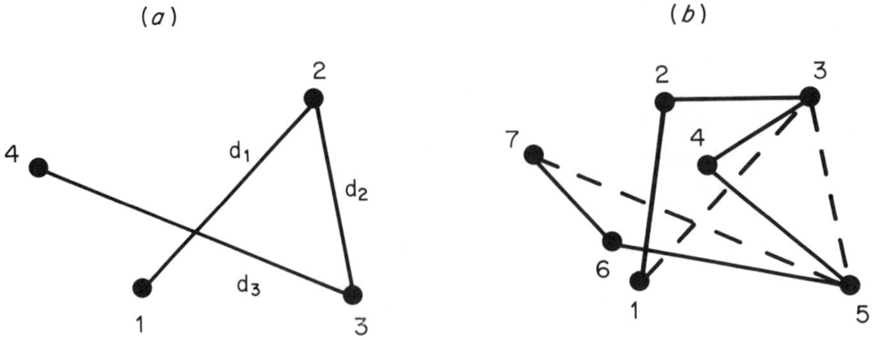

Fig. 2.3. (*a*) Positions of a Brownian particle in four sequential frames filmed at n frames per second.
(*b*) Positions of the same particle filmed at $2n$ frames per second.
The dashed "trajectory" in (b) is the same as the "trajectory" in (a).

$$u_{app} = \frac{1}{3}(d_1 + d_2 + d_3)n \tag{2.1}$$

If the same particle had been filmed at a speed of $2n$ frames per second, the resulting "trajectory" would be that shown in Fig. 2.3b. The increased resolution in time reveals that the particle does not move in a straight line in between two consecutive positions, but makes a "jump" in between. As can be seen from Fig. 2.3b, the apparent velocity of the particle has increased to

$$u_{app} = \frac{1}{3} n \sum_{i=1}^{6} d_i, \tag{2.2}$$

where d_i is the distance between consecutive positions. It follows that the apparent velocity depends on the time scale in which the motion is observed. As we will show later, the average distance $\bar{d}$ travelled during the time interval t between frames is proportional to $t^{1/2}$, and hence is inversely proportional to $n^{1/2}$:

$$\bar{d} \propto t^{1/2} \propto n^{-1/2} \tag{2.3}$$

Therefore

$$u_{app} \propto n^{1/2} \tag{2.4}$$

i.e. the apparent velocity of a Brownian particle approaches infinity as the film speed approaches infinity. Another way of expressing this is to say that the apparent distance a particle travels depends on the magnification used in observing the particle. If the length of the trajectory that a particle follows in a given period of time is measured with a given length scale ℓ, then the total (apparent) length L of the trajectory is found to vary as

$$L \propto \ell^{1-D} \tag{2.5}$$

where D is called the fractal dimension of the trajectory (Mandelbrot, 1975). When a particle moves along a straight line, $D = 1$ and L is constant. When $D > 1$, $L \to \infty$ as $\ell \to 0$. Because for a Brownian particle

$$L \propto u_{app} \propto n^{1/2} \tag{2.6}$$

and

$$\ell \propto \bar{d} \propto n^{-1/2} \tag{2.7}$$

it follows that

$$L \propto \ell^{-1} \tag{2.8}$$

Therefore, the fractal dimension of a Brownian particle is 2, despite the fact that the motion occurs in a space for which $D = 3$. The name fractal is due to the fact that, in general, D can have any value, not just integral values as for the usual spatial dimensions. Fractals are often encountered in colloid and surface science. For example, aggregates or clusters of collodial particles have a fractal nature, rough surfaces have a dimension $2 < D < 3$, as does a fluid front advancing in a porous medium.

Brownian trajectories are an example of self-similar phenomena. If a trajectory is observed at a given magnification over a period of time, a trajectory like the one shown in Fig. 2.4a is seen. If the detailed section PQ of Fig. 2.4a is magnified, as shown in Fig. 2.4b, a qualitatively similar trajectory is observed. The fact that the apparent velocity of a Brownian particle depends on the way it is observed, can be stated differently by saying that the trajectory of a Brownian particle is a non-differential curve. It is impossible to determine the tangent of the trajectory at a given point because it changes magnitude and direction as the time of observation is changed. This explains the failure of early efforts by Exner (1900) who tried to measure the velocity of a Brownian particle. It is clear that the velocity of a Brownian particle is not a physically meaningful concept. What is meaningful instead is the displacement of a particle from a given initial position or orientation. This was first realized independently by Einstein (1905) and Smoluchowski (1906).

The question arises as to whether the apparent velocity of a Brownian particle keeps increasing with film speed or if there exists a film speed beyond

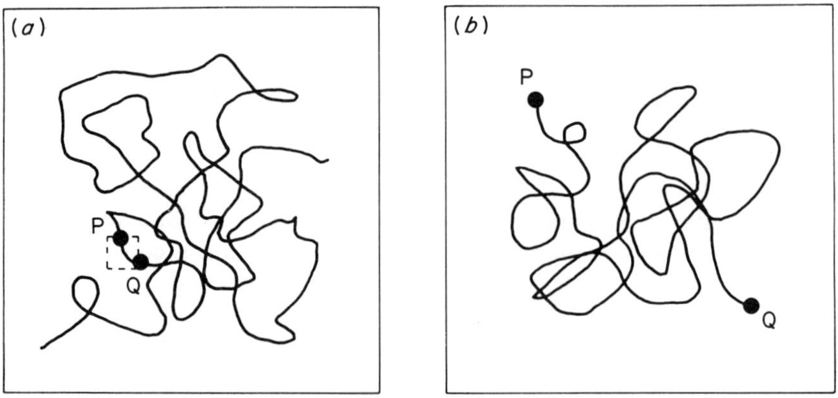

Fig. 2.4. Brownian motion as a self-similar phenomenon. The section PQ of the Brownian trajectory observed at a certain magnification in (*a*) is shown magnified in (*b*). This enlarged trajectory is qualitatively similar in form to the trajectory shown in (*a*).

which the velocity of the particle becomes well defined. If such a velocity exists, then a single jump of a colloidal particle observed between two consecutive frames filmed at n frames per second consists in reality of a large number of much smaller irregular jumps. How big (or rather how small) would such a jump have to be and how much time would it take? To answer these questions let us consider a colloidal particle which, due to collisions with fluid molecules, acquires an initial velocity u_0 at a time $t=0$. When the particle moves through the liquid it experiences a drag force given by Stokes law:

$$\boldsymbol{F}_{\text{drag}} = -f\boldsymbol{u} \tag{2.9}$$

where f is the friction coefficient of the particle. We have seen that for a spherical particle $f=6\pi\mu a$ (cf. eq (1.68)). According to Newton's second law, the acceleration $\boldsymbol{a}$ due to the application of a force $\boldsymbol{F}$ on a particle of mass m is

$$\boldsymbol{F} = m\boldsymbol{a} \tag{2.10}$$

Combining eqs (2.9) and (2.10) yields the force balance for the particle

$$ma = -fu \tag{2.11}$$

Considering motion in the X direction only and using the following notations

$$a = \frac{\mathrm{d}u}{\mathrm{d}t} = \frac{\mathrm{d}^2x}{\mathrm{d}t^2} = \ddot{x} \text{ and } u = \frac{\mathrm{d}x}{\mathrm{d}t} = \dot{x}$$

eq (2.11) can be rewritten as

$$\ddot{x} = -\beta\dot{x} \tag{2.12}$$

where $\beta = f/m$ is the friction coefficient per unit mass of the particle. The solution of eq (2.12) for a particle at the origin and moving with velocity u_0 at $t=0$ is

$$x(t) = \frac{u_0}{\beta}(1 - \mathrm{e}^{-\beta t}) \tag{2.13}$$

and its velocity equals

$$u = u_0\mathrm{e}^{-\beta t} \tag{2.14}$$

It can be seen that β has the dimension of inverse time, so that β^{-1} can be regarded as a characteristic relaxation time. Loosely speaking, this relaxation time can be identified as the time of a single "jump". Suppose the particle comes to rest after travelling a distance ℓ, then according to eq (2.13):

$$\ell = x(t=\infty) = u_0/\beta \tag{2.15}$$

Again loosely speaking, ℓ can be equated with the length of a jump. Therefore, in an approximate sense, Brownian motion can be thought of as a series of β jumps per second of length ℓ. An estimate of u_0 (and hence of ℓ) can be made from the equipartition of energy (in one dimension):

$$\frac{1}{2}m\overline{u_0^2} = \frac{1}{2}kT \tag{2.16}$$

Therefore

$$u_0 \simeq \sqrt{\overline{u_0^2}} = \sqrt{kT/m} \tag{2.17}$$

or

$$\ell \simeq \frac{\sqrt{mkT}}{f} \tag{2.18}$$

Table 2.1 gives estimates of β^{-1} and ℓ for spherical particles suspended in air (aerosols) or in water (hydrosols). It is assumed that the density of the particles is 1 kg/dm^3.

As can be seen from Table 2.1, for all practical purposes, β^{-1} is extremely short compared with a typical observation time. To observe the "true" jumps of a 100-nm particle suspended in water, it must be filmed with a film speed of more than a billion frames per second. Taking a frame whose

TABLE 2.1
Estimates of the duration and length of jumps of Brownian particles

a	Aerosols[a]		Hydrosols[b]	
(μm)	β^{-1} (s)	ℓ (nm)	β^{-1} (s)	ℓ (nm)
10	10^{-3}	35	2×10^{-5}	0·7
1	10^{-5}	14	2×10^{-7}	0·3
0·1	10^{-7}	3·5	2×10^{-9}	0·07
0·01	10^{-9}	1·4	2×10^{-11}	0·03

[a] $\mu \simeq 2 \times 10^{-5}$ Pa s.
[b] $\mu \simeq 1 \times 10^{-3}$ Pa s.

length is about 1 cm, this corresponds to moving the film through a camera at about the speed of light, while at the same time observing the position of the particle with sub-Angstrom precision. Because of the extremely large number of collisions per second with fluid molecules, the true motion is not really a series of jumps, but a very irregular motion having the same kinetic energy as that of the fluid molecules (i.e. that given by eq (2.16)).

2.1.2 Brownian motion as a random walk

2.1.2.1 Translational Brownian motion

We have seen that Brownian motion can be regarded as a series of jumps of irregular magnitude and in arbitrary directions. A jump in an arbitrary direction can be decomposed into three equivalent independent jumps along the three axes of a Cartesian-coordinate system. Therefore it is useful and simpler to look at jumps in one dimension along a line. Let us assume further that each jump is of the same size and has an equal probability of being in a forward or backward direction. We can then ask what the probability is that the particle will be at a certain position after N jumps. This problem is usually referred to as the *random-walk* problem (Chandrasekhar, 1943).

After N jumps the particle could be at any one of the points

$$-N,\ -N+1,\ .\ .\ .\ .\ .,\ -2,\ -1,\ 0,\ 1,\ 2,\ .\ .\ .\ .\ .\ N-1,\ N$$

The probability $W(m,N)$ of the particle arriving at point m in N jumps can be expressed as

$$W(m,N) = n_s w \tag{2.19}$$

where w is the probability of any given sequence of jumps and n_s the number of sequences leading to m after N jumps. As the probability of going forward (or backward) in any one jump is $\frac{1}{2}$, it follows that

$$w = (\tfrac{1}{2})^N \tag{2.20}$$

To calculate n_s, suppose first that $m > 0$. Then to arrive at position m, at least m positive jumps are needed. The remaining $N-m$ jumps must be half in the positive direction and half in the negative direction. Therefore, there must be $\frac{1}{2}(N-m)$ jumps in the negative direction and $m+\frac{1}{2}(N-m) = \frac{1}{2}(N+m)$ jumps in the positive direction for the particle to arrive at m in N jumps. The same result is obtained when $m < 0$. Hence the number of possible sequences, n_s, leading to m after N jumps, is the number of ways $\frac{1}{2}(N+m)$

jumps can be chosen from N jumps. From combinatorial theory, the number of ways in which n objects can be chosen from s objects is

$$\frac{s!}{n!(s-n)!}$$

and so the number of ways in which $\frac{1}{2}(N+m)$ jumps can be chosen from N jumps is

$$n_s = \frac{N!}{[\frac{1}{2}(N+m)]![N-\frac{1}{2}(N+m)]!} = \frac{N!}{[\frac{1}{2}(N+m)]![\frac{1}{2}(N-m)]!} \tag{2.21}$$

Combining eqs (2.19) and (2.21) gives:

$$W(m,N) = \frac{N!}{[\frac{1}{2}(N+m)]![\frac{1}{2}(N-m)]!}\left(\frac{1}{2}\right)^N \tag{2.22}$$

Equation (2.22) is an example of a *Bernouillian distribution*. The case of physical interest is when $N \gg 1$ and $m \ll N$, i.e. the particle makes a very large number of jumps (β per second), but it does not move very far from its initial position. Taking the natural logarithm of eq (2.22) and making use of Stirling's approximation valid for large n:

$$\ln n! = (n+\tfrac{1}{2})\ln n - n + \tfrac{1}{2}\ln 2\pi + \ldots \tag{2.23}$$

and of the series expansion

$$\ln\left(1 \pm \frac{m}{N}\right) = \pm\frac{m}{N} - \tfrac{1}{2}\left(\frac{m}{N}\right)^2 \pm \ldots \tag{2.24}$$

leads to

$$\ln W(m,N) = -\tfrac{1}{2}\ln N + \tfrac{1}{2}\ln 2 - \tfrac{1}{2}\ln\pi - \frac{m^2}{2N} \tag{2.25}$$

or

$$W(m,N) = \left(\frac{2}{\pi N}\right)^{1/2} \mathrm{e}^{-m^2/2N} \tag{2.26}$$

Therefore, for large N and $m/N \ll 1$, the Bernouillian distribution reduces to a *Gaussian distribution*.

Instead of considering the discrete variables m and N, it is convenient to express the results in terms of the probability $W(x,t)\mathrm{d}x$, i.e. of finding the particle at a distance x (more precisely between x and $x+\mathrm{d}x$) at time t. This probability is related to $W(m,N)$ by

$$W(x,t)\,\mathrm{d}x = W(m,N)\frac{\mathrm{d}x}{2\ell} \tag{2.27}$$

where

$$x = m\ell \tag{2.28}$$

ℓ being the length of a jump, with the number of jumps per unit time given by

$$n = N/t \tag{2.29}$$

The factor 2 in eq (2.27) arises because m can take only even or odd values depending on whether N is even or odd. From eqs (2.27) to (2.29) it follows that

$$W(x,t) = \frac{1}{(2\pi n\ell^2 t)^{1/2}}\,\mathrm{e}^{-x^2/2n\ell^2 t} \tag{2.30}$$

From eq (2.30) it can be shown that

$$\int_{-\infty}^{\infty} W(x,t)\,\mathrm{d}x = 1 \tag{2.31}$$

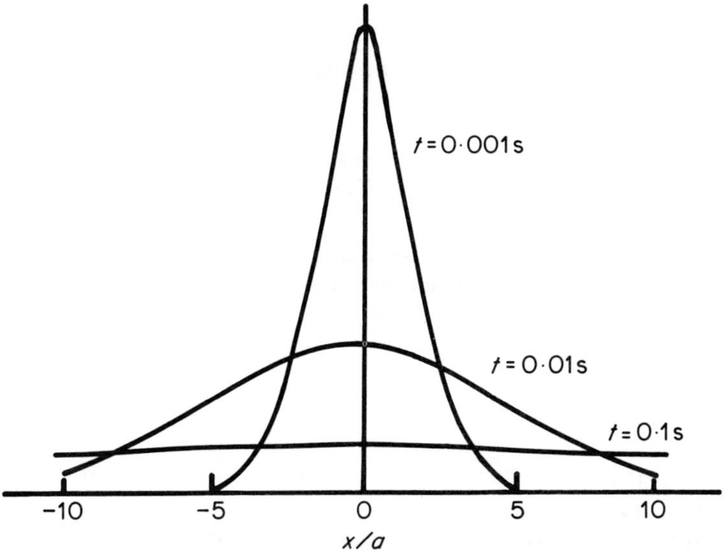

Fig. 2.5. The distribution of particles undergoing a random walk, initially at the origin, for various times t. $n\ell = 2{\cdot}5 \times 10^{-12}\ \mathrm{m^2\ s^{-1}}$, $a = 0.1\ \mu\mathrm{m}$.

which means that it is certain that the particle will be found "somewhere" on the line.

The time development of a distribution of an ensemble of non-interacting particles, initially at the origin, each undergoing a simple random walk, is shown in Fig. 2.5. From the Gaussian distribution (eq (2.30)) one can find that the average displacement at any time is zero:

$$\langle x \rangle = \int_{-\infty}^{\infty} x W(x,t)\, dx = 0 \tag{2.32}$$

while the mean absolute displacement is

$$\langle |x| \rangle = 2 \int_{0}^{\infty} x W(x,t)\, dx = \ell \sqrt{\frac{2nt}{\pi}} \tag{2.33}$$

and the mean-square displacement

$$\langle x^2 \rangle = \int_{-\infty}^{\infty} x^2 W(x,t)\, dx = n \ell^2 t \tag{2.34}$$

If Brownian motion is interpreted as a series of β jumps per second of length, ℓ, (given by eq (2.18)) then the mean-sqare displacement is found to be

$$\langle x^2 \rangle \simeq \frac{f}{m} \frac{m kT}{f^2} = \frac{kT}{f} t \tag{2.35}$$

However, the analogy between Brownian motion and a random walk is not exact and the correct result, as will be shown in Section 2.1.3, for the mean-square displacement is twice as large:

$$\langle x^2 \rangle = 2 \frac{kT}{f} t \tag{2.36}$$

Nevertheless, the analogy is close and provides a convenient intuitive way of interpreting Brownian motion.

The result (eq (2.30)) can be readily generalized to three dimensions. If the jumps are completely random, then there is no correlation between the displacements in each of the three spatial directions and we can write:

$$W(x_1,x_2,x_3,t)\,dx_1dx_2dx_3 = W(x_1,t)\,dx_1\,W(x_2,t)\,dx_2\,W(x_3,t)\,dx_3$$

$$= \frac{1}{(2\pi n\ell^2 t)^{3/2}} \exp\left[-\frac{(x_1^2+x_2^2+x_3^2)}{2n\ell^2 t}\right] dx_1dx_2dx_3 \qquad (2.37)$$

from which it follows that

$$\langle x_i x_j \rangle = n\ell^2 t\mathbf{I} \qquad (2.38)$$

and

$$\langle r^2 \rangle = \langle x_1^2 \rangle + \langle x_2^2 \rangle + \langle x_3^2 \rangle = 3n\ell^2 t$$

2.1.2.2. Rotary Brownian motion

Besides undergoing a perpetual irregular translational motion, colloidal particles also change their orientation ceaselessly in a similarly irregular manner. A simple case to consider is the rotary motion of an axisymmetric particle, such as a rod, confined to a plane (see Fig. 2.6).

In applying the random-walk model to such a particle, it is assumed that

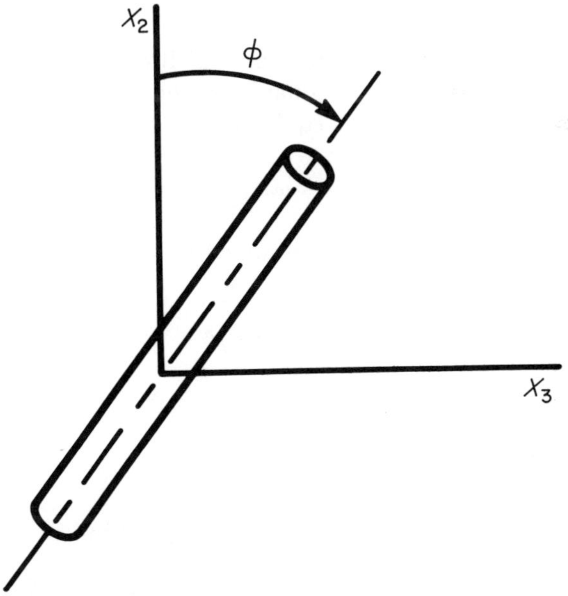

Fig. 2.6. Orientation of a rod-shaped particle. Rotary Brownian motion causes variations in the angle ϕ.

the particle makes n angular jumps per second of magnitude ϵ. Therefore for a particle initially oriented at an angle ϕ_0:

$$W(\phi,t)=\frac{1}{(2\pi n\epsilon^2 t)^{1/2}}\,\mathrm{e}^{-\Delta\phi^2/2n\epsilon^2 t} \tag{2.40}$$

where $\Delta\phi=\phi-\phi_0$.

For rotational motion, n and ϵ can be estimated from a torque balance on a particle. The analogue of Newton's second law for rotational motion is

$$L=I\ddot{\phi} \tag{2.41}$$

where L is the total torque exerted on the particle, I is its moment of inertia, and $\ddot{\phi}=\mathrm{d}^2\phi/\mathrm{d}t^2$ its angular acceleration. Hence a torque balance yields (cf. eq (2.11)):

$$I\ddot{\phi}=-f_{\mathrm{r}}\dot{\phi} \tag{2.42}$$

where f_{r} is the rotary friction coefficient and $\dot{\phi}$ the angular velocity. Equation (2.42) can be rewritten as

$$\ddot{\phi}=-\beta_{\mathrm{r}}\dot{\phi} \tag{2.43}$$

with $\beta_{\mathrm{r}}=f_{\mathrm{r}}/I$. By analogy with translational Brownian motion, it follows that a particle executes about β_{r} angular jumps per second. The magnitude of the jump can be found from (cf. eq (2.15))

$$\epsilon=\omega_0/\beta_{\mathrm{r}} \tag{2.44}$$

where ω_0 is the angular velocity determined by the equipartition of kinetic and thermal energy:

$$\tfrac{1}{2}I\overline{\omega_0^2}=\tfrac{1}{2}kT \tag{2.45}$$

and therefore

$$\epsilon\simeq\frac{\sqrt{IkT}}{f_{\mathrm{r}}} \tag{2.46}$$

For spherical particles, $f_{\mathrm{r}}=8\pi\mu a^3$ and $I=\frac{2}{5}ma^2$ and hence $\beta_{\mathrm{r}}^{-1}=(9/20)\beta^{-1}$ and $\epsilon=0.58\ \ell/a$ rad. Estimates of β^{-1} and ℓ were given earlier in Table 2.1. For a particle with a 100-nm radius in water, $\beta_{\mathrm{r}}\simeq10^{-9}$ s and $\epsilon\simeq4\times10^{-4}$ rad ($\simeq0{\cdot}02°$). For a doublet of touching spheres of radius a,

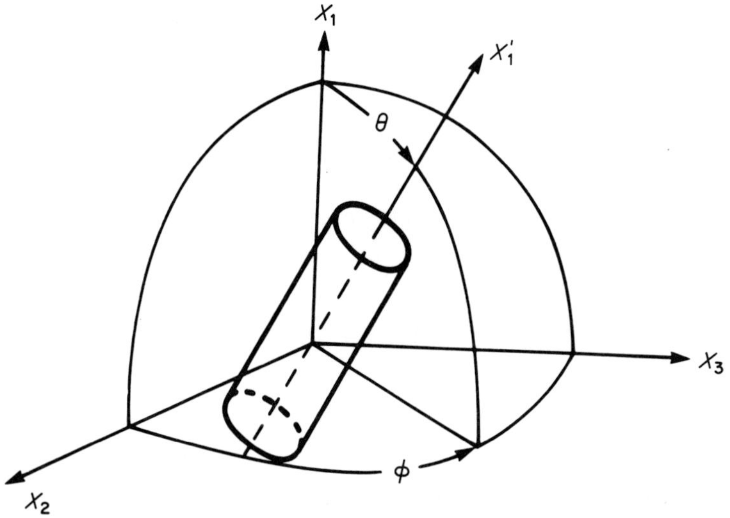

Fig. 2.7. The orientation of an axisymmetric particle can be described by the polar angles θ and ϕ.

$f_r = 29 \cdot 92\ \pi\mu a^3$ and the moment of inertia about a symmetry axis perpendicular to the axis of revolution is $I = (12/5)ma^2$, whence $\epsilon \simeq \ell/a \simeq 0 \cdot 01°$.

From eq (2.40) it is found that

$$\langle \Delta\phi^2 \rangle = n\epsilon^2 t \tag{2.47}$$

Substituting the estimates of $n = \beta_r$ and ϵ into eq (2.47) results in $\langle \Delta\phi^2 \rangle \simeq (kT/f_r)t$. The correct result is

$$\langle \Delta\phi^2 \rangle = 2\frac{kT}{f_r}t \tag{2.48}$$

The orientation of axisymmetric particles in three dimensions can be described by the polar angles θ and ϕ (see Fig 2.7). It can be readily shown that

$$\langle \Delta\theta^2 \rangle = 2\frac{kT}{f_r}t \tag{2.49a}$$

and

$$\langle (\sin\theta\ \Delta\phi)^2 \rangle = 2\frac{kT}{f_r}t \tag{2.49b}$$

or, defining $e_i = \Delta\theta$ and $e_j = \sin\theta\, \Delta\phi$:

$$\langle e_i e_j \rangle = 2\frac{kT}{f_r} t\boldsymbol{I} \tag{2.49c}$$

Besides randomly changing the angles θ and ϕ, an axisymmetric particle also executes a random motion about its axis of revolution. This motion, however, does not change the particle orientation. Again the mean-square angular displacement is given by

$$\langle \Delta\gamma^2 \rangle = 2\frac{kT}{f_s} t \tag{2.50}$$

where f_s is the spin friction coefficient.

2.1.3 Langevin theory of Brownian motion

2.1.3.1 Translational Brownian motion

The random trajectories of Brownian particles are due to random fluctuations in the forces that colliding fluid molecules collectively exert on them. A fluctuating force $R(t)$ can be said to be acting on each particle. Because $R(t)$ is a random force, its instantaneous magnitude cannot be predicted but only its statistical average or the probability of it having a certain magnitude. As such, $R(t)$ is not an ordinary function of time, but a stochastic function. Formally (Uhlenbeck and Ornstein, 1930), the force balance on a colloidal particle can be written as (cf. eq (2.11)):

$$m\ddot{x} = -f\dot{x} + R(t) \tag{2.51}$$

or

$$\ddot{x} = -\beta\dot{x} + A(t) \tag{2.52}$$

where $A(t) = R(t)/m$.

Equation (2.52) is commonly referred to as the *Langevin equation*. It can be solved formally by treating $A(t)$ as a regular function. Integrating once yields:

$$\dot{x} = u = u_0 e^{-\beta t} + e^{-\beta t} \int_0^t e^{\beta t} A(t)\, dt \tag{2.53}$$

Integrating a second time gives

$$x = u_0 \int_0^t \mathrm{e}^{-\beta t}\mathrm{d}t + \int_0^t f(t)\mathrm{e}^{-\beta t}\mathrm{d}t \tag{2.54}$$

where

$$f(t) = \int_0^t \mathrm{e}^{\beta t} A(t)\,\mathrm{d}t$$

Integrating the second term on the right-hand side of eq (2.54) by parts results in

$$x = \frac{u_0}{\beta}\left(1 - \mathrm{e}^{-\beta t}\right) - \frac{\mathrm{e}^{-\beta t}}{\beta}\int_0^t \mathrm{e}^{\beta t} A(t)\,\mathrm{d}t + \frac{1}{\beta}\int_0^t A(t)\,\mathrm{d}t \tag{2.55}$$

Integrals such as those in eqs (2.53) to (2.55) have no real meaning because $A(t)$ is not well defined. However, average values of these integrals can be obtained. Because of the random nature of the force $A(t)$, we have

$$\langle A(t) \rangle = 0 \tag{2.56}$$

and

$$\langle A(t_1)A(t_2) \rangle = K\,\delta(t_1 - t_2) \tag{2.57}$$

Equation (2.56) states that the average force is zero and eq (2.57) indicates that there is no correlation between the random force at time t_1 and the force at another time t_2. (When $t_1 \neq t_2$, the Dirac delta function $\delta(t_1 - t_2) = 0$. When $t_1 = t_2$, $\delta(t_1 - t_2) = 1$.) K is a constant to be determined.

Squaring eq (2.55) and averaging results in:

$$\langle x^2 \rangle = \frac{u_0^2}{\beta^2}(1 - \mathrm{e}^{-\beta t})^2 - \frac{\mathrm{e}^{-2\beta t}}{\beta^2}\int_0^t\int_0^t \mathrm{e}^{\beta(t_1 + t_2)}\langle A(t_1)A(t_2)\rangle\,\mathrm{d}t_1\mathrm{d}t_2$$

$$+ \frac{1}{\beta^2}\int_0^t\int_0^t \langle A(t_1)A(t_2)\rangle\,\mathrm{d}t_1\mathrm{d}t_2$$

$$- \frac{2\mathrm{e}^{-\beta t}}{\beta^2}\int_0^t\int_0^t \mathrm{e}^{\beta t_1}\langle A(t_1)A(t_2)\rangle\,\mathrm{d}t_1\mathrm{d}t_2 \tag{2.58}$$

In writing eq (2.58) terms containing the factor $\langle A(t) \rangle$ have been omitted, because their contribution is zero according to the condition given by eq (2.56). The integrals in eq (2.58) can be evaluated by changing to the new variables $v = t_1 + t_2$ and $w = t_1 - t_2$. For example, evaluating the third integral on the right-hand side of eq (2.58) yields

$$\frac{1}{\beta^2}\int_0^t\int_0^t \langle A(t_1)A(t_2)\rangle \mathrm{d}t_1\mathrm{d}t_2 = \frac{1}{\beta^2}\int_0^{2t}\mathrm{d}v\int_{-\infty}^{\infty}K\,\delta(w)\mathrm{d}w = \frac{Kt}{\beta^2} \tag{2.59}$$

in which we have used eq (2.57), and, because $\delta(w) = 0$ for $w \neq 0$, we can integrate from minus to plus infinity. Evaluating the other integrals appearing in eq (2.58) in a similiar manner gives:

$$\langle x^2 \rangle = \frac{u_0^2}{\beta^2}(1 - \mathrm{e}^{-\beta t})^2 + \frac{Kt}{\beta^2} + \frac{K}{2\beta^3}(-3 + 4\mathrm{e}^{-\beta t} - \mathrm{e}^{-2\beta t}) \tag{2.60}$$

As $t \to \infty$, eq (2.60) reduces to

$$\langle x^2 \rangle = \frac{Kt}{\beta^2} \tag{2.61}$$

Comparing this result to eq (2.36) yields

$$K = \frac{2fkT}{m^2} \tag{2.62}$$

The same result for K can be found by averaging the velocity u (eq (2.53)) and comparing with eq (2.16). As $t \to 0$, eq (2.60) reduces to

$$\langle x^2 \rangle = u_0^2 t^2 \tag{2.63}$$

Therefore, on average, the particle moves with uniform velocity u_0.

The asymptotic results of eqs (2.62) and (2.63) show that the time β^{-1} may be interpreted as the decay time of the velocity autocorrelation function. At times $t \gg \beta^{-1}$, particle velocities are uncorrelated and a particle moves according to a random walk. For times $t \ll \beta^{-1}$ the particle velocities at two different times are correlated.

2.1.3.2. Rotary Brownian motion

The same analysis as described above for translational Brownian motion can be applied to rotary Brownian motion. The fundamental equation is again the Langevin equation, which takes the form

$$\ddot{\phi} = -\beta_r \dot{\phi} + A_r(t) \tag{2.64}$$

where $A_r(t) = L_r(t)/I$, $L_r(t)$ being a random torque exerted on the particle by the fluid molecules. As before

$$\langle A_r(t) \rangle = 0 \tag{2.65}$$

and

$$\langle A_r(t_1)\, A_r(t_2) \rangle = K_r \delta(t_1 - t_2) \tag{2.66}$$

Hence

$$\langle \Delta\phi^2 \rangle = \frac{\omega_0^2}{\beta_r^2}\left(1 - e^{-\beta_r t}\right)^2 + \frac{K_r t}{\beta_r^2} + \frac{K_r}{2\beta_r^3}\left(-3 + 4e^{-\beta_r t} - e^{-2\beta_r t}\right) \tag{2.67}$$

For $t \gg \beta_r^{-1}$:

$$\langle \Delta\phi^2 \rangle = \frac{K_r t}{\beta} = 2\frac{kT}{f_r} t \tag{2.68}$$

2.1.4 Brownian motion and diffusion

Diffusion can be defined as the spontaneous equalization of the concentration of colloidal (or other) particles as a result of the chaotic Brownian motion of each particle. Let us consider two containers with number concentrations of n_1 and n_2, respectively, divided by a surface A, with $n_1 > n_2$ (see Fig. 2.8). Each particle on each side of the dividing surface A undergoes random Brownian displacement. Each particle is equally likely to move to the right or to the left. Obviously, a particle in compartment I does not "know" that it must diffuse to compartment II. However, because there are more particles in compartment I than in compartment II, at the interface between them there will be more particles jumping from compartment I toward compartment II than vice versa and so, on average, there will be a net transport of particles from higher to lower concentration. In other words, diffusion only occurs when concentration gradients exist in a dispersion. The net number of particles, j, diffusing through area A per unit time is simply

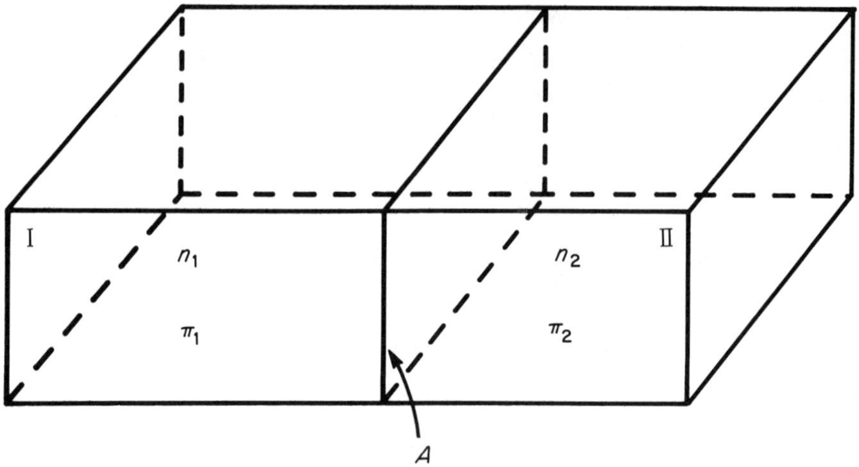

Fig. 2.8. Two compartments I and II, divided by a surface A. The concentration and osmotic pressure in I and II are denoted by n_1 and π_1, and n_2 and π_2, respectively.

proportional to the concentration gradient and the surface area A. In one dimension:

$$j = -DA\frac{dn}{dx} \tag{2.69}$$

where D is a proportionality constant called the diffusion constant (its dimensions are $m^2\ s^{-1}$). The minus sign in eq (2.69) occurs because $dn/dx < 0$ at surface A.

The number of particles passing a unit surface per unit time is called a flux, denoted by the symbol J. Hence

$$J = -D\frac{dn}{dx} \tag{2.70}$$

More generally, in three dimensions

$$J = -D\nabla n \tag{2.71}$$

which is known as *Fick's first law of diffusion.*

As a result of the Brownian motion of each particle, there exists a net transport through the dividing surface. The system appears to behave as if there was a fictitious force acting on the particles in compartment I, pushing them into compartment II. This fictitious force is often called a Brownian force, a random force or a thermodynamic force. This force should not be confused with the truly random force acting on Brownian particles as described in Section 2.1.3. This fictitious force manifests itself as a pressure difference between compartments I and II, commonly referred to as the *osmotic pressure*. According to *van 't Hoff's law*, the change in the osmotic pressure $\Delta\pi$ is related to the difference in the concentration of colloidal particles on either side of the surface:

$$\Delta\pi = kT\Delta n \tag{2.72}$$

Thus the fictitious force per unit volume can be written as

$$F = -\frac{\Delta\pi}{\Delta x} = -kT\frac{\mathrm{d}n}{\mathrm{d}x} \tag{2.73}$$

and the random force per particle as

$$F_{\mathrm{r}} = -\frac{kT}{n}\frac{\mathrm{d}n}{\mathrm{d}x} \tag{2.74}$$

which vanishes when the concentration is the same everywhere. Hence, as a result of Brownian motion, it is as if there is a deterministic force proportional to the gradient of log n acting on each particle. As the total force on each particle must be zero, this force must balance the Stokes drag force:

$$-\frac{kT}{n}\frac{\mathrm{d}n}{\mathrm{d}x} = fu \tag{2.75}$$

where f is the friction coefficient, or

$$un = -\frac{kT}{f}\frac{\mathrm{d}n}{\mathrm{d}x} \tag{2.76}$$

The product un is just the flux of particles passing through the surface A, moving with velocity u; thus

$$J = -\frac{kT}{f}\frac{\mathrm{d}n}{\mathrm{d}x} \tag{2.77}$$

Comparing this result with eq (2.70) shows that

$$D = \frac{kT}{f} \tag{2.78}$$

and the famous Stokes–Einstein relation is recovered. Thus diffusion can be considered to be a phenomenon in which each particle experiences a thermodynamic force and, as a result, moves with a Brownian velocity u, given by eq (2.76). Again, this Brownian velocity must not be confused with the true motion of a colloidal particle, to which, as we have seen, a velocity cannot be ascribed.

From a knowledge of the number of particles passing through a given surface area A, a mass balance equation can be derived for a volume element ΔV enclosed by six surface areas (see Fig. 2.9). First consider the fluxes in the X_1 direction only. Consider a unit area at x_1 and at $x_1 + \Delta x_1$, then $\Delta V = A\Delta x_1 = \Delta x_1$. Suppose that there is a concentration gradient at x_1, as well as a flow of particles due to fluid motion or caused by an external field, then the number of particles entering the volume element ΔV per unit time is

$$J_1 = J_p(x_1) + J_0(x_1) \tag{2.79}$$

where J_p is the flux due to Brownian motion and J_0 that due to the flow or the external field. The subscripts "p" and "o" refer to *perikinetic* motion (i.e.

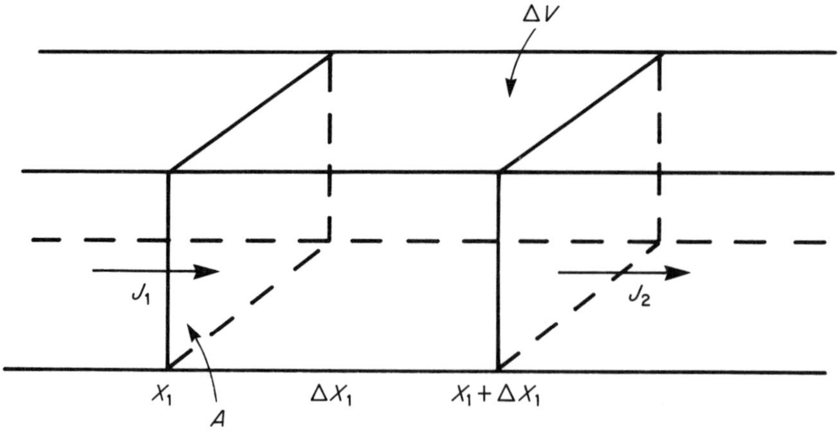

Fig. 2.9. The volume element ΔV enclosed by six surface areas. The fluxes J_1 and J_2 passing through two areas in the X_1 direction are shown. Similar fluxes pass through areas in the X_2 and X_3 directions.

random) and *orthokinetic* motion (i.e. caused by deterministic forces). $J_0(x_1)$ can be expressed as (cf. eqs (2.76) and (2.77))

$$J_0(x_1) = n(x_1)\,u(x_1) \tag{2.80}$$

where $u(x_1)$ is the deterministic velocity in the absence of Brownian motion. Using Fick's first law for the flux at x_1 gives:

$$J_1 = -D\left(\frac{\mathrm{d}n}{\mathrm{d}x_1}\right)_{x_1} + n(x_1)\,u(x_1) \tag{2.81}$$

and similarly for the flux at $x_1 + \Delta x_1$

$$J_2 = -D\left(\frac{\mathrm{d}n}{\mathrm{d}x_1}\right)_{x_1+\Delta x_1} + n(x_1+\Delta x_1)\,u(x_1+\Delta x_1) \tag{2.82}$$

where the subscripts 1 and 2 refer to the values at x_1 and $x_1 + \Delta x_1$, respectively. If $J_1 \neq J_2$, an accumulation or depletion of particles occurs in the volume element ΔV. If the change in the number of particles per unit volume is denoted by Δn, then in a time interval Δt the accumulation in ΔV is $\Delta V(\Delta n/\Delta t)$.
Therefore

$$\Delta V\frac{\Delta n}{\Delta t} = J_1 - J_2 = -\Delta J = -\Delta V\frac{\Delta J}{\Delta V} = -\Delta V\frac{\Delta J}{\Delta x} \tag{2.83}$$

or, in the limit of small ΔV

$$\frac{\partial n}{\partial t} = -\frac{\partial J}{\partial x_1} = -\frac{\partial}{\partial x_1}\left(D\frac{\partial n}{\partial x_1}\right) + \frac{\partial}{\partial x_1}(un) \tag{2.84}$$

The same reasoning applies to the difference in the fluxes in the X_2 and X_3 directions. Thus eq. (2.84) can be generalized to

$$\frac{\partial n}{\partial t} = -\nabla\cdot\boldsymbol{J} \tag{2.85}$$

or

$$\frac{\partial n}{\partial t} = \nabla\cdot(D\nabla n) - \nabla\cdot(un) \tag{2.86}$$

which is known as the *convective-diffusion equation* introduced in Chapter 1 (eq (1.1)). For the case in which D is a constant and $\boldsymbol{u}=0$, eq (2.86) reduces to *Fick's second law of diffusion*:

$$\frac{\partial n}{\partial t}=D\nabla^2 n \tag{2.87}$$

Similar reasoning in orientational space results in the rotary convective-diffusion equation

$$\frac{\partial p}{\partial t}=\nabla\cdot(D_r\nabla p)-\nabla\cdot(\omega p) \tag{2.88}$$

p being the orientation distribution function, D_r the rotary diffusion constant, and ω the deterministic angular velocity. When D_r is a constant and $\omega=0$, eq (2.88) reduces to the rotary-diffusion equation

$$\frac{\partial p}{\partial t}=D_r\nabla^2 p \tag{2.89}$$

It is of interest to look at the solutions of eqs (2.87) and (2.89) for a point source at the origin (corresponding to a large number, N, of non-interacting particles initially at the origin). For translational diffusion in one dimension the solution is

$$n(x,t)=\frac{1}{2(\pi Dt)^{1/2}}\,\mathrm{e}^{-x^2/4Dt} \tag{2.90}$$

Comparing this result with the random-walk model (cf. eq (2.30)), it can be seen that the random walk can be regarded as a diffusion process with the diffusion constant $D=\frac{1}{2}n\ell^2$. From eq (2.90) it follows that the mean-square displacement is given by

$$\langle x^2\rangle=\int_{-\infty}^{\infty}x^2 n(x,t)\,\mathrm{d}x=2Dt \tag{2.91}$$

in agreement with eqs (2.34) and (2.36).

Similarly, for rotational diffusion (in one dimension, $\phi=0$ at $t=0$):

$$p(\phi, t) = \frac{1}{2(\pi D_r t)^{1/2}} e^{-\phi^2/4D_r t} \tag{2.92}$$

and

$$\langle \phi^2 \rangle = 2D_r t \tag{2.93}$$

Table 2.2 gives the values of the translational and rotational diffusion constants for spheres, together with typical root mean square (r.m.s.) displacements.

TABLE 2.2
Translational and rotational diffusion constants for spheres, together with typical r.m.s. displacements in 0·1 s (in water at room temperature)

a	D_t ($m^2\,s^{-1}$)	$\bar{\Delta}_t$ (μm)	D_r (s^{-1})	$\bar{\Delta}_r$ (°)
1 nm	$2{\cdot}2 \times 10^{-10}$	6·6	$1{\cdot}7 \times 10^{8}$	914×360
10 nm	$2{\cdot}2 \times 10^{-11}$	2·1	$1{\cdot}7 \times 10^{5}$	29×360
0·1 μm	$2{\cdot}2 \times 10^{-12}$	0·66	$1{\cdot}7 \times 10^{2}$	329
1 μm	$2{\cdot}2 \times 10^{-13}$	0·21	$1{\cdot}7 \times 10^{-1}$	10
10 μm	$2{\cdot}2 \times 10^{-14}$	0·07	$1{\cdot}7 \times 10^{-4}$	0·3

2.1.5 Brownian motion of arbitrarily shaped rigid particles

For non-spherical particles the diffusion can no longer be described by a simple constant D. Instead the diffusion is anisotropic and the particle is more likely to diffuse more rapidly in some preferred direction (depending on its geometry). For example, a cylindrical particle diffuses faster in the direction of its axis of revolution than in a direction perpendicular to it, because of a difference in hydrodynamic drag. In such cases the diffusion must be described by a diffusion tensor rather than a diffusion constant. When there is no coupling between translational and rotational motion, the diffusion tensor can be found from the resistance tensor, via a generalized Stokes–Einstein equation (Happel and Brenner, 1983):

$$\boldsymbol{F} = -\boldsymbol{f}_t \cdot \boldsymbol{u} \tag{2.94}$$

$$\mathbf{L} = -\boldsymbol{f}_r \cdot \boldsymbol{\omega} \tag{2.95}$$

$$\boldsymbol{D}_t : \boldsymbol{f}_t = kT \tag{2.96}$$

and

$$\boldsymbol{D}_{\mathrm{r}}:\boldsymbol{f}_{\mathrm{r}}=kT \tag{2.97}$$

Similarly, the diffusion constants appearing in the convective-diffusion equation must be replaced by tensors:

$$\frac{\partial n}{\partial t}=\nabla\cdot(\boldsymbol{D}_{\mathrm{t}}\cdot\nabla n)-\nabla\cdot(\boldsymbol{u}n) \tag{2.98}$$

$$\frac{\partial p}{\partial t}=\nabla\cdot(\boldsymbol{D}_{\mathrm{r}}\cdot\nabla p)-\nabla\cdot(\omega p) \tag{2.99}$$

The symmetry of a particle determines how many non-zero components the diffusion tensors contain. For spheres, each tensor contains three non-zero components which are all equal; i.e.

$$\boldsymbol{D}_{\mathrm{t}}=\begin{pmatrix} D & 0 & 0 \\ 0 & D & 0 \\ 0 & 0 & D \end{pmatrix}=D\boldsymbol{I} \tag{2.100}$$

where D is the usual diffusion constant of a sphere.

For triaxial particles (e.g. ellipsoids):

$$\boldsymbol{D}_{\mathrm{t}}=\begin{pmatrix} D_1 & 0 & 0 \\ 0 & D_2 & 0 \\ 0 & 0 & D_3 \end{pmatrix} \tag{2.101}$$

where the three components correspond to diffusion along the three principal axes. The same is also true for the rotary-diffusion tensor. The additional symmetry of a body of revolution (spheroids, rods, disks, etc.) results in $D_2=D_3$.

Off-diagonal components in the diffusion or resistance tensor correspond to coupling between motion in orthogonal directions. For example, a force exerted on a particle in a given direction can result in a velocity in a different direction. This leads to off-diagonal contributions to the diffusion, characterized by

$$\langle x_i x_j\rangle=2D_{ij}t \tag{2.102}$$

Because $D_{ij}=D_{ji}$, $\boldsymbol{D}$ can have at most six independent non-zero components.

Values of the components of $\boldsymbol{D}_{\mathrm{t}}$ and $\boldsymbol{D}_{\mathrm{r}}$ of ellipsoidal particles have been calculated by Perrin (1934) and Brenner (1974); they can be written as

$$D_1^{\mathrm{t}} = \frac{kT}{16\pi\mu}(\chi + \alpha_1 a_1^2) \tag{2.103a}$$

$$D_2^{\mathrm{t}} = \frac{kT}{16\pi\mu}(\chi + \alpha_2 a_2^2) \tag{2.103b}$$

$$D_3^{\mathrm{t}} = \frac{kT}{16\pi\mu}(\chi + \alpha_3 a_3^2) \tag{2.103c}$$

$$D_1^{\mathrm{r}} = \frac{3kT}{16\pi\mu}\left(\frac{\alpha_2 a_2^2 + \alpha_3 a_3^2}{a_2^2 + a_3^2}\right) \tag{2.104a}$$

$$D_2^{\mathrm{r}} = \frac{3kT}{16\pi\mu}\left(\frac{\alpha_3 a_3^2 + \alpha_1 a_1^2}{a_3^2 + a_1^2}\right) \tag{2.104b}$$

$$D_3^{\mathrm{r}} = \frac{3kT}{16\pi\mu}\left(\frac{\alpha_1 a_1^2 + \alpha_2 a_2^2}{a_1^2 + a_2^2}\right) \tag{2.104c}$$

where a_1, a_2 and a_3 are the three semi-axes of the ellipsoid,

$$\alpha_i = \int_0^\infty \frac{\mathrm{d}\lambda}{(a_i^2 + \lambda)\Delta(\lambda)}$$

$$\chi = \int_0^\infty \frac{\mathrm{d}\lambda}{\Delta(\lambda)}$$

and

$$\Delta(\lambda) = [(a_1^2 + \lambda)(a_2^2 + \lambda)(a_3^2 + \lambda)]^{1/2}$$

Values of D_i^{t} and D_i^{r} are shown in Fig. 2.10.

For arbitrarily shaped particles, a coupling exists between translational and rotational motion, i.e. besides causing translation, a force exerted on a particle will also cause it to rotate. Similarly, exerting a torque causes it to translate. In such cases

$$\boldsymbol{F} = -(\boldsymbol{f}_{\mathrm{t}} \cdot \boldsymbol{u} + \boldsymbol{c} \cdot \boldsymbol{\omega}) \tag{2.105}$$

$$\boldsymbol{L} = -(\boldsymbol{c}^{\dagger} \cdot \boldsymbol{u} + \boldsymbol{f}_{\mathrm{r}} \cdot \boldsymbol{\omega}) \tag{2.106}$$

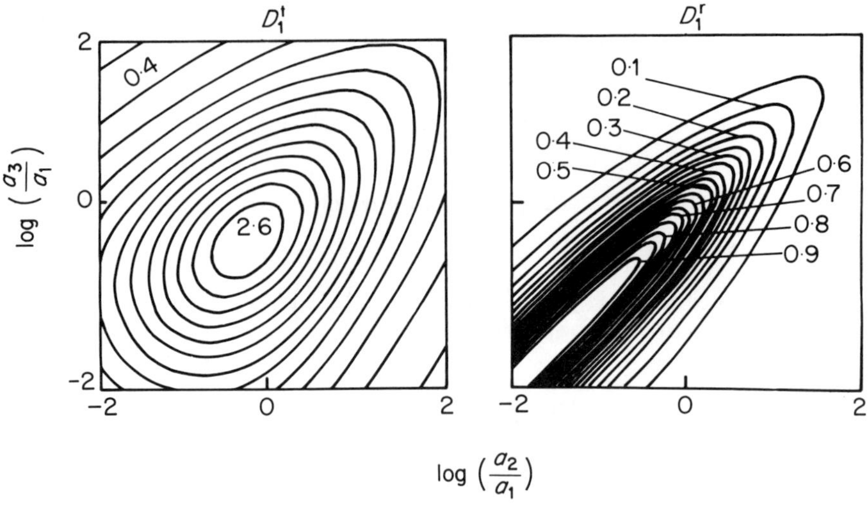

Fig. 2.10. (*a*) Translational-diffusion coefficients D_1^t of ellipsoids of various shapes. The values on the graph indicate the value of $D_1^t/(2.66D_s)$, D_s being the diffusion constant of a sphere of the same volume. The lines are 0·2 units apart.
(*b*) Rotary-diffusion coefficients D_1^r of ellipsoids of various shapes. Values are for $3D_r^r/2D_r^s$, D_r^s being the rotary-diffusion constant of a sphere of the same volume. Values of D_2^t, D_3^t, D_2^r and D_3^r can be obtained by permutation of axis ratios.

where $\boldsymbol{c}$ is a coupling tensor and $\boldsymbol{c}^\dagger$ its transpose. Equations (2.105) and (2.106) can be combined as

$$\begin{pmatrix} \boldsymbol{F} \\ \boldsymbol{L} \end{pmatrix} = -\begin{pmatrix} \boldsymbol{f}_t & \boldsymbol{c} \\ \boldsymbol{c}^\dagger & \boldsymbol{f}_r \end{pmatrix} \bullet \begin{pmatrix} \boldsymbol{u} \\ \boldsymbol{\omega} \end{pmatrix} \tag{2.107a}$$

or more concisely

$$\tilde{\boldsymbol{F}} = -\mu \boldsymbol{R} \cdot \boldsymbol{U} \tag{2.107b}$$

Equation (2.107b) can be considered as a generalized version of the Stokes drag law. Here $\boldsymbol{F}$ is a generalized force-torque vector (six components), $\boldsymbol{U}$ a generalized velocity (six components), and $\boldsymbol{R}$ is called the grand resistance tensor (6 × 6 components). It can be shown (Brenner, 1964) that of the 36 components, 21 are independent. $\boldsymbol{R}$ is related to the grand diffusion tensor $\boldsymbol{D}$ by the generalized Stokes–Einstein relation

$$\boldsymbol{D} = \boldsymbol{R}^{-1} \frac{kT}{\mu} \tag{2.108}$$

As a result, the translational and rotational convective-diffusion equations are also coupled and can be written as:

$$\frac{\partial n}{\partial t} = \nabla \cdot (\boldsymbol{D} \cdot \nabla n) - \nabla \cdot (\boldsymbol{U} n) \tag{2.109}$$

where the operator ∇ must now be expressed in a six-dimensional coordinate system.

To describe the diffusion of arbitrarily shaped particles exactly is a demanding task. No results are available for particles more complex than ellipsoids. The common approach is to compare Brownian particles of odd shapes with spheres or spheroids and define equivalent apparent diffusion constants.

2.1.6. Brownian motion of emulsion droplets

The problem of finding the friction coefficient for a liquid sphere can be solved in a way similar to the friction coefficient of a solid sphere derived in Chapter 1 (Section 1.3.4.3). In the case of a liquid sphere the fluid flow field inside the drop must be solved and the no-slip boundary condition replaced by the continuity of the normal and tangential stresses and velocities at the drop/liquid interface. This leads to (Rybczynski, 1911; Hadamard, 1911, 1912)

$$F = -6\pi\mu a u \frac{p + \frac{2}{3}}{p + 1} \tag{2.110}$$

where p is the viscosity ratio

$$p = \mu_i / \mu \tag{2.111}$$

μ being the viscosity of the medium and μ_i that of the droplet. It follows from the Stokes–Einstein relation that the diffusion constant of a droplet is given by

$$D = \frac{kT}{6\pi\mu a} \frac{p + 1}{p + \frac{2}{3}} \tag{2.112}$$

Notice that as $p \to \infty$, eq (2.112) reduces to the result for a solid sphere. As $p \to 0$ (corresponding to an air bubble suspended in a liquid), eq (1.112) becomes

$$D = \frac{kT}{4\pi\mu a} \tag{2.113}$$

In the derivation of eq (2.111) it is assumed that the liquid inside the drop is freely mobile. This is not always the case, especially when surfactants are present at the drop/liquid interface to stabilize the drop. Boussinesq (1913) assumed that a surface viscosity μ_s can be ascribed to the drop surface. Including this in the analysis leads to

$$D = \frac{kT}{6\pi\mu a} \frac{\mu_s/a + 3\mu + 3\mu_i}{\mu_s/a + 2\mu + 3\mu_i} \tag{2.114}$$

When $\mu_s = 0$, eq (2.112) is recovered. When $\mu_s \to \infty$, the result for a solid sphere is obtained, which is independent of the viscosity ratio p.

2.2 Brownian motion of electrically charged colloidal particles

2.2.1 Qualitative remarks

When Brownian particles possess an electrical charge, they are surrounded by a diffuse ionic double layer. Here we are interested in the way the ionic double layer affects the Brownian motion of the particle. As most colloidal particles are charged, especially in polar media, this problem is of a fundamental importance in colloid science. Despite its importance, this problem has received relatively little attention. In fact, more papers exist in the literature describing secondary effects, such as the effects of double layers on the concentration dependence of diffusion coefficients, than on the primary effect which double layers have on isolated single particles. As such the problem is only partly solved. In many instances these effects are often ignored in studies which deal with the electrolyte dependence of particles such as proteins, or particles with adsorbed polymer, etc., where changes in diffusion constants are commonly ascribed to changes in polymer configurations and the structure of the liquid near the interface. Although such phenomena certainly can occur, it is useful to know the magnitude of the contribution of the double layer to changes in the diffusion coefficient for ideal smooth colloidal particles. In Section 2.1.2 we saw that the Brownian motion of a colloidal particle can be modelled approximately by a random walk in which the particle executes β jumps per second of length ℓ given by eqs (2.12) and (2.15). What happens to the double layer during such a jump?

If the double layer remains undeformed (spherical), then during a jump each ion in the double layer will be displaced by a distance of order ℓ. On average, it will take a time t_i to relax back to the new equilibrium position where (cf. eq (2.91))

$$t_i = \ell^2/2D_i \tag{2.115}$$

where D_i is the diffusion constant of ion i. As long as this time is short compared to the time of a jump (β^{-1}), then the double layer remains essentially in its steady-state configuration. Hence when

$$t_i \ll \beta^{-1} \tag{2.116}$$

the ions in the double layer can follow the jump. Inserting eq (2.15) for ℓ, using the Stokes–Einstein relation (eq (2.78)) for D_i and assuming that the friction coefficient for an ion is approximately given by Stokes law, i.e. $f_i \simeq 6\pi\mu a_i$, a_i being the ion radius, the condition described by eq (2.116) can be written as

$$a_i \ll a \tag{2.117}$$

Because the radius of an ion is usually much smaller than the radius of a colloidal particle, eq (2.117) is almost always satisfied.

It can be assumed that the larger the velocity of a jump, the larger the deformation of the double layer will be. As a result of double-layer deformation, the double layer can be regarded as a dipole. This introduces a new length scale into the problem, namely the distance d between the centres of positive and negative charges. It can be anticipated that d increases with increasing velocity u_0 of the jump. As soon as $d > \ell$, the relaxation time for the ions is

$$t_i = d^2/2D_i \tag{2.118}$$

As $u_0 \to \infty$, $d \to \infty$, because all ions surrounding the particle will be stripped away. It follows that at high velocities the double layer is unable to follow the particle and, therefore, non-steady conditions would be expected.

Because the time between jumps is extremely short, when u_0 is large there is never time to establish a steady-state ion distribution corresponding to u_0. Instead, the deviations from the undisturbed distribution will remain extremely small. This is represented schematically in Fig. 2.11 which shows fluctuations in the number density of ions in a volume V in the double layer. During each jump of time, β^{-1}, the ion distribution starts to approach the steady-state distribution corresponding to the direction and velocity of the jump. Because this direction and velocity is continuously changing, the "target" steady state is continuously changing as well. As a result the ion distribution fluctuates about its equilibrium distribution. Hence, despite the

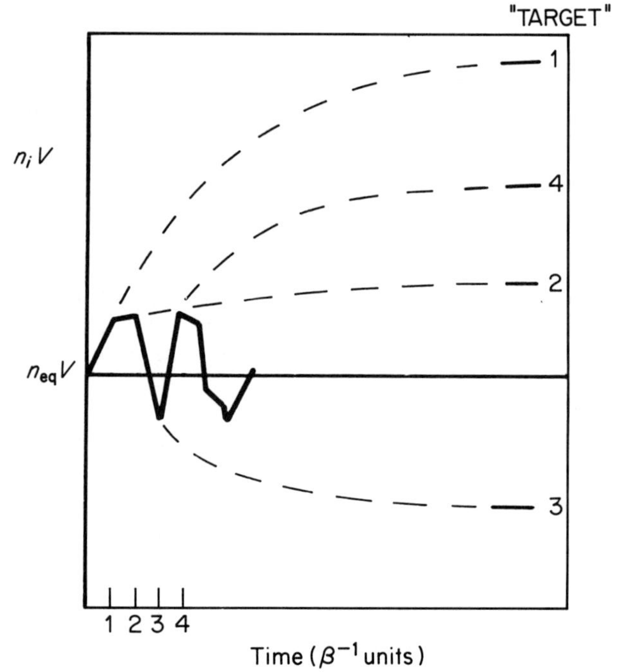

Fig. 2.11. Fluctuations in the number of ions in a volume V within the double layer as a function of time. In each jump the ion concentration moves towards the new steady-state ion distribution corresponding to the direction and velocity of the jump. Before reaching this distribution, a new jump occurs and the ion concentration varies accordingly. Because the jumps are random the net result is that the ion concentration fluctuates about the equilibrium concentration, n_{eq}.

fact that u_0 can be large for a Brownian particle, on average the ion distribution remains the same as the equilibrium one. The only difference is an increase in local fluctuations in ion concentration.

The time needed to reach a target steady state is about u_0/a. Hence during a time β^{-1} the change in concentration is about

$$\frac{\langle \Delta n_i^2 \rangle^{1/2}}{n_i} \simeq \frac{u_0}{a\beta} \tag{2.119}$$

Even without jumps, the ion concentration in volume V in the double layer is continuously changing due to translational Brownian motion of the ions. From fluctuation theory it can be shown that the amplitude of these fluctuations is

$$\frac{\langle \Delta n_i^2 \rangle}{n_i} \simeq \frac{1}{V} \tag{2.120}$$

These fluctuations approach infinity as $V \to 0$, contrary to the fluctuations caused by Brownian jumps.

These arguments suggest that despite the continuous changes in ion concentrations the double layer can be considered, on average, as having an equilibrium ion distribution. As a result we can expect that the Stokes–Einstein relation

$$Df = kT \tag{2.121}$$

also holds for electrically charged particles surrounded by a double layer, with f determined by the friction coefficient of a charged sphere. Experiments reported later bear out this conclusion.

Therefore, in order to determine D, we can consider the problem of the friction coefficient of a charged sphere surrounded by an equilibrium ion distribution.

2.2.2 Friction coefficient of a charged sphere surrounded by an electrical double layer

2.2.2.1. General theory

Consider a spherical particle moving with constant velocity u^0 (without Brownian motion) through a quiescent fluid. Let the particle be surrounded by an electrical double layer (see Fig. 2.12).

The distribution of the ions of species i around the particle is governed by the convective-diffusion equation (eq (2.86)):

$$\nabla \cdot (D_i \nabla n_i - n_i \boldsymbol{u}_i) = 0 \tag{2.122}$$

where n_i is the concentration of ion i, and $\boldsymbol{u}_i$ is its velocity due to hydrodynamic and electric forces:

$$\boldsymbol{u}_i = \boldsymbol{u}_{\text{hydr}} + \boldsymbol{u}_{\text{el}} \tag{2.123}$$

where (cf. eq (1.85))

$$\boldsymbol{u}_{\text{el}} = -\frac{z_i e}{f_i} \nabla \psi \tag{2.124}$$

ψ being the potential, f_i the friction coefficient of the ion, z_i its valency, and e the proton charge.

Combining eq (2.122) with eq (2.124) yields

$$\nabla \cdot \left(D_i \nabla n_i + \frac{n_i z_i e}{f_i} \nabla \psi - n_i \boldsymbol{u}_{\text{hydr}}\right) = 0 \tag{2.125}$$

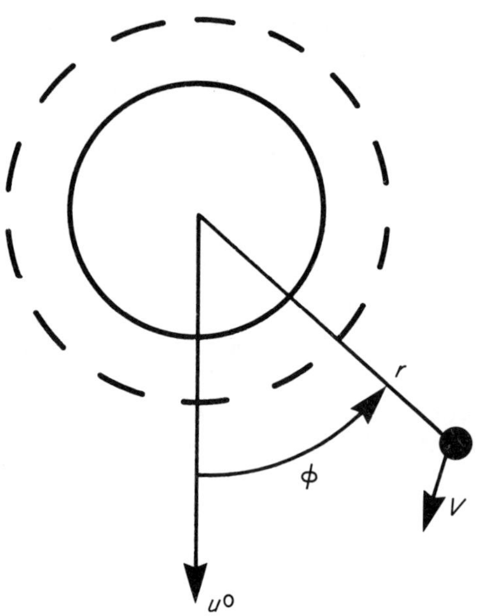

Fig. 2.12. A sphere with a double layer moving through a quiescent fluid. The disturbance flow $\boldsymbol{v}$ around the sphere depends on the coordinates (r, ϕ).

The potential ψ can be found from the Poisson equation (cf. eq (1.88)):

$$\nabla^2 \psi = -\frac{\rho}{\epsilon} \tag{2.126}$$

where ρ is the charge density given by

$$\rho = \sum_i n_i z_i e \tag{2.127}$$

The hydrodynamic velocity, $\boldsymbol{u}_{\text{hydr}}$, is assumed to be equal to the fluid-flow velocity and is given by the creeping-flow equation, including a term for the external force (cf. eq (1.106))

$$\mu \nabla^2 \boldsymbol{u}_{\text{hydr}} - \nabla P = \rho \nabla \psi \tag{2.128}$$

Finally, we require the continuity equation (eq (1.17))

$$\nabla \cdot \boldsymbol{u}_{\text{hydr}} = 0 \tag{2.129}$$

Equations (2.125) to (2.129), together with the appropriate boundary conditions, determine the ion, the potential and the pressure distribution around a particle as well as the hydrodynamic flow field. From the flow field the drag force on the particle and hence the friction coefficient can be determined.

Before discussing these equations in some detail, it is useful to express them in a non-dimensional form. For the sake of simplicity we consider here only symmetrical ions for which

$$n_+^\infty = n_-^\infty = n_\infty \tag{2.130}$$

and

$$z_+ = -z_- = z \tag{2.131}$$

where n_∞ is the ion concentration in the bulk, far from the particle. The theory can be readily adapted for different electrolytes.

Let us define the following dimensionless variables:

$$\Phi = \frac{ze\psi}{kT} \tag{2.132}$$

$$n^\pm = \frac{n_\pm}{n_\infty} \tag{2.133}$$

$$u = \frac{u_{\text{hydr}}}{u^0} \tag{2.134}$$

$$p = \frac{a^2 f_1 P}{\mu kT} \tag{2.135}$$

and

$$r = \frac{r^*}{a} \tag{2.136}$$

where r^* is the dimensional distance to the sphere centre. Substituting them into eqs (2.125) to (2.129) results in

$$\nabla \cdot (\nabla n^+ + n^+ \nabla \Phi - n^+ \mathrm{Pe}\boldsymbol{u}) = 0 \tag{2.137}$$

$$\nabla \cdot (\nabla n^- - n^- \nabla \Phi - n^- q \mathrm{Pe}\boldsymbol{u}) = 0 \tag{2.138}$$

$$\nabla^2 \Phi = -\nu(n^+ - n^-) \tag{2.139}$$

$$\mathrm{Pe}\nabla^2 \boldsymbol{u} - \nabla p = b_1 \nu (n^+ - n^-) \nabla \Phi \tag{2.140}$$

$$\nabla \cdot \boldsymbol{u} = 0 \tag{2.141}$$

where

$$\mathrm{Pe} = \frac{u^0 a}{D_1} \tag{2.142}$$

D_1 being the diffusion coefficient of the positive ion,

$$\nu = \tfrac{1}{2}(\kappa a)^2 \tag{2.143}$$

κ^{-1} being the double-layer thickness (cf. eq (1.94)),

$$q = \frac{D_1}{D_2} = \frac{b_2}{b_1} \tag{2.144}$$

D_2 being the diffusion coefficient of the negative ion, and

$$b_i = \frac{\epsilon k T f_i}{(ez)^2 \mu} \tag{2.145}$$

Equations (2.137) to (2.139) are three second-order equations, eq (2.140) is a second-order vector equation (because of axisymmetry, each vector has only two components), and eq (2.41) is a first-order equation. Hence 11 boundary conditions are required to specify the problem completely. These are:

$$
\begin{array}{llll}
\text{(i)} & n^+ = 1 & \text{at} & r = \infty \\
\text{(ii)} & n^- = 1 & \text{at} & r = \infty \\
\text{(iii)} & \Phi = 0 & \text{at} & r = \infty \\
\text{(iv)} & \Phi = \Phi_s & \text{at} & r = 1 \\
\text{(v)} & u_r = 0 & \text{at} & r = \infty \\
\text{(vi)} & u_\phi = 0 & \text{at} & r = \infty \\
\text{(vii)} & u_r = \cos\phi & \text{at} & r = 1 \\
\text{(viii)} & u_\phi = \sin\phi & \text{at} & r = 1 \\
\text{(ix)} & \boldsymbol{r} \cdot (\nabla n^+ + n^+ \nabla\Phi - n^+ \mathrm{Pe}\boldsymbol{u}) = 0 & \text{at} & r = 1 \\
\text{(x)} & \boldsymbol{r} \cdot (\nabla n^- - n^- \nabla\phi - n^- q\mathrm{Pe}\boldsymbol{u}) = 0 & \text{at} & r = 1 \\
\text{(xi)} & p = \text{constant} & \text{at} & r = \infty
\end{array}
\tag{2.146}
$$

Boundary conditions (i) and (ii) state that far from the sphere the ion concentration is uniform; conditions (iii) and (iv) state that far from the particle the potential is zero, while at the surface of the particle

$$\Phi_s = \frac{ze\psi_s}{kT} \tag{2.147}$$

ψ_s being the surface potential. Because the underlying equations that describe the motion of a charged sphere in an electric field are the same as those discussed above, we can identify ψ_s with the zeta-potential (see Section 3.3). Conditions (v) and (vi) state that far from the sphere the fluid velocity is zero, while (vii) and (viii) are the no-slip boundary conditions at the sphere surface. Conditions (ix) and (x) are the no-penetration (or no-flux) boundary conditions ensuring that no ions penetrate the solid sphere.

Equations (2.137) to (2.141), together with the boundary conditions in eq (2.146), specify the problem completely. It can be seen that the solution depends on five dimensionless parameters Pe, ν, b_1, b_2 and Φ_s, which all have a well-defined physical meaning: Pe is a Péclet number which indicates the ratio of fluid flow to the diffusivity of ion 1; ν is a measure of the thickness of the double layer; and b_1 and b_2 are the scaled friction coefficients of ions 1 and 2, respectively. Because f_i is approximately proportional to μa_i, b_i can also be regarded as a dimensionless ion radius, scaled by a naturally occurring length scale $(e^2/\epsilon kT)$. Finally, Φ_s represents the dimensionless surface potential. Typical ranges of the parameters are:

$$\begin{aligned} 10^{-3} &< \nu < 10^4 \\ 10^{-1} &< b_i < 1 \\ 0 &< \mathrm{Pe} < 10^4 \\ -10 &< \Phi^s < 10 \end{aligned}$$

Values of b_i determined from limiting conductivities for various ions are shown in Table 2.3.

In considering the problem of the Brownian motion of a sphere with a double layer, we have argued that, on average, the ion distribution approximates to an equilibrium distribution at Pe = 0, despite the fact that Pe for a single jump is usually much larger than one. Hence the limit Pe = 0 is the most important limit for the diffusion of charged spheres. Before considering this limit it is illuminating to examine qualitatively the set of eqs (2.137) to (2.141) and to see how the five dimensionless parameters affect the solution.

TABLE 2.3
Values of scaled friction coefficient of various ions[a]

Cation	b_+	Anion	b_-
H^+	0·056	OH^-	0·098
K^+	0·263	Br^-	0·248
NH_4^+	0·264	I^-	0·252
Ag^+	0·312	Cl^-	0·254
Na^+	0·386	NO_3^-	0·270
Li^+	0·500	ClO_4^-	0·285
$(CH_3CH_2)_4N^+$	0·599	CHO_3^-	0·435
Ba^{2+}	0·077	$CH_3CO_2^-$	0·473
Ca^{2+}	0·081	$CH_2ClCO_2^-$	0·488
Sr^{2+}	0·081	$C_2H_5CO_2^-$	0·540
Mg^{2+}	0·092	$C_3H_7CO_2^-$	0·593
La^{3+}	0·030	$C_6H_5CO_2^-$	0·599
		SO_4^{2-}	0·060
		$Fe(CN)_6^{3-}$	0·021
		$Fe(CN)_6^{4-}$	0·011

[a] The friction coefficient is calculated from the limiting conductance Λ_i° of ion i from the relation $f_i = Ne^2 z_i/\Lambda_i^\circ$, where N is Avogadro's number. Most Λ_i° values are taken from MacInnes (1939).

(a) The effect of double-layer thickness

From eq (2.140) it can be seen that as $\nu \to 0$, i.e. when the double-layer thickness extends to infinity, that

$$\mathrm{Pe}\nabla^2 \boldsymbol{u} - \nabla p = 0$$

which is identical to the creeping-flow equation. Thus the flow around the sphere is u_0, given by eq (1.67). Hence the drag is the same as in the absence of a double layer:

$$f = f_0 = 6\pi\mu a$$

When ν is small but non-zero, we might try to make an expansion in ν. However, the solution of eq (2.137) for Pe = 0 and small Φ_s (cf. eq (1.96)) is

$$\Phi_0 = \Phi_s \frac{e^{-\tau(r-1)}}{r} \tag{2.148}$$

where $\tau = \kappa a$, which for small τ reduces to

$$\Phi_0 = \frac{\Phi_s}{r} - \Phi_s \tau \frac{r-1}{r} + \ldots \qquad (2.149)$$

The first term on the right-hand side of eq (2.149) satisfies the boundary conditions at $r=1$ and $r=\infty$, but the second term does not. This suggests that an expansion must be made in $\nu^{1/2}$ (or τ), rather than in ν. Because the first-order term does not satisfy the boundary condition at infinity, an expansion in $\nu^{1/2}$ must be a singular perturbation expansion, where an inner expansion near the sphere and an outer expansion far from the sphere must be made separately and, subsequently, the two solutions must be matched in an intermediate regime (Proudman and Pearson, 1962). We will not try to pursue this matter here, but it suggests that the first-order correction to the friction coefficient is of the order $\nu^{1/2}$. Hence for small ν

$$f = f_0(1 + A\nu^{1/2}) \qquad (2.150)$$

where A is a function of Pe, Φ_s and b_i.

When $\nu \to \infty$, i.e. when $\kappa^{-1} \to 0$, then $n^+ = n^- = 1$, except for a small region near the sphere surface. Outside this region the flow equals the Stokes flow. In the limit $\nu \to \infty$, this region becomes negligible and hence $f = f_0$. An expansion suggests that as $\nu \to \infty$

$$f = f_0(1 + B\nu^{-1}) \qquad (2.151)$$

Because $f \to f_0$ either as $\kappa a \to 0$ or $\kappa a \to \infty$, this suggests that appreciable deviations from Stokes flow occur only when κa is of the order of unity, i.e. when the double-layer thickness is comparable to the particle radius. For a given set of parameters Pe, Φ_s and b_i, the friction coefficient first increases, passes through a maximum and decreases to f_0 as κa increases.

(b) The effect of surface potential

When the surface potential is very small, then the gradient $\nabla\psi$ will also be small. As $\nabla\psi \to 0$, eq (2.140) reduces to the regular creeping-flow equation and hence $u \to u_{st}$, where u_{st} is the solution for a neutral sphere. The limit $\Phi_s \to 0$ corresponds to the absence of a double layer and hence $f \to f_0$ in this limit.

For small Φ_s, an expansion in Φ_s can be made from which it can be readily shown that, up to order Φ_s^2:

$$f = f_0(1 + C\Phi_s^2) \tag{2.152}$$

The fact that the first-order correction to the friction coefficient is of the order Φ_s^2, rather than of the order Φ_s, follows directly from symmetry. A negatively charged particle surrounded by a 1-1 electrolyte must behave the same way as a positively charged particle (of the same charge). If the $0(\Phi_s)$ term is non-zero, this would not be true.

In the limit $\Phi_s \to \infty$, the particle accumulates an infinite number of counterions around it and pushes all the co-ions to infinity. It is obvious that an infinite amount of energy would be required for a sphere to attract an infinite number of ions and keep them around it against the hydrodynamic forces that want to convect them away. Hence we anticipate that as $\Phi_s \to \infty$, then $f_0 \to \infty$. This can also be seen from eqs (2.137) to (2.140). As $\Phi_0 \to \infty$, n becomes of the form $\exp[\pm\phi_s f(r,\Phi)]$ and a regular expansion (in, for example, Φ_s^{-1}) is impossible. These exponential terms enter the right-hand side of eq (2.140), implying that as $\Phi_s \to \infty$, $u \to \infty$ and hence $f \to \infty$.

(c) The effect of ion size

The ion size enters into the problem via the two parameters, b_1 and q, or equivalently b_1 and b_2. From an expansion in the parameter b_1, it can be shown that there is a correction of order b_1 to the zero-order solution. We have normalized our equations with respect to the positive ion. We could equally well have chosen the negative ion as a reference. If this is done, an expansion in b_2 results in a correction of order b_2 to the zero-order solution. When $b_1 = b_2 = 0$, Stokes flow is recovered, and hence as $b_i \to 0$

$$f = f_0(1 + D_1 b_1 + D_2 b_2) \tag{2.153}$$

It should be noted that as $b_1 \to 0$, Pe $\to 0$, except when $u^0 \to \infty$. If $b_i \to \infty$, while at the same time Pe $\to \infty$, then $f \to f_0$.

(d) The effect of fluid velocity

It is easily shown that as Pe $\to \infty$, then $f \to f_0$. In this case all the ions are stripped away from the vicinity of the particle. Of more interest is the limit Pe $\to 0$, which, as we have discussed, is likely to apply to the Brownian motion of a particle in a quiescent medium. It can readily be seen that for small Pe a non-zero contribution to the friction coefficient exists due to the presence of the double layer. Therefore, it is most likely that f decreases monotonically from its value at Pe $= 0$, approaching f_0 as Pe $\to \infty$.

Before discussing the important case of Pe $= 0$ in more detail, we present

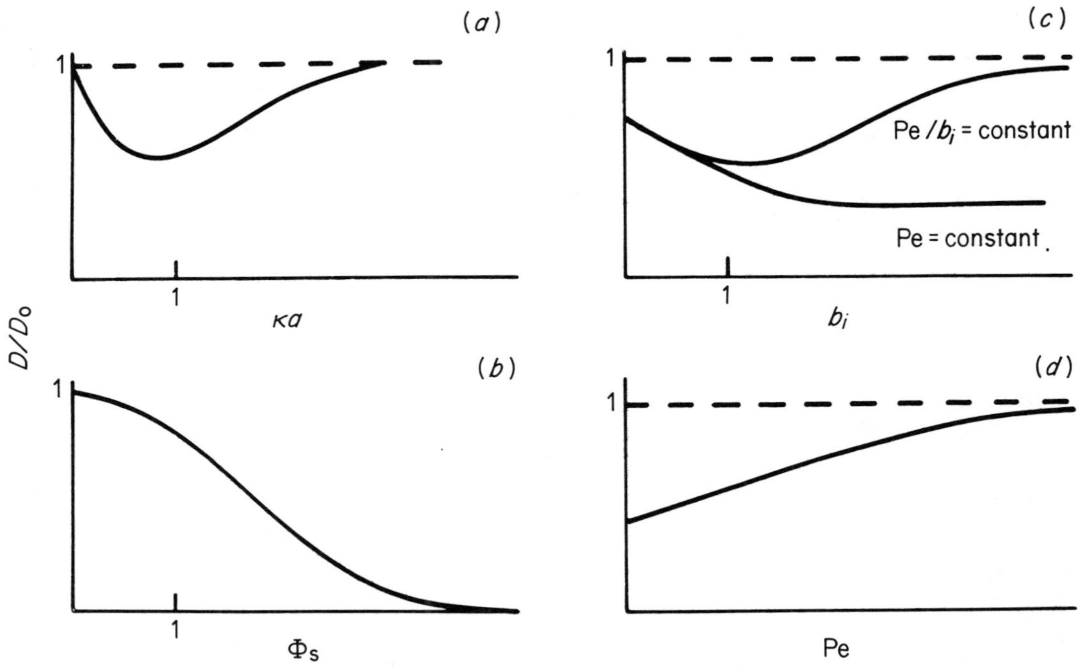

Fig. 2.13. Schematic representations of the trends in the diffusion coefficient D as suggested by the asymptotic solutions (eqs (2.150) to (2.153)), as a function of electrolyte concentration (κa), surface potential (Φ_s), ion size (b_i), and fluid velocity (Pe).

schematically the various trends in the diffusion coefficient D corresponding to the effects outlined above, assuming that the Stokes–Einstein relation $Df=kT$ holds. These are shown in Fig. 2.13, which shows D/D_0 as a function of κa, Φ_s, b_i and Pe. The values 1 on the horizontal axes are approximate. The minimum in Fig. 2.13a, the limiting values in Fig. 2.13c and the initial value in Fig. 2.13d depend on the magnitude of the remaining parameters. In Fig. 2.13d the first-order correction of f is of the order of Pe and hence the initial part of the curve D/D_0 vs. Pe is linear. The sign of the slope can only be determined from the solution of the set of eqs (2.137) to (2.141). It can be seen from Fig. 2.13 that, in general, D/D_0 can have any value between 0 and 1. As we will show later, however, low values of D/D_0 can only be achieved with unrealistically high surface potentials. Under realistic conditions D/D_0 is in the range 0.85–1.

2.2.2.2 Low Péclet number solutions

When Pe $\ll$ 1, an expansion in the parameter Pe can be made:

$$
\begin{aligned}
n^+ &= n_0^+ + \mathrm{Pe}\, n_1^+ \\
n^- &= n_0^- + \mathrm{Pe}\, n_1^- \\
\Phi &= \Phi_0 + \mathrm{Pe}\, \Phi_1 \\
\boldsymbol{u} &= \boldsymbol{u}_0 + \mathrm{Pe}\, \boldsymbol{u}_1 \\
p &= p_0 + \mathrm{Pe}\, p_1
\end{aligned}
\tag{2.154}
$$

Substituting the expansion eq (2.154) into eqs (2.137) to (2.141) and collecting zero-order terms results in:

$$\nabla\cdot(\nabla n_0^+ + n_0^+ \nabla\Phi) = 0 \tag{2.155}$$

$$\nabla\cdot(\nabla n_0^- - n_0^- \nabla\Phi) = 0 \tag{2.156}$$

$$\nabla^2\Phi_0 = -\nu(n_0^+ - n_0^-) \tag{2.157}$$

These three equations describe the ion and potential distributions around an undisturbed colloidal particle. The solution of eqs (2.155) and (2.156) is the well-known Boltzmann distribution (cf. eq (1.87)):

$$n_0^\pm = \mathrm{e}^{\mp\Phi_0} \tag{2.158}$$

with Φ_0 determined by eq (2.157):

$$\frac{d^2\Phi_0}{dr^2}+\frac{2}{r}\frac{d\Phi_0}{dr}=-\nu(e^{-\Phi_0}-e^{\Phi_0}) \tag{2.159}$$

subjected to the boundary conditions $\Phi_0=\Phi_s$ when $r=1$ and $\Phi_0\to 0$ as $r\to\infty$.

For low potentials, $\Phi_s \ll 1$, the solution of eq (2.159) is (cf. eq 1.96)):

$$\Phi_0=\Phi_s\frac{e^{-\tau(r-1)}}{r} \tag{2.160}$$

For surface potentials which are not small, eq (2.159) must be solved numerically. This can readily be done by realizing that as $r\to\infty$ the solution is similar to eq (2.160), with Φ_s replaced by an integration constant K which must be determined. The constant K can be found by iteration: an arbitrary choice of K is made, the equation integrated backwards to $r=1$, and the resulting value of Φ found at $r=1$. An algorithm can readily be devised such that for subsequent choices of K the solution converges to the value $\Phi_0=\Phi_s$ at $r=1$.

Substituting the expansion given by eq (2.154) into eq (2.137) to (2.141) and collecting terms of order Pe, gives:

$$\nabla\cdot(\nabla n_1^+ + n_1^+\nabla\Phi_0 + n_0^+\nabla\Phi_1 - n_0^+\boldsymbol{u}_0)=0 \tag{2.161}$$

$$\nabla\cdot(\nabla n_1^- - n_1^-\nabla\Phi_0 - n_0^-\nabla\Phi_1 - q\,n_0^-\boldsymbol{u}_0)=0 \tag{2.162}$$

$$\nabla^2\Phi_1=-\nu(n_1^+-n_1^-) \tag{2.163}$$

$$\nabla^2\boldsymbol{u}_0-\nabla p_0=\nu b_1(n_0^+-n_0^-)\nabla\Phi_1+(n_1^+-n_1^-)\nabla\Phi_0 \tag{2.164}$$

$$\nabla\cdot\boldsymbol{u}_0=0 \tag{2.165}$$

The first thing to notice is that the zero-order solution for $\boldsymbol{u}_0$ is different from the Stokes solution for a neutral sphere because of the occurrence of the force terms on the right-hand side of eq (2.164). This implies that the presence of the double layer contributes a term of order unity to the flow around the particle and hence to the friction or diffusion coefficient. This is so despite the fact that as Pe → 0 the double layer is undisturbed. At first sight this seems curious because the right-hand side of eq (2.164) is determined by terms of O(Pe), which vanish in the limit Pe → 0. Thus the equations seem to contradict the famous philosophical theorem which states that an infinitesimal cause cannot lead to a finite effect. However, this paradoxical behaviour arises from our particular choice to "non-dimensionalize" the governing equations. It can be seen from these equations that Pe is present only in the combination Peu. Hence the equations could have been non-dimensionalized by defining a non-dimensional velocity $\tilde{u} = \text{Pe}u = au_0/D_1$. In this case Pe disappears from the differential equations but enters again via boundary conditions (vii) and (viii) of eq (2.146), which must be replaced by $\tilde{u}_r = \text{Pe} \cos\phi$ and $\tilde{u}_\phi = \text{Pe} \sin\phi$ at $r = 1$. With this choice of parameters, an expansion in Pe results in

$$\tilde{u} = \tilde{u}_0 + \text{Pe}\,\tilde{u}_1$$

from which it follows that $\tilde{u}_0 = 0$ (i.e. when Pe $= 0$ there is no flow) and $\tilde{u}_1$ will be given by eq (2.164). In this case all the first-order effects are of the order Pe.

Progress in trying to solve the set of eqs (2.161) to (2.165) can be made by considering small potentials only ($\Phi_s \ll 1$) and making an expansion in terms of Φ_s:

$$\begin{aligned} n_1^{\pm} &= \Phi_s m_1^{\pm} \cos\phi + \ldots \\ \Phi_1 &= \Phi_s \chi_1 \cos\phi + \ldots \\ \boldsymbol{u} &= \boldsymbol{u}_0^{\text{st}} + \Phi_s^2 \boldsymbol{u}_2 + \ldots \\ p &= p_0 + \Phi_s^2 p_2 + \ldots \end{aligned} \tag{2.166}$$

where the terms $\Phi_s \boldsymbol{u}_1$ and $\Phi_s p_1$ have been omitted, because it can readily be shown from eqs (2.163) and (2.164) that they are zero. This also follows directly from the symmetry between positive and negative particles. The dependence of eq (2.166) on the angle ϕ is suggested by the geometry of the problem.

Substituting the expansion given by eq (2.166) into eqs (2.161) to (2.165) and collecting terms of order Φ_s, the following differential equations are obtained for $m_1^{\pm}$ and χ_1:

$$\left(\frac{\mathrm{d}^2}{\mathrm{d}r^2} + \frac{2}{r}\frac{\mathrm{d}}{\mathrm{d}r} - \frac{1}{r^2}\right)(m_1^+ + \chi_1) = f(r) \tag{2.167}$$

$$\left(\frac{\mathrm{d}^2}{\mathrm{d}r^2} + \frac{2}{r}\frac{\mathrm{d}}{\mathrm{d}r} - \frac{1}{r^2}\right)(m_1^- - \chi_1) = -q f(r) \tag{2.168}$$

and

$$\left(\frac{\mathrm{d}^2}{\mathrm{d}r^2}+\frac{2}{r}\frac{\mathrm{d}}{\mathrm{d}r}\right)\chi_1=-\nu(m_1^+-m_1^-) \tag{2.169}$$

where

$$f(r)=\left(\frac{\mathrm{d}u_r^{\mathrm{st}}/\cos\phi}{\mathrm{d}r}\right)\left(\frac{\mathrm{d}(\Phi_0/\Phi_s)}{\mathrm{d}r}\right)$$

or

$$f(r)=\frac{3}{2r^4}\left(1-\frac{1}{r^2}\right)(1+\tau r)\mathrm{e}^{-\tau(r-1)} \tag{2.170}$$

Equations (2.167) to (2.169) can be solved by using the method of variations of parameters (Boyce and Diprima, 1977).

The creeping-flow equation of the order of Φ_s^2 becomes:

$$\nabla^2 \boldsymbol{u}_2-\nabla p_2=b_1\nu\left[2w\,\nabla\chi_1+(m_1^+-m_1^-)w'\right]\cos\phi \tag{2.171}$$

where $w=\Phi_0/\Phi_s$ and the prime denotes differentiation with respect to r. The terms on the right-hand side of eq (2.171) are all known functions of r and ϕ, so that this equation can be solved. From the flow field and ion and potential distributions we can determine the hydrodynamic and electrical stresses (cf. eq (1.106)) exerted on the sphere from which the drag coefficient can be calculated. This calculation is not trivial, but has been performed in a somewhat different manner by Ohshima *et al.* (1984) with the result that (for symmetrical electrolytes):

$$f=f_0\left[1+\bar{b}\Phi_s^2\tilde{f}(\kappa a)\right] \tag{2.172}$$

where $\bar{b}=(b_1+b_2)/2$ and $\tilde{f}(\kappa a)$ is a function of κa only, given by

$$\begin{aligned}\frac{2}{3}\tilde{f}(\kappa a)= {} & \mathrm{e}^{2\kappa a}[3E_4(\kappa a)-5E_6(\kappa a)]^2+8\mathrm{e}^{\kappa a}\left[E_3(\kappa a)-E_5(\kappa a)\right]- \\ & \mathrm{e}^{2\kappa a}\left[4E_3(2\kappa a)+3E_4(2\kappa a)-7E_8(2\kappa a)\right]\end{aligned} \tag{2.173}$$

with

$$E_n(\kappa a)=(\kappa a)^{n-1}\int_{\kappa a}^{\infty}\frac{\mathrm{e}^{-x}}{x^n}\,\mathrm{d}x$$

It can readily be shown that $\tilde{f}(\kappa a)=0$ for $\kappa a=0$ or $\kappa a=\infty$ and that it has a maximum near $\kappa a=1$. Thus eq (2.172) correctly describes the behaviour

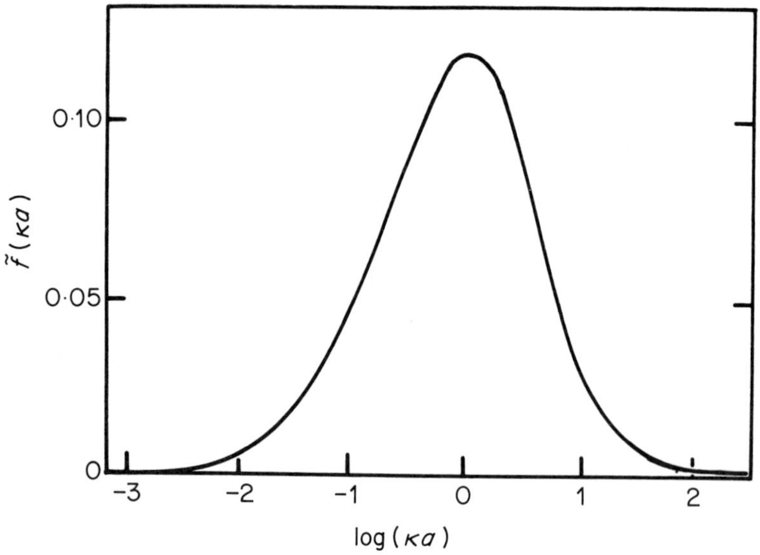

Fig. 2.14. The function $\tilde{f}(\kappa a)$, describing the electrolyte dependence on the friction or diffusion coefficient of a charged sphere.

predicted in Fig. 2.13a. The function $\tilde{f}(\kappa a)$ is shown in Fig. 2.14. As it should be, eq (2.172) is of the form of eqs (2.152) and (2.153).

Assuming the Stokes–Einstein relation holds, the diffusion constant of a charged sphere can be expressed as

$$D = D_0[1 - \bar{b}\Phi_s^2 \tilde{f}(\kappa a)] \quad (2.174)$$

Another expansion that can be performed on eqs (2.161) to (2.165) is an expansion in ν^{-1}. This has also been done by Ohshima *et al.* (1984), with the result:

$$f = f_0 \left[1 + \frac{12}{(\kappa a)^2}(b_1 G_+^2 + b_2 G_-^2)\right] \quad (2.175)$$

with

$$G_\pm = \ln \frac{1 + \exp(\mp |\Phi_s|/2)}{2}$$

As $\Phi_s \to 0$, eq (2.175) becomes of the form of eq (2.172). Notice also that it is of the form of eq (2.151). As $\Phi \to \infty$, the correction term approaches infinity. Hence $f \to \infty$, as predicted above. As $\kappa a \to \infty$ and $\Phi_s \to 0$, eqs (2.172) and (2.175) become identical.

It is possible to solve the set of eqs (2.161) to (2.165) numerically. Making the substitutions

$$
\begin{aligned}
n_1^+ &= g_1\cos\phi & u_r^0 &= g_4\cos\phi \\
n_1^- &= g_2\cos\phi & u_\phi^0 &= g_5\sin\phi \\
\Phi_1 &= g_3\cos\phi & p_0 &= g_6\cos\phi
\end{aligned}
$$

into eqs (2.161) to (2.165) results in a series of five ordinary second-order differential equations and one first-order equation, or equivalently, a series of 11 coupled first-order equations for the functions $g_i(r)$. These equations are subject to the boundary conditions given by eq (2.146) and can be solved using standard methods. The difficulty is that some of the boundary conditions are specified at $r=1$, while others are specified at $r=\infty$. One way of approaching this is to make initial guesses of the unknown boundary conditions at $r=1$ (or $r=\infty$) and establish the correct values by trial and error. It is possible to write algorithms that produce increasingly better guesses and converge to the correct solution. These equations have been solved in a somewhat different manner by Ohshima *et al.* (1984). Results of their calculations are presented in Fig. 2.15, which shows D/D_0 vs. κa for various values of Φ_s and b_i. Fig. 2.16 shows experimental results by Schumacher and

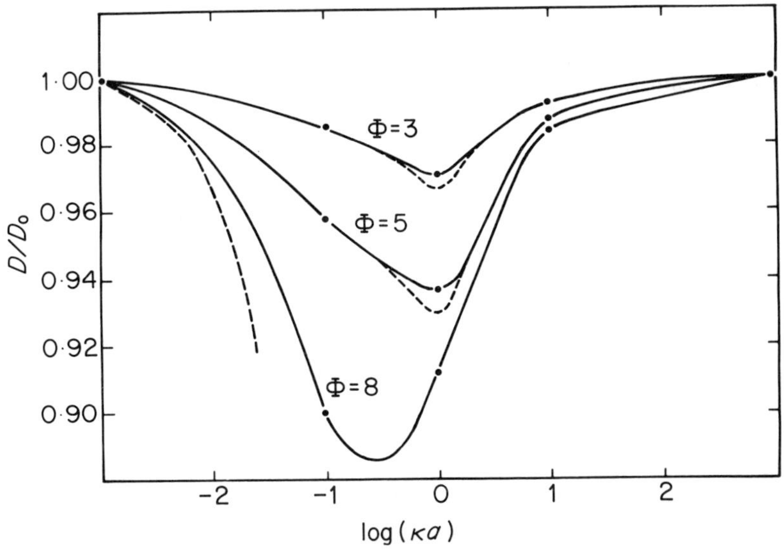

Fig. 2.15. D/D_0 from numerical calculations of the set of governing equations as a function of κa for various values of Φ_s (points are those calculated by Ohshima *et al.*, 1984). $b_1=0.176$; $b_2=0.169$.

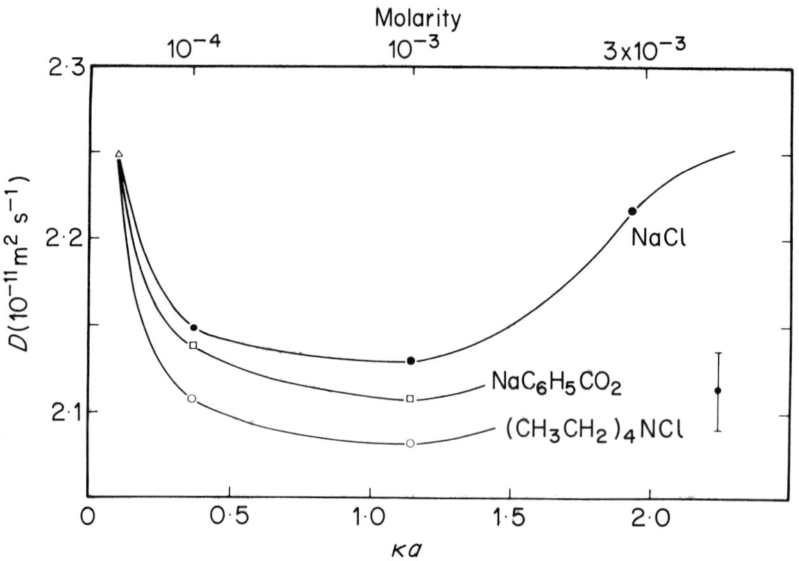

Fig. 2.16. Experimental observations of the diffusion coefficient of gold particles ($a \simeq 10$ nm) suspended in aqueous electrolyte solutions, obtained by photon correlation spectroscopy (after Schumacher and van de Ven, 1987).

van de Ven (1987) for gold particles of 10-nm radius suspended in aqueous electrolyte solutions. It can be seen that there is a good agreement between theory and experiment.

2.3 Brownian motion of particles with adsorbed polymer

2.3.1 Composite-sphere model

To calculate the diffusion coefficient of a sphere coated with polymer is a complicated problem. In principle the segment density distribution of the polymer at the surface must be known and the flow field around and through this layer of adsorbed polymer calculated. This can be done approximately in various ways. First, we will consider a homogeneous polymer layer and in the next section we will discuss the problem of a particle coated with a more realistic layer. A sphere with a homogeneous polymer layer can be thought of as a solid sphere surrounded by a porous shell, and an important parameter in describing the flow around such a sphere is the permeability of this shell

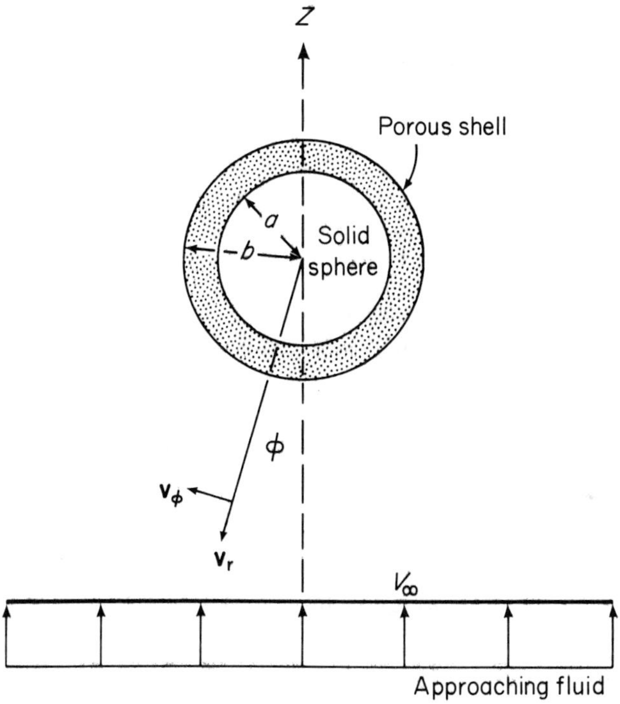

Fig. 2.17. Coordinate system for axisymmetric flow relative to an isolated composite sphere.

which, in principle, can be found from the segment density at the surface.

To calculate the diffusion coefficient of a sphere with adsorbed polymer, the friction coefficient of such a sphere when it is kept fixed in a uniform flow of velocity v_∞ must be determined. The geometry of the system is shown schematically in Fig. 2.17. The radius of the solid sphere is denoted by a, and that of the composite sphere by b, so that the thickness of the layer is $d = b - a$. The governing equations for $r > b$ are the well-known creeping-flow and continuity equations:

$$\mu \nabla^2 v = \nabla p \tag{2.176a}$$

and

$$\nabla \cdot v = 0 \tag{2.176b}$$

where v is the velocity vector and p the pressure.

In the region $a \leqslant r \leqslant b$ we require (Masliyah *et al.*, 1987) instead of eq (2.176a):

$$-\frac{\mu}{k}v+\mu\nabla^2 v=\nabla p \tag{2.177}$$

where k is the permeability of the porous shell. Equation (2.177) is an extension of *Darcy's law*:

$$\nabla p=-\frac{\mu}{k}v \tag{2.178}$$

which on its own is incompatible with the no-slip boundary condition at $r=a$. Equation (2.177) is usually referred to as the *Brinkman equation*. The term $\mu\nabla^2 v$ is important only in a thin boundary layer near the solid sphere (of order $\sqrt{k}$), while eq (2.178) holds in the remaining part of the shell.

The appropriate boundary conditions for the system of eqs (2.176) to (2.177) are:

$$v_r(a)=v_\phi(a)=0 \tag{2.179}$$

$$v_r(b) \text{ and } v_\phi(b) \text{ are continuous at } r=b \tag{2.180}$$

$$\Pi_{rr} \text{ and } \Pi_{r\phi} \text{ are continuous at } r=b \tag{2.181}$$

$$v_r(\infty)=-v_\infty\cos\phi;\ v_\phi(\infty)=v_\infty\sin\phi \tag{2.182}$$

where Π_{rr} and $\Pi_{r\phi}$ denote the normal and tangential components of the stress tensor, respectively. Equation (2.179) expresses the no-slip condition at the surface $r=a$, eq (2.180) expresses the continuity of normal and tangential velocities, and eq (2.181) that of normal and tangential stress components at the interface $r=b$. Finally, eq (2.182) states that far from the sphere the flow field must approach the uniform flow velocity v_∞. By defining a stream function ψ (cf. eq (1.41)), the creeping-flow and the Brinkman equations can be written as:

$$E^4\psi=0,\ r\geqslant b \tag{2.183}$$

and

$$E^4\psi-k^{-1}E^2\psi=0,\ a\leqslant r\leqslant b \tag{2.184}$$

where E is as defined by eq (1.43).

The general solution of eqs (2.183) and (2.184) is:

$$\psi(\xi,\phi)=-\frac{kv_\infty}{2}(A\xi^{-1}+B\xi+C\xi^2+D\xi^4)\sin^2\phi,\ \beta\leqslant\xi \tag{2.185}$$

and

$$\psi(\xi,\phi) = -\frac{k\nu_\infty}{2}\left[E\xi^{-1}+F\xi^2+G(\xi^{-1}\cosh\xi-\sinh\xi)\right.$$
$$\left.+H(\xi^{-1}\sinh\xi-\cosh\xi)\right]\sin^2\phi,\ \alpha\leqslant\xi\leqslant\beta \qquad (2.186)$$

where $\xi=r/\sqrt{k}$, $\alpha=a/\sqrt{k}$ and $\beta=b/\sqrt{k}$. The integration constants A, B, C, D, E, F, G and H can be found from the eight boundary conditions, eqs (2.179) to (2.182) (Masliyah *et al.*, 1987). The flow field can be determined from the stream function and the friction coefficient of the sphere can be calculated from the flow field. The result is:

$$f=6\pi\mu b\lambda \qquad (2.187)$$

where λ is a correction factor accounting for the porosity of the shell, given by

$$\lambda=\frac{A_1}{A_2\beta(\alpha\sinh\beta-\cosh\alpha)} \qquad (2.188)$$

where

$A_1 = (\alpha^4 + 2\alpha\beta^3 + 3\alpha^2)\cosh\alpha + 3\alpha^2(\cosh\beta - \beta\sinh\beta - \alpha\sinh\alpha) + \cosh(\beta-\alpha)\left[(\alpha^3+2\beta^2+3\alpha)(\alpha\beta\sinh\beta - \alpha\cosh\beta - \beta\cosh\alpha) + 3\alpha^2\beta\sinh\alpha\right] + \sinh(\beta-\alpha)\left[(\alpha^3+2\beta^2+3\alpha)\cosh\alpha + 3\alpha^2(\alpha\beta\sinh\beta - \alpha\cosh\beta - \sinh\alpha)\right]$

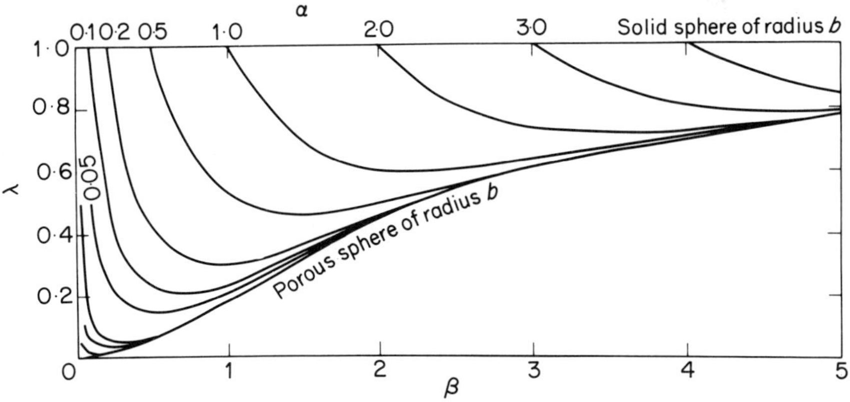

Fig. 2.18. Correction factor, λ, describing the friction coefficient of a composite sphere, as a function of β (the dimensionless radius of the composite sphere) for various values of α (the dimensionless radius of the core).

and

$$A_2 = -6\alpha + (3\alpha + 3\beta + \alpha^3 + 2\beta^3)\cosh(\beta - \alpha) + 3(\alpha^2 - 1)\sinh(\beta - \alpha)$$

The correction factor λ is plotted in Fig. 2.18 as a function of β for various values of α. It can be seen that when $\alpha = \beta$ (i.e. $a = b$ and $d = 0$), $\lambda = 1$. When $\alpha = 0$, we have a porous sphere of radius b in which case λ varies from 0 for $\beta = 0$ to 1 for $\beta = \infty$ (corresponding to a solid sphere of radius b). The diffusion coefficient of a sphere of radius a with an adsorbed polymer layer of thickness d depends on λ:

$$D = \frac{aD_0}{(a+d)\lambda} \tag{2.189}$$

where D_0 is the diffusion coefficient of a solid sphere of radius a. To evaluate λ it is necessary to find an expression for the permeability, k, of the polymer layer. The flow through a porous polymer layer can be approximated by that of a flow through an array of cylindrical particles normal to their axes, in which case the permeability is related to the cylinder radius, R, and the volume fraction, c, by (Happel, 1959):

$$k = -\frac{R^2}{8c}\left[\ln c + \frac{1 - c^2}{1 + c^2}\right] \tag{2.190}$$

In the case of polymers, R may be equated with the average radius of the polymer chain and c with the volume fraction of the polymer.

Equations (2.187) to (2.190) determine fully the diffusion coefficient of a sphere with a uniform layer of polymer. These equations were tested experimentally by measuring the sedimentation velocity of a large sphere of 3·18 mm diameter, to which up to 22 threads of length 5·1 mm and diameter 0·4 mm had been attached. The results are shown in Fig. 2.19. It can be seen that the agreement between theory and experiment is reasonable, taking into consideration the fact that the theory contains no adjustable parameters. A similar model for the friction coefficient of hairy spheres is given by Sasaki (1985).

2.3.2 Multiple-layer model

In most cases the configuration of polymers adsorbed on surfaces is rather complex. As discussed in Chapter 1, Section 1.4.5, we usually distinguish between polymer segments adsorbed in trains, loops and tails. The diffusion constant of a colloidal particle with adsorbed polymer depends on the way in

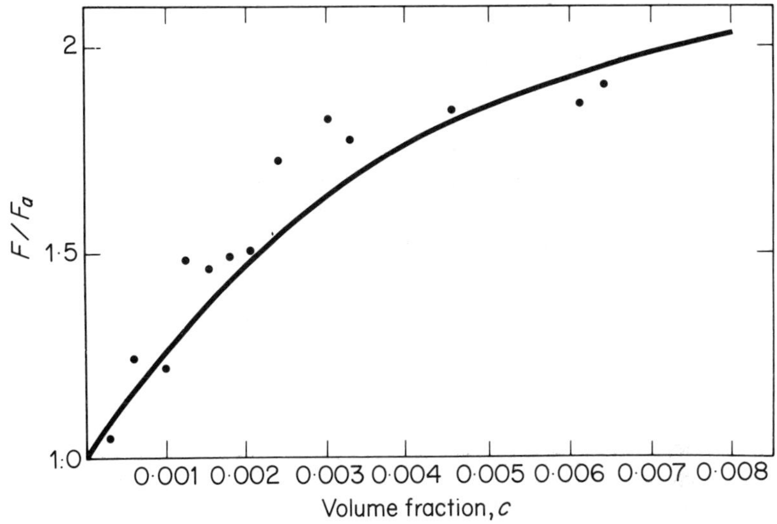

Fig. 2.19. Ratio of the drag force on a sphere of radius a and a sphere of radius a with a layer of threads of thickness d, as a function of the volume fraction of threads (after Masliyah and van de Ven, 1986). (———) Composite-sphere model; •, experimental.

which the polymer is adsorbed. Knowledge of the segment-density distribution at the surface can be found from lattice models (Scheutjens and Fleer, 1982) or Monte Carlo simulations (Napper, 1983). Alternatively, it is possible to describe the structure of adsorbed polymers as a "self-similar" grid (de Gennes, 1987). Assuming a certain segment-density profile, the effective hydrodynamic radius (i.e. the diffusion constant) can be calculated from a parallel-plate model.

Consider the two parallel plates with adsorbed polymers on their inner surface, as shown schematically in Fig. 2.20. The basic idea is to compare the volume flux through a parallel-plate channel having polymer adsorbed on its inner walls with that through a parallel-plate channel with no polymer and a slightly smaller gap. The gap is chosen such that the two fluxes are identical. The thickness, δ, of the polymer layer can then be found from the relation $L = L' + 2\delta$, where L and L' are the gap width of the channel with and without polymer, respectively. The liquid (with polymer) between the plates can be divided into a certain number of layers, M, of which the thickness, ℓ, can be taken as equal to the cross-section of a polymer chain. Assuming the concentration of polymer is known in each layer and has an associated permeability k_i, we can solve the Brinkman equation (eq (2.177)) for each

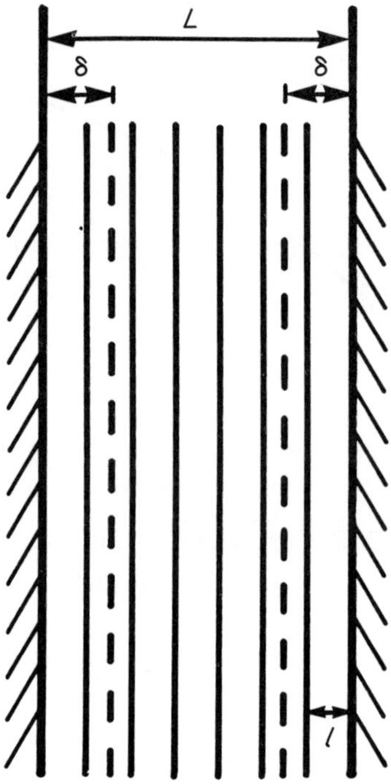

Fig. 2.20. Schematic representation of the parallel-plate method for calculating the hydrodynamic layer thickness, δ. The flow in between two plates covered with polymer is compared to the flow in a smooth channel with the plates closer together by a distance of 2δ.

layer, with the boundary conditions for the velocity $v(y)$ between layers i and $i+1$ $(y=x/\ell)$

$$v_{i+1}(0)=v_i(1) \tag{2.191}$$

and

$$(\mathrm{d}v_{i+1}/\mathrm{d}y)_{y=0}=(\mathrm{d}v_i/\mathrm{d}y)_{y=1} \tag{2.192}$$

with the result (Cohen Stuart *et al.*, 1984):

$$v_i=A_i\sinh\left(\frac{y\ell}{\sqrt{k_i}}\right)+B_i\cosh\left(\frac{y\ell}{\sqrt{k_i}}\right)+C_i \tag{2.193}$$

where

$$A_i=\frac{\sqrt{k_i}}{\ell}\left(\frac{\mathrm{d}v_i}{\mathrm{d}y}\right)_{y=0};\ B_i=v_i(0)-C_i^2;\ C_i=k_i\frac{\nabla p}{\mu}$$

From the knowledge of the solution in each layer, a solution for the flow between two plates can be built up numerically which satisfies the conditions

$$v_1(0) = v_M(1) = 0 \tag{2.194}$$

and

$$(\mathrm{d}v_{M/2}/\mathrm{d}v)_{y=0} = 0 \tag{2.195}$$

where M is the total number of layers (i.e. $M\ell = L$). The flux, j, of liquid passing between the two plates is given by

$$j = \sum_{i=1}^{M} \int_0^1 v_i \mathrm{d}y \tag{2.196}$$

This flux may be compared with the flux between two plates at a distance $L - 2\delta$ apart:

$$j = 2 \int_{\delta}^{L/2} v' \mathrm{d}x \tag{2.197}$$

where v' is the Poiseuille flow field between the plates. Equating eqs (2.196) and (2.197) yields a value for the hydrodynamic layer thickness, δ. Cohen Stuart *et al.* (1984) assumed the following simple relation for the permeability k_i

$$k_i = K \frac{1 - c_i}{c_i} \tag{2.198}$$

K being a constant for a given polymer and c_i the volume fraction of polymer in layer i. Their calculations show that δ depends strongly on the chain length of the polymer, on the Flory–Huggins parameter, χ (cf. eq (1.134)), and on the constant K; it was found to be rather insensitive to the adsorption energy, χ_s. This is because the amount of polymer in tails is hardly affected by χ_s and it turns out that δ is mainly determined by the tails. This is illustrated in Fig. 2.21 which shows the hydrodynamic thickness, δ, as a function of the chain length (n being the number of segments in the chain). It can be shown that approximately

$$\delta \propto n^{0 \cdot 8} \tag{2.199}$$

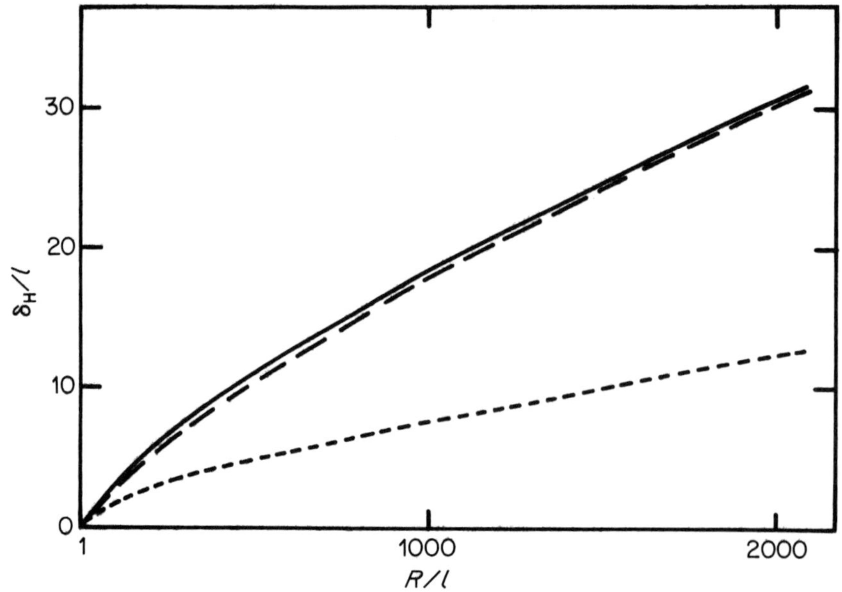

Fig. 2.21. Hydrodynamic layer thickness, δ_H, as a function of chain length, R. Segmental adsorption energy $\chi_s = 1$, θ solvent ($\chi = 0 \cdot 5$), equilibrium volume fraction of polymer $\phi = 10^{-3}$. (———) Complete segment density profile; (— — — —) only tails; (-----) only loops and trains. Distances are rendered dimensionless by the diameter, ℓ, of the polymer chain (after Cohen Stuart *et al.*, 1984).

Since the radius of gyration, R_g, of a polymer chain is proportional to $n^{0 \cdot 6}$, it follows that for short chains δ is smaller, while for large chains δ is larger than the free-coil diameter.

In applying the above results to polymer adsorption on spherical colloidal particles, it may be assumed that the effective radius of the particle is increased by δ and hence

$$D = \frac{kT}{6\pi\mu(\mathrm{a} + \delta)} \tag{2.200}$$

Experiments using poly(ethylene) oxide adsorbed on poly(styrene) latex confirm the theoretical predictions (Cohen Stuart *et al.*, 1984). An example is shown in Fig. 2.22, which shows measured values of R_g and δ as a function of chain length. It can be seen that $\delta \propto M^{0 \cdot 8}$ and $R_g \propto M^{0 \cdot 6}$ (M being the molecular weight of the polymer), as predicted by theory.

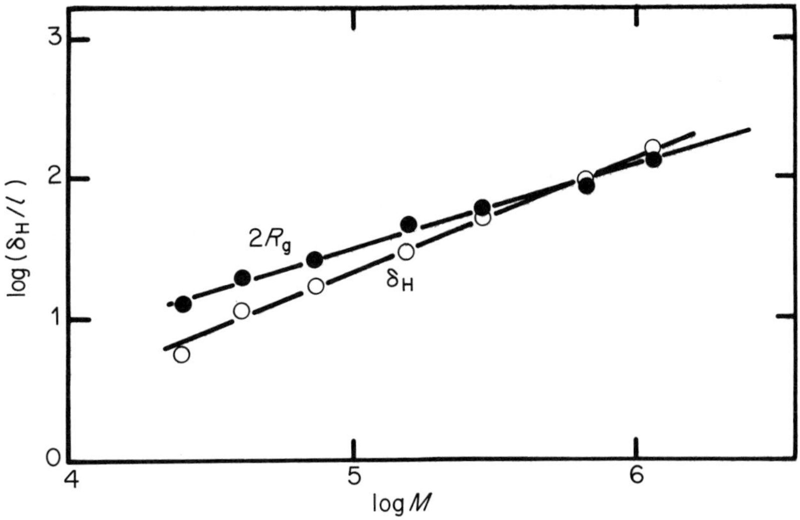

Fig. 2.22. (○) The plot of log (δ_H) for poly(ethylene) oxide adsorbed from water on poly(styrene) latex as a function of log(M). (●) For comparison, $2R_g$ (from light scattering) is also shown (after Cohen Stuart *et al.*, 1984).

2.4 Experimental techniques

2.4.1 Direct microscopic observations

The most direct way of obtaining the diffusion coefficients of colloidal particles is to observe their motion through a microscope. The first person to make such quantitative observations was Perrin (1913), who received a Nobel prize for his work on Brownian motion. The novelty of this early work lies in the fact that Avogadro's number can be determined from diffusion experiments. As we have seen, the mean-square displacement is proportional to the Boltzmann constant, k, which is related to Avogadro's number, N_{AV}, by

$$R = kN_{AV} \tag{2.201}$$

where R is the universal gas constant ($R = 8 \cdot 31$ J mol^{-1} K^{-1}), which was already known from experiments on gases. Avogadro's number gives the number of atoms or molecules in a mole of a substance and from its value the mass and the size of a single atom can be estimated. For example, from the molar mass of water (ca. 18g) and Avogadro's number ($N_{AV} = 6 \cdot 02 \times 10^{23}$) the mass of one water molecule is approximately 3×10^{-23} g, and from the volume of 1 mol of water (ca. 18 cm³) the volume of 1 water molecule is about 3×10^{-23} cm³ or 30 Å³, corresponding to an equivalent radius (allowing for free volume) of about 2 Å. Being able to determine the mass and size of atoms and molecules was considered solid evidence for their existence, an issue still hotly debated at the beginning of this century. Perrin (1913) obtained a value of $N_{AV} = 6 \cdot 7 \times 10^{23}$, by measuring both the translational and the rotational Brownian motion. This value is close to the presently accepted value.

Nowadays, the motion of Brownian particles can be observed and recorded on film or video. Interest in obtaining diffusion constants in this way has recently been revived (Vadas *et al.*, 1976; Suda and Imai, 1985). Results for latex spheres and doublets of equal spheres suspended in water are shown in Table 2.4 and it may be seen that theory and experiment agree in most cases to within a few percent. Vadas *et al.* (1976) also measured the probability density function, according to theory given by (cf. eq (2.90)):

$$n(x,t) = \frac{1}{2\sqrt{\pi D t}} e^{-x^2/4Dt} \tag{2.202}$$

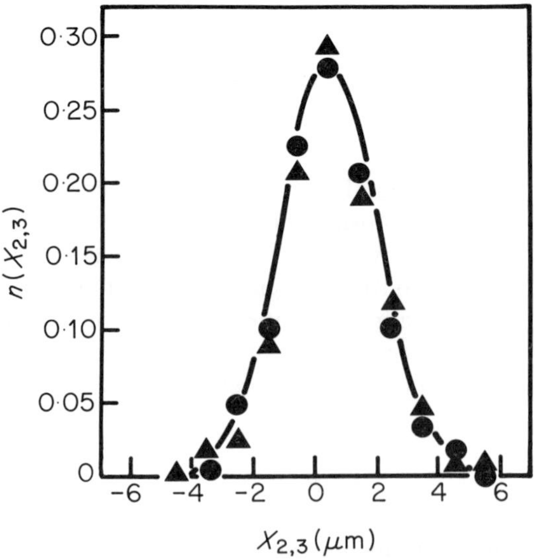

Fig. 2.23. Translational Brownian diffusion of a 1·19-μm sphere. Plot of the probability density distribution function with the solid line calculated from eq (2.202). The points are experimental values for translational motion along the X_2 (●) and X_3 (▲) axes, respectively; $t = 0 \cdot 040$ s (after Vadas *et al.*, 1976).

TABLE 2.4

Observed diffusion coefficients of spheres and doublets (after Vadas *et al.*, 1976)

Particle	Diameter of sphere (μm)	Δt (s)	D_1 (exp.) (10^{-11} m^2 s^{-1})	D_1 (theory)[a] (10^{-11} m^2 s^{-1})	D_2 (exp.) (10^{-11} m^2 s^{-1})	D_2 (theory)[b] (10^{-11} m^{2}s^{-1})
Sphere	1·19	0·020	4·26	4·12	$D_1 = D_2$	
Sphere	2·07	0·020	2·47	2·43	$D_1 = D_2$	
Doublet	1·19	0·060	3·18	3·18	2·98	2·83
Doublet	2·07	0·10	1·91	1·83	1·60	1·63
Doublet	2·07	0·20	2·13	1·83	1·70	1·63

[a]For a sphere: $D_1 = D_2 = kT/6\pi\mu a$. [b]For a doublet: $D_1 = kT/24.32\ \mu a$; $D_2 = kT/27.31\ \mu a$.

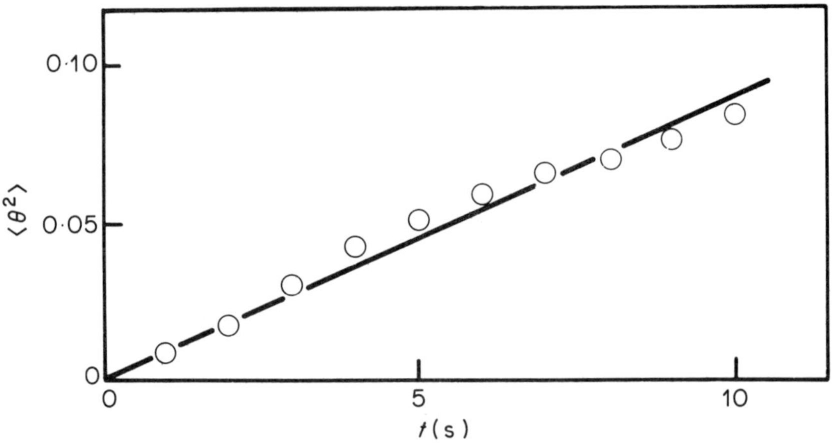

Fig. 2.24. Mean-square displacement $\langle \theta^2 \rangle$ in the polar angle, θ, as a function of time, measured on individual tungsten oxide particles suspended in glycerol/water (after Sudsa and Imai, 1985).

Their observations are shown in Fig. 2.23 for the translational motion along the X_2 and X_3 axes, respectively, for a latex sphere of diameter $1 \cdot 19\ \mu$m and are in excellent agreement with the predicted Gaussian distribution.

Figure 2.24 shows the observations of Suda and Imai (1985) on the rotational Brownian motion of plate-like tungsten oxide particles suspended in a 50% glycerol/water mixture. The polar angles, θ, were determined from the projected length, ℓ', of the particle (of length ℓ) onto a screen by

$$\ell' = \ell \sin\theta \tag{2.203}$$

From the slope of the plot of $\langle \theta^2 \rangle$ versus time t, the rotary diffusion coefficient can be determined; again a fair agreement with theory is found.

The most serious disadvantage of the direct method for determining diffusion coefficients is that it is tedious and time consuming. However, with today's technology in automated film and video analysis, this disadvantage is rapidly disappearing. The method may well become superior to other methods, especially for particles of $> \sim 0 \cdot 1\ \mu$m in size. The main advantage is that one can keep track of a whole distribution of particles and, furthermore, one is able to determine all the relevant diffusion constants for non-spherical particles. Because all particles are observed individually, there is no problem in finding the distribution in diffusion constants for polydisperse dispersions.

2.4.2 Diffusion experiments

Instead of looking at the motion of individual particles, the diffusion constants may be determined by studying the diffusion of a large number of particles. There are two ways in which diffusion experiments are usually performed: (i) starting from an initial non-uniform concentration distribution and observing the evolution towards an equilibrium or steady state; and (ii) maintaining a steady state and measuring the corresponding concentration profile. In both cases the experimental concentration profiles are compared with solutions of the (convective) diffusion equation. Obviously these kinds of experiment can be performed for a large variety of geometries. Here we will consider a simple example of each of the two cases mentioned.

As an example of evolution towards an equilibrium configuration, consider a dispersion of neutrally buoyant monodisperse particles of number concentration n_0 initially placed into contact with a volume of the suspending medium (containing no particles). This is shown schematically in Fig. 2.25. The subsequent evolution of the particles is described by the diffusion equation

$$\frac{\partial n}{\partial t} = D \frac{\partial^2 n}{\partial z^2} \tag{2.204}$$

where z is the vertical coordinate. The problem can be simplified by letting the distances to the top and bottom surface, ℓ_1 and ℓ_2, be large and restricting the problem to times $t \ll \ell_i^2/2D$, i.e. times for which diffusing particles are still far from the surfaces $z = +\ell_1$ and $z = -\ell_2$. In this case the initial condition can be taken to be

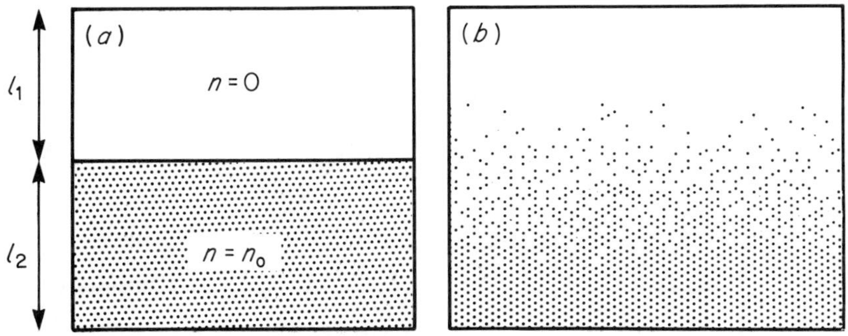

Fig. 2.25. The initial particle distribution in a diffusion experiment (a) and the distribution after some time t (b).

$$\begin{aligned} n(z,0) &= n_0 \qquad -\infty < z \leqslant 0 \\ n(z,0) &= 0 \qquad 0 < z < \infty \end{aligned} \tag{2.205}$$

and the boundary conditions are:

$$\begin{aligned} n(t,\infty) &= 0 \\ n(t,-\infty) &= n_0 \end{aligned} \tag{2.206}$$

Equation (2.204) together with the conditions given by eqs (2.205) and (2.206) give the simple solution (Crank, 1956):

$$n(z,t) = \tfrac{1}{2} n_0 \operatorname{erfc}\left(\frac{z}{2\sqrt{Dt}}\right) \tag{2.207}$$

where erfc is the complimentary error function. The solution is shown in Fig. 2.26. By comparing the experimentally obtained diffusion profiles with Fig. 2.26, a value of D is obtained. So far, such diffusion methods have found more application to molecular solutions than to colloidal dispersions because diffusion is a rather slow process in dispersions. For example, in an aqueous dispersion of 10-nm radius particles, it takes about 6 h to attain a r.m.s. displacement of 1 mm. Therefore, after 6 h the concentration profile extends for only a few millimetres beyond the initial boundary at $z = 0$. The

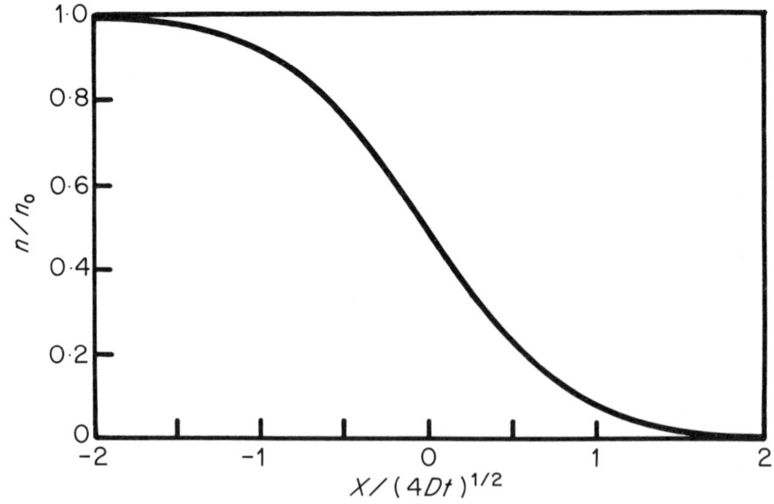

Fig. 2.26. Concentration–distance curve for an extended source of infinite extent.

experimental observation of such thin boundaries is difficult. One way of observing them is by using Schlieren optics (see e.g. Hiemenz, 1977). Nowadays it might be better to observe such profiles by turbidity measurements; owing to recent technological advances, photoreceptors with a resolution of 10 μm exist, allowing the accurate determination of concentration profiles which extend over only a few millimeters. In principle, diffusion experiments can be extended to all kinds of geometries, provided that solutions of the diffusion equation can be found.

As an example of a steady-state profile we will discuss convective diffusion in two concentric tubes (van de Ven, 1975). The steady-state profile is governed by

$$D\nabla^2 n - \nabla\cdot(\boldsymbol{u}n)=0 \tag{2.208}$$

Equation (2.208) has been solved numerically by Mashelkar and Venkatsubramanian (1975) with the following boundary conditions: at $z=0$ (exit of inner tube of radius ℓ)

$$\begin{array}{ll} n=n_0 & 0<r<\ell \\ n=0 & \ell<r<R \end{array} \tag{2.209}$$

R being the inner radius of the outer tube. At $z=\infty$,

$$n=\text{constant} \tag{2.210}$$

the constant determined by the relative fluxes of liquid entering via the inner and outer tube. Furthermore, it was assumed that at $z>0$ the flow was given by a fully developed Poiseuille flow:

$$u=u_{\mathrm{m}}\left[1-(r/R)^2\right] \tag{2.211}$$

Rewriting eq (2.208) with u as given by eq (2.211) and non-dimensionalizing the equation by dividing distances by R, gives

$$\nabla^2 n-\mathrm{Pe}\;\nabla\cdot\left[\{1-r^2\}n\right]=0 \tag{2.212}$$

where

$$\mathrm{Pe}=u_{\mathrm{m}}R^2/D \tag{2.213}$$

Solutions of this equation with the above boundary conditions for various Péclet numbers (Pe) are shown in Fig. 2.27. Measuring the concentration profile for such a geometry experimentally yields Pe, from which D can be calculated.

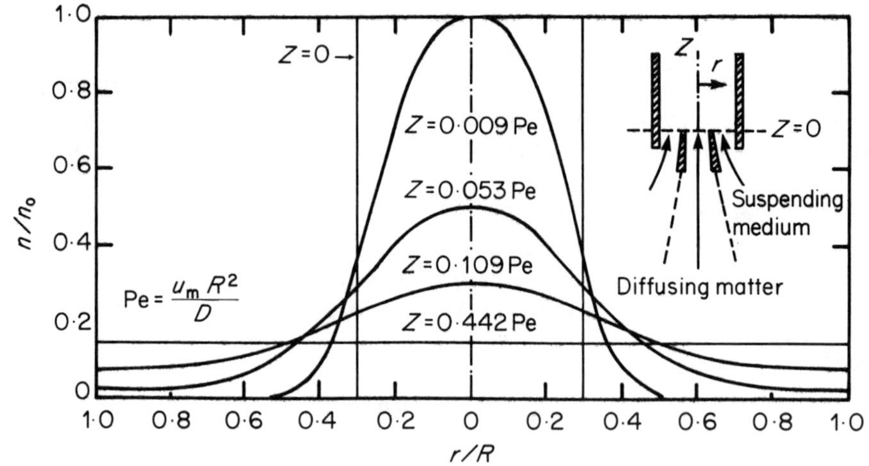

Fig. 2.27. Dimensionless radial concentration profiles of Brownian particles (or molecules) at an initial concentration n_0 diffusing from a centrally placed micropipette, of radius $0 \cdot 3R$, into the suspending medium undergoing Poiseuille flow for various distances, z, downstream of the infusion site. Final $n/n_0 = 0 \cdot 14$ (after Bell *et al.*, 1984).

2.4.3 Photon correlation spectroscopy

2.4.3.1 Monodisperse dispersions

When a light beam travels through a colloidal dispersion, the suspended particles scatter and sometimes absorb a certain amount of the light. The scattered light contains information about the sizes and shapes of the suspended particles. Almost all modern light-scattering experiments are performed using monochromatic laser light. The intensity of the light, scattered by a single particle, observed at a given angle, is determined mainly by the volume of the particle and the wavelength of the light. For the light beam traversing a dilute suspension, the total amount of light scattered in a certain direction is simply the sum of the contribution of all the particles in the beam. Colloidal particles suspended in a fluid undergo translational and rotational Brownian motion and this causes small fluctuations in the intensity and wavelength of the scattered light. Thus the measurement of these fluctuations yields information about the diffusion coefficients and may be made by counting the number of scattered photons in regular, short time intervals, over times sufficiently long to obtain proper statistical results. This technique is called photon correlation spectroscopy (PCS), also referred to as quasi-elastic light scattering (QELS).

The principle of a PCS measurement is shown schematically in Fig. 2.28a. A laser beam is passed through a suspension and the intensity of the scattered light is measured at a fixed scattering angle, θ_s. The scattered light impinges upon a photomultiplier and the output goes to an autocorrelator and a recorder or visual-display monitor. The method shown in Fig. 2.28a is usually referred to as the homodyne method. In certain cases it is advantageous to split the incoming beam and recombine it with the scattered light before reaching the detector. This is the so-called heterodyne method, shown schematically in Fig. 2.28b.

Instead of specifying the scattering geometry by the scattering angle θ_s, it is often convenient to express it in terms of the wave vector $\boldsymbol{q}$, defined as

$$\boldsymbol{q} = \boldsymbol{K}_i - \boldsymbol{K}_s \tag{2.214}$$

where $\boldsymbol{K}_i$ is the wave vector of the incoming light ($\boldsymbol{K}_i = 2\pi n/\lambda_i$), and $\boldsymbol{K}_s$ is that of the scattered light ($\boldsymbol{K}_s = 2\pi n/\lambda_f$), n being the refractive index of the

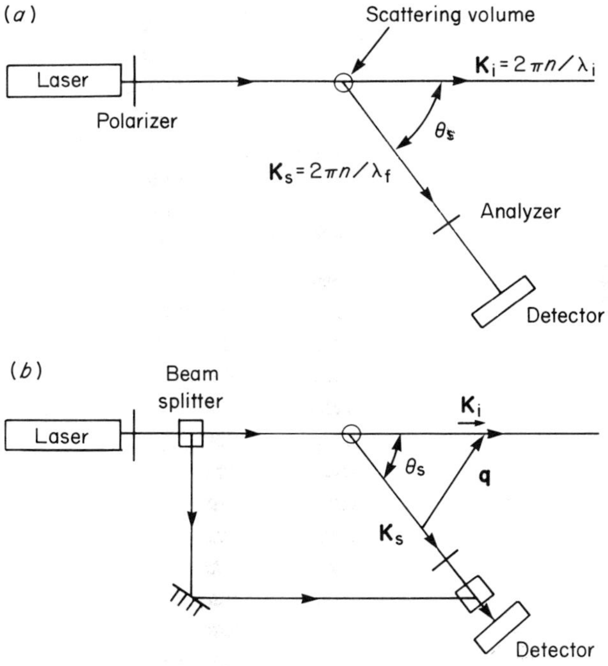

Fig. 2.28. Scheme to show the principle of a PCS measurement: (a) homodyne method; (b) heterodyne method. $\boldsymbol{K}_i$, wave vector of the incoming beam, $\boldsymbol{K}_i = 2\pi n/\lambda_i$; $\boldsymbol{K}_s$, wave vector of the scattering beam, $\boldsymbol{K}_s = 2\pi n/\lambda_f$; scattering vector, $\boldsymbol{q} = \boldsymbol{K}_i - \boldsymbol{K}_s$, $q = |\boldsymbol{q}| = 4\pi n/\lambda_i \sin(\theta_s/2)$; θ_s, scattering angle.

medium and λ_i the wavelengths of the incoming and scattered light. For small colloidal particles, the wavelength of the scattered light is almost equal to that of the incoming light and hence $\lambda_i \simeq \lambda_f = \lambda$. From the definition of $\boldsymbol{q}$ it can be shown that

$$q = |\boldsymbol{q}| = \frac{4\pi n}{\lambda} \sin\left(\frac{\theta_s}{2}\right) \tag{2.215}$$

The photons falling on the photo tube of the detector cause a current output which is proportional to the square of the incident electric field, which in turn is proportional to the intensity of the light, i.e. $i(t) \propto E(t)^2$, where i is the current output of the detector. The output is passed through an autocorrelator (a hardwired computer), which calculates the autocorrelation function $\langle i(t)\, i(0) \rangle$.

In homo- and heterodyne experiments two autocorrelation functions are usually defined (Berne and Pecora, 1976). The intensity measured in a heterodyne experiment is

$$I_1(t) \equiv \langle E_s^*(0)\; E_s(t) \rangle \tag{2.216}$$

and in a homodyne experiment

$$I_2(t) \equiv \langle |E_s^*(0)|^2 |E_2(t)|^2 \rangle \tag{2.217}$$

where E_s is the scattered electric field strength and E_s^* its complex conjugate. In the homodyne method, the scattered light impinges directly on the photon detector and, because the current is proportional to E_s^2, we have

$$\langle i(t)\, i(0) \rangle = BI_2(t) \tag{2.218}$$

where B is a constant depending on the detection unit. If the scattering volume consists of a large number of statistically independent subregions governed by Gaussian statistics, the autocorrelation functions are related to each other by

$$I_2(t) = |I_1(0)|^2 + |I_1(t)|^2 \tag{2.219}$$

In the heterodyne method, a small portion of the unscattered light, having an electric field $E_0(t)$, is mixed with the scattered light E_s, and hence the autocorrelation becomes

$$\langle i(t)\, i(0) \rangle = B\langle |E_0(t) + E_s(t)|^2 |E_0(0) + E_s(0)|^2 \rangle \tag{2.220}$$

Usually one chooses $|E_0(0)| \gg E_s$ and eq (2.220) reduces to

$$\langle i(t)\, i(0) \rangle = B\langle I_0^2 + 2I_0 \operatorname{Re} I_1(t) \rangle \tag{2.221}$$

where $I_0 = \langle | E_0 |^2 \rangle$ and Re I_1 is the real part of $I_1(t)$. From eqs (2.218) to (2.221) it can be seen that if the approximation of independent Gaussian subregions holds, the homo- and heterodyne methods yield identical information, embodied in the function $I_1(t)$. To obtain useful information from light-scattering experiments, this function must be calculated for the systems one is interested in.

When the scattering is due to spherical particles, it can be shown that (Berne and Pecora, 1976)

$$\begin{aligned} I_{VV}(q,t) &= \alpha^2 F_1(q,t) \\ I_{VH}(q,t) &= I_{HV}(q,t) = 0 \\ I_{HH}(q,t) &= \cos^2\theta_s I_{VV}(q,t) \end{aligned} \tag{2.222}$$

where V and H stand for the polarization directions (vertical or horizontal) of the incoming light (first symbol) and scattered light (second symbol). Eq (2.222) refers to the autocorrelation function I_1, and α is the polarizability of the particle. The polarizability factor arises because light scattering can be thought of as light emitted by oscillating dipoles induced by the electric field, whose moments are proportional to α. The function $F_1(q,t)$ is given by

$$F_1(q,t) = \langle N \rangle F_s(q,t) \tag{2.223}$$

where $\langle N \rangle$ is the average number of particles in the illuminated volume and F_s is a volume average, given by

$$F_s(q,t) = \langle \exp i\boldsymbol{q} \cdot [\boldsymbol{r}(t) - \boldsymbol{r}(0)] \rangle \tag{2.224}$$

$\boldsymbol{r}$ being the position of a particle and $i = \sqrt{-1}$. It can be seen that in order to obtain I_1, eq (2.224), which expresses the fact that I_1 is determined by the fluctuations in location $\boldsymbol{r}$ of the (independently) scattering particles, must be evaluated. In deriving eq (2.224) it is assumed that the number of particles in the scattering volume is large enough for fluctuations in the number of particles to contribute negligibly to the fluctuations in the scattered light. For small scattering volumes, these fluctuations in number must also be taken into account. For spherical colloidal particles, these fluctuations in location are due to translational Brownian motion and are governed by the diffusion equation

$$\frac{\partial n}{\partial t} = D \nabla^2 n \tag{2.225}$$

The Fourier transform of eq (2.225) is

$$\frac{\partial F_s(q,t)}{\partial t} = -q^2 D F_s(q,t) \tag{2.226}$$

with $F_s(q,t)$ given by eq (2.224). The solution of eq (2.226) subject to $F(q,0)=1$, is

$$F_s(q,t) = \exp(-q^2 Dt) \tag{2.227}$$

Combining eqs (2.223) and (2.227) yields the heterodyne autocorrelation function for a suspension of non-interacting spherical particles:

$$F_1(q,t) = \langle N \rangle \exp(-q^2 Dt) \tag{2.228}$$

From eq (2.219) it follows that the homodyne autocorrelation function given by eq (2.222), with F_1 replaced by F_2, is

$$F_2(q,t) = \langle N \rangle^2 (1 + \exp - 2q^2 Dt) \tag{2.229}$$

It follows that D can be obtained by measuring $I_1(t)$ or $I_2(t)$ with monochromatic light of a given wavelength, λ, and at a fixed scattering angle, θ_s.

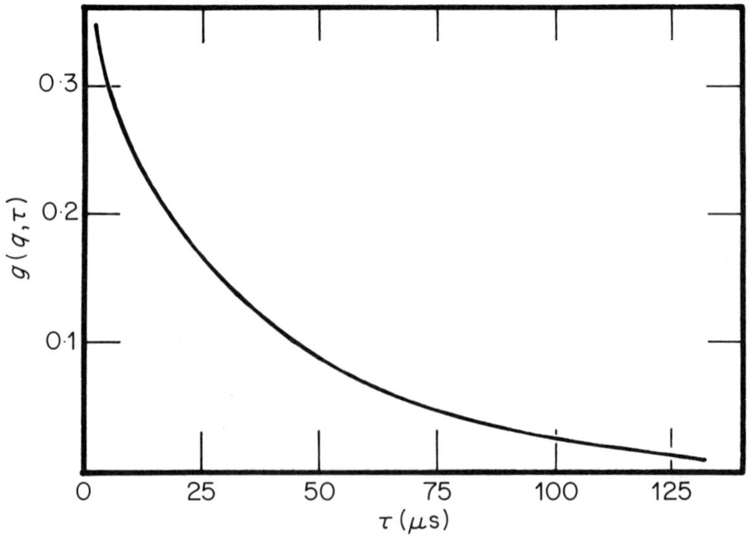

Fig. 2.29. Experimental homodyne time correlation function $g(q,\tau) = \langle i(t+\tau)\, i(t) \rangle - \langle i \rangle^2$ for spherical gold particles of radius 10 nm (after Schumacher and van de Ven, 1987).

An example of the results of a homodyne experiment on spherical gold particles is shown in Fig. 2.29.

For non-spherical particles the polarizability is no longer a scalar as for spheres (cf. eq (2.222)), but becomes a tensor. For axisymmetric particles

$$\boldsymbol{\alpha} = \begin{pmatrix} \alpha_{\parallel} & 0 & 0 \\ 0 & \alpha_{\perp} & 0 \\ 0 & 0 & \alpha_{\perp} \end{pmatrix} \tag{2.230}$$

and eq (2.222) becomes

$$\begin{aligned} I_{\mathrm{VV}}(\boldsymbol{q},t) &= \langle N\rangle\langle\alpha_{11}(0)\,\alpha_{11}(t)\rangle F_{\mathrm{s}}(\boldsymbol{q},t) \\ I_{\mathrm{VH}}(\boldsymbol{q},t) &= \langle N\rangle\langle\alpha_{12}(0)\,\alpha_{12}(t)\rangle F_{\mathrm{s}}(\boldsymbol{q},t) \end{aligned} \tag{2.231}$$

where α_{ij} are the components of the polarization tensor $\boldsymbol{\alpha}$ in space-fixed coordinates, with the X_1 axis coinciding with the vertical direction, i.e. the direction in which the incoming light is polarized, found from

$$\alpha_{ij} = \beta_{im}\,\beta_{jm}\,\alpha'_{mn} \tag{2.232}$$

where $\alpha'_{mn} = \boldsymbol{\alpha}$ are the components of $\boldsymbol{\alpha}$ given by eq (2.230) and β_{ij} is the transformation matrix given by

$$\beta_{ij} = \begin{pmatrix} \cos\theta & -\sin\theta & 0 \\ \sin\theta\cos\phi & \cos\theta\cos\phi & -\sin\phi \\ \sin\theta\sin\phi & \cos\theta\sin\phi & \cos\phi \end{pmatrix} \tag{2.233}$$

θ and ϕ being the polar angles describing the orientation of the scattering axisymmetric particles with respect to the space-fixed coordinates X_i. From eqs (2.232) and (2.233) it follows that

$$\alpha_{11} = \alpha_{\parallel}\cos^2\theta + \alpha_{\perp}\sin^2\theta = \alpha + \frac{1}{3}\,\beta(3\cos^2\theta - 1) \tag{2.234}$$

and

$$\alpha_{12} = (\alpha_{\parallel} - \alpha_{\perp})\,\sin\theta\cos\theta\cos\phi = \beta\sin\theta\cos\theta\cos\phi \tag{2.235}$$

where

$$\alpha = \frac{1}{3}(\alpha_{\parallel} + 2\alpha_{\perp}) \tag{2.236}$$

and

$$\beta = \alpha_{\parallel} - \alpha_{\perp} \tag{2.237}$$

α is the isotropic part of $\boldsymbol{\alpha}$ and β is the optical anisotropy. To obtain I_{VV} or I_{VH} the α_{ij} values in eq (2.231) must be averaged over all the orientations for a given time, t. The fluctuations in the orientation distribution are governed by the rotary-diffusion equation

$$\frac{\partial p}{\partial t} = D_r \nabla^2 p \tag{2.238}$$

subject to the normalization condition

$$\int_0^{\pi} \int_0^{2\pi} p(\theta,\phi;t) \sin\theta \mathrm{d}\theta \mathrm{d}\phi = 1 \tag{2.239}$$

The general solution of eq (2.238) is (Vadas *et al.*, 1976)

$$p(\theta,\phi;t) = \sum_{m,n} (A_{\mathrm{mn}} \sin m\phi + B_{mn} \cos m\phi) P_n^m(\cos\theta) \mathrm{e}^{-n(n+1)D_r t} \tag{2.240}$$

where A_{mn} and B_{mn} are determined by the distribution at time $t = 0$, and P_n^m are Legendre polynomials. With the aid of eq (2.240), the averages in eq (2.231) can be determined, resulting in

$$I_{\mathrm{VV}}(q,t) = \langle N \rangle \left(\alpha^2 + \frac{4}{45} \beta^2 \mathrm{e}^{-6D_r t} \right) \mathrm{e}^{-q^2 Dt} \tag{2.241}$$

$$I_{\mathrm{VH}}(q,t) = \frac{1}{15} \langle N \rangle \beta^2 \mathrm{e}^{-6D_r t} \mathrm{e}^{-q^2 Dt} \tag{2.242}$$

Equations (2.241) and (2.242) allow the rotary-diffusion coefficient, D_r, to be determined from PCS measurements. The translational-diffusion coefficient in this case equals

$$D = \frac{1}{3} (D_{\parallel} + 2D_{\perp}) \tag{2.243}$$

where $D_{\parallel}$ and $D_{\perp}$ are translational-diffusion coefficients for the motion along and perpendicular to the particle's axis of symmetry.

The results for axisymmetric particles can be readily extended to ellipsoidal (or other triaxial particles such as brick or ribbon shaped). In this case

$$\alpha = \begin{pmatrix} \alpha_1 & 0 & 0 \\ 0 & \alpha_2 & 0 \\ 0 & 0 & \alpha_3 \end{pmatrix} \tag{2.244}$$

and the orientation of a particle is described by the three Euler angles (θ, ϕ and ψ), with the corresponding transformation matrix

$$\beta_{ij} = \begin{pmatrix} \cos\theta & -\sin\theta\cos\psi & \sin\theta\cos\psi \\ \sin\theta\cos\phi & -\sin\phi\sin\psi + \cos\theta\cos\phi\cos\psi & -\sin\phi\cos\psi - \cos\theta\cos\phi\sin\psi \\ \sin\theta\sin\phi & \cos\phi\sin\psi + \cos\theta\sin\phi\cos\psi & \cos\phi\cos\psi - \cos\theta\sin\phi\sin\psi \end{pmatrix} \quad (2.245)$$

from which it follows that

$$\alpha_{11} = \alpha_1\cos^2\theta + \alpha_2\sin^2\theta\cos^2\psi + \alpha_3\sin^2\theta\sin^2\psi \quad (2.246)$$

and

$$\alpha_{12} = \alpha_1\sin\theta\cos\theta\cos\phi + \alpha_2\sin\theta\cos\psi(\sin\phi\sin\psi - \cos\theta\cos\phi\cos\psi) + \alpha_3\sin\theta\sin\psi(\sin\phi\cos\psi + \cos\theta\cos\phi\sin\psi) \quad (2.247)$$

The rotary-diffusion equation becomes

$$\frac{\partial p}{\partial t} = \nabla\cdot(\boldsymbol{D}_{\mathrm{r}}\cdot\nabla p) \quad (2.248)$$

which has the general solution (Jayasuriya and van de Ven, 1985):

$$p = \sum_{n,k,m} A\{n,2k,2m\}\exp(-ik\psi)\,P^n_{k,m}(\cos\theta)\exp(-im\phi)\mathrm{e}^{-\kappa^2 t} \quad (2.249)$$

where $\boldsymbol{P}^n_{k,m}$ are generalized harmonic functions and $A\{n,2k,2m\}$ are coefficients that can be found from the initial distribution; κ^2 are the eigen values.

Using the orientation distribution (eq (2.249)), the averages in eq (2.231) can be obtained with α_{11} and α_{12} given by eqs (2.246) and (2.247). When the principal axes of the rotary-diffusion tensor $\boldsymbol{D}_{\mathrm{r}}$ and the polarizability tensor $\boldsymbol{\alpha}$ coincide, the result for I_{VH} is (Berne and Pecora, 1976):

$$I_{\mathrm{VH}}(q,t) = \frac{1}{15}\langle N\rangle \sum_{i=1}^{2} A_i\exp(-q^2D + f_i)t \quad (2.250)$$

where

$$D = \frac{1}{3}(D_1^{\mathrm{t}} + D_2^{\mathrm{t}} + D_3^{\mathrm{t}}) \quad (2.251)$$

D_i^{t} being the components of the translational diffusion tensor and

$$f_1 = 6D_{\mathrm{r}} - 2\Delta \quad (2.252)$$

$$f_2 = 6D_r + 2\Delta \qquad (2.253)$$

where

$$D_r = \frac{1}{3}(D_1^r + D_2^r + D_3^r) \qquad (2.254)$$

and

$$\Delta = \left[(D_1^r - D_2^r)^2 + (D_1^r - D_3^r)(D_2^r - D_3^r)\right]^{1/2} \qquad (2.255)$$

For ellipsoidal particles which are not too far from axisymmetry, i.e. when $2D_1^r > D_2^r + D_3^r \gg D_2^r - D_3^r$, where we have chosen $D_1^r > D_2^r > D_3^r$, f_1 reduces to

$$f_1 \simeq 3(D_2^r + D_3^r) \qquad (2.256)$$

which means that at long periods of time such a body behaves like an axisymmetric body with an effective diffusion coefficient given by (Jayasuriya and van de Ven, 1985)

$$D_{\text{eff}}^r = \frac{D_2^r + D_3^r}{2} \qquad (2.257)$$

An alternative to measuring autocorrelation functions is to measure the corresponding frequency spectrum. This is simply the Fourier transform of the autocorrelation function. For a heterodyne autocorrelation function obtained from the scattering by monodisperse spherical particles, the spectrum is a Lorentzian:

$$F_1(q, \omega) = \frac{\langle N \rangle}{\pi} \frac{q^2 D}{\omega^2 + (q^2 D)^2} \qquad (2.258)$$

ω being the frequency. Similar spectra can be obtained for other correlation functions.

2.4.3.2 Polydisperse dispersions

For dilute polydisperse systems, the intensity of the scattered light is the sum of the light scattered by the individual particles. In the case of spheres made of the same material but having a discrete size distribution, the scattering function is

$$I_{VV} = \alpha^2 \sum_{i=1}^{n} \langle N_i \rangle \, e^{-q^2 D_i t} \qquad (2.259)$$

where N_i is the number of particles of size i and having a diffusion constant D_i; n is the total number of different particle sizes. For a continuous size distribution, the above summation can be replaced by an integration:

$$I_{VV} = \alpha^2\langle N\rangle \int_0^\infty f(a)\,\mathrm{e}^{-q^2 D(a)t}\,\mathrm{d}a$$

$$= \alpha^2\langle N\rangle \bar{F}_s \tag{2.260}$$

where $f(a)$ is the particle-size distribution and $\bar{F}_s$ is the weighted average of F_s defined by eq (2.227). Alternatively, we can write

$$\bar{F}_s = \int_0^\infty f(a)\,\mathrm{e}^{-q^2 D(a)t}\mathrm{d}a = \int_0^\infty g(D)\,\mathrm{e}^{-q^2 Dt}\mathrm{d}D \tag{2.261}$$

where $g(D)$ is the distribution of the diffusion coefficients.

Experimentally, the problem consists of determining $f(a)$ or $g(D)$ from a measurement of $\bar{F}_s$. This is not usually a simple problem, especially when the degree of polydispersity is large. There are several ways to approach this problem. First, a certain distribution can be assumed, e.g. a normal (i.e. Gaussian) or a log-normal distribution. Such a distribution contains two unknown parameters, the average radius, $\bar{a}$, and the standard deviation, σ. After performing the integration in eq (2.261) for the chosen distribution, we obtain a relation of the form

$$\bar{F}_s = f(\bar{a},\sigma) \tag{2.262}$$

and from the best fit between the experimental and theoretical data we can estimate $\bar{a}$ and σ. The problem with this approach is that the distribution is not known *a priori*.

A second method that can be used is to expand $\bar{F}_s$ about its mean value (Brown *et al.*, 1975). Let $q^2 D = \Gamma$, then we can write

$$\bar{\Gamma} = \int \Gamma G(\Gamma)\,\mathrm{d}\Gamma \tag{2.263}$$

and

$$\mathrm{e}^{-\Gamma t} = \mathrm{e}^{-\bar{\Gamma}t}\mathrm{e}^{-(\Gamma-\bar{\Gamma})t}$$

$$= \mathrm{e}^{-\bar{\Gamma}t}\left[1-(\Gamma-\bar{\Gamma})t+\frac{1}{2}(\Gamma-\bar{\Gamma})t^2-\frac{1}{6}(\Gamma-\bar{\Gamma})t^3+\ldots\right] \tag{2.264}$$

Hence

$$\bar{F}_s = \int e^{-\Gamma t} G(\Gamma)\, d\Gamma = e^{-\bar{\Gamma} t}\left(1 + \frac{1}{2}\mu_2 t^2 - \frac{1}{6}\mu_3 t^3 + \ldots\right) \tag{2.265}$$

where the moments μ_i are given by

$$\mu_i = \int (\Gamma - \bar{\Gamma})^i G(\Gamma)\, d\Gamma \tag{2.266}$$

Again it is possible to obtain $\bar{\Gamma}$, μ_2, μ_3, etc., from the comparison of experiment with theory. The advantage of this method is that no *a priori* knowledge of the size distribution is required. One problem is to determine how many moments are required to describe the distribution. This can be solved by doing experiments at different sample times, e.g. as $t \to 0$, higher order moments become negligible (Brown *et al.*, 1975).

A third method that can be used is to calculate the size distribution directly from the measured autocorrelation function. The integrals appearing in eqs (2.261) and (2.263) are so-called Fredholm integrals which can be transformed into matrix equations. This can be achieved by approximating the continuous distribution by a discrete one, consisting of M distinct sizes varying from a_1 to a_M (see Fig. 2.30a), with $a_{i+1} = a_i + \Delta a$. Such a distribution $\bar{F}_s$ is given by

$$F_s = \sum_{i=1}^{M} f(a_i)\, e^{-q^2 D_i t} \Delta a \tag{2.267}$$

where $f(a_i)\Delta a$ is the fraction of particles of size a_i. Note that for a given i, D_i is known (from the Stokes–Einstein equation), but $f(a_i)$ is not.

Next we choose M points on the experimental $\bar{F}_s$ versus t curve (Fig. 2.30b). For each point i, we know the corresponding value of $\bar{F}_s(t_i)$. Hence, we have M equations with M unknowns:

$$\begin{pmatrix} F_s(t_1) \\ \vdots \\ F_s(t_i) \\ \vdots \\ F_s(t_M) \end{pmatrix} = \begin{pmatrix} e^{-q^2 D_1 t_1} & e^{-q^2 D_2 t_1} & \cdots\cdots & e^{-q^2 D_M t_1} \\ \\ e^{-q^2 D_1 t_i} & e^{-q^2 D_2 t_i} & \cdots\cdots & e^{-q^2 D_M t_i} \\ \\ e^{-q^2 D_1 t_M} & e^{-q^2 D_2 t_M} & \cdots\cdots & e^{-q^2 D_M t_M} \end{pmatrix} \bullet \begin{pmatrix} f(a_i) \\ \vdots \\ f(a_i) \\ \vdots \\ f(a_M) \end{pmatrix} \Delta a$$

or

$$\boldsymbol{F}_s = \boldsymbol{E} \cdot \boldsymbol{f}\, \Delta a \tag{2.268}$$

Notice that the vector $\boldsymbol{F}_s$ is known (from experiment), the matrix $\boldsymbol{E}$ is known (its components can be calculated from D_i and t_i), but the vector $\boldsymbol{f}$ is unknown. The vector $\boldsymbol{f}$ is the size distribution which we want to determine. Inverting eq (2.268) yields

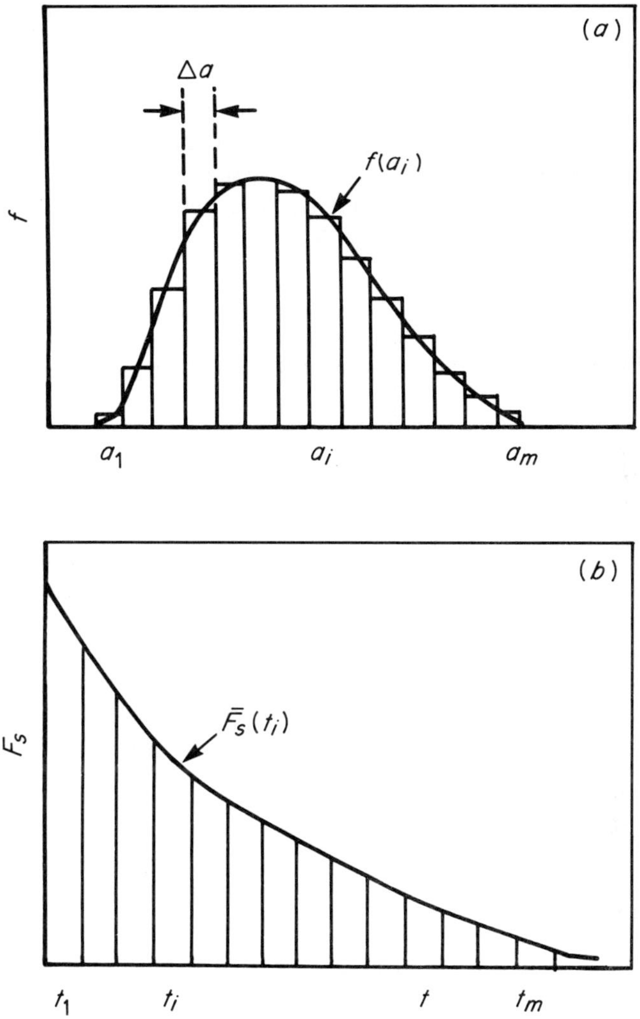

Fig. 2.30. Determination of the unknown particle-size distribution (a) from the experimentally obtained autocorrelation function $\bar{F}_s(t)$ shown in (b).

$$f = \boldsymbol{E}^{-1} \cdot \boldsymbol{F}_s / \Delta a \tag{2.269}$$

This method is quite general and applies to all experimental properties that can be measured as a function of time. The difficulty with this method is that the inversion of eq (2.268) is non-unique; many distributions can give rise to a very similar $F_s(t)$ behaviour. These problems can be overcome by imposing constraints on the problem. By minimizing a modified weighted sum of the residuals, together with the constraint $f(a) > 0$, it is possible to find a unique regularized solution for $f(a)$ (Provencher, 1979).

The method can also be readily extended to non-spherical particles in which there are variations in both size and shape. For example, for axisymmetric particles of constant $\alpha_\perp$ and $\alpha_\parallel$, the heterodyne autocorrelation function is proportional to (cf. eq (2.242))

$$A(t) = \mathrm{e}^{-6D_r t}\mathrm{e}^{-q^2 Dt} \tag{2.270}$$

For a polydisperse system there is a variation in the two principal axes, a and b, and hence

$$\bar{A} = \int_0^\infty \int_0^\infty f(a,b)\, \mathrm{e}^{-6D_r t}\mathrm{e}^{-q^2 Dt}\mathrm{d}a\,\mathrm{d}b \tag{2.271a}$$

or

$$\bar{A} = \sum_i \sum_j f_{ij}(a,b)\, \mathrm{e}^{-6D^r_{ij}t}\mathrm{e}^{-q^2 D_{ij}t}\Delta a\,\Delta b \tag{2.271b}$$

Choosing M intervals for a, N for b and MN points on the experimental $\bar{A}$ versus t curve, MN equations are obtained with MN unknowns which can be written as

$$\boldsymbol{A} = \boldsymbol{E}\cdot\boldsymbol{f}\,\Delta a\,\Delta b \tag{2.272}$$

which upon inversion yields

$$\boldsymbol{f} = \boldsymbol{E}^{-1}\boldsymbol{A}/(\Delta a\,\Delta b) \tag{2.273}$$

2.4.4 Relaxation experiments

If a suspension of non-spherical particles is perturbed, its random distribution of orientations is likely to become non-random. If the perturbation is removed, rotary Brownian motion will cause the orientation distribution to relax back to its initial random distribution. In principle, by observing the system the time development of any property of a suspension that is orientation dependent can be followed. It has been shown (Jayasuriya and van de Ven, 1985) that not all properties show the same relaxation behaviour. As an example we will discuss the relaxation obtained in light-scattering or light-transmittance experiments as these are probably the most widely used techniques. When

$$2\pi \ell (m^2 - 1) \ll \lambda \tag{2.274}$$

where ℓ is the dimension of a suspended particle, m the relative refractive index of the medium with respect to the particle, and λ the wavelength of the incident light, then the scattering by the suspended particles can be described by the Rayleigh–Debye theory. Consider a vertically polarized incident light beam travelling along the X_2 axis and scattering in the X_2X_3 plane at an angle θ_s (with X_3 axis). The intensity of the light scattered by a spheroidal particle of axis ratio r_p and orientation (θ,ϕ) is given by (Okagawa and Mason, 1977)

$$I_s(\theta_s;\theta,\phi) = KP(\theta_s;\theta,\phi) \tag{2.275}$$

where K is a constant which depends on particle volume and the experimental conditions. $P(\theta_s;\theta,\phi)$ is a form factor given by

$$P(\theta_s) = \left[\frac{3(\sin u - u\cos u)}{u^3}\right]^2 \tag{2.276}$$

with

$$u = q(\cos^2\alpha + r_p^{-2}\sin^2\alpha)^{1/2} \tag{2.277}$$

where q is the scattering vector and

$$\cos\alpha = f(\theta_s,\phi)\sin\theta$$

where

$$f(\theta_s,\phi) = \sin\phi\cos\left(\frac{\theta_s}{2}\right) - \cos\phi\sin\left(\frac{\theta_s}{2}\right) \tag{2.278}$$

In a suspension with an orientation distribution $p(\theta,\phi;t)$, the intensity of the light scattered at an angle θ_s is given by

$$I_s(\theta) = K\int_0^{\pi}\int_0^{2\pi} P(\theta_s;\theta,\phi)\,p(\theta,\phi;t)\sin\theta\,d\theta\,d\phi \tag{2.279}$$

To evaluate the integral in eq (2.279) we expand the form factor

$$P(\theta_s) = k_1 + k_2 f^2(\theta_s,\phi)\sin^2\theta + k_3 f^4(\theta_s,\phi)\sin^4\theta + \ldots \tag{2.280}$$

where the k_i values are constants which depend on q and r_p.

With the aid of the known function $p(\theta,\phi;t)$ (cf. eq (2.240)) and the expansion eq (2.280), the integral occurring in eq (2.279) can be evaluated term by term, to give (Jayasuriya and van de Ven, 1985):

$$\Delta I_s(\theta_s)=\Delta I_s^0-\sum_{n=1}^{\infty} a_n e^{-2n(2n+1)D_r t} \tag{2.281}$$

ΔI_s^0 being the difference in scattered intensity between the times $t=0$ and $t=\infty$. The coefficients a_n depend on the orientation distribution at time $t=0$. It follows from eq (2.281) that the scattering intensity relaxes to the intensity for a random distribution with an infinite number of relaxation times. At large times the only remaining relaxation time occurs at $n=1$ and hence

$$\Delta I_s(\theta_s, t\rightarrow\infty)=\Delta I_s^0-a_1 e^{-6D_r t} \tag{2.282}$$

Equations (2.281) and (2.282) also apply to the intensity of the transmitted light (with I_s replaced by the intensity of the transmitted light I_t). In the absence of absorption, the intensity of the transmitted light is the difference

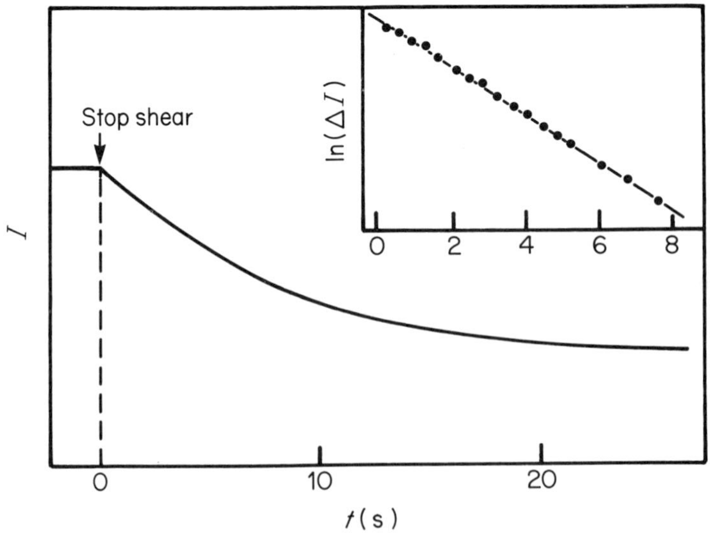

Fig. 2.31. Relaxation in intensity (I) of transmitted light through a suspension of non-spherical latex particles. The orientation distribution was disturbed by a shear flow and at time $t=0$ the shear was switched off. Inset: same relaxation on logarithmic scale (after Jayasuriya and van de Ven, 1985).

between the intensity of the incident light and the light scattered in all directions. Because for all directions the scattering intensity varies according to eq (2.281), the light transmittance follows the same dependency as the scattered intensity.

The same analysis applies to ellipsoidal or other triaxial particles. When they are not too far from axisymmetry, these particles behave as spheroids with an effective rotary-diffusion constant given by

$$D_r = (D_2 + D_3)/2 \tag{2.283}$$

From eqs (2.281) and (2.282) it can be seen that the rotary-diffusion constant, D_r, may be obtained from measurements of relaxation in light scattering or transmittance. The result of a typical relaxation experiment is given in Fig. 2.31. The data are replotted on a logarithmic scale in the inset of the figure, showing that after a few seconds the relaxation is described by a single relaxation time ($\tau = 1/6D_r$). These experiments were performed on triaxial latex particles with $a_1 = 470$, $a_2 = 330$ and $a_3 = 250$ nm. The experimental value for D_r ($2{\cdot}4 \times 10^{-2}\,\mathrm{s}^{-1}$) is close to the value calculated on the assumption that the particles are ellipsoids ($D_r = 2{\cdot}0 \times 10^{-2}\,\mathrm{s}^{-1}$).

In essence, rotary-relaxation experiments are very similar to translational-diffusion experiments where the particles diffuse towards an equilibrium distribution. For rotary diffusion, however, the time scales are usually much shorter and more readily observable.

Chapter 3

Non-interacting Collodial Particles Subjected to an External Field

3.1 Colloidal particles subjected to shear fields

3.1.1 A neutral sphere in simple shear

A simple but representative case of a particle subjected to a hydrodynamic flow field is the motion of a spherical particle in a simple shear flow. The behaviour of particles of a more complex geometry in other flow fields is often qualitatively similar to the motion of a sphere and hence this case merits special attention. The motion of a sphere in simple shear and the corresponding flow field around it has been solved by Cox *et al.* (1968). Consider a sphere in a simple shear flow as shown schematically in Fig. 3.1. The undisturbed flow field is given by

$$v_1 = v_2 = 0, \quad v_3 = Gx_2 \tag{3.1}$$

where G is the rate of shear.

It is convenient to express the distances in dimensionless units by dividing them by the radius a of the sphere. When $\mathrm{Re} \ll 1$, the flow field around the sphere is governed by the creeping-flow equation

$$\mu\nabla^2 v = \nabla p \tag{3.2}$$

and the continuity equation

$$\nabla \cdot v = 0 \tag{3.3}$$

The appropriate boundary conditions are: (i) as $r \to \infty$ the flow velocity approaches its undisturbed value given by eq. (3.1); (ii) at $r = 1$ (i.e. on the sphere surface) the no-slip boundary condition applies:

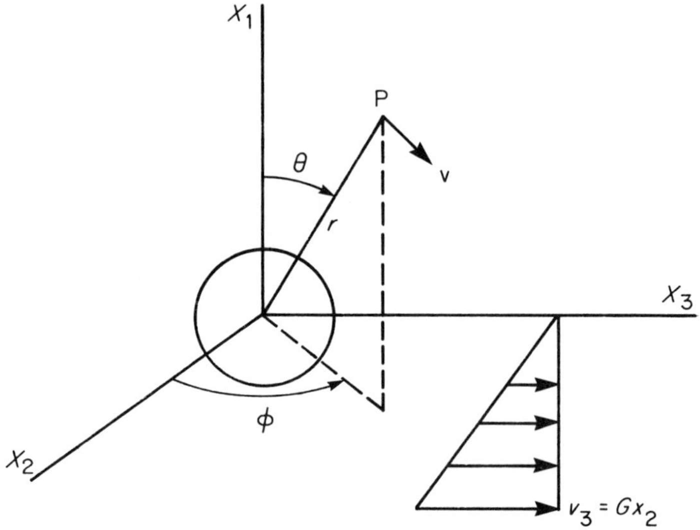

Fig. 3.1. Geometry of a sphere subjected to a simple shear flow given by $v_3 = Gx_2$. The presence of the sphere perturbs the simple shear locally in the neighbourhood of the sphere. The velocity, $\boldsymbol{v}$, of the fluid can be expressed in terms of the polar coordinates (r, θ, ϕ).

$$v_1 = 0,\ v_2 = -\omega x_3,\ v_3 = \omega x_2 \text{ at } r = 1 \tag{3.4}$$

Here ω is the angular velocity of the sphere about the vorticity-axis X_1, which is determined by the condition that the net torque on the sphere is zero. The solution of eqs (3.2) and (3.3) with the above boundary conditions is:

$$v_i = Gx_2\delta_{i3} + \frac{G}{2}\left[\frac{5(1-r^2)\,x_i x_2 x_3}{r^7} - \frac{\delta_{i2}x_3 + \delta_{i3}x_2}{r^5}\right] \tag{3.5}$$

where δ_{ij} is the Kronecker delta.

From the condition of zero torque it can be shown that

$$\omega = \frac{1}{2}G \tag{3.6}$$

By expressing the velocity in terms of polar coordinates

$$\boldsymbol{v} = \begin{pmatrix} v_r \\ v_\theta \\ v_\phi \end{pmatrix} = \begin{pmatrix} dr/dt \\ r d\theta/dt \\ r\sin\theta d\phi/dt \end{pmatrix} \tag{3.7}$$

we obtain

$$\frac{dr}{dt} = G\,A(r)\sin^2\theta\,\sin 2\phi \tag{3.8a}$$

$$\frac{d\theta}{dt} = \frac{1}{4}\,G\,B(r)\sin 2\theta\,\sin 2\phi \tag{3.8b}$$

$$\frac{d\phi}{dt} = \frac{1}{2}\,G[1 + B(r)\cos 2\phi] \tag{3.8c}$$

with

$$A(r) = \frac{r}{2} - \frac{5}{4r^2} + \frac{3}{4r^4} \tag{3.9a}$$

and

$$B(r) = 1 - \frac{1}{r^5} \tag{3.9b}$$

The set of eqs (3.8) describes the streamlines around a sphere subjected to shear. They can also be considered to describe the trajectories of small tracer particles around the sphere.

Dividing eq (3.8a) by eq (3.8b) and integrating, results in

$$\cos\theta = C\,f(r) \tag{3.10}$$

with

$$f(r) = \left(r^3 - \frac{5}{2} + \frac{3}{2r^2}\right)^{-1/3} \tag{3.11}$$

where C is a constant of integration. Similarly, dividing eq (3.8c) by eq (3.8b) and integrating using eq (3.10) leads to

$$\sin\theta\cos\phi = \pm f(r)\,\sqrt{E + g(r)} \tag{3.12}$$

with

$$g(r) = \int_r^\infty y^{-3} f(y)\,dy \tag{3.13}$$

where E is a second constant of integration. Alternatively, because $x_1 = r\cos\theta$ and $x_2 = r\sin\theta\cos\phi$, we can write:

$$x_1 = C\, r f(r) \tag{3.14}$$

$$x_2 = \pm r f(r) \sqrt{E + g(r)} \tag{3.15}$$

Equations (3.14) and (3.15) describe a set of surfaces of revolution about the X_1 and X_2 axes, respectively. The streamlines are parameterized by the constants C and E which are the trajectory or streamline constants. There are two different types of streamline depending on the value of E. For $E > 0$, streamlines are open and extend from $-\infty$ to $+\infty$ in the X_3 direction. When $-g(1) < E < 0$, the streamlines are closed. The open and closed streamlines are separated by a surface of limiting streamlines for which $E = 0$. A schematic representation of the various types of streamline is given in Fig. 3.2 for the case $C = 0$. When $C = 0$ all trajectories are in the X_2X_3 plane. A photograph of a sphere subjected to a simple shear surrounded by dye is shown in Fig. 3.3. Initially, the dye was placed all around the sphere, but outside the region of closed streamlines, the dye was convected away; it remained only in the region of closed streamlines.

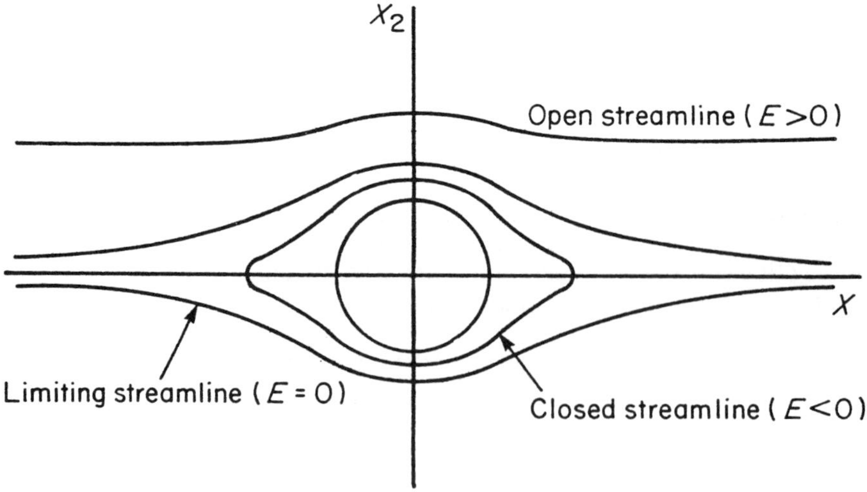

Fig. 3.2. Scheme to show the various types of streamline around a sphere subjected to simple shear. The limiting streamline ($E = 0$) separates open streamlines from closed ones. The sphere rotates with an angular velocity $\omega = \frac{1}{2} G$ about the vorticity axis X_1 (directed towards the viewer).

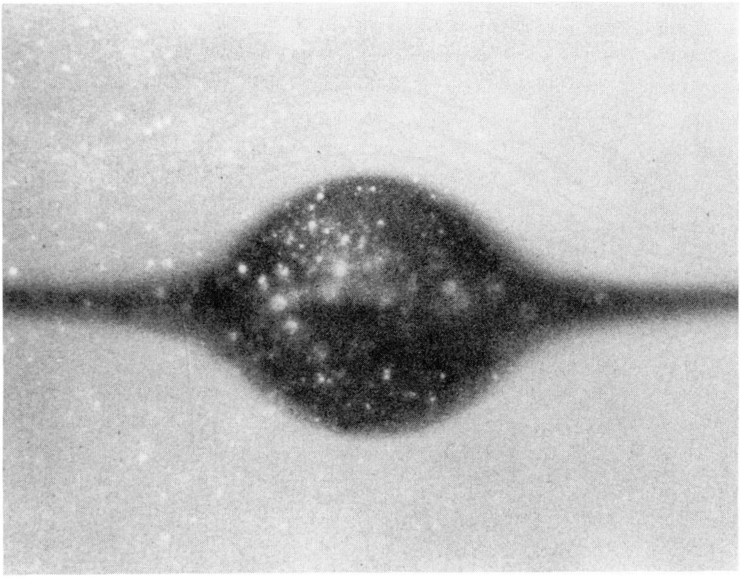

Fig. 3.3. Photograph of a sphere in simple shear, surrounded by dye. The dye remains only in the region of the closed streamlines, the contours of which can be clearly seen.

The surface of limiting streamlines is presented in Fig. 3.4, where the projection of limiting streamlines is shown for various values of the trajectory constant C. From the trajectory equation it can readily be shown that the total volume enclosed by the surfaces of limiting streamlines is infinite. It is of interest to note that a fluid element on an open streamline (or a small tracer particle with radius $b \ll a$) can never approach closer than the minimum distance between the streamline $C = E = 0$ and the sphere. This minimum distance can be found from eq (3.15) it occurs when

$$f(r_{\min})\sqrt{g}(r_{\min}) = 1 \qquad (3.16)$$

From the functions $f(r)$ and $g(r)$ defined above, it follows from eq (3.16) that

$$r_{\min} = 1 \cdot 157 \qquad (3.17)$$

Hence a small particle or a fluid element arriving from infinity can never approach to within distances of less than about 16% of the sphere radius. This fact has important ramifications for the mass and heat transfer to such a sphere.

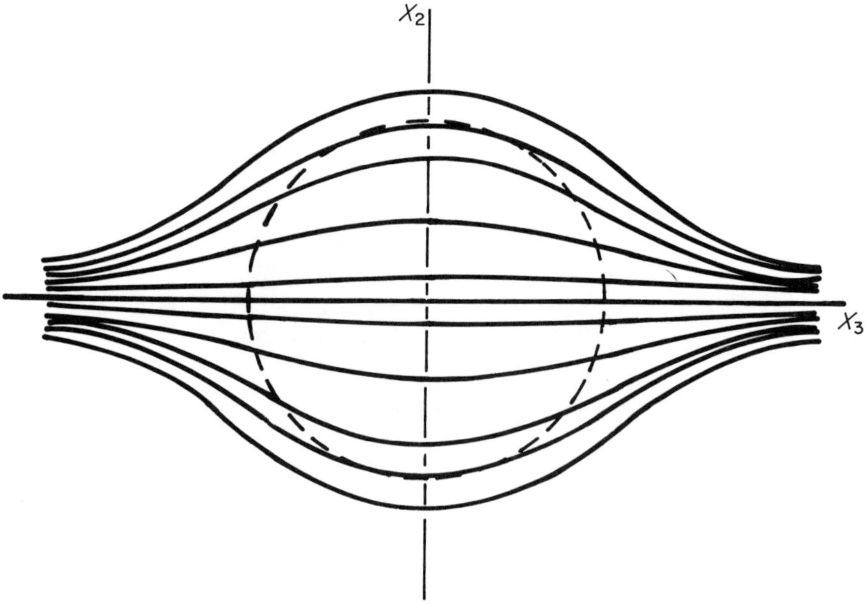

Fig. 3.4. Projections of limiting streamlines for various values of the streamline constant C (after Cox *et al.*, 1968).

3.1.2 A charged sphere in simple shear

An electrical double layer surrounding a particle subjected to a simple shear affects the flow around a charged sphere in a similar way to that of a uniform flow around it, as discussed in Chapter 2 (Section 2.2.2). The flow disturbs the equilibrium distribution of ions around the particles and the basic equations describing the fluid flow and the distribution of ions are the same as those presented in Chapter 2. The only difference is that some of the boundary conditions must be changed, especially conditions (v) to (viii) of eq (2.146). These four conditions must be replaced by six different ones, describing the three fluid velocity components at the sphere surface and at infinity. Specifically, it is required that the flow field approaches an undisturbed simple shear flow as $r \to \infty$:

$$
\begin{aligned}
v_r &= \tfrac{1}{2} G\, r \sin^2\theta \sin 2\phi \\
v_\theta &= \tfrac{1}{4} G\, r \sin 2\theta \sin 2\phi \qquad (3.18) \\
v_\phi &= \tfrac{1}{2} G\, r \sin\theta\, (1 + \cos 2\phi)
\end{aligned}
$$

and the no-slip boundary condition at the sphere surface is required: at $r=1$

$$\begin{aligned} v_r &= 0 \\ v_\theta &= 0 \\ v_\phi &= \omega \sin\theta \end{aligned} \tag{3.19}$$

Most of the conclusions reached in earlier sections about the solutions of the convective-diffusion equation, the Poisson equation, and the creeping-flow and the continuity equations for a sphere in a uniform flow, also apply to a sphere in a simple shear flow. The flow field and the potential and ion distributions are determined by five dimensionless parameters (for symmetrical electrolytes): Pe, κa, Φ_s, b_1 and b_2, with Pe now defined as

$$\text{Pe} = \frac{Ga^2}{D_1} \tag{3.20}$$

Again it is possible to make expansions in each of these parameters.

Let us consider what happens to the ion distribution when Pe varies from zero to infinity. As Pe $\rightarrow \infty$ (i.e. as flow dominates all Brownian motion), the double layer will take the form of the region of closed streamlines. When Pe $=0$ (diffusion-controlled conditions), the ion distribution is symmetric around the sphere. At small but non-zero Péclet numbers, an expansion in Pe can be made:

$$\begin{aligned} n^+ &= e^{-\Phi_0} + \text{Pe}\, n_1^+ \\ n^- &= e^{\Phi_0} + \text{Pe}\, n_1^- \\ \Phi &= \Phi_0 + \text{Pe}\, \Phi_1 \end{aligned} \tag{3.21}$$

and subsequently, an expansion in $\Phi_s (\Phi_s \ll 1)$:

$$\begin{aligned} n_1^+ &= \Phi_s\, m_1^+ \\ n_1^- &= \Phi_s\, m_1^- \\ \Phi_1 &= \Phi_s\, \chi_1 \end{aligned} \tag{3.22}$$

This results in the following equations for m_1^+, m_1^- and χ_1:

$$\nabla^2(m_1^+ + \chi_1) = f(r) \sin^2\theta \sin 2\phi \tag{3.23}$$

$$\nabla^2(m_1^- - \chi_1) = -q\, f(r) \sin^2\theta \sin 2\phi \tag{3.24}$$

$$\nabla^2\chi_1 = -\nu(m_1^+ - m_1^-) \tag{3.25}$$

where

$$f(r) = A(r)\ w'$$

where $A(r)$ is given by eq (3.9a) and $w' = \mathrm{d}w/\mathrm{d}r$ with $w = \Phi_0/\Phi_s$.

Now, writing

$$\begin{aligned} m_1^+ &= g_1 \sin^2\theta \sin 2\phi \\ m_1^- &= g_2 \sin^2\theta \sin 2\phi \\ \chi_1 &= s \sin^2\theta \sin 2\phi \end{aligned} \tag{3.26}$$

and integrating eq (3.23) results in (Booth, 1950)

$$s + g_1 = \frac{F'(\tau,\tau)}{3\tau} + \frac{F(\tau,\tau r)}{\tau^2} \tag{3.27}$$

where $F' = \mathrm{d}F/\mathrm{d}x$ with $x = \tau r (\tau = \kappa a)$ and

$$\begin{aligned} F(\tau,x) = \frac{\tau\,\mathrm{e}^{\tau-x}}{5(1+\tau)}\Bigg[5\left(1+\frac{3}{x}+\frac{6}{x^2}+\frac{6}{x^3}\right) - \frac{5\tau^3}{2}\left(-\frac{x}{8}+\frac{1}{8}-\frac{1}{4x}+\frac{3}{4x^2}\right. \\ \left.+\frac{2}{x^3}+\frac{x^2\,Ei(x)\mathrm{e}^x}{8}\right) + \frac{\tau^5}{4}\left(-\frac{x}{24}+\frac{1}{24}-\frac{1}{12x}+\frac{1}{4x^2}-\frac{1}{x^3}\right. \\ \left.+\frac{5}{x^4}+\frac{x^2\,Ei(x)\mathrm{e}^x}{24}\right)\Bigg] \end{aligned} \tag{3.28}$$

with

$$Ei(x) = \int_x^\infty \frac{\mathrm{e}^{-u}}{u}\,\mathrm{d}u$$

Similarly,

$$s - g_2 = q(s + g_1) \tag{3.29}$$

By adding eq (3.27) to eq (3.28) and substituting into eq (3.25), using eq (3.26), the potential, s, can be found from

$$\left(\frac{\mathrm{d}}{\mathrm{d}r^2} + \frac{2}{r}\frac{\mathrm{d}}{\mathrm{d}r} - \frac{6}{r^2} - \tau^2\right) s = -(1+q)\,h(\tau,r) \tag{3.30}$$

where

$$h(\tau,r) = F(\tau,\tau r) + \frac{\tau}{3r^3} F'(\tau,\tau) \tag{3.31}$$

Hence, to the first order in Pe and at small surface potentials

$$\begin{aligned} n^+ &= e^{-\Phi_0} + \text{Pe}\ \Phi_s\, g_1(r) \sin^2\theta \sin 2\phi \\ n^- &= e^{\Phi_0} + \text{Pe}\ \Phi_s\, g_2(r) \sin^2\theta \sin 2\phi \\ \Phi &= \Phi_0\ \text{Pe} + \Phi_s s(r) \sin^2\theta \sin 2\phi \end{aligned} \tag{3.32}$$

with the functions g_1, g_2 and s being determined by eqs (3.27) to (3.31), respectively. These functions depend on ν (or τ) and b_1 and b_2 (or q). From the form of the solution it can be seen that the maximum ion density occurs when $\phi = 45°$. As noted above, when $\text{Pe} = \infty$ the maximum ion density occurs when $\phi = 90°$. It is therefore reasonable to assume that the angle at which the maximum ion density occurs varies gradually from 45° to 90° as Pe varies from 0 to ∞. This is shown schematically in Fig. 3.5.

The flow field around the sphere can be found from (cf. eq (2.171)):

$$\nabla^2 v - \nabla p = b_1 \nu [2w\, \nabla \chi_1 + (m_1^+ - m_1^-)\, w'] \tag{3.33}$$

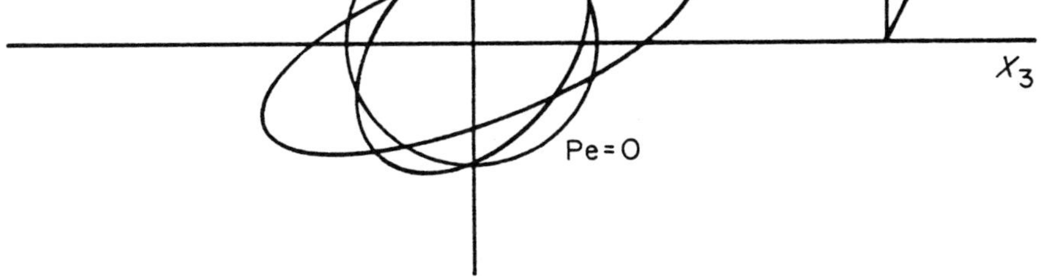

Fig. 3.5. Schematic ion distributions around a sphere in simple shear for various Péclet numbers, Pe: $\text{Pe}_2 > \text{Pe}_1 > 0$. The X_1 axis is directed into the page. The angle ϕ is measured from the X_2 axis.

together with the continuity equation $\nabla \cdot v = 0$. Equation (3.33), with the usual boundary conditions of no-slip at the sphere surface and that of an undisturbed shear field at infinity, was solved by Booth (1950), with the exception of the explicit solution for the angular dependence and the omission of a constant angular-velocity term. Taking these into account, the solution can be writen as

$$v = (\mathrm{d}r/\mathrm{d}t,\ r\mathrm{d}\theta/\mathrm{d}t,\ r\sin\theta\ \mathrm{d}\phi/\mathrm{d}t)$$

with (van de Ven, 1988b)

$$\frac{\mathrm{d}r}{\mathrm{d}t} = G\,\tilde{A}(r)\sin^2\theta\,\sin 2\phi \tag{3.34a}$$

$$\frac{\mathrm{d}\theta}{\mathrm{d}t} = \tfrac{1}{4}G\,\tilde{B}(r)\sin 2\theta\,\sin 2\phi \tag{3.34b}$$

$$\frac{\mathrm{d}\phi}{\mathrm{d}t} = \tfrac{1}{2}G\,[1 + \tilde{B}(r)\cos 2\phi] \tag{3.34c}$$

in which

$$\tilde{A}(r) = A(r) + \bar{b}\,\Phi_s^2\,a(\tau, r) \tag{3.35a}$$

$$\tilde{B}(r) = B(r) + \bar{b}\,\Phi_s^2\,b(\tau, r) \tag{3.35b}$$

where $\bar{b} = (b_1 + b_2)/2$ and

$$a(\tau, r) = -\frac{Q(\kappa a)}{2r^2}\left(1 - \frac{1}{r}\right) \tag{3.35c}$$

$$b(\tau, r) = \frac{2}{r}\int_1^r r\,U(\tau, r)\,\mathrm{d}r \tag{3.35d}$$

$$U(\tau, r) = \frac{\mathrm{e}^{\tau}(1+\tau)^2}{5\tau^2}\int_r^\infty \frac{\tau u + 1}{u^2}\,\mathrm{e}^{-\tau u}\,h(\tau, \tau u)\left(\frac{r^3}{u^3} - \frac{u^2}{r^2}\right)\mathrm{d}u \tag{3.35e}$$

and

$$Q(\tau) = -U(\tau, 1) \tag{3.35f}$$

It can be shown from the functions a and b that as $r \to 1$ or $r \to \infty$, a and $b \to 0$. Thus at both the sphere surface and at infinity, the flow field

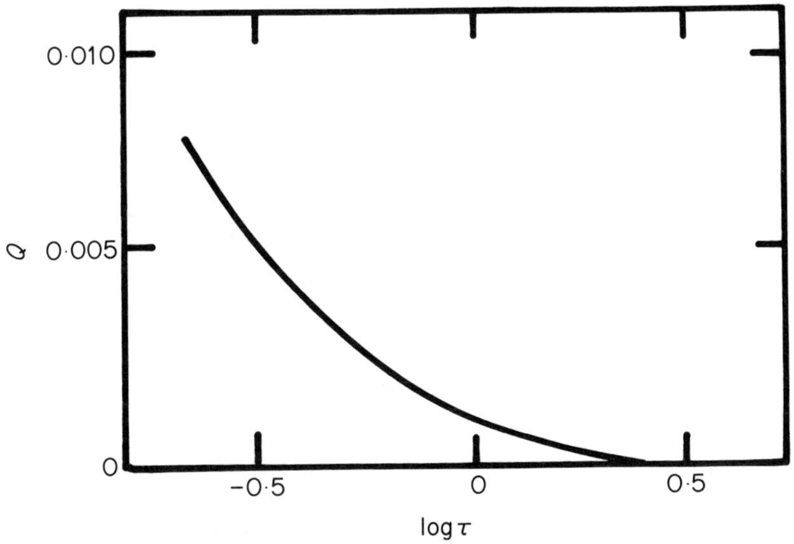

Fig. 3.6. The function Q, appearing in eq (3.35c) as a function of $\tau = \kappa a$.

is the same as that for a neutral sphere. The function $Q(\kappa a)$ is shown in Fig. 3.6.

It can be seen from eqs (3.34) that the flow field around a charged sphere is very similar to that of a neutral sphere (cf. eq (3.8)). The streamline equations can be integrated analogously:

$$x_1 = C\, r\, \tilde{f}(r) \tag{3.36}$$

$$x_2 = \pm r\, \tilde{f}(r) \sqrt{\tilde{g}(r)} + E \tag{3.37}$$

where C and E are streamline constants and

$$\tilde{f}(r) = f(r)\, [1 + \bar{b}\, \Phi_s^2\, h_1(\tau)] \tag{3.38}$$

where

$$h_1(\tau) = -\int_r^{\infty} \left[\frac{a(\tau, r)\, B(r)}{4A(r)^2} + \frac{b(\tau, r)}{2A(r)}\right] \mathrm{d}r$$

and

$$\tilde{g}(r) = g(r) + b\, \Phi_s^2\, h_2(\tau) \tag{3.39}$$

with

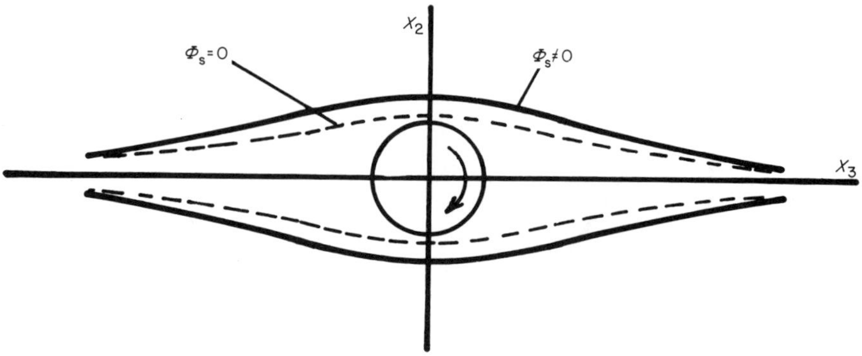

Fig. 3.7. Schematic comparison of limiting streamlines around a neutral ($\Phi_s=0$) and a charged sphere.

$$h_2(\tau)=\int_r^\infty \frac{1-B}{2Af^2}\left(2h_1+\frac{b}{1-B}-\frac{a}{2A}\right)\mathrm{d}r$$

It can be shown that both h_1 and $h_2 > 0$. It follows from eqs (3.36) to (3.39) that the surface of limiting streamlines is farther removed from a charged sphere than from a neutral sphere (shown schematically in Fig. 3.7) and thus a fluid element passes a charged sphere at a larger distance than a neutral one. The minimum distance to which a fluid element (or a small tracer particle) can approach the sphere is increased to:

$$r_{\min}=1\cdot 157+\bar{b}\,\Phi_s^2\,h_3(\tau) \tag{3.40}$$

with

$$h_3(\tau)=\left(\frac{2h_1}{f(r_0)}+\frac{h_2}{2g(r_0)}\right)\Big/\left(\frac{g'(r_0)}{2g(r_0)}-\frac{f'(r_0)}{f(r_0)}\right)$$

where $r_o=1\cdot 157$. It can be shown that $h_3(\tau)>0$.

Because $\tilde{B}(r)\rightarrow 0$ as $r\rightarrow 1$, it follows directly that the angular velocity of the sphere is unaffected by the presence of the double layer (when $\phi_s \ll 1$). Hence, also in the presence of a double layer

$$\omega=\tfrac{1}{2}G \tag{3.41}$$

and the period of rotation, T, about the vorticity axis equals

$$T=\frac{4\pi}{G} \tag{3.42}$$

Because of symmetry, terms with odd powers in Φ_s must vanish and it can be anticipated that the first-order correction to T due to double layers will be of the order Φ_s^4:

$$T = T_0[1 + 0(\Phi_s^4)] \tag{3.43}$$

with T_0 given by eq (3.42).

3.1.3 Diffusion of spheres in simple shear

In the previous chapter we saw that the Brownian motion of a colloidal particle can be regarded as a random walk. When a Brownian particle is subjected to an external field such as, for example, a simple shear flow, a deterministic motion is superimposed on this random walk. This is represented schematically in Fig. 3.8.

The diffusion of non-interacting spherical particles in a simple shear flow is described by the convective-diffusion equation

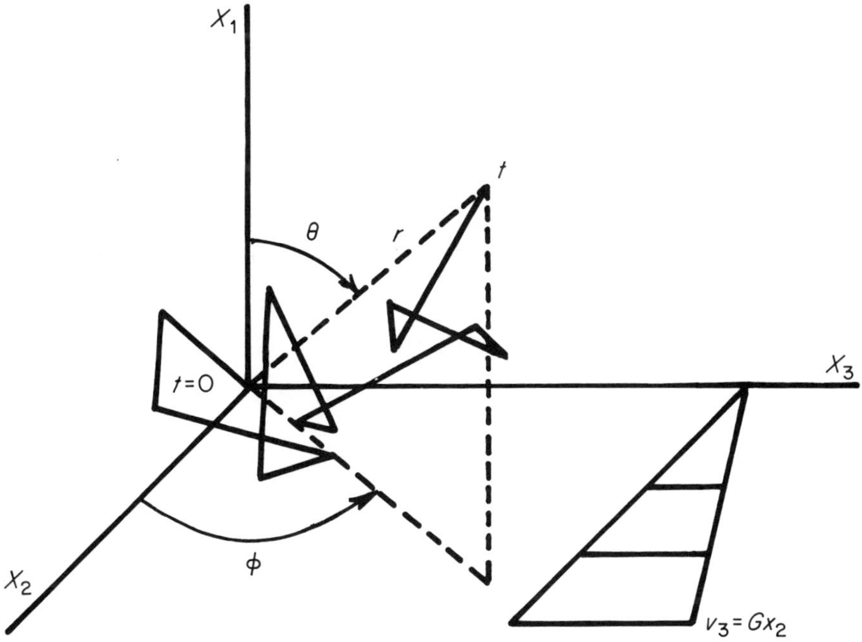

Fig. 3.8. Random walk superimposed on a simple shear flow of gradient G. The probability of finding a particle at position (r, θ, ϕ) at time t, when it was at the origin at time $t = 0$, is described by the convective-diffusion equation.

$$\frac{\partial n(x_1,x_2,x_3,t)}{\partial t} = \nabla\cdot(\boldsymbol{D}\cdot\nabla n) - \nabla\cdot(\boldsymbol{u}n) \tag{3.44}$$

Here x_i are the Cartesian coordinates (cf. Fig. 3.8); n is the number concentration or, alternatively, the probability of finding a particle at position (x_1,x_2,x_3) at time t, $\boldsymbol{D}$ is the diffusion tensor, and $\boldsymbol{u}$ is the deterministic translational velocity of the sphere. The diffusion tensor of a sphere does not always reduce to a simple constant. It has been shown that the ion distribution around electrically charged spheres in simple shear becomes asymmetric. Because the double layer affects the friction coefficient of the particle, its diffusion will no longer be isotropic and hence

$$\boldsymbol{D} = \begin{pmatrix} D_1 & 0 & 0 \\ 0 & D_2 & 0 \\ 0 & 0 & D_3 \end{pmatrix} \tag{3.45}$$

From the ion distribution around particles with low surface potentials, it may be anticipated that

$$D_i = D_0[1 - \bar{b}\,\Phi_s^2 f_i(\kappa a)] \tag{3.46}$$

where the functions $f_i(\kappa a)$ are, as yet, unknown functions of κa. When $\mathrm{Pe} \ll 1$, the principal axes of $\boldsymbol{D}$ coincide with the axes of an ellipsoidal particle aligned along $\theta = 90°$, $\phi = 45°$. For other Péclet numbers, one of the principal axes is at an angle ϕ_m in the range $45° < \phi_m < 90°$. It is often convenient to express $\boldsymbol{D}$ in space-fixed coordinates. Writing $\boldsymbol{D} = D'_{mn}$ in body-fixed coordinates, and denoting the components in space-fixed coordinates by D_{ij} gives

$$D_{ij} = \alpha_{im}\,\alpha_{jn}\,D'_{mn} \tag{3.47}$$

where α_{ij} is the transformation matrix

$$\alpha_{ij} = \begin{pmatrix} 0 & -1 & 0 \\ \cos\phi_m & 0 & -\sin\phi_m \\ \sin\phi_m & 0 & \cos\phi_m \end{pmatrix} \tag{3.48}$$

From eqs (3.47) and (3.48) it follows that

$$D_{ij} = \begin{pmatrix} D_2 & 0 & 0 \\ 0 & D_1\cos^2\phi_m + D_3\sin^2\phi_m & (D_1 - D_3)\sin\phi_m\cos\phi_m \\ 0 & (D_1 - D_3)\sin\phi_m\cos\phi_m & D_1\sin^2\phi_m + D_2\cos^2\phi_m \end{pmatrix} \tag{3.49}$$

Note that when $D_1 = D_2 = D_3 = D$, the off-diagonal terms vanish and all the diagonal terms become equal to D.

The velocity $\boldsymbol{u}$ of an isolated sphere far from bounding walls is the same as that of the simple shear flow:

$$u_1 = u_2 = 0, \; u_3 = Gx_2$$

and so, for isolated spheres, eq (3.44) can be rewritten

$$\frac{\partial n}{\partial t} = \nabla \cdot (\boldsymbol{D} \cdot \nabla n) - Gx_2 \frac{\partial n}{\partial x_3} \tag{3.50}$$

We will first look at the solution of eq (3.50) for neutral spheres, which simplifies to

$$\frac{\partial n}{\partial t} = D\,\nabla^2 n - G\,x_2 \frac{\partial n}{\partial x_3} \tag{3.51}$$

The solution of this equation for a point source (corresponding to a large number of non-interacting particles at the origin) has been given by Elrick (1962):

$$n^*(x_1^*, x_2^*, x_3^*, t^*) = \frac{3^{1/2} M}{4(\pi t^*)^{3/2}(t^{*2} + 12)^{1/2}} \times \exp\left[-\left\{ \frac{3x_3^* - \frac{1}{2}t^* x_2^*}{t^*(t^{*2} + 12)} + \frac{x_2^{*2}}{4t} + \frac{x_1^{*2}}{4t} \right\} \right] \tag{3.52}$$

where

$$n^* = (G/D)^{-3/2} n, \; x_i^* = (G/D)^{1/2} x_i \text{ and } t^* = Gt$$

M is the total amount of diffused matter:

$$M = \int_{-\infty}^{\infty} \int_{-\infty}^{\infty} \int_{-\infty}^{\infty} n^* \, \mathrm{d}x_1^* \, \mathrm{d}x_2^* \, \mathrm{d}x_3^*$$

Equation (3.52) gives the probability of finding a particle at position (x_1, x_2, x_3) at time t, which was at the origin at time $t = 0$. From eq (3.52), the components of the mean-square displacement tensor are found to be

$$\langle x_i^* x_j^* \rangle = \frac{1}{M} \int_{-\infty}^{\infty} \int_{-\infty}^{\infty} \int_{-\infty}^{\infty} x_i^* x_j^* n^* \, dx_1^* \, dx_2^* \, dx_3^* \tag{3.53}$$

Carrying out the integrations and converting back to dimensional parameters gives (van de Ven, 1977):

$$\langle x_i x_j \rangle = \begin{pmatrix} 2Dt & 0 & 0 \\ 0 & 2Dt & DGt^2 \\ 0 & DGt^2 & 2Dt + \frac{2}{3} DG^2 t^3 \end{pmatrix} \tag{3.54}$$

The scalar mean-square displacement is the trace of the matrix:

$$\langle r^2 \rangle = 6Dt + \frac{2}{3} DG^2 t^3 \tag{3.55}$$

It can be seen that at short periods of time, Einstein's result $\langle r^2 \rangle = 6Dt$ is recovered, while at long time periods $\langle r^2 \rangle \propto t^3$. At first sight the term t^3 might seem surprising because when $\mathrm{Pe} = 0$, $\langle x_3^2 \rangle \propto t$ and when $\mathrm{Pe} = \infty$, $\langle x_3^2 \rangle = 0$ (or $\langle x_3^2 \rangle \propto t^2$ for a space-fixed observer). However, there is a coupling between diffusion in the X_2 direction and convection in the X_3 direction. For diffusion in the X_2 direction, $\langle x_2^2 \rangle \propto t$, for convection in the X_3 direction, $dx_3/dt \propto x_2 \propto t^{1/2}$, from which it follows that $\langle x_3^2 \rangle \propto t^3$.

It must be remembered that the results expressed by eqs (3.54) and (3.55) are valid for particles initially at the origin, or more generally for observers moving with the velocity of the particles at time $t = 0$.

In interpreting the solution (eq (3.52)) of the convective-diffusion equation, it is useful to transform it into spherical polar coordinates (see Fig. 3.8):

$$n^*(r^*, \theta, \phi, t^*) = \frac{3^{1/2} M}{4(\pi t^*)^{3/2} (t^{*2} + 12)^{1/2}} \times \exp\left[-r^{*2} \left\{ f(\phi, t^*) \sin^2\theta + \frac{\cos^2\theta}{4t^*} \right\} \right] \tag{3.56}$$

where

$$f(\phi, t^*) = \frac{3(\cos\phi - \frac{1}{2} t^* \sin\phi)^2}{t^*(t^{*2} + 12)} + \frac{\sin^2\phi}{4t^*} \tag{3.57}$$

From eqs (3.56) and (3.57) it can be seen that the distribution of particles is Gaussian in every direction, with a spread of

$$2\text{Pe}[f(\phi, t^*) \sin^2\theta + \cos^2\theta/(4t^*)]^{1/2}$$

Here Pe is the translational Péclet number, defined as

$$\text{Pe} = \frac{G\,a^2}{D} \tag{3.58}$$

The probability $p(\phi)\mathrm{d}\phi$ of finding a particle between ϕ and $\phi + \mathrm{d}\phi$ is given by

$$p(\phi) = \frac{1}{M}\int_0^\infty \int_0^\pi r^2 n^* \sin\theta \, \mathrm{d}r^* \mathrm{d}\theta \tag{3.59}$$

or

$$p(\phi) = \frac{3^{1/2}}{4\pi t^*(t^{*2} + 12)^{1/2} f(\phi, t^*)} \tag{3.60}$$

It can be readily checked that

$$\int_0^{2\pi} p(\phi)\, \mathrm{d}\phi = 1$$

From eq (3.60) it follows that $p(\phi)$ has a maximum value at

$$\phi_{\max} = \tfrac{1}{2} \tan^{-1}\left(\frac{3}{t^*}\right) \tag{3.61}$$

Hence, $\phi_{\max}$ changes from 45° when $t = 0$ to 90° when $t = \infty$. This is illustrated in Fig. 3.9, where $p(\phi)$ is plotted for various values of t^*.

We now return to the problem of the diffusion of charged spheres. In this case it is necessary to solve eq (3.50). This can be done in the way described by Foister and van de Ven (1980). Here we give only the results for the mean-square displacements:

$$\langle x_i x_j \rangle = \begin{pmatrix} \langle x_1^2 \rangle & 0 & 0 \\ 0 & \langle x_2^2 \rangle & \langle x_2 x_3 \rangle \\ 0 & \langle x_2 x_3 \rangle & \langle x_3^2 \rangle \end{pmatrix} \tag{3.62}$$

where

$$\langle x_1^2 \rangle = 2D_2 t \tag{3.63a}$$

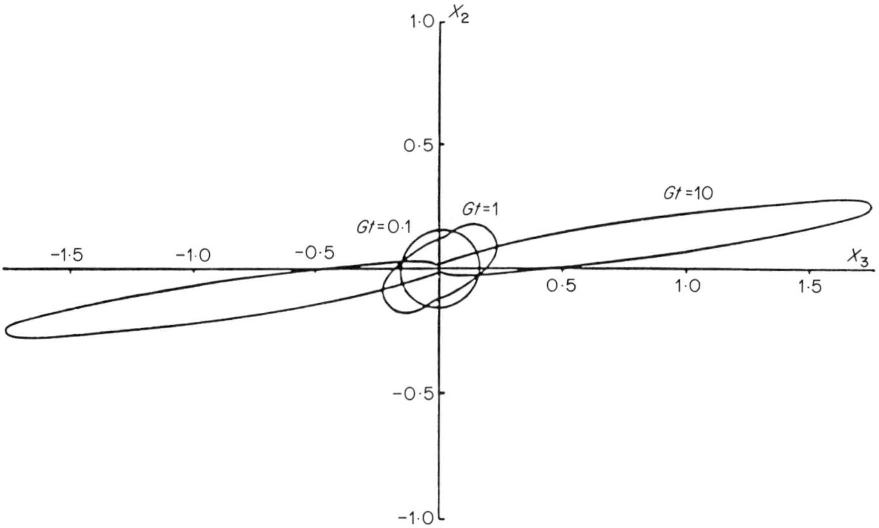

Fig. 3.9. Orientation distribution function $p(\phi)$ of particles undergoing a random walk in a simple shear flow for various values of Gt; $p(\phi)d\phi$ is the probability of finding a particle between ϕ and $\phi+d\phi$ when it was at the origin at time $t=0$ (after van de Ven, 1977).

$$\langle x_2^2\rangle = 2(D_1\cos^2\phi_m + D_3\sin^2\phi_m)\, t \tag{3.63b}$$

$$\langle x_2 x_3\rangle = \sin 2\phi_m(D_1 - D_3)t + (D_1\cos^2\phi_m + D_3\sin^2\phi_m)\, Gt^2 \tag{3.63c}$$

and

$$\langle x_3^2\rangle = 2(D_1\sin^2\Phi_m + D_3\cos^2\phi_m)t + \tfrac{1}{2}\sin 2\phi_m(D_1 - D_3)Gt^2 + \tfrac{2}{3}(D_1\cos^2\phi_m + D_3\sin^2\Phi_m)\, G^2t^3 \tag{3.63d}$$

It can be readily seen that when $D_1 = D_2 = D_3$, eqs (3.63) reduce to eq (3.54). At short times, $\langle r^2\rangle$ reduces to

$$\langle r^2\rangle = 6D_{\rm eff}t \tag{3.64}$$

with $D_{\rm eff} = \dfrac{1}{3}(D_1 + D_2 + D_3)$, and at large times to

$$\langle r^2\rangle = \frac{2}{3} D_{\rm eff}\, G^2 t^3 \tag{3.65}$$

with $D_{\rm eff} = D_1\cos^2\phi_m + D_3\sin^2\phi_m$.

3.1.4 An infinite cylinder in simple shear

The motion of a cylinder in shear flow is of interest for several reasons. Firstly, the streamlines can be readily calculated for certain orientations of the cylinder, and secondly, its motion also resembles the motion of other axi-symmetric bodies such as spheroids (ellipses of revolution). Here we will consider the simple case of an infinitely long cylinder aligned with the vorticity-axis X_1, subjected to a simple shear (see Fig. 3.10). The flow field around the cylinder must satisfy the creeping-flow and the continuity equations

$$\mu\nabla^2 v = \nabla p \tag{3.66}$$

$$\nabla \cdot v = 0 \tag{3.67}$$

with the no-slip boundary condition at the cylinder surface:

$$v_2 = -\omega x_3;\ v_3 = \omega x_2 \text{ at } r = 1 \tag{3.68}$$

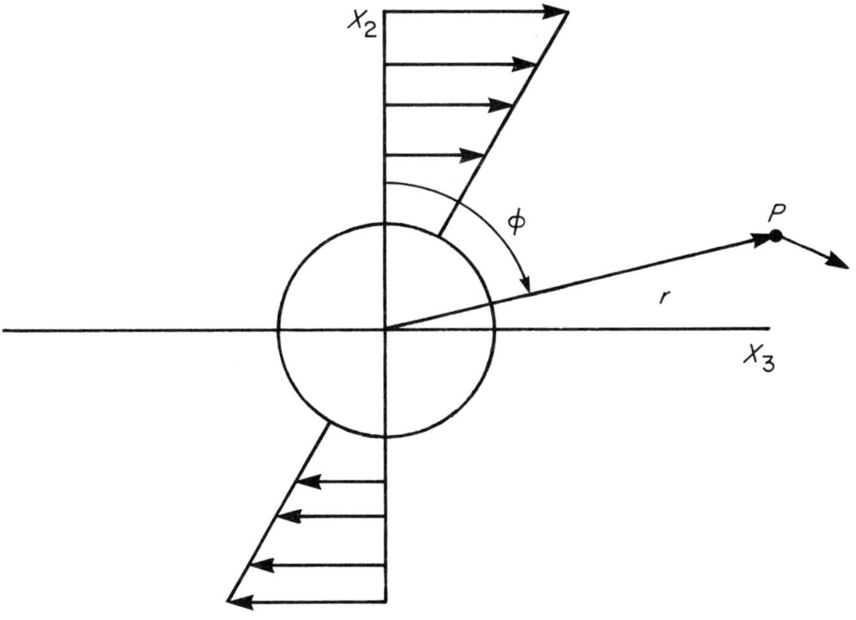

Fig. 3.10. A schematic representation of a cylinder aligned along the vorticity axis X_1 (directed towards the viewer) subjected to a simple shear flow of gradient G. The undisturbed shear flow $v_3 = Gx_2$ is shown. The cylinder causes a disturbance flow **v** which can be expressed in terms of the polar coordinates (r, ϕ).

where ω is the angular velocity of the cylinder about the X_1 axis, and where the flow must approach the undisturbed shear flow far from the surface of the cylinder:

$$v_2 = 0;\ v_3 = Gx_2 \text{ at } r = \infty \tag{3.69}$$

The solution of eqs (3.66) and (3.67) which satisfies the conditions given by eqs (3.68) and (3.69) is (Cox *et al.*, 1968):

$$v_r = \mathrm{d}r/\mathrm{d}t;\ v_\phi = r\mathrm{d}\phi/\mathrm{d}t$$

with

$$\frac{\mathrm{d}r}{\mathrm{d}t} = G\,A(r)\,\sin 2\phi \tag{3.70a}$$

$$\frac{\mathrm{d}\phi}{\mathrm{d}t} = \tfrac{1}{2}G\,[1 + B(r)\cos 2\phi] \tag{3.70b}$$

with

$$A(r) = \frac{r}{2} - \frac{1}{r} + \frac{1}{2r^3} \tag{3.71a}$$

$$B(r) = 1 - \frac{1}{r^4} \tag{3.71b}$$

From the zero-torque condition it follows that, as in the case of spheres, the angular velocity is

$$\omega = \tfrac{1}{2}G \tag{3.72}$$

Notice the similarity between eqs (3.70) and (3.8). Equations (3.71) can be integrated, analogously to eqs (3.8), with the result

$$x_2 = \pm r\,f(r)\sqrt{E + g(r)} \tag{3.73}$$

with

$$f(r) = 2^{1/2}[(r - r^{-1})]^{-1} \tag{3.74a}$$

and

$$g(r) = r^{-2} \tag{3.74b}$$

It appears that the streamlines around an infinitely long cylinder aligned along the vorticity axis are qualitatively similar to those around a sphere.

They consist of open streamlines characterized by values of the streamline constant, E, larger than one, and of closed streamlines characterized by $-1 < E < 0$. $E = 0$ corresponds to the limiting streamline.

The distance of closest approach between the limiting streamline and the cylinder can be found from eq (3.73); it is the solution of

$$f(r_{\min})\sqrt{g(r_{\min})} = 1 \tag{3.75}$$

Substituting eq (3.74) into eq (3.75) and solving gives

$$r_{\min} = 1 \cdot 30 \tag{3.76}$$

Hence a streamline coming from infinity can never approach a cylinder to within distances of less than 30% of the cylinder radius (cf. 16% for spheres).

It is also possible to examine the effects of an electrical double layer surrounding the cylinder and results very similar to those obtained for a sphere may be expected.

3.1.5 Spheroids in simple shear

Colloidal dispersions encountered in practical applications often contain particles which are not spherical. Non-spherical (or anisometric) particles behave differently from spheres, and it is important to understand the effects that the particle shape has on a variety of dispersion properties. One class of particles for which exact results can be found are spheroidal particles, i.e. particles with the shape of an ellipse of revolution. Furthermore, it can be shown that a large class of particles, such as those having an axisymmetric shape (rods, disks, cylinders, etc.), behave in ways qualitatively similar to those of spheroids.

Consider a spheroidal particle whose major axis is at an arbitrary orientation (θ, ϕ) in a simple shear flow which is given by (see Fig. 3.11)

$$v_1 = v_2 = 0; \; v_3 = Gx_2$$

The flow field around a spheroid in an instantaneously fixed position in space can be calculated using the creeping-flow equation

$$\mu \nabla^2 v' = \nabla p' \tag{3.77}$$

where the prime indicates that the velocity and the pressure are expressed in particle-fixed coordinates. These particle-fixed coordinates, x_i', are related to the space-fixed coordinates, x_i, by

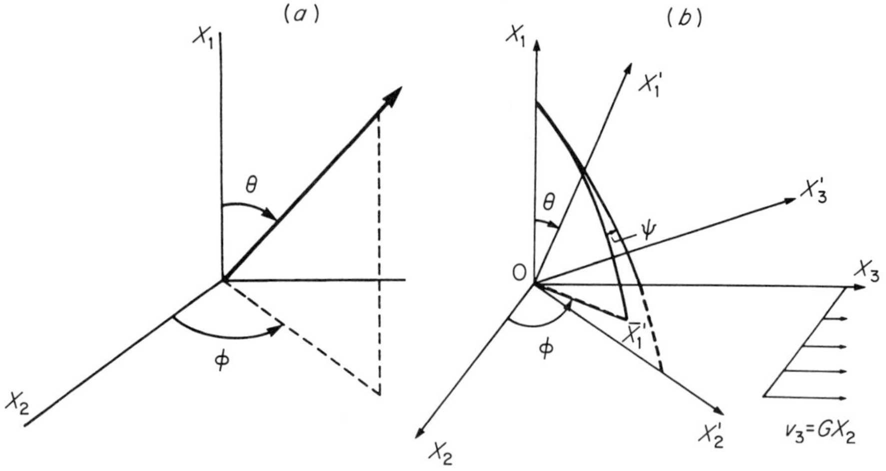

Fig. 3.11. The coordinate system used to describe the rotation of a spheroid in simple shear. (*a*) The orientation of the axis of revolution X_1' is described by the polar angles (θ, ϕ). (*b*) To describe the spin of the spheroid about the X_1' axis, the third Euler angle ψ is required.

$$x_i = \alpha_{ij} x_j' \tag{3.78}$$

with the transformation matrix α_{ij} given by

$$\alpha_{ij} = \begin{pmatrix} \cos\theta & -\sin\theta\cos\psi & \sin\theta\sin\psi \\ \sin\theta\cos\phi & -\sin\phi\sin\psi + \cos\theta\cos\phi\cos\psi & -\sin\phi\cos\psi - \cos\theta\cos\phi\sin\psi \\ \sin\theta\sin\phi & \cos\phi\sin\psi + \cos\theta\sin\phi\cos\psi & \cos\phi\cos\psi - \cos\theta\sin\phi\sin\psi \end{pmatrix} \tag{3.79}$$

Here ψ is the third Euler angle (besides θ and ϕ) (see Fig. 3.11). This angle is irrelevant in determining the orientation of the spheroid (which depends on θ and ϕ only), but is important in describing the spin of the particle about the X_1' axis.

Besides eq (3.77), the velocity must also satisfy the continuity equation

$$\nabla \cdot v' = 0 \tag{3.80}$$

Equations (3.77) and (3.80) are subject to the usual boundary conditions: (i) at the surface the no-slip boundary condition applies, i.e. at the surface, which is defined by

$$\frac{x_1'^2}{a^2} + \frac{x_2'^2 + x_3'^2}{b^2} = 1 \tag{3.81}$$

where a and b are the semi-axes of the spheroid, we must have

$$
\begin{aligned}
v_1' &= \omega_2 x_3' - \omega_3 x_2' \\
v_2' &= \omega_3 x_1' - \omega_1 x_3' \\
v_3' &= \omega_1 x_2' - \omega_2 x_1'
\end{aligned}
\tag{3.82}
$$

where the ω_i are the angular velocities of the spheroid about its axes X_i'. (ii) Far from the surface, the fluid velocity must match the undisturbed fluid velocity, which in particle-fixed coordinates is given by

$$v_i' = G\, B_{ij} x_j' \tag{3.83a}$$

with

$$B_{ij} = \begin{pmatrix} \alpha_{21}\alpha_{31} & \alpha_{22}\alpha_{31} & \alpha_{23}\alpha_{31} \\ \alpha_{21}\alpha_{32} & \alpha_{22}\alpha_{32} & \alpha_{23}\alpha_{32} \\ \alpha_{21}\alpha_{33} & \alpha_{22}\alpha_{33} & \alpha_{23}\alpha_{33} \end{pmatrix} \tag{3.83b}$$

The solution of eqs (3.77) and (3.80) with the boundary conditions given by eqs (3.82) and (3.83) has been given by Jeffery (1922), and is

$$
\begin{aligned}
\frac{v_1'}{G} = {} & x_1'[\alpha_{21}\alpha_{31} + \beta'(W - V) - 2(\alpha + 2\beta)A] \\
& + x_2'[\alpha_{22}\alpha_{31} + \beta' T^* - 2(\beta - \alpha)H] \\
& + x_3'[\alpha_{23}\alpha_{31} + \beta' S - 2(\beta - \alpha)G^*] \\
& - \frac{2x_1' P^2}{(a^2+\lambda)\Delta}\left[\tilde{R} + \frac{x_2'^2\, \tilde{W}}{(b^2+\lambda)^2} - \frac{x_3'^2\, \tilde{V}}{(b^2+\lambda)^2}\right]
\end{aligned}
\tag{3.84a}
$$

$$
\begin{aligned}
\frac{v_2'}{G} = {} & x_1'[\alpha_{21}\alpha_{32} + \beta' T^* + 2(\beta - \alpha)H] \\
& + x_2'[\alpha_{22}\alpha_{32} + \alpha' U - \beta' W - 2(\alpha + 2\beta)B^*] \\
& + x_3'[\alpha_{23}\alpha_{32} + \alpha' R] \\
& - \frac{2x_2' P^2}{(b^2+\lambda)\Delta}\left[\tilde{R} - \frac{x_1'^2\, \tilde{W}}{(a^2+\lambda)^2} + \frac{x_3'^2\, \tilde{U}}{(b^2+\lambda)^2}\right]
\end{aligned}
\tag{3.84b}
$$

$$
\begin{aligned}
\frac{v_3'}{G} = {} & x_1'[\alpha_{21}\alpha_{33} + \beta' S - 2(\alpha - \beta)G^*] \\
& + x_2'[\alpha_{22}\alpha_{33} + \alpha' R] \\
& + [x_3'\, \alpha_{23}\alpha_{33} + \beta' V - \alpha' U - 2(\alpha + 2\beta)C^*] \\
& - \frac{2x_3'\, P^2}{(b^2+\lambda)\Delta}\left[\tilde{R} + \frac{x_1'^2\, \tilde{V}}{(a^2+\lambda)^2} - \frac{x_2'^2\, \tilde{U}}{(b^2+\lambda)^2}\right]
\end{aligned}
\tag{3.84c}
$$

where

$$A=\frac{\alpha_{21}\alpha_{31}}{6\beta_0''}$$

$$B^*=\frac{\beta_0''(2\alpha_{22}\alpha_{32}-\alpha_{23}\alpha_{33})-\alpha_0''\alpha_{21}\alpha_{31}}{6\beta_0''(2\alpha_0''+\beta_0'')}$$

$$C^*=-(A+B^*)$$

$$F=\frac{\alpha_{22}\alpha_{33}+\alpha_{23}\alpha_{32}}{8\alpha_0'b^2}$$

$$G^*=\frac{\alpha_{23}\alpha_{31}+\alpha_{21}\alpha_{33}}{4\beta_0'(a^2+b^2)}$$

$$H=\frac{\alpha_{21}\alpha_{32}+\alpha_{22}\alpha_{31}}{4\beta_0'(a^2+b^2)}$$

$$R=-4Fb^2$$

$$S=-2G^*(a^2+b^2)$$

$$T^*=-2H(a^2+b^2)$$

$$U=2(B^*-C^*)b^2$$

$$V=2(b^2C^*-a^2A)$$

$$W=-(U+V)$$

$$\begin{aligned}\tilde{R}=&\frac{x_2'x_3'}{(b^2+\lambda)^2}[R+4(b^2+\lambda)F]\\&+\frac{x_1'x_3'}{(a^2+\lambda)(b^2+\lambda)}[S+2(a^2+b^2+2\lambda)G^*]\\&+\frac{x_1'x_2'}{(a^2+\lambda)(b^2+\lambda)}[T^*+2(a^2+b^2+2\lambda)H]\end{aligned}$$

$$\tilde{U}=U-2(b^2+\lambda)(B^*-C^*)$$

$$\tilde{V}=V-2(b^2+\lambda)C^*+2(a^2+\lambda)A$$

$$\tilde{W}=W-2(a^2+\lambda)A+2(b^2+\lambda)B^*$$

$$\frac{1}{P^2}=\frac{x_1'^2}{(a^2+\lambda)^2}+\frac{x_2'^2+x_3'^2}{(b^2+\lambda)^2}$$

$$\Delta=(b^2+\lambda)(a^2+\lambda)^{1/2}$$

where λ is the positive root of $\dfrac{x_1'^2}{a^2+\lambda}+\dfrac{x_2'^2+x_3'^2}{b^2+\lambda}=1$

$$\alpha=\int_\lambda^\infty \frac{d\lambda}{(a^2+\lambda)\Delta}=p-\frac{2}{s^2(a^2+\lambda)^{1/2}} \quad \text{with } s=(|a^2-b^2|)^{1/2}$$

and

$$p=\frac{2}{s^3}\left[-\tan^{-1}\frac{(a^2+\lambda)^{1/2}}{s}+\frac{\pi}{2}\right] \qquad \text{if } a/b<1$$

$$p=\frac{1}{s^3}\ln\frac{(a^2+\lambda)^{1/2}+s}{(a^2+\lambda)^{1/2}-s} \qquad \text{if } a/b>1$$

$$\beta=\int_\lambda^\infty \frac{d\lambda}{(b^2+\lambda)\Delta}=\frac{1}{(b^2+\lambda)(a^2+\lambda)^{1/2}}-\frac{\alpha}{2}$$

$$\alpha'=\int_\lambda^\infty \frac{d\lambda}{(b^2+\lambda)^2\Delta}=\frac{(a^2+\lambda)^{1/2}}{2s^2\,(b^2+\lambda)^2}-\frac{3\beta}{4s^2}$$

$$\beta'=\int_\lambda^\infty \frac{d\lambda}{(a^2+\lambda)(b^2+\lambda)\Delta}=\frac{\beta-\alpha}{s^2}$$

$$\alpha''=\int_\lambda^\infty \frac{\lambda d\lambda}{(b^2+\lambda)^2\Delta}=\beta-\alpha'$$

$$\beta''=\int_\lambda^\infty \frac{\lambda d\lambda}{(a^2+\lambda)(b^2+\lambda)\Delta}=\alpha-\beta'$$

The corresponding integrals with the lower limit of integration replaced by zero are denoted by α_0, α_0', α_0'', β_0, β_0' and β_0''.

The velocity field (eq. (3.84)) is rather complicated and we must resort to a numerical evaluation to appreciate its nature. We will return to this a little later on. From the flow field we can calculate the torque exerted by the fluid on the particle. The particle rotates in such a way that the net torque exerted on it is zero. This results in

$$\omega_1 = \dot{\phi}\cos\theta + \dot{\psi}$$
$$\omega_2 = \dot{\theta}\sin\psi - \dot{\phi}\sin\theta\cos\psi \qquad (3.85)$$
$$\omega_3 = \dot{\theta}\cos\psi + \dot{\phi}\sin\theta\sin\psi$$

with

$$\dot{\theta} = \frac{\mathrm{d}\theta}{\mathrm{d}t} = \frac{1}{4} BG \sin 2\theta \sin 2\phi \qquad (3.86a)$$

$$\dot{\phi} = \frac{\mathrm{d}\phi}{\mathrm{d}t} = \frac{1}{2} G(1 + B\cos 2\phi) \qquad (3.86b)$$

$$\dot{\psi} = \frac{\mathrm{d}\psi}{\mathrm{d}t} = -\frac{1}{2} GB \cos\theta\cos 2\phi \qquad (3.86c)$$

where

$$B = \frac{(a/b)^2 - 1}{(a/b)^2 + 1} \qquad (3.87a)$$

For prolate spheroids $a/b > 1$ and $0 < B < 1$; for oblate spheroids $a/b < 1$ and $-1 < B < 0$. Equations (3.86) are usually referred to as *Jeffery's equations*. It can be shown that all axisymmetric bodies rotate in the same way, with the modification that the axis ratio a/b be replaced by an equivalent axis ratio r_e. Thus Jeffery's equations apply to all axisymmetric particles with the shape factor B given by

$$B = \frac{r_e^2 - 1}{r_e^2 + 1} \qquad (3.87b)$$

Notice the similarity between Jeffery's equation and the equations describing the streamlines around spheres in simple shear (eqs (3.8)). As mentioned above, changes in ψ do not affect the particle orientation and are irrelevant to most properties of a suspension containing axisymmetric particles.

From eqs (3.86a) and (3.86b) it can be seen that the motion of a spheroid in simple shear is periodic. These equations can readily be integrated with the result

$$\tan\theta = C(r_e \sin^2\tau + \cos^2\tau)^{1/2} \qquad (3.88)$$

and

$$\tan\phi = r_e \tan\tau \qquad (3.89)$$

where

$$\tau = \frac{2\pi t}{T} + \kappa \tag{3.90}$$

T is the period of rotation about the vorticity axis (X_1 axis), given by

$$T = \frac{2\pi}{G}(r_e + r_e^{-1}) \tag{3.91}$$

C and κ are integration constants, usually referred to as the orbit constant and the phase angle, respectively, defined as

$$C = (\tan\theta)_{\tau=0} \tag{3.92}$$

i.e. C is determined by the value of θ when $\tau = 0$, and

$$\kappa = \tan^{-1}\left(\frac{\tan\phi_0}{r_e}\right) \tag{3.93}$$

ϕ_0 being the angle ϕ at time $t = 0$.

The orbit constant C determines the trajectory that the tip of the spheroid (i.e. the point $x_1' = a$, $x_2' = x_3' = 0$) traces out in space. When $C = 0$, $\theta = 0$ and the

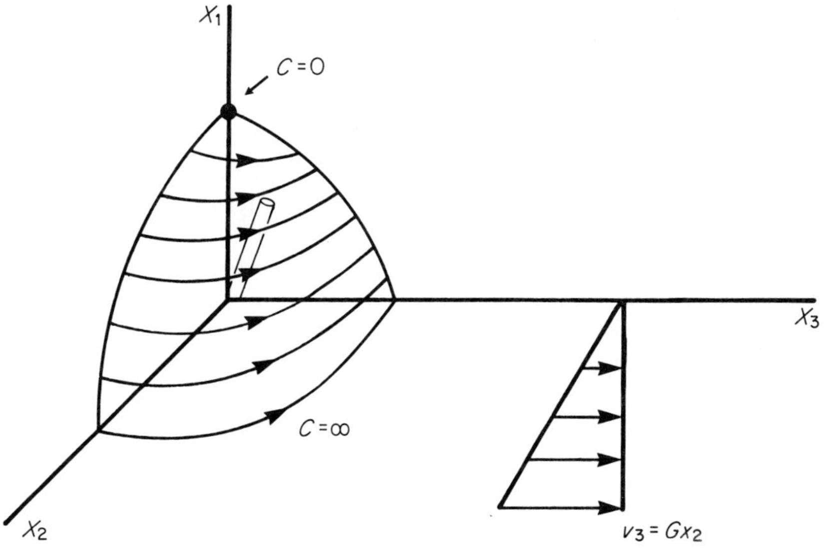

Fig. 3.12. Scheme to show the typical orbits for prolate spheroids in simple shear.

spheroid is aligned with the X_1 axis and the only motion is a spin about this axis. When $C=\infty$, $\theta=90°$ and the spheroid's X_1' axis lies in the X_2X_3 plane, and rotates about the vorticity-axis X_1. Various orbits, corresponding to different values of the orbit constant C, are shown in Fig. 3.12.

The rotation of a cylinder in simple shear is demonstrated in Fig. 3.13, which shows several frames of a cine film made by Mason and co-workers. It is apparent from Fig. 3.13 and from eqs (3.86a) and (3.86b) that the rotational motion of a spheroid is non-uniform with time. Prolate spheroids spend more time with their axis of revolution aligned with the flow than perpendicular to it. In contrast, oblate spheroids spend more time with their axis of revolution perpendicular to the flow. As we will discuss later, this non-uniformity in angular velocity has many important consequences. This non-uniform motion can also be seen in Fig. 3.14, which shows experimental measurements of the angle ϕ versus time, t, for two different axisymmetric bodies, together with the relevant theoretical curves. Fig. 3.15 shows how the period of rotation T varies as a function of axis ratio for various axisymmetric bodies.

Apart from knowing how spheroids rotate, it is also of interest to look at the velocity field around them. Of special interest are the trajectories of fluid elements, as these are the limiting cases of the motion of small tracer particles around or near to the rotating spheroids. These trajectories are called path lines, rather than streamlines, because of their non-steady nature. Equations (3.84) describe the flow field around a spheroid at orientation (θ, ϕ), but because the particle is rotating according to eqs (3.86), this flow field is changing continuously. The modification of eqs (3.84) to account for the rotation was made by Adamczyk and van de Ven (1983). In a time interval Δt an element of fluid (or a small tracer particle with radius $r \ll b$) moves by an amount $v'\Delta t$ with respect to a spheroid at an orientation (θ, ϕ, ψ). During this time interval, the spheroid rotates by an amount $(\Delta\theta, \Delta\phi, \Delta\psi)$. The new position of the fluid element must be expressed in terms of a particle-fixed coordinate system that rotates along with the spheroid. This can be achieved by two simultaneous transformations; first by transforming the position of the tracer into space-fixed coordinates, using the transformation matrix α_{ji} evaluated at time t, and subsequently transforming the space-fixed coordinates into rotating particle-fixed coordinates using α_{ij} evaluated at $t+\Delta t$. As a result the motion of a fluid element with respect to particle-fixed coordinates v_{iJ}' rotating according to Jeffery's equations is given by

$$v_i' = v_{iJ}' + \beta_{iJ} x_j' \tag{3.94}$$

where v_{iJ}' is the fluid velocity of a spheroid given by eqs (3.84), but now θ, ϕ and ψ are functions of time; furthermore

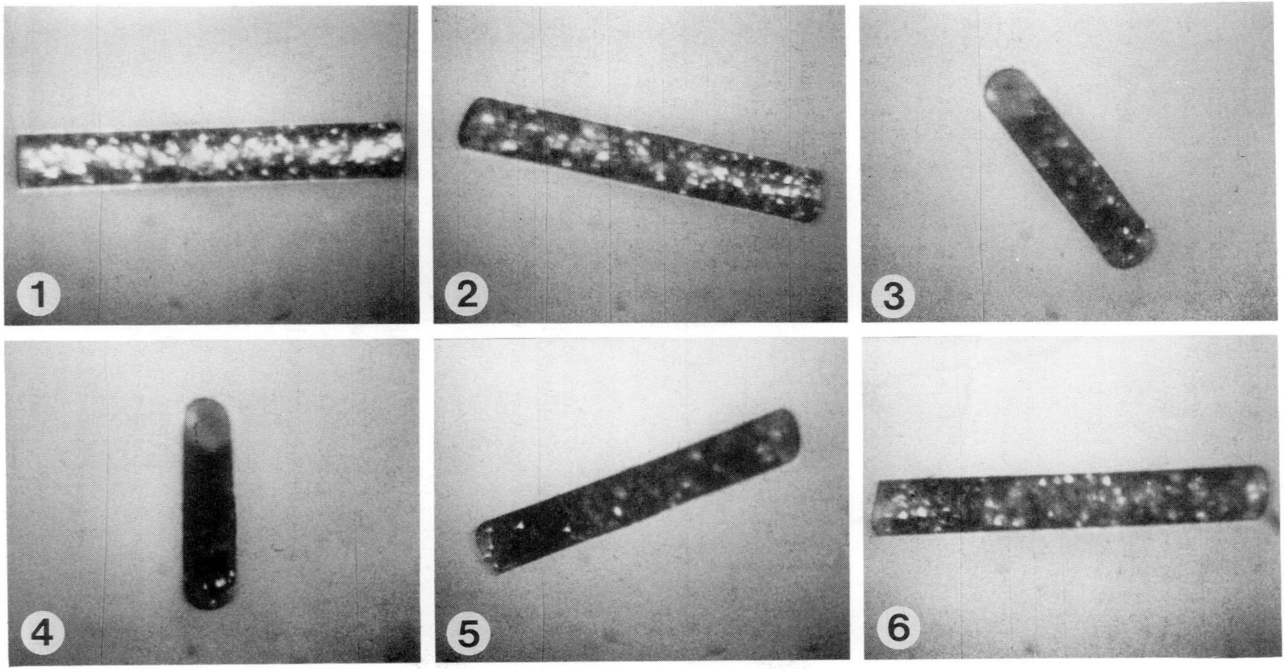

Fig. 3.13. Selected frames of film showing the rotation of a rod in simple shear. The time intervals between the frames are 5·8, 2·9, 0·8, 2·5 and 6·3 s. $T/2 = 18{\cdot}2$ s.

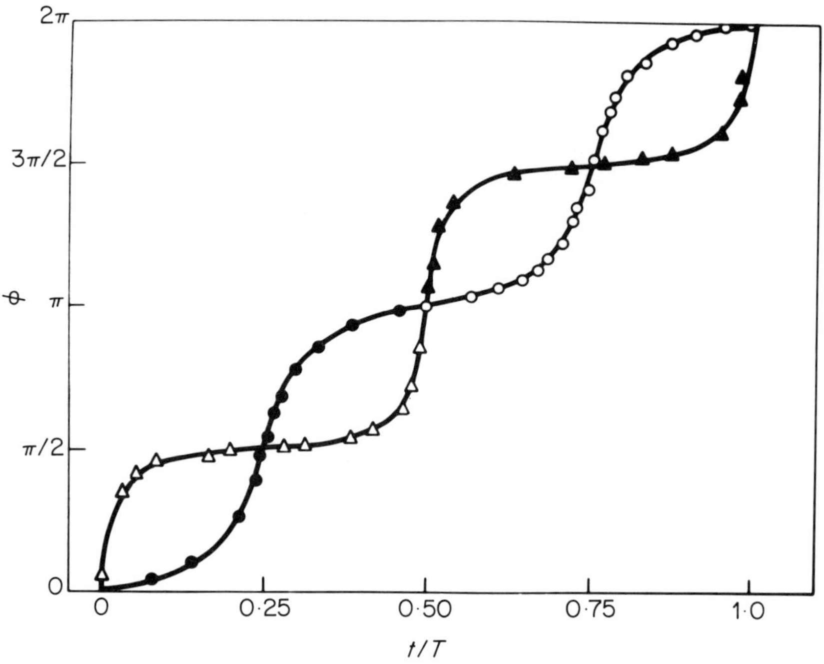

Fig. 3.14. Measured angles ϕ versus t/T for a rod, $r_e = 8 \cdot 2$ (▲) and a disk $r_e = 0 \cdot 25$ (○). The points are experimental and the lines are calculated from theory. The graph also illustrates that the angular velocities for rods and disks are $\pi/2$ out of phase with each other (after Karnis *et al.*, 1966).

$$\beta_{k\ell} = \alpha_{j\ell} \frac{\mathrm{d}\alpha_{jk}}{\mathrm{d}t} \tag{3.95}$$

where α_{ij} is a function of time, again because θ, ϕ and ψ are functions of time. It can be shown that

$$\beta_{ij} = \begin{pmatrix} 0 & \omega_3 & -\omega_2 \\ -\omega_3 & 0 & \omega_1 \\ \omega_2 & -\omega_1 & 0 \end{pmatrix} \tag{3.96}$$

where ω_i is given by eq (3.85).

Equation (3.94) expresses the flow field in rotating-particle-fixed coordinates. This equation has been solved numerically by Adamczyk and van de Ven (1983). To obtain the path lines in space-fixed coordinates, the solution must be transformed into space-fixed coordinates:

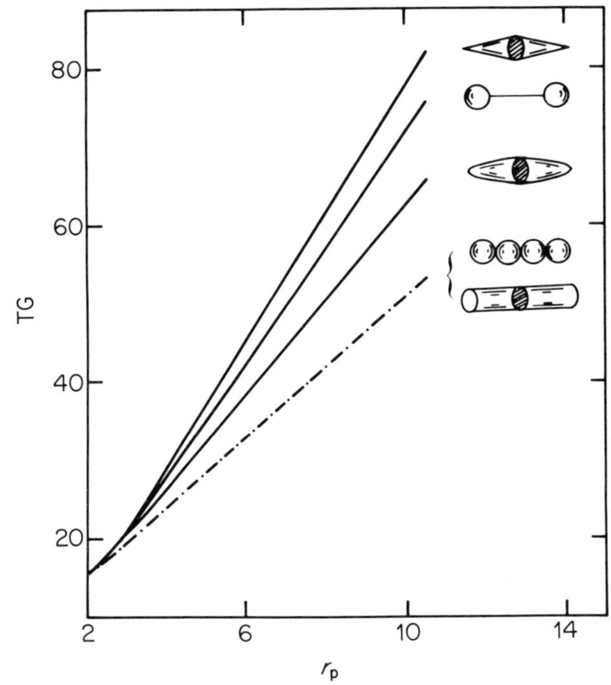

Fig. 3.15. Plot of the theoretically calculated values of the period of rotation, T, for a symmetrical double cone, rigid dumbbell and spheroid against the particle axis ratio, r_p (——), compared with the best-fit experimental curve (-·-·-·) for cylinders and linear chains of spheres with equal diameters (after Takamura *et al.*, 1981c).

$$x_i(t) = \alpha_{ij}(t)\, x_j'(t) \tag{3.97}$$

When $r_e = 1$ or $r_e = \infty$ with $\theta = 90°$, the problem reduces to that of finding the streamlines around a sphere or an infinitely long cylinder aligned along the vorticity axis, problems which were discussed earlier (Sections 3.1.1 and 3.1.4). For spheroids aligned along the vorticity axis, the path lines reduce to streamlines which are qualitatively similar. The streamlines are of two kinds, open and closed, separated by a surface of limiting streamlines. Examples of closed streamlines are shown in Fig. 3.16, which shows the maximum and minimum distances d^*_{max} and d^*_{min}, normalized with the semi-axis b, to which a closed streamline approaches the surface of the spheroid. In all cases, d_{max} tends to infinity at a limiting value of d_{min}, corresponding to the closest approach of the limiting streamline. For comparison, the limiting streamlines for spheres ($d^*_{min} = 0 \cdot 157$) and for cylinders ($d^*_{min} = 0 \cdot 30$) are included.

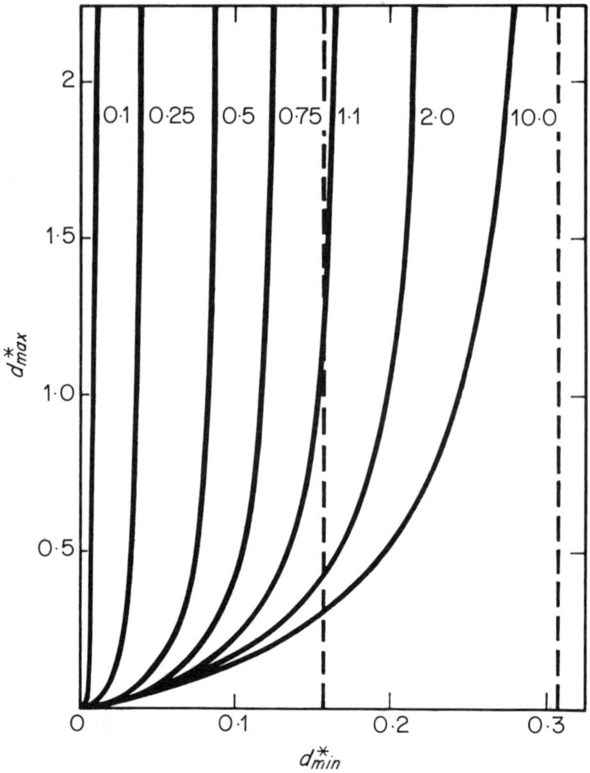

Fig. 3.16. Values of d^*_{min} and d^*_{max} for closed orbits in the equatorial plane for spheroids aligned along the vorticity axis and having various particle axis ratios, r_{p}. The dashed lines denote the asymptotic values of d^*_{min} for spheres and cylinders, respectively (after Adamczyk and van de Ven, 1983).

For spheroids not aligned along the vorticity axis, the path lines are more complex; nevertheless, similarities with the streamlines around spheres do exist. For arbitrary orientations of the spheroid there are open trajectories, which in general are asymmetric, and are referred to as single-pass path lines. Instead of closed streamlines, there exist permanent orbits which are, in general, neither closed nor periodic. An example is given in Fig. 3.17 which shows the distance between a fluid element and the surface of the spheroid as a function of time. The distance remains finite and its variations are non-periodic. Finally, instead of a surface of limiting streamlines, for spheroids there is a region of transient orbits. These are characterized by the property that the fluid elements arrive from infinity, orbit the spheroid several times and subsequently separate towards infinity. Two examples are shown in

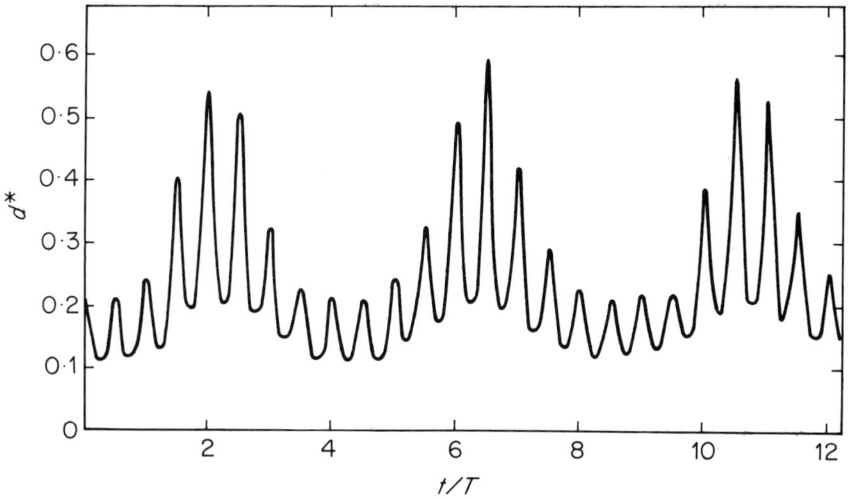

Fig. 3.17. Apparently permanent non-closed orbits in the equatorial plane ($\theta = 90°$) for spheroids of axis ratio $r_p = 2$. The distance between the position of the fluid element and the surface of the spheroid around which it is in orbit is plotted as a function of time. Initial conditions: $\phi_0 = 0$, $x_1' = x_2' = 0$, $x_3'/b = 1{\cdot}21$ (after Adamczyk and van de Ven, 1983).

Fig 3.18. The region of transient orbits separates the single-pass path lines from the permanent orbits as can be seen from Fig 3.19 which presents the number of orbits, as a function of distance from the spheroid's surface.

The similarities and differences between the streamlines around spheres (and spheroids aligned along the vorticity axis) and path lines around arbitrarily oriented spheroids is summarized in Table 3.1. When spheroids are surrounded by ionic double layers, effects similar to those for spheres may be expected. When the ratio of the double-layer thickness to the minor semi-axis is of the order of unity, deviations from Jeffery's equations may be expected, resulting in, for example, variations in the period, T. For spheres it could be demonstrated that at low surface potentials the double layer does not affect the period (up to the order of Φ_s^2) (cf. eq (3.43)). There are no particular reasons to believe that the same will be true for spheroids and at low potentials it is likely that

$$T = T_0[1 + \bar{b}\,\Phi_s^2 f(r_e, \tau)] \tag{3.98}$$

where T_0 is the period of a neutral spheroid, given by eq (3.91) and $f(r_e, \tau)$ is an unknown function of particle shape and double-layer thickness (or electrolyte concentration). These aspects deserve further investigation.

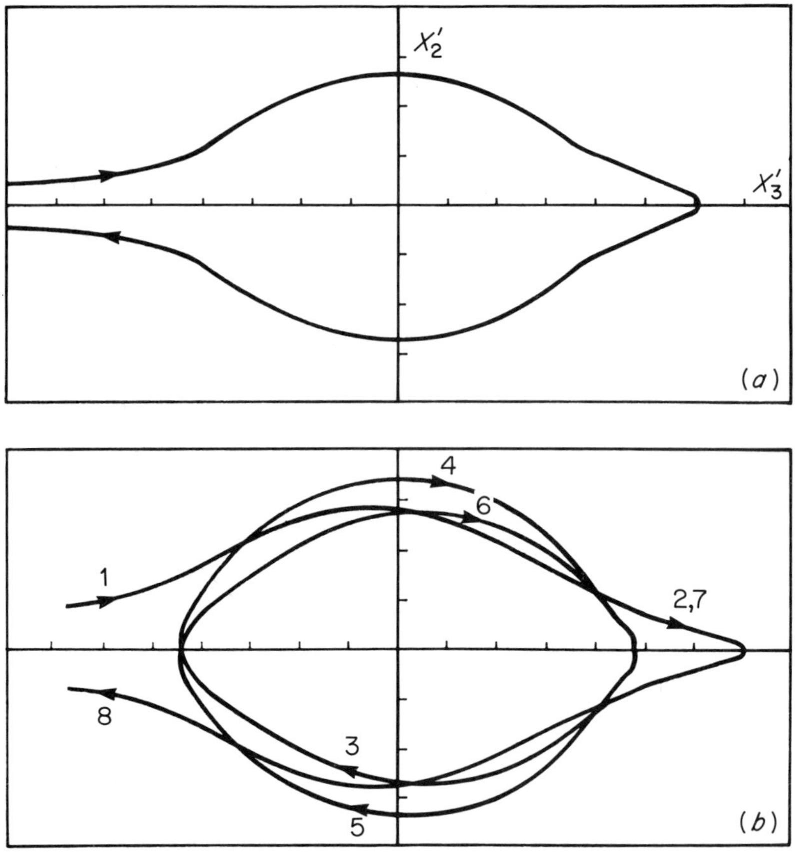

Fig. 3.18. Examples of transient orbits in the equatorial plane ($\theta = 90°$) for spheroids of axis ratio $r_p = 2$, initially aligned along the X_2' axis ($\phi = 0$).
(*a*) Initial conditions: $x_1' = x_2' = 0$, $x_3'/b = 3$ (path of approach and recession is symmetric).
(*b*) Initial conditions: $x_1' = 0$, $x_2'/b = 0{\cdot}426$, $x_3/b = 3{\cdot}238$. The numbers indicate the course the pathline takes while orbiting the spheroid. For the region 2,7 the positions differ by less than the thickness of the line. Notice that here the pathline is also symmetrical about the X_3' axis. One division on the axis corresponds to $b/2$ (after Adamczyk and van de Ven, 1983).

3.1.6 Diffusion of spheroids in simple shear

3.1.6.1 Rotary diffusion

Colloidal particles subjected to simple shear will undergo translational and rotary Brownian motion at the same time. Brownian motion causes random variations in the motion and orientations of spheroids and their trajectories

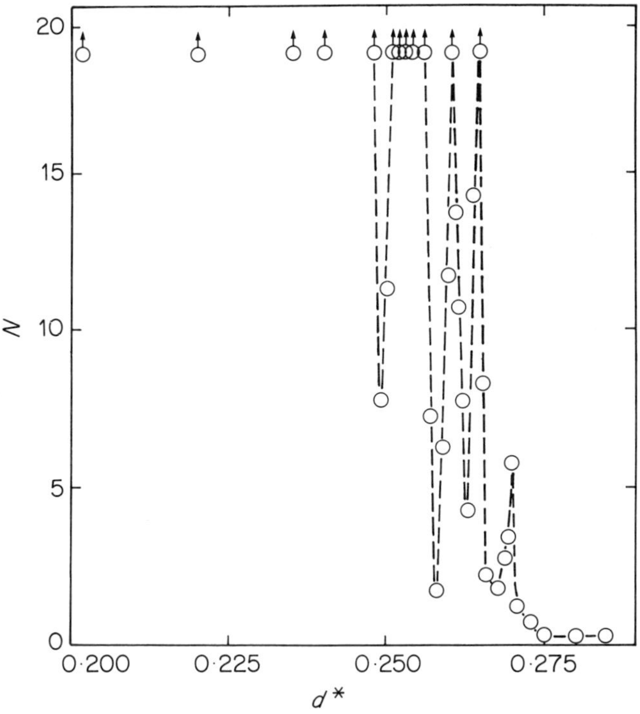

Fig. 3.19. An example of a transition region consisting of transient orbits for spheroids of $r_p = 2$ with initial orientation $\theta = 30°$, $\phi_0 = 0°$. The number of orbits before separating towards inifinity, N, is plotted as a function of the initial distance, d (non-dimensionalized by b) from the surface. The distance d is taken along the X_2 axis which initially lies in the $X_1'X_2'$ plane (after Adamczyk and van de Ven, 1983).

TABLE 3.1
Comparison of the flow fields around rotating spheres and spheroids in simple shear

Spheres, aligned spheroids	Arbitrarily oriented spheroids
Open (symmetric) streamlines	Open (asymmetric) single-pass path lines
Surface of limiting streamlines	Region of transient orbits
Closed (periodic) streamlines	Permanent (mainly non-closed, non-periodic) orbits

and orbits will no longer be fully deterministic. Instead, we must deal with stochastic trajectory and orbit equations. Translational diffusion will not affect the rotary motion of the particles, because for simple shear the shear rate is the same for all streamlines. Rotary diffusion will affect the translational motion because the translational motion of a spheroid depends on its orientation. For this reason rotary Brownian motion will be discussed first.

The probability of finding a spheroid in a given orientation is governed by the rotary convective-diffusion equation (cf. eq (2.99)):

$$\frac{\partial p}{\partial t} = \nabla \cdot (\boldsymbol{D}_{\mathrm{r}} \cdot \nabla p) - \nabla \cdot (\omega p) \tag{3.99}$$

For axisymmetric particles subjected to simple shear the angular velocity vector ω is given by eq (3.85) and $\boldsymbol{D}_{\mathrm{r}}$ is the rotary-diffusion tensor. For an electrically charged particle surrounded by a double layer it can be anticipated that

$$\boldsymbol{D}_{\mathrm{r}} = \begin{pmatrix} D_1 & 0 & 0 \\ 0 & D_2 & 0 \\ 0 & 0 & D_3 \end{pmatrix} \tag{3.100}$$

where the three principal diffusion constants depend on $\bar{b}$, Φ_{s}^2, a function of r_{e} and τ and also on the orientation (θ, ϕ). As no theory for evaluating D_i is available, we will deal here with neutral spheroids for which

$$\boldsymbol{D}_{\mathrm{r}} = \begin{pmatrix} D_{\mathrm{s}} & 0 & 0 \\ 0 & D_{\mathrm{r}} & 0 \\ 0 & 0 & D_{\mathrm{r}} \end{pmatrix} \tag{3.101}$$

where D_{s} is the diffusion constant for rotary motion about the X_1' axis (see Fig. 3.11) and D_{r} is the diffusion constant for rotary motion about the X_2' or X_3' axes. Because motion about the X_1' axis does not affect the orientation of the particle, the component D_{s} is irrelevant in determining the orientation distribution function, p. Hence without loss of generality, we can choose $\psi = 90°$ in eq (3.85) and eq (3.99) reduces to

$$\frac{\partial p}{\partial t} = D_{\mathrm{r}} \nabla^2 p - \nabla \cdot (\omega p) \tag{3.102}$$

with

$$\boldsymbol{\omega}=\begin{pmatrix}\mathrm{d}\theta/\mathrm{d}t\\ \sin\theta\mathrm{d}\phi/\mathrm{d}t\end{pmatrix} \tag{3.103}$$

with $\mathrm{d}\theta/\mathrm{d}t$ and $\mathrm{d}\phi/\mathrm{d}t$ given by Jeffery's equations (eqs (3.86)).

No general solutions of eq (3.102) exist, but several asymptotic solutions as well as numerical solutions are available. Because $\boldsymbol{\omega}$ is proportional to the shear rate, G, it follows from eq (3.102) that its solution depends on the rotary Péclet number, $\mathrm{Pe_r}$, defined as

$$\mathrm{Pe_r}=G/D_\mathrm{r} \tag{3.104}$$

When $\mathrm{Pe_r}\to\infty$, diffusion is negligible and the particle rotates according to Jeffery's equations. When $\mathrm{Pe_r}\to 0$, Brownian motion is dominant, all orientations are equally probable and p is constant. Exact solutions are available for large and small Péclet numbers.

(a) Large rotary Péclet numbers

Several papers in the literature deal with various aspects of the solution of eq (3.102) for large Péclet numbers, restricting themselves to short times, slender bodies, steady states, etc. (Hinch and Leal 1973; Cohen and Leal, 1978; van de Ven *et al.*, 1981a). Here we will first give the solution for short times and subsequently the steady-state solution which is reached at long times.

When $\mathrm{Pe_r}\gg 1$, the deviations from Jeffery's equations are initially small and it is convenient to express them in a coordinate system that rotates according to Jeffery's equations. Let us define the deviations from Jeffery's motion as

$$\Delta\theta=\theta-\theta_\mathrm{s};\ \Delta\phi=\phi-\phi_\mathrm{s}$$

where $(\theta_\mathrm{s},\phi_\mathrm{s})$ is the orientation of the rotating coordinate system. Then, up to first order, it follows from eqs (3.86) that

$$\frac{\mathrm{d}\Delta\theta}{\mathrm{d}t}=\tfrac{1}{2}BG\cos2\theta_\mathrm{s}\sin2\phi_\mathrm{s}\Delta\theta+\tfrac{1}{2}BG\sin2\theta_\mathrm{s}\cos2\phi_\mathrm{s}\Delta\phi \tag{3.105a}$$

$$\frac{\mathrm{d}\Delta\phi}{\mathrm{d}t}=-BG\sin2\phi_\mathrm{s}\Delta\phi \tag{3.105b}$$

Defining $\theta^* = \mathrm{Pe}_r^{1/2}\Delta\theta$, $\phi^* = \mathrm{Pe}_r^{1/2}\sin\theta_s\Delta\phi$ and $t^* = Gt$, eq (3.102) can be rewritten as (neglecting terms of $0(\mathrm{Pe}_r^{-1})$):

$$\frac{\partial p}{\partial t^*} = \frac{\partial^2 p}{\partial \theta^{*2}} + \frac{\partial^2 p}{\partial \phi^{*2}} - \boldsymbol{\nabla}\cdot(\boldsymbol{\omega}^* p) \tag{3.106}$$

with

$$\boldsymbol{\omega}^* = \begin{pmatrix} \frac{1}{2}B\cos 2\theta_s \sin 2\phi_2 & B\cos 2\phi_s \cos\theta_s \\ 0 & -B\sin 2\phi_s \end{pmatrix} \cdot \begin{pmatrix} \theta^* \\ \phi^* \end{pmatrix} \tag{3.107}$$

With the aid of Jeffery's equations, the angles θ_s and ϕ_s appearing in eq (3.107) can be converted into functions of time. The solution of eq (3.106) is a generalized Gaussian (van de Ven *et al.*, 1981a):

$$p = \frac{(\beta_{11}\beta_{22} - \beta_{12}^2)^{1/2}}{2\pi} \exp[-(\tfrac{1}{2}\beta_{11}\theta^{*2} + \tfrac{1}{2}\beta_{22}\phi^{*2} + \beta_{12}\theta^*\phi^*)] \tag{3.108}$$

where β_{ij} are components of a matrix $\boldsymbol{\beta}$. The components of the inverted matrix $\boldsymbol{\beta}^{-1}$ correspond to the second moments of the orientation distribution:

$$\beta_{11}^{-1} = \langle\theta^{*2}\rangle = \mathrm{Pe}_r \langle\Delta\theta^2\rangle \tag{3.109a}$$

$$\beta_{22}^{-1} = \langle\phi^{*2}\rangle = \mathrm{Pe}_r \sin^2\theta_s \langle\Delta\phi^2\rangle \tag{3.109b}$$

$$\beta_{12}^{-1} = \langle\theta^*\phi^*\rangle = \mathrm{Pe}_r \sin\theta_s \langle\Delta\theta\Delta\phi\rangle \tag{3.109c}$$

The above components are related to each other by the following set of equations

$$\frac{\mathrm{d}\langle\theta^{*2}\rangle}{\mathrm{d}t^*} = 2A_{11}\langle\theta^{*2}\rangle + 2A_{12}\langle\theta^*\phi^*\rangle + 2 \tag{3.110a}$$

$$\frac{\mathrm{d}\langle\phi^{*2}\rangle}{\mathrm{d}t^*} = 2A_{22}\langle\phi^{*2}\rangle + 2 \tag{3.110b}$$

$$\frac{\mathrm{d}\langle\theta^*\phi^*\rangle}{\mathrm{d}t^*} = (A_{11} + A_{22})\langle\theta^*\phi^*\rangle + A_{12}\langle\phi^{*2}\rangle \tag{3.110c}$$

Here A_{ij} are the components of the matrix appearing in eq (3.107). The set of eqs (3.110) determines the second moments of the orientation distribution when $\mathrm{Pe}_r \gg 1$. Denoting θ^* by e_i and ϕ^* by e_j, one can write

$$e_ie_j = \begin{pmatrix} \langle\theta^{*2}\rangle & \langle\theta^*\phi^*\rangle \\ \langle\theta^*\phi^*\rangle & \langle\phi^{*2}\rangle \end{pmatrix}$$

Equation (3.110b) can be integrated immediately with the result that, for $r_e > 1$:

$$\langle\phi^{*2}\rangle = \frac{(r_e^2+1)^2}{4r_e(r_e^2\sin^2\tau+\cos^2\tau)^2}\left[(2+B^2)\frac{2\pi t}{T} - 2B(\sin 2\tau - \sin 2\kappa) + \tfrac{1}{4}B^2(\sin 4\tau - \sin 4\kappa)\right] \quad (3.111)$$

For oblate spheroids, r_e must be replaced by r_e^{-1}.

It can be readily seen that as $t \to 0$,

$$\sin^2\theta_s\langle\Delta\phi^2\rangle \to 2D_r t \quad (3.112a)$$

The moments $\langle\theta^{*2}\rangle$ and $\langle\theta^*\phi^*\rangle$ can be found by integrating eq (3.110) numerically. For small periods of time,

$$\langle\Delta\theta^2\rangle = 2D_r t \quad (3.112b)$$

$$\sin\theta_s\langle\Delta\theta\Delta\phi\rangle = \frac{1+B-S}{S(1+C^2S)^{1/2}}D_r G t^2 \quad (3.112c)$$

where C is the orbit constant and $S = r_e^2\sin^2\kappa + \cos^2\kappa$.

Examples of $\langle\phi^{*2}\rangle$, $\langle\theta^{*2}\rangle$ and $\langle\theta^*\phi^*\rangle$ are given in Fig. 3.20. It can be seen that the short-time behaviour predicted by eqs (3.112) is obeyed. As $t \to \infty$, eqs (3.110) no longer apply because $\Delta\phi$ and $\Delta\theta$ are no longer small.

Besides the second moments of the orientation distribution, it is also of interest to examine the variations in the period of rotation, T. These can be found from variations in the angle ϕ which can be obtained by expanding the rotary convective-diffusion equation, taking the quadratic terms $\Delta\theta^2$ and $\Delta\phi^2$ into account. In this case eq (3.106) is again obtained, but with ω^* given by

$$\omega^* = \begin{pmatrix} a_1\theta^* + a_2\phi^* + \mathrm{Pe}_r^{-1/2}(a_3\theta^*\phi^* + a_4\theta^{*2} + a_5\phi^{*2}) \\ b_1\phi^* + \mathrm{Pe}_r^{-1/2}\, b_2\phi^{*2} \end{pmatrix} \quad (3.113)$$

The coefficients a_i and b_i are functions of t^*; they can be found from Jeffery's equations and are similar to the coefficients A_{ij} in eq (3.107).

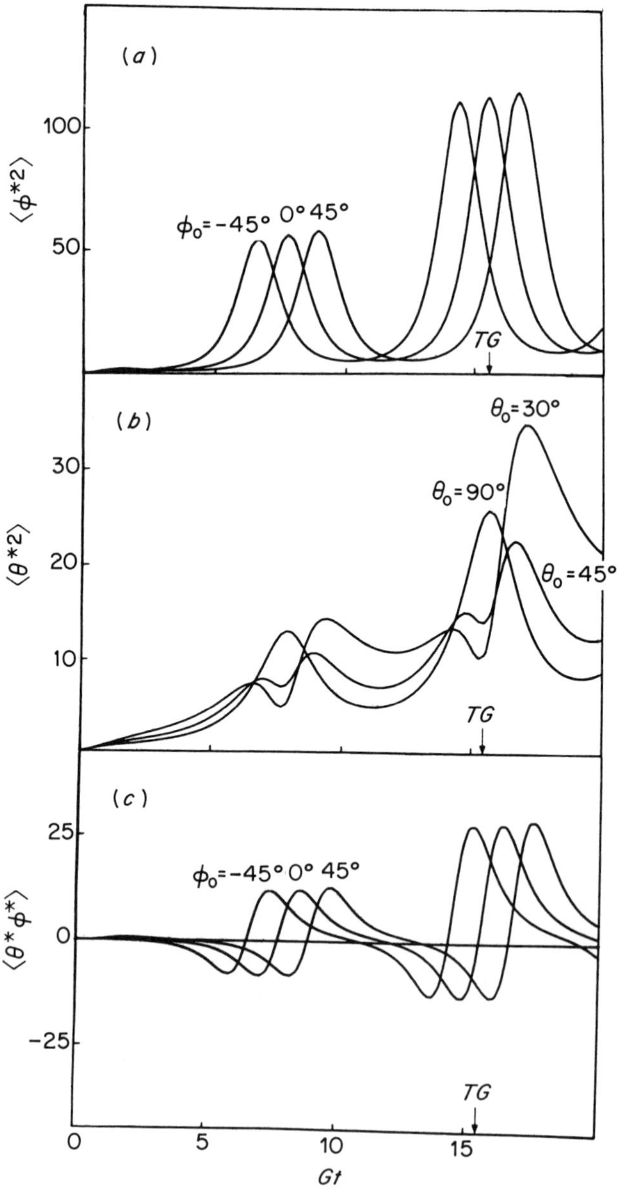

Fig. 3.20. Examples of second moments of the orientation distribution function:
(a) $\langle \phi^{*2} \rangle$; $C=1$.
(b) $\langle \theta^{*2} \rangle$; arbitrary C.
(c) $\langle \theta^* \phi^* \rangle$; $\phi_0 = 0$.
$r_e = 1 \cdot 9817$ (corresponding to two equal-sized spheres in contact). After van de Ven *et al.* (1981a).

Substituting eq (3.113) into eq (3.106), multiplying by $\langle\phi^*\rangle$ and integrating over the whole orientation space yields (for $r_e > 1$)

$$\frac{d\langle\phi^*\rangle}{dt^*} = \frac{-B\, r_e \sin 2\tau}{r_e^2\sin^2\tau + \cos^2\tau}\langle\phi^*\rangle + \frac{B}{\mathrm{Pe}_r^{1/2}}\frac{r_e^2\sin^2\tau - \cos^2\tau}{r_e^2\sin^2\tau + \cos^2\tau}\langle\phi^{*2}\rangle \tag{3.114}$$

Integrating eq (3.114) yields

$$\langle\phi^*\rangle = \frac{B(r_e^2+1)^3}{4\mathrm{Pe}_r^{1/2}\, r_e(r_e^2\sin^2\tau + \cos^2\tau)}\left[\frac{g(t^*)\sin 2\tau + 3r_e^2\sin^2\tau}{r_e^2\sin^2\tau + \cos^2\tau} - \frac{\sin^2\tau}{r_e}\right] \tag{3.115}$$

with

$$g(t^*) = -\frac{r_e + r_e^{-1}}{2}\left[2B\sin 2\kappa - \tfrac{1}{4}B^2\sin 4\kappa + (2+B^2)\frac{2\pi t}{T}\right]$$

From eq (3.115) the average deviation of the azimuthal angle is found to be

$$\langle\Delta\phi\rangle = \frac{h_\phi(t^*)}{\mathrm{Pe}_r\sin\theta_s} \tag{3.116}$$

where $h_\phi(t^*) = \mathrm{Pe}_r^{1/2}\langle\phi^*\rangle$.

It is clear that changes in $\langle\Delta\phi\rangle$ will cause changes in the period of rotation, T, about the vorticity axis. To proceed further we must clarify what is meant by the period of rotation of a particle subjected to Brownian motion in shear flow. In the absence of Brownian motion, the period is defined as the time it takes for the particle to rotate from $\phi = 0$ to $\phi = 2\pi$. A Brownian particle, however, can cross the angle $\phi = 2\pi$ several times and the definition of $\bar{T}$ becomes imprecise. Several ways of defining an average period $\bar{T}$ are possible. Here we define $\bar{T}$ in terms of an ensemble of particles which are aligned at (θ_0, ϕ_0) (at time $t = 0$), as the time at which the ensemble average has increased by 2π, i.e. $\langle\phi\rangle = \phi_0 + 2\pi$. Note that this ensemble average is different from the time average obtained from the average period of rotation of a single particle undergoing a large number of rotations, since the ensemble average depends on the initial orientation of the particles in the ensemble, while the time average is independent of the initial orientation.

According to eq (3.116), at time $t = T$:

$$\langle\phi\rangle = \phi_0 + 2\pi + \frac{h_\phi(GT)}{\mathrm{Pe}_r\sin\theta_0}$$

and hence, using eq (3.86b)

$$\bar{T} = T - \frac{2h_\phi(GT)}{\mathrm{Pe_r}\sin\theta_0(1 + B\cos 2\phi_0)} \tag{3.117}$$

Figure 3.21 shows $\mathrm{Pe_r}\sin\theta_0(\bar{T} - T)G$ as a function of the initial orientation ϕ_0. It is interesting to note that the average period of rotation depends on the

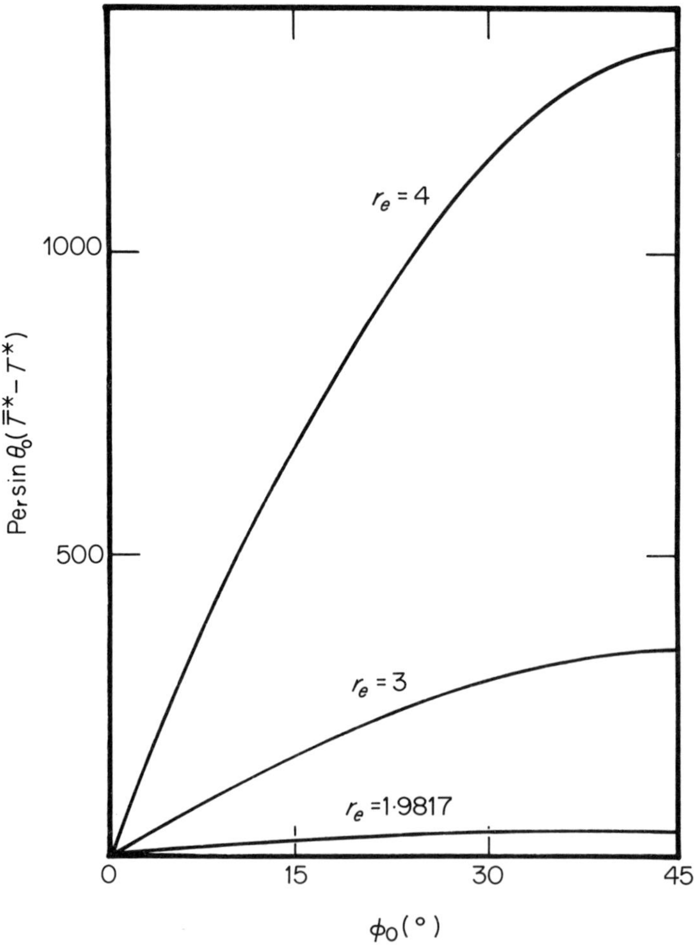

Fig. 3.21. The change in the average period of rotation, $\bar{T}^*$, due to rotary Brownian motion as a function of the initial orientation, ϕ_0, for various values of r_e; values for $45° < \phi_0 < 90°$ can be obtained by symmetry about $\phi_0 = 45°$. For negative ϕ_0 values, the magnitude of the change is identical to positive ϕ_0 values, but with reversed sign (after van de Ven *et al.*, 1981a).

way in which it is observed, i.e. on what value is taken for the initial orientation, ϕ_0. The reason is that the motion of the spheroid is non-uniform. (For a sphere $h_\phi(T)=0$ and $\bar{T}=T$.) The time-averaged period of rotation is ill-defined because the period of rotation is undefined when the orbit constant $C=0$ ($\theta_0=0$), i.e. when the particle is aligned with the vorticity axis. For values of θ near 0°, small changes in θ can cause very large changes in $\Delta\phi$. (When $\theta=0°$, $\Delta\phi=\infty$.) Given sufficient time, a Brownian particle will eventually get arbitrarily close to orientations near $\theta=0°$ and variations in ϕ and T will become unbounded. From Fig. 3.21 it can be seen that variations in $\bar{T}$ become very large as the equivalent axis ratio r_e increases.

Besides the average period of rotation $\bar{T}$, the spread in the distribution of periods is also of interest. From the theory outlined above it can be shown that when $r_e > 1$ the standard deviation, σ_T, is given by (van de Ven *et al.*, 1981a)

$$\sigma_T^2=\frac{3r_e^4+2r_e^2+3}{\sin^2\theta_0(1+B)^2}D_rT\left(1-\frac{\langle\Delta\theta(T)\,\Delta\phi(T)\rangle^2}{\langle\Delta\theta(T)^2\rangle\langle\Delta\phi(T)^2\rangle}\right) \tag{3.118}$$

from which it follows that the spread in the distribution of periods also depends on the initial orientation of the particles.

From the theory outlined above all kinds of orientational averages can be evaluated, some of which will be encountered later in this book. For example,

$$\langle\sin^2\theta\rangle=\sin^2\theta_s+\sin 2\theta_s\langle\Delta\theta\rangle+\cos 2\theta_s\langle\Delta\theta^2\rangle+\ldots \tag{3.119}$$

$\langle\Delta\theta^2\rangle$ can be found from eq (3.110), while $\langle\Delta\theta\rangle$ can be obtained in the same way as $\langle\Delta\phi\rangle$ (cf. eq (3.115)):

$$\frac{d\langle\theta^*\rangle}{dt^*}=a_1\langle\theta^*\rangle+a_2\langle\phi^*\rangle+\mathrm{Pe}_r^{-1/2}\times[a_3\langle\theta^*\phi^*\rangle+a_4\langle\theta^{*2}\rangle+a_5\langle\phi^*\rangle] \tag{3.120}$$

which yields

$$\langle\Delta\theta\rangle=\frac{h_\theta(t^*)}{\mathrm{Pe}_r}$$

where $h_\theta(t^*)$ can be found from a numerical integration of eq (3.120). Hence eq (3.119) can be rewritten as

$$\langle\sin^2\theta\rangle=\sin^2\theta_s+\mathrm{Pe}_r^{-1}[\sin 2\theta_s h_\theta(t^*)+\cos 2\theta_s\langle\theta^{*2}\rangle]+\ldots \tag{3.121}$$

Similar expansions can be made for all orientational averages.

So far we have been concerned with solving eq (3.102) for short time periods. To be more precise, the theory holds when

$$D_r t \ll r_e^{-4}\sin^2\theta_0 = \frac{C^2}{r_e^4(C^2+1)} \quad \text{when } r_e > 1$$

and

$$D_r t \ll r_e^4 \sin^2\theta_0 \qquad\qquad \text{when } r_e < 1$$

At long time periods, the orientation distribution function attains a steady-state value ($\partial p/\partial t = 0$). In the absence of Brownian motion, the distribution of orbits is fully determined by the initial orientations of the particles. Each particle rotates with its initial orbit constant C. Because each particle rotates with a period T, independent of the phase angle κ, the distribution of the angles ϕ for an ensemble of particles oscillates in time with a period $\frac{1}{2}T$ (for particles with fore–aft symmetry, which find themselves in the same orientation after half a period). Brownian motion will cause a drift in both C and κ values. The effect of Brownian motion on the κ values causes a damping of the amplitude of oscillation of period $\frac{1}{2}T$. As $t \to \infty$, the distribution in κ values, or alternatively in ϕ angles, reaches a steady state. This steady state can readily be obtained from Jeffery's equations. The probability of finding a particle between ϕ and $\phi + \mathrm{d}\phi$ is proportional to the time, t, a particle spends in that interval

$$p_\infty(\phi)\mathrm{d}\phi = k\,\mathrm{d}t \tag{3.123}$$

where $p_\infty(\phi)$ is the steady-state distribution in ϕ for $\mathrm{Pe_r} = \infty$, and k is a constant of proportionality which can be found from the normalization condition

$$\int_0^{2\pi} p_\infty(\phi)\,\mathrm{d}\phi = 1 \tag{3.124}$$

Hence, from eqs (3.123) and (3.124):

$$p_\infty(\phi)\,\mathrm{d}\phi = \frac{2\mathrm{d}\phi}{TG(1 + B\cos 2\phi)} \tag{3.125}$$

When $\mathrm{Pe_r} = \infty$, no changes in orbit constant C occur. Therefore orientational averages in any property $S(\theta, \phi)$ can be found from

$$\int_0^{\pi}\int_0^{2\pi} S(\theta,\phi)\,p(\theta,\phi)\,\sin\theta \mathrm{d}\theta \mathrm{d}\phi = \int_0^{\infty}\int_0^{2\pi} f(C)\,S(C,\phi)\,p_{\infty}(\phi)\,\mathrm{d}C\mathrm{d}\phi \qquad (3.126)$$

$f(C)$ being the distribution of orbit constants.

The steady-state orientation distribution is independent of κ, and without loss of generality, we can choose $\kappa = 0$ in eqs (3.88) and (3.90) and write

$$\tan\theta = Cr_e(r_e\cos^2\phi + \sin^2\phi)^{-1/2} \qquad (3.127)$$

With the aid of this equation, every function of θ can be converted to a function of C and ϕ. As an example consider $\langle\cos^2\theta\rangle$:

$$\cos^2\theta = \frac{r_e^2\cos^2\phi + \sin^2\phi}{C^2r_e^2 + r_e^2\cos^2\phi + \sin^2\phi}$$

and hence for a given orbit with an orbit constant C:

$$\langle\cos^2\theta\rangle = \int_0^{2\pi} \frac{r_e^2\cos^2\phi + \sin^2\phi}{C^2r_e^2 + r_e^2\cos^2\phi + \sin^2\phi}\,\frac{2\mathrm{d}\phi}{TG(1 + B\cos 2\phi)}$$
$$= [(C^2r_e^2 + 1)(C^2 + 1)]^{-1/2} \qquad (3.128)$$

For a suspension with a distribution of orbit constants $f(C)$:

$$\langle\cos^2\theta\rangle = \int_0^{\infty} [(C^2r_e^2 + 1)(C^2 + 1)]^{-1/2} f(C)\,\mathrm{d}C \qquad (3.129)$$

where $f(C)$ depends on the rotary Péclet number, Pe_r. When $\mathrm{Pe}_r = \infty$, $f(C)$ is solely determined by the initial distribution of the orbit constants. The distribution $f(C)$ has been determined by Leal and Hinch (1971) for large but finite rotary Péclet numbers. In essence they solved the steady-state convective-diffusion equation in the limit $\mathrm{Pe}_r = \infty$:

$$\nabla\cdot(\omega p) = 0 \qquad (3.130)$$

This equation possesses an infinite number of solutions, each corresponding to a different distribution of C and κ values. Brownian motion causes a steady-state distribution to be reached as $t \to \infty$ and this imposes a condition on eq (3.130). This condition can be found by integrating the equation

$$D_r\nabla^2 p = \nabla\cdot(\omega p) \tag{3.131}$$

over a region A of the unit sphere on which p is defined:

$$D_r\int\int_A \nabla^2 p \, dA = \int\int_A \nabla\cdot(\omega p)\, dA \tag{3.132}$$

Applying the divergence theorem:

$$D_r\int \nabla p\cdot \boldsymbol{n}\, dl = \int(\omega p)\cdot\boldsymbol{n}\, dl \tag{3.133}$$

where $\boldsymbol{n}$ is the outward normal of the bounding surface. Choosing the bounding curve of domain A to be a single Jeffery orbit of orbit constant C for which $\omega\cdot\boldsymbol{n}=0$, one obtains

$$\int_C \frac{\partial p}{\partial n}\, dl = 0 \tag{3.134}$$

which is independent of D_r. The solution of eq (3.130), subject to eq (3.134), determines the orientation distribution in the limit $\mathrm{Pe}_r \to \infty$. It follows from these equations that the final distribution is independent of D_r. However, D_r does determine how long it takes for the steady-state solution to be reached. As $D_r \to 0$, the time for reaching the steady-state solution approaches infinity. The solution of eq (3.130), subject to eq (3.134), is given by Leal and Hinch (1971). Expressing p in the naturally occurring non-orthogonal coordinates C and τ (cf. eqs (3.88) and (3.89)) yields

$$p(\theta,\phi)\,\sin\theta d\theta d\phi = p(C,\tau)dCd\tau = f(C)g(C,\tau)dCd\tau \tag{3.135}$$

Thus, as a first approximation, $p(C,\tau)$ is a product of a distribution around each of the various orbits described by $g(C,\tau)$ (cf. eq (3.126)) and a distribution in the orbit constants C. The function $f(C)$ is given by

$$f(C) = C_n C[(HC^4 + KC^2 + M)F(C, r_e)]^{-3/4} \tag{3.136}$$

with

$$H = r_e^2 + 1;\; K = (r_e^2 + 14 + r_e^{-2})/4;\; M = 1 + r_e^{-2}$$

and

$$F(C,r_e)=\left[\frac{2MC^2+K-(K^2-4HM)^{1/2}}{2HC^2+K+(K^2-4HM)^{1/2}}\right]^{(4-K)/(K^2-4HM)^{1/2}} \qquad K^2>4HM$$

$$=\exp\left[\frac{2(K-4)}{2HC^2+K}\right] \qquad \text{for } K^2=4HM$$

$$=\exp\left[\frac{2(4-K)}{(4HM-K^2)^{1/2}}\right]\tan^{-1}\left[\frac{2HC^2+K}{(4HM-K^2)^{1/2}}\right] \qquad \text{for } K^2<4HM \tag{3.137}$$

C_n is a normalization constant determined by

$$\int_0^\infty f(C)\,\mathrm{d}C=1$$

Equation (3.137) can be integrated analytically for $r_e\to 0$, $r_e=1$ and $r_e\to\infty$:

$$f(C)=C(C^2+1)^{-3/2} \qquad r_e=1$$

$$f(C)=4C(4C^2+1)^{-3/2} \qquad r_e\to 0$$

$$f(C)=4C\,r_e^2(4C^2r_e^2+1)^{-3/2} \qquad r_e\to 0$$

The distribution in orbit constants for various axis ratios is shown in Fig. 3.22. The function $f(C)$ permits the determination of orientational averages in the limit $\mathrm{Pe}_r\to\infty$, an example of which has been given in eq (3.129).

(b) Small rotary Péclet numbers

The solution of the rotary convective-diffusion equation (eq (3.102)) in the limit of $\mathrm{Pe}_r\to 0$ has been given by Zuzovsky *et al.* (1980b), using the method of multiple time scales. In polar coordinates, eq (3.102) becomes

$$\frac{\partial p}{\partial t}=D_r\left[\frac{1}{\sin\theta}\frac{\partial}{\partial\theta}\left(\sin\theta\frac{\partial p}{\partial\theta}\right)+\frac{1}{\sin^2\theta}\frac{\partial^2 p}{\partial\phi^2}\right]$$
$$-G\left[\frac{1}{\sin\theta}\frac{\partial}{\partial\theta}\left(\frac{1}{4}Bp\sin\theta\sin 2\theta\sin 2\phi\right)\right.$$
$$\left.+\frac{\partial}{\partial\phi}\left\{\frac{1}{2}p(1+B\cos 2\phi)\right\}\right] \tag{3.138}$$

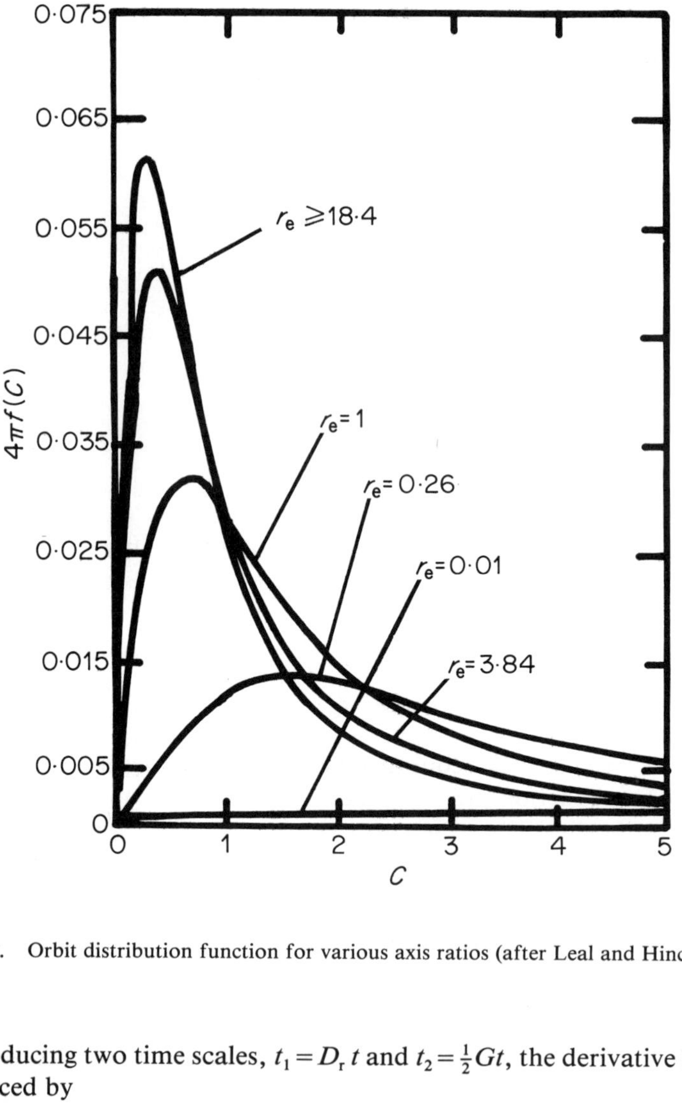

Fig. 3.22. Orbit distribution function for various axis ratios (after Leal and Hinch, 1971).

By introducing two time scales, $t_1 = D_r t$ and $t_2 = \frac{1}{2}Gt$, the derivative $\partial/\partial t$ can be replaced by

$$\frac{\partial}{\partial t} = D_r \frac{\partial}{\partial t_1} + \frac{G}{2} \frac{\partial}{\partial t_2}$$

The solution can be expressed as a power series in $\mathrm{Pe_r}$:

$$p = p_0 + \mathrm{Pe_r} p_1 + \mathrm{Pe_r^2} p_2 + \ldots \quad (3.139)$$

Solving for the zero-order solution yields:

$$p_0 = \sum_{m,n} (A_{mn}\sin m\phi + B_{mn}\cos m\phi)\, P_n^m(\cos\theta)\, e^{-n(n+1)t_1} \tag{3.140}$$

with

$$\left.\begin{matrix} A_{mn} \\ B_{mn} \end{matrix}\right\} = \frac{(n-m)!\,(2n+1)}{(n+m)!\,\pi} \int_0^{\pi}\int_0^{2\pi} p(t=0) \left\{\begin{matrix} \sin m\phi \\ \cos m\phi \end{matrix}\right\}$$

$$\times P_n^m(\cos\theta)\,\sin\theta\, d\theta\, d\phi \tag{3.141}$$

and when $m \neq 0$ and

$$B_{on} = \frac{2n+1}{2\pi} \int_0^{\pi}\int_0^{2\pi} p(t=0)\, P_n^0(\cos\theta)\,\sin\theta\, d\theta\, d\phi$$

For an initially random orientation distribution, p_0 reduces to $p_0 = \frac{1}{4}\pi$. The first-order correction, p_1, is given by

$$p_1 = \frac{B}{16\pi} \sin^2\theta\, [\sin 2\phi - \sin 2(\phi - t_2)\cdot e^{-6t_1}] \tag{3.142}$$

The term p_2 has also been evaluated. The expression for p_2 is complicated and can be found in Zuzovsky *et al.* (1980b).

As $t \to \infty$, the solution reduces to

$$p = \frac{1}{4\pi} + \mathrm{Pe_r}\, \frac{B}{16\pi} \sin^2\theta \sin 2\phi + \mathrm{Pe_r^2} \left[\frac{B}{96\pi} \sin^2\theta \cos^2\phi \right.$$

$$\left. + \frac{B^2}{256\pi} \{\sin^4\theta\,(1-\cos^4\phi) - 8/15\} \right] \tag{3.143}$$

From the solution it can be seen that, as time progresses, an initially isotropic distribution changes to one in which the most probable orientation is 45° (and 135°) (up to first order). Obviously, the steady-state solution is independent of the initial distribution.

(c) Intermediate Péclet numbers

For intermediate Péclet numbers, no analytical results for the orientation distribution p are available. Numerical results are given by Sheraga (1955) for Péclet numbers up to 60. As noted by Brenner (1974), at higher Péclet numbers the numerical methods employed become inaccurate. For Péclet numbers $\mathrm{Pe_r} > 60$, results can be obtained by performing dynamic simulation calculations (van de Ven and Mason, 1976b; Zuzovsky *et al.*, 1980a).

In dynamic simulation calculations, Jeffery's equations (eqs (3.86a) and (3.86b)) are integrated with the addition of a term allowing for the stochastic rotary motion of the particle:

$$\frac{\Delta\theta}{\Delta t} = \tfrac{1}{4}BG \sin 2\theta \sin 2\phi + R_1/\Delta t \tag{3.144a}$$

$$\frac{\Delta\phi}{\Delta t} = \tfrac{1}{2}G\,(1 + B\cos 2\phi) + R_2/\Delta t \tag{3.144b}$$

R_1 and R_2 are random angular displacements, chosen (using random-number generator techniques) from a Gaussian distribution of zero mean and standard deviation, $2D_r\Delta t$, in the variables $\Delta\theta$ and $\sin\theta\Delta\phi$. The time interval Δt must be chosen as small compared to G^{-1}, but large compared to the characteristic time of the decay in the autocorrelation of the angular velocity (cf. Section 2.1.3.2):

$$I/f_r \ll \Delta t \ll G^{-1} \tag{3.145}$$

where I is the moment of inertia and f_r the rotary friction coefficient.

By integrating eqs (3.144), stochastic trajectories can be generated. By generating a large number of trajectories and averaging, the orientation distribution can be obtained. The variations in $p(\phi)$ as a function of $\mathrm{Pe_r}$ are shown schematically in Fig. 3.23. The probability varies from a constant when $\mathrm{Pe_r} = 0$ to one which is maximum for orientations in the direction of flow and minimum in the direction perpendicular to the flow, when $\mathrm{Pe_r} = \infty$.

3.1.6.2 Translational diffusion

The translational diffusion of axisymmetric particles is governed by the translational convective-diffusion equation

$$\frac{\partial n}{\partial t} = \nabla\cdot(\boldsymbol{D}\cdot\nabla n) - \nabla\cdot(\boldsymbol{u}n) \tag{3.146}$$

where $\boldsymbol{D}$ is the translational-diffusion tensor. Contrary to rotary diffusion in simple shear, which is unaffected by translational diffusion, the

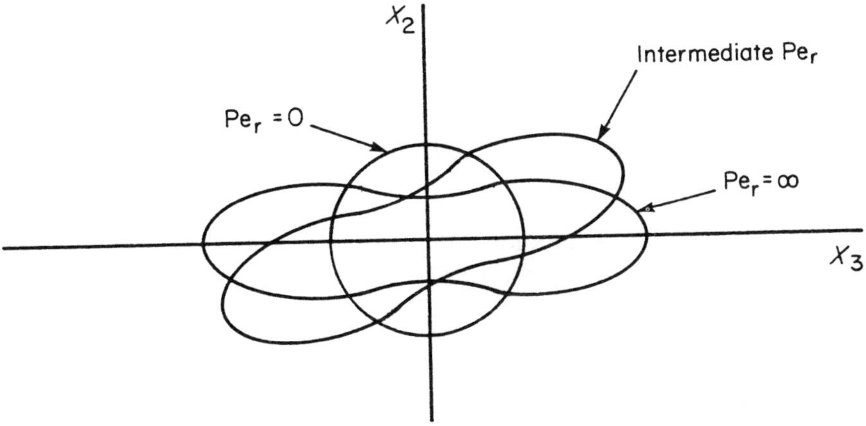

Fig. 3.23. Steady-state solutions of the rotary convective-diffusion equation for spheroids (schematic). The distribution changes gradually from random ($Pe_r = 0$) until the case where the probability of finding particles aligned with the flow is maximum ($Pe_r = \infty$).

translational diffusion depends on the rotary diffusion. The translational diffusion depends on the orientation of the particles, which in turn is affected by rotary diffusion. Hence, in general, $\boldsymbol{D}$ is a function of the rotary Péclet number, Pe_r.

The components of the tensor $\boldsymbol{D}$ in space-fixed coordinates can be found from the diffusion tensor in body-fixed coordinates, represented as

$$D'_{mn} = \begin{pmatrix} D_1 & 0 & 0 \\ 0 & D_2 & 0 \\ 0 & 0 & D_2 \end{pmatrix} \tag{3.147}$$

where D_1 is the component parallel to the axis of revolution X'_1 and D_2 is along the axis perpendicular to the axis of revolution. In space-fixed coordinates, the diffusion tensor is given by

$$D_{ij} = \alpha_{im} \alpha_{jn} D'_{mn} \tag{3.148}$$

where α_{ij} is the transformation matrix (cf. eq (3.79) with $\psi = 0$).

$$\alpha_{ij} = \begin{pmatrix} \cos\theta & -\sin\theta & 0 \\ \sin\theta\cos\phi & \cos\theta\cos\phi & -\sin\phi \\ \sin\theta\sin\phi & \cos\theta\sin\phi & \cos\phi \end{pmatrix} \tag{3.149}$$

It follows from eqs (3.148) and (3.149) that D_{ij} is a function of the angles θ and ϕ, which are determined by Jeffery's equations and by rotary diffusion. Hence, in general, n in eq (3.146) is not only a function of x_1, x_2, x_3 and t, but also of θ and ϕ. Thus eq (3.146) represents an equation in an abstract five-dimensional space. The equation reduces to an equation in ordinary three-dimensional space in two cases: (i) When D_{ij} can be expressed as a function of t only; this will be the case when $\mathrm{Pe}_r = \infty$ and all components of D_{ij} can be converted to functions of time via Jeffery's equations. In this case

$$\frac{\partial n}{\partial t} = D_{ij}(t)\frac{\partial^2 n}{\partial x_i \partial x_j} - Gx_2\frac{\partial n}{\partial x_3} \tag{3.150}$$

(ii) When the orientation distribution is independent of time, i.e. when a steady-state orientation distribution has been reached. It can be shown by averaging eq (3.146) that, in this case

$$\frac{\partial n}{\partial t} = \overline{D_{ij}}\frac{\partial^2 n}{\partial x_i \partial x_j} - Gx_2\frac{\partial n}{\partial x_3} \tag{3.151}$$

where

$$\overline{D_{ij}} = \int_0^{\pi}\int_0^{2\pi} D_{ij}\,p(\theta,\phi)\sin\theta \mathrm{d}\theta \mathrm{d}\phi \tag{3.152}$$

The formal general solution of eq (3.146) is a generalized Gaussian:

$$n = B(t)\exp\left(-\tfrac{1}{2}\beta_{k\ell}x_k x_\ell\right) \tag{3.153}$$

with

$$B(t) = (2\pi)^{-3/2}\det(\beta^{-1})^{-1/2}$$

and

$$\frac{\mathrm{d}\beta_{k\ell}^{-1}}{\mathrm{d}t} = 2D_{k\ell} + A_{ki}\beta_{\ell i}^{-1} + A_{\ell i}\beta_{ik}^{-1} \tag{3.154}$$

where $A_{ij} = G\,\delta_{i3}\delta_{j2}$.

The coefficients of the inverse of matrix β are the second moments of the probability distribution:

$$\beta_{k\ell}^{-1} = \langle x_k x_\ell \rangle \tag{3.155}$$

Using eqs (3.148) and (3.149), eq (3.154) can be rewritten as

$$\begin{aligned}
\frac{\mathrm{d}\langle x_1^{*2}\rangle}{\mathrm{d}t^*} &= q-(q-1)\sin^2\theta \\
\frac{\mathrm{d}\langle x_2^{*2}\rangle}{\mathrm{d}t^*} &= 1+(q-1)\sin^2\theta\cos^2\phi \\
\frac{\mathrm{d}\langle x_3^{*2}\rangle}{\mathrm{d}t^*} &= 1+(q-1)\sin^2\theta\sin^2\phi + 2\langle x_2^* x_3^*\rangle \\
\frac{\mathrm{d}\langle x_1^* x_2^*\rangle}{\mathrm{d}t^*} &= (q-1)\sin\theta\cos\theta\cos\phi \\
\frac{\mathrm{d}\langle x_1^* x_3^*\rangle}{\mathrm{d}t^*} &= (q-1)\sin\theta\cos\theta\sin\phi + \langle x_1^* x_2^*\rangle \\
\frac{\mathrm{d}\langle x_2^* x_3^*\rangle}{\mathrm{d}t^*} &= (q-1)\sin^2\theta\sin\phi\cos\phi + \langle x_2^{*2}\rangle
\end{aligned} \tag{3.156}$$

Here $x_i^* = x_i\sqrt{G/2D_2}$, $t^* = Gt$ and $q = D_1/D_2$.

When dealing with a steady-state orientation distribution, the orientational terms appearing in eqs (3.156) must be replaced by their orientational averages, e.g.

$$\langle \sin^2\theta \rangle = \int_0^\pi \int_0^{2\pi} \sin^2\theta \, p(\theta,\phi) \sin\theta \mathrm{d}\theta \mathrm{d}\phi \tag{3.157}$$

Integrating eqs (3.156), in the case of a steady-state orientation distribution function, results in

$$\begin{aligned}
\langle x_1^{*2}\rangle &= [q-(q-1)\langle\sin^2\theta\rangle]\, t^* \\
\langle x_2^{*2}\rangle &= [1+(q-1)\langle\sin^2\theta\cos^2\phi\rangle]\, t^* \\
\langle x_3^{*2}\rangle &= [1+(q-1)\langle\sin^2\theta\sin^2\phi\rangle]\, t^* + (q-1) \\
&\quad \times \langle\sin^2\theta\sin\phi\cos\phi\rangle t^{*2} + \tfrac{1}{3}[1+(q-1)\langle\sin^2\theta\cos^2\phi\rangle]\, t^{*3} \\
\langle x_1^* x_2^*\rangle &= (q-1)\langle\sin\theta\cos\theta\cos\phi\rangle\, t^*
\end{aligned} \tag{3.158}$$

$$\langle x_1^* x_3^* \rangle = (q-1)\,\langle \sin\theta\cos\theta\sin\phi \rangle\, t^* + \tfrac{1}{2}(q-1)\,\langle \sin\theta\cos\theta\cos\phi \rangle\, t^{*2}$$
$$\langle x_2^* x_3^* \rangle = (q-1)\,\langle \sin^2\theta\sin\phi\cos\phi \rangle\, t^* + \tfrac{1}{2}[1 + (q-1)\,\langle \sin^2\theta\cos^2\phi \rangle]\, t^{*2}$$

The orientational averages occurring in eqs (3.158) can be evaluated when $\mathrm{Pe_r} \ll 1$, using eq (3.143), resulting in (up to order $\mathrm{Pe_r^2}$):

$$\langle x_1^2 \rangle = 2D_{\mathrm{eff}}t - 2(D_1 - D_2)\,\frac{\mathrm{Pe_r^2}B}{90}\left(1 - \frac{11B}{112}\right) \qquad t \equiv 2D_a t$$
$$\langle x_2^2 \rangle = 2D_{\mathrm{eff}}t + 2(D_1 - D_2)\,\frac{\mathrm{Pe_r^2}B}{120}\left(1 - \frac{29B}{168}\right) \qquad t \equiv 2D_b t$$
$$\langle x_3^2 \rangle = 2D_c t + (D_1 - D_2)\,\frac{\mathrm{Pe_r}Bt}{30} + \frac{2}{3}\,D_b G^2 t^3$$
$$\langle x_1 x_2 \rangle = \langle x_1 x_3 \rangle = 0 \tag{3.159}$$
$$\langle x_2 x_3 \rangle = 2(D_1 - D_2)\,\frac{\mathrm{Pe_r}Bt}{30} + D_b G t^2$$

where

$$D_c = D_{\mathrm{eff}} + (D_1 - D_2)\,\frac{\mathrm{Pe_r^2}B}{72}\left(1 - \frac{11B}{280}\right)$$

and

$$D_{\mathrm{eff}} = \frac{1}{3}\,(D_1 + 2D_2) \tag{3.160}$$

It can be seen that when $\mathrm{Pe_r} = 0$ the mean-square displacements are equal to those of a sphere with a diffusion constant given by eq (3.160). The effect of flow is to decrease $\langle x_1^2 \rangle$ and increase the other moments, because for rod-shaped particles $B > 0$ and $D_1 > D_2$, while for disk-shaped particles $B < 0$ and $D_1 < D_2$. For rods $1 < D_1/D_2 < 2$, while for disks $2/3 < D_1/D_2 < 1$ (Cerda and van de Ven, 1983).

It is also possible to evaluate the orientational averages in eq (3.158) when $\mathrm{Pe_r} \to \infty$ (again for a steady-state orientation distribution function), using eqs (3.125), (3.126), (3.127) and (3.136). In this case the mean-square displacements must be evaluated numerically. As a final example, we consider the case when $\mathrm{Pe_r} \to \infty$, but for short time periods. In this case the rotary Brownian motion can be neglected and the rotary motion is given by Jeffery's equations. Expressing the orientational functions in eqs (3.156) as functions of time and integrating, yields

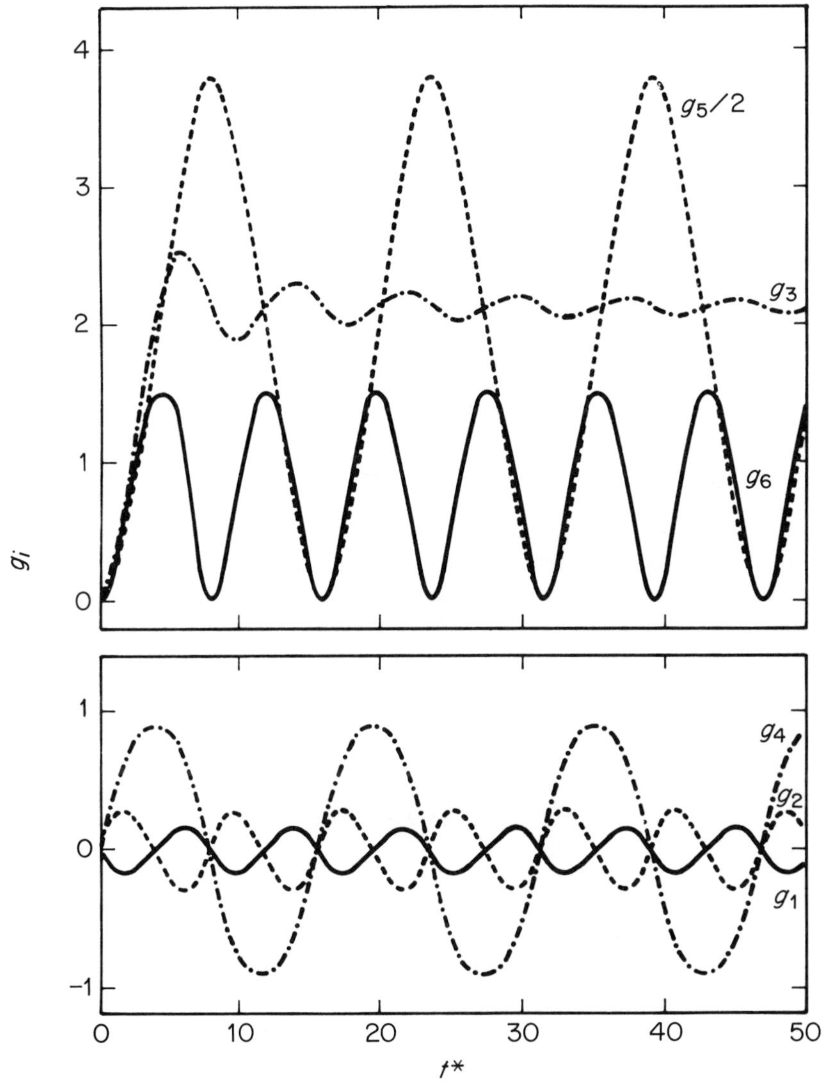

Fig. 3.24. The functions $g_i(t)$ as a function of dimensionless time t^* for touching doublets of spheres ($r_e = 1 \cdot 9817$); $\kappa = 0$ and $C = 1$ (after Cerda and van de Ven, 1983).

$$\begin{aligned}
\langle x_1{}^{*2}\rangle &= qt^* - (q-1)\,[c_1 t^* + g_1(t^*)] \\
\langle x_2^{*2}\rangle &= t^* + (q-1)\,[c_2 t^* + g_2(t^*)] \\
\langle x_3^{*2}\rangle &= t^* + \tfrac{1}{3}t^{*3} + (q-1)\,[g_3(t^*)\,t^* + \tfrac{1}{3}c_2 t^{*3}] \\
\langle x_1^* x_2^*\rangle &= (q-1)\,g_4(t^*) \\
\langle x_1^* x_3^*\rangle &= (q-1)\,g_5(t^*) \\
\langle x_2^* x_3^*\rangle &= \tfrac{1}{2}t^{*2} + (q-1)\,[\tfrac{1}{2}c_2 t^{*2} + g_6(t^*)]
\end{aligned} \tag{3.161}$$

Here

$$c_1 = 1 - \frac{1}{\sqrt{(C^2 r_e^2 + 1)(C^2+1)}}$$

$$c_2 = \frac{C^2}{C^2 + 1 + \sqrt{(C^2 r_e^2 + 1)(C^2+1)}}$$

and the functions $g_i(t^*)$ are periodic functions of t^* of order unity. An example for a doublet of touching spheres with phase angle $\kappa = 0$ and orbit constant $C = 1$ is given in Fig. 3.24. At large t (but still satisfying eq (3.122)), eqs (3.161) reduce to

$$\begin{aligned}
\langle x_1^2\rangle &= 2[D_1 - (D_1 - D_2)c_1]\,t \\
\langle x_2^2\rangle &= 2[D_2 + (D_1 - D_2)c_2]\,t \\
\langle x_3^2\rangle &= \tfrac{2}{3}[D_1 + (D_1 - D_2)c_2]\,G^2 t^3 \\
\langle x_2 x_3\rangle &= [D_1 - (D_1 - D_2)c_2]G t^2
\end{aligned} \tag{3.162}$$

It can be seen that eqs (3.162) and the long-time-period behaviour of eqs (3.159) are qualitatively similar. When $r_e = \infty$, $c_1 = 1$ and $c_2 = 0$, the obvious results are

$$\langle x_1^2\rangle = \langle x_2^2\rangle = 2D_2 t\ ;\ \langle x_3^2\rangle = \tfrac{2}{3}D_2 G^2 t^3\ ;\ \langle x_2 x_3\rangle = D_2 G t^2$$

When $C = 0$, then $c_1 = 0$ and $c_2 = 0$ and we have

$$\langle x_1^2\rangle = 2D_1 t\ ;\ \langle x_2^2\rangle = 2D_2 t\ ;\ \langle x_3^2\rangle = \tfrac{2}{3}D_2 G^2 t^3\ ;\ \langle x_2 x_3\rangle = D_2 G t^2$$

An example of the observed moment $\langle x_3^2\rangle$ of doublets of spheres in shear flow is shown in Fig. 3.25, showing the transition of the linear to the t^3 dependence.

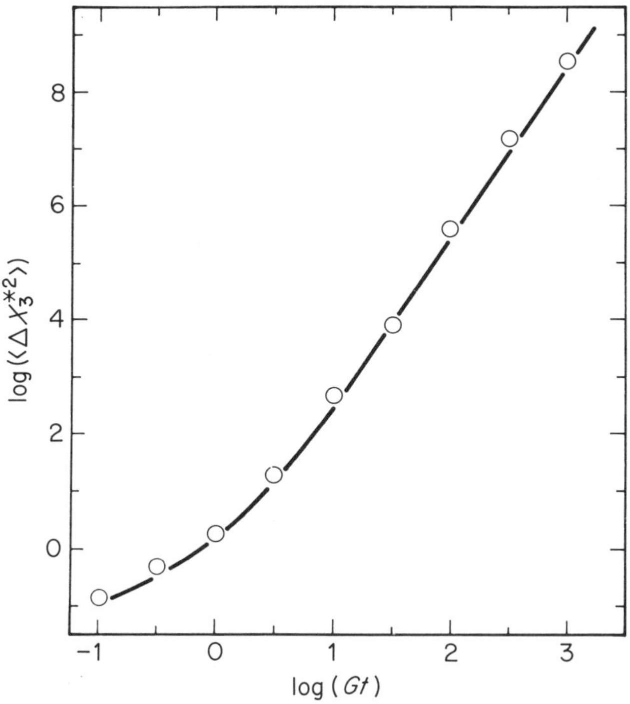

Fig. 3.25. Mean-square displacement in the direction of flow as a function of time for doublets of latex spheres subjected to Poiseuille flow. The solid line is calculated from eq (3.161) for orbit constant $C = 1$. Small oscillations superimposed on the line shown (as predicted by eq (3.161)) are smaller than the thickness of the line (after Cerda and van de Ven, 1983).

3.1.7 Ellipsoids in simple shear

The motion of ellipsoids, or of all triaxial particles in general such as brick- or ribbon-shaped ones, is qualitatively different from the motion of axi-symmetric particles and thus is of particular interest in understanding the motion of particles with more complex shapes. The rotary motion is of special interest because the translational motion is simply rectilinear (in the absence of diffusion): the particle centre moves with the same velocity as the fluid flow. The rotary motion, in the absence of rotary Brownian motion, can be determined from the flow field around an ellipsoid in simple shear, as calculated by Jeffery (1922). Results of such calculations can be found in the papers by van de Ven (1975), Gierszewski and Chaffey (1978), Harris *et al.* (1979), and Hinch and Leal (1979).

The motion depends on three shape factors B_i ($i=1$–3) given by

$$B_i=\frac{r_i^2-1}{r_i^2+1} \tag{3.163}$$

where, for ellipsoids, $r_1=a_2/a_3$, $r_2=a_1/a_3$ and $r_3=a_1/a_2$. For other triaxial particles three equivalent axis ratios can be defined. For ellipsoids, $r_2/r_1=r_3$, but this is not necessarily true for all triaxial particles. The equations of motion can be written as

$$\frac{\mathrm{d}\theta}{\mathrm{d}t}=\tfrac{1}{4}B_3G\sin2\theta\sin2\phi-\tfrac{1}{2}(B_3-B_2)G\sin\theta\sin\psi\, f(\theta,\phi,\psi) \tag{3.164a}$$

$$\frac{\mathrm{d}\phi}{\mathrm{d}t}=\tfrac{1}{2}G(1+B_3\cos2\phi)+\tfrac{1}{2}(B_3-B_2)G\cos\psi\, f(\theta,\phi,\psi) \tag{3.164b}$$

$$\frac{\mathrm{d}\psi}{\mathrm{d}t}=-\tfrac{1}{2}G[B_3\cos\theta\cos2\phi+(B_3-B_2)\cos\theta\cos\psi\, f(\theta,\phi,\psi)$$
$$+B_1\{\tfrac{1}{2}\sin2\phi\sin2\psi\,(1+\cos^2\theta)-\cos\theta\cos2\phi\cos2\psi\}] \tag{3.164c}$$

where

$$f(\theta,\phi,\psi)=\cos\theta\sin2\phi\sin\psi-\cos2\phi\cos\psi \tag{3.164d}$$

We note that for spheroidal particles ($a_2=a_3$), $B_2=B_3$ and $B_1=0$, in which case eqs (3.164) reduce to the classical equations of Jeffery (eqs (3.86)).

In general, the motion of an ellipsoid in simple shear is non-periodic, except for three particular orientations in which one of the principal axes of the ellipsoid is aligned with the vorticity axis X_1 (see Fig. 3.26). In such a case,

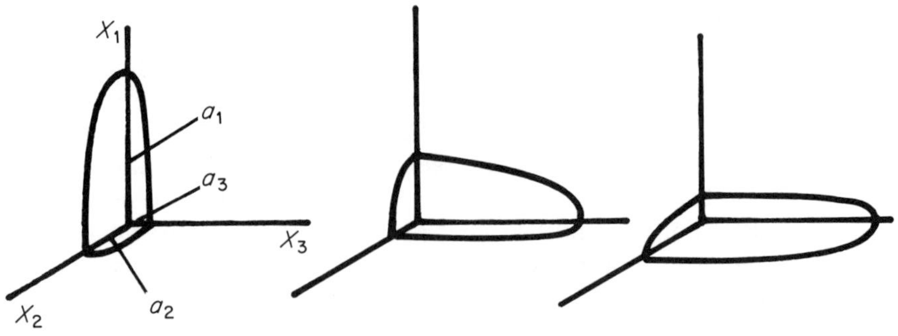

Fig. 3.26. Three orientations of an ellipsoid for which the rotation about the X_1 axis is periodic.

either $\theta = 0$, or $\theta = \pi/2$ and $\psi = 0$, or $\theta = \pi/2$ and $\psi = \pi/2$. The three orientations for which the motion is periodic are shown in Fig. 3.26 and their periods are given by

$$T = \frac{2\pi}{G}(r_i + r_i^{-1}) \tag{3.165}$$

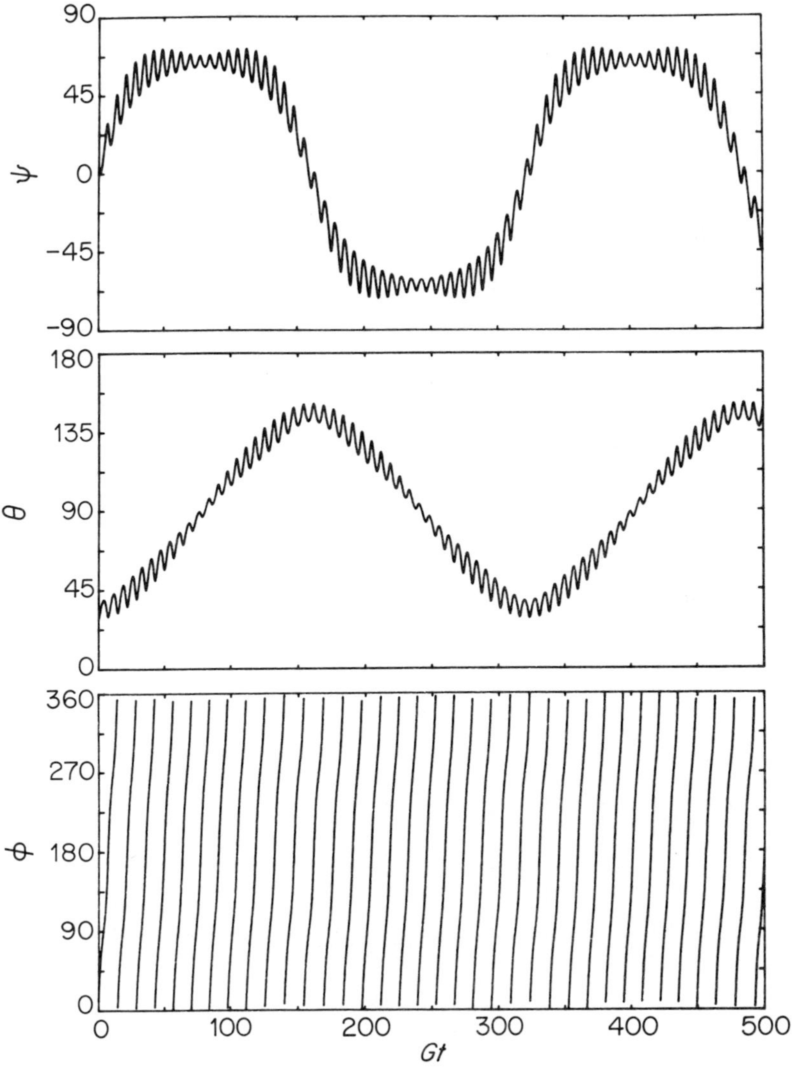

Fig. 3.27. Variations in ψ, θ and ϕ for an ellipsoid subjected to a simple shear flow. $a_1/a_2 = 1 \cdot 42$; $a_1/a_3 = 1 \cdot 88$.

For other orientations of the ellipsoid, the equations of motion must be integrated numerically. An example of variations in the three Euler angles with time is given in Fig. 3.27.

Despite the fact that the motion of an ellipsoid is not periodic, one can define either a time-averaged period of rotation for a single ellipsoid with a given initial orientation, or one can define an ensemble-averaged period of rotation for an ensemble with a given initial orientation distribution. An example of a distribution of periods for a single ellipsoid is given in Fig. 3.28. For the particular axis ratios chosen, the three periodic orbits have periods of $TG = 13 \cdot 334$, $17 \cdot 803$ and $23 \cdot 368$. Therefore, all the possible periods must be in the range $13 \cdot 334 \leqslant TG \leqslant 23 \cdot 368$. It can be seen from Fig. 3.28 that the ellipsoid does not sample all the possible orientations, but remains in a region of orientation space for which $15 \cdot 03 < TG < 18 \cdot 06$, with an average period of $\bar{T}G = 16 \cdot 55$. Because an average period can be assigned to each initial orientation, an ensemble average can also be assigned for a given distribution of orientations. For a random distribution and for particles not too far from axisymmetry, the average period is approximately given by (Jayasuriya and van de Ven, 1986a):

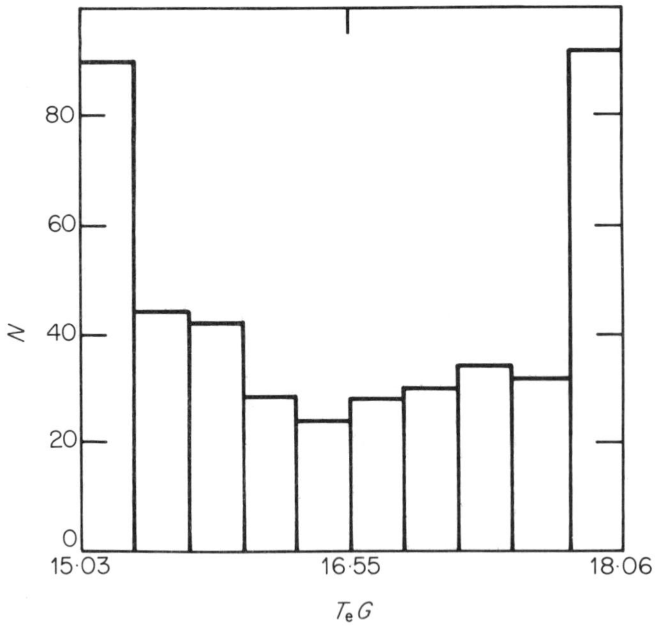

Fig. 3.28. Histogram of periods of rotation for 500 rotations executed by a single ellipsoid ($r_1 = 2 \cdot 420$, $r_2 = 3 \cdot 428$) having the initial orientation $\theta = 30°$, $\phi = 40°$, $\psi = 50°$ (after Jayasuriya and van de Ven, 1986a).

TABLE 3.2
Comparison between the rotary motion of spheroids and ellipsoids in simple shear

Spheroids	Ellipsoids
Periodic motion	Non-periodic motion
Period of rotation about vorticity axis independent of initial orientation	Distribution in *TG* dependent on initial orientation
Tip of spheroid traces out a closed orbit on a sphere of radius *a* around the origin	Orbit remains in a restricted region of orientation space

$$\bar{T}=\frac{2\pi}{G}(r_e+r_e^{-1}) \tag{3.166}$$

where

$$r_e=\frac{a_1}{\frac{1}{2}(a_2+a_3)}$$

A comparison between the motion of spheroids and ellipsoids is given in Table 3.2.

3.1.8 A drop in simple shear

A drop subjected to a simple shear behaves, in general, rather differently from a solid sphere. A drop in shear can deform and eventually, at high deformation, break up into smaller droplets. The theory for small deformations was first worked out by Taylor (1934) and was generalized by Cox (1969). The behaviour of the drop depends on two dimensionless parameters: the viscosity ratio, p, defined as

$$p=\frac{\mu_1}{\mu_0} \tag{3.167}$$

and the capillary number Ca, defined as

$$Ca=\frac{\mu_0 Ga}{\sigma} \tag{3.168}$$

where μ_1 is the viscosity of the drop, μ_0 of the suspending medium, G is the rate of shear, a the undeformed drop radius and σ the interfacial tension.

For drops of colloidal size $Ca \ll 1$. For example, for a particle of radius $a = 100$ nm in water $Ca \simeq 10^{-8}G$ seconds. As G is usually smaller than 10^4 s^{-1}, $Ca \ll 1$ in water. For drops of low viscosity suspended in high-viscosity oils (e.g. water droplets suspended in heavy oils), Ca can be much larger. For example, when $\mu_0 = 10$ Pa s, $\sigma \simeq 10$ mN m^{-1} and $a \simeq 1\mu$m, $Ca \simeq 10^{-3}G$ s, which becomes of order unity as $G \simeq 10^3$ s^{-1}. Hence for oil-in-water emulsions, $Ca \ll 1$. For water-in-oil emulsions $Ca \ll 1$, except when $p \gg 1$ and at high shear rates. It can be shown that for both conditions deformations of the drop are small, and the general theory by Cox (1969) for small deformations is adequate for drops of colloidal size. In essence, he solved the creeping-flow and continuity equations with the following boundary conditions:

(i) as $r \to \infty$, the flow field approaches the undisturbed shear field $v_1 = v_2 = 0$, $v_3 = Gx_2$;
(ii) the normal and tangential components of the flow inside and outside the drop are continuous at the drop surface;
(iii) the difference in the normal components of the stresses are balanced by interfacial-tension forces on the deformed surface, which is represented as an expansion in a small parameter, ϵ, about the undisturbed spherical shape.

Cox then solved the governing equations for a drop in a time-dependent shear flow in which the shear rate, $G(t)$, is a function of time. He considered

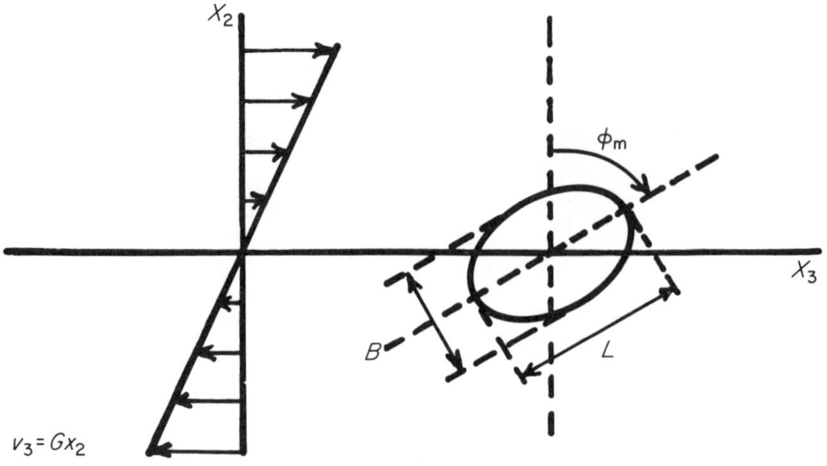

Fig. 3.29. Schematics of a deformed drop of major axis L and minor axis B in a simple shear of gradient G.

the special case of a drop in simple shear characterized by $G=0$, $t<0$ and $G=1$ for $t \geqslant 0$, corresponding to the onset of a sudden shearing motion with a constant gradient at time $t=0$. The drop deforms into a spheroid and, after going through a transient regime, it attains a steady-state orientation. The deformation, E, of the drop is defined by

$$E=\frac{L-B}{L+B}=\frac{r_e-1}{r_e+1} \tag{3.169}$$

where L and B are the major and minor axes respectively of the spheroidal drop and r_e is its axis ratio (see Fig. 3.29).

An example of variations in E and the angle of orientation ϕ_m is shown in Fig. 3.30. It can be seen that E and ϕ_m go through a series of damped

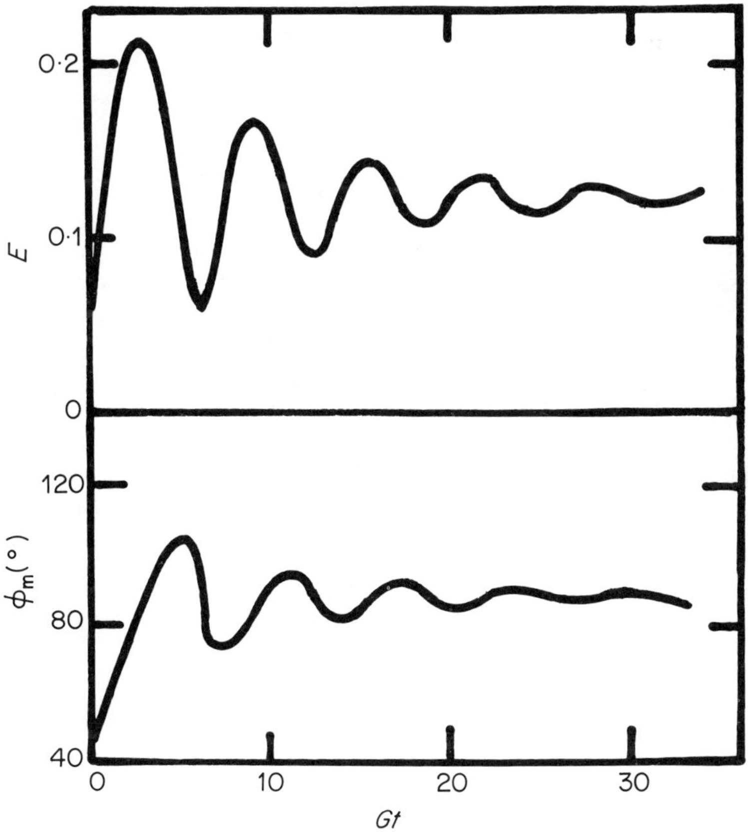

Fig. 3.30. Variations of ϕ_m and E with time for the case $p=10$, $Ca=1$ (after Cox, 1969).

oscillations prior to reaching the steady state. Such transient behaviour has been observed by Torza *et al.* (1971b). The steady-state deformation is given by

$$E = \frac{5(19p+16)}{4(p+1)[(20/Ca)^2+(19p)^2]^{1/2}} \qquad (3.170)$$

and

$$\phi_m = \frac{\pi}{4} + \tfrac{1}{2}\tan^{-1}\left[\frac{19}{20}\,p\,Ca\right] \qquad (3.171)$$

It can readily be seen that ϕ_m varies from 45°, when $Ca=0$ or $p=0$, to 90° when Ca or $p \to \infty$. When $Ca \ll 1$, $\phi_m = 45°$ and eq (3.170) reduces to

$$E = \frac{19p+16}{16(p+1)}\,Ca \qquad (3.172a)$$

When $p \gg 1$, $\phi_m = 90°$ and eq (3.170) reduces to

$$E = 5/4p \qquad (3.172b)$$

Equation (3.172a) applies to all oil-in-water emulsions, to water-in-oil emulsions at low rates of shear and water in low-viscosity oil emulsions. Equation (3.172b) applies to water in high-viscosity oil emulsions at high rates of shear. The full eqs (3.170) and (3.171) must be used for water in high-viscosity oil emulsions at intermediate shear rates. This is shown schematically in Fig. 3.31 which shows a plot of E versus p for various values of Ca.

The flow field around a deformed drop is rather complex, but in the limit of $E \to 0$, it becomes simple. It can be shown (Cox, 1969; Torza *et al.*, 1971b) that in this case the streamline trajectories outside the drop are given by

$$\frac{dr}{dt} = G\,A(r)\,\sin^2\theta\sin 2\phi \qquad (3.173a)$$

$$\frac{d\theta}{dt} = \tfrac{1}{4}G\,B(r)\,\sin 2\theta\sin 2\phi \qquad (3.173b)$$

$$\frac{d\phi}{dt} = \tfrac{1}{2}G\,[1 + B(r)\cos 2\phi] \qquad (3.173c)$$

where

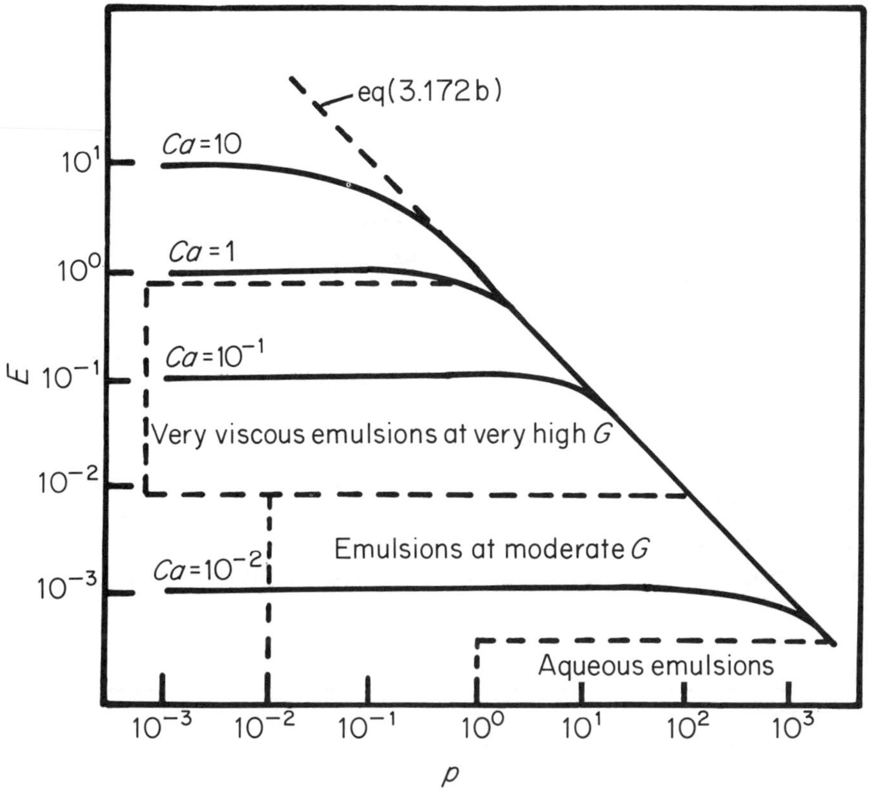

Fig. 3.31. Drop deformation, E, as a function of the viscosity ratio, p, for various capillary numbers, Ca. The regions of typical E and p values for various emulsions are shown.

$$A(r) = \frac{r}{2} - \frac{5p+2}{4(1+p)r^2} + \frac{3p}{4(1+p)r^4} \tag{3.174a}$$

$$B(r) = 1 - \frac{p}{(1+p)r^5} \tag{3.174b}$$

Notice that as $p \to \infty$ these equations reduce to those describing the streamlines around a solid sphere (eqs (3.8) and (3.9)). The flow inside the drop is given by the same set of equations, but with

$$A(r) = \frac{3r(r^2-1)}{4(1+p)} \tag{3.175a}$$

and

$$B(r) = \frac{5r^2 - 3}{2(1+p)} \tag{3.175b}$$

Particles exist that are intermediate between solid particles and fluid drops. An example is a human red blood cell. At high shear rates it deforms into a spheroid with a rotating membrane and attains a steady-state orientation, while at low shear rates it behaves as a rigid disk and rotates according to Jeffery's equations (Goldsmith and Marlow, 1972). For larger than colloidal-size drops, it is possible that break-up occurs at a critical number, Ca_c. Emulsion droplets usually cannot be broken because, for them, $Ca \ll Ca_c$. However, the value of Ca_c is of importance in two aspects: (i) in emulsification, where large drops are broken up into smaller emulsion droplets; and (ii) in the coalescence of unstable emulsion droplets, leading to larger drops. The coalescence proceeds until the drops are large enough to be broken apart and a dynamic equilibrium will be established between coalescence and break-up. The value of Ca_c has been determined experimentally by various people (Rallison, 1984); the results are presented in Fig. 3.32. It has

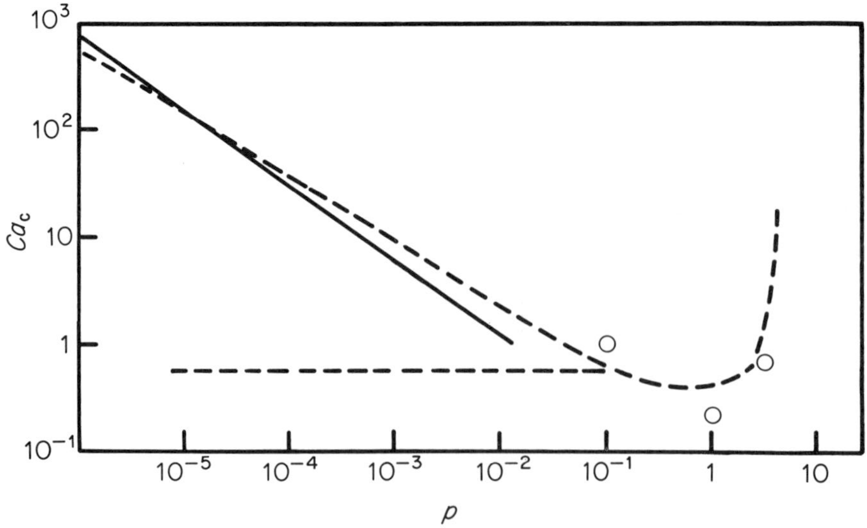

Fig. 3.32. Variation in Ca_c with p for simple shear flows. The dashed lines are experimental results from Grace (1971). Horizontal branches burst by tip streaming and the main branch by drop fracture. The solid line is from the asymptotic theory ($p \to 0$) of Hinch and Acrivos (1980). Open circles are the values derived using the theory of Barthes-Biesel and Acrivos (1973) (after Rallison, 1984).

been found experimentally (Rumscheidt and Mason, 1962), that drops can break up in several ways, depending on the value of the viscosity ratio, p. At low viscosity ratios, drops break up by a mechanism known as tip streaming. Small droplets are ejected from the pointed tips of the drop, which form at an advanced state of deformation. At intermediate values of p, drops break into two more-or-less equal fragments. At still higher viscosity ratios the drop deforms into a long thin thread which is subsequently broken up by a Rayleigh instability into a number of roughly equal-sized drops, together with a large number of much smaller satellite drops.

Drops containing surfactant molecules at their surface behave differently from pure drops. If the only effect of surfactants was to lower the interfacial tension of a drop, then two drops with identical p and Ca values, but one with surfactant and the other without, should deform in exactly the same way. It has been observed (Smith and van de Ven, 1985a) that this is not the case. An example of the break-up of a water drop with surfactant suspended in oil in simple shear is shown in Fig. 3.33. Surfactants affect the drops in two ways: (i) by changing the boundary conditions at the drop–fluid interface, and (ii) by Marangoni effects caused by an uneven distribution of surfactants which results from the fluid flow.

Little is known about the effects of double layers on drops subjected to

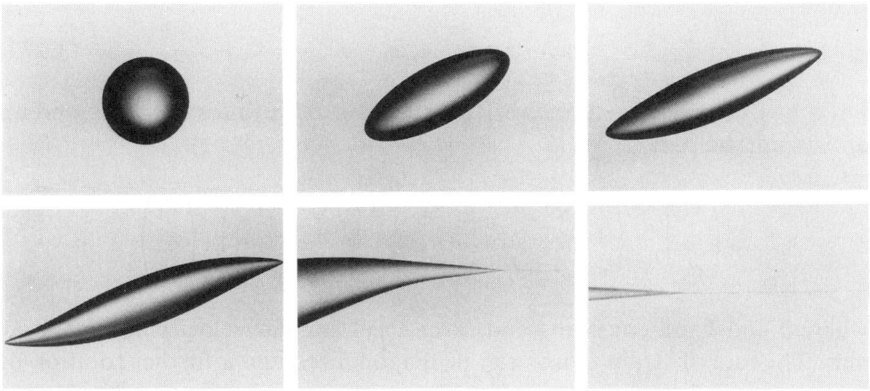

Fig. 3.33. Deformation and rupture of a drop of saturated TRS 1080 solution (mixed alkyl- and aryl-sulphonates) in silicone oil. Initial drop radius is 1·17 mm. Rupture occurs by the expulsion of extremely small droplets (ca. 10 μm) from the drop tips (last frame, barely visible). Rates of shear: $G = 0, 0{\cdot}14, 0{\cdot}18, 0{\cdot}35, 0{\cdot}42$, and $0{\cdot}49\ s^{-1}$ (after Smith and van de Ven, 1985a).

shear. As in the case of spheres and spheroids, one can predict that translational and rotary diffusion will be affected by double layers. In the absence of double layers, the description of the diffusion process is rather straightforward. Rotary Brownian motion will cause a spread about the orientation ϕ_m. Translational diffusion is rather similar to the diffusion of spheroidal particles, i.e. it will be given by the set of eqs (3.158). However, the orientational averages are different because the droplets do not rotate. For drops surrounded by double layers, the direction of the maximum ion density will usually not coincide with the direction of orientation ϕ_m. However, for most practical situations, both angles will be 45° when Pe $\ll$ 1 and $Ca \ll 1$. Water drops surrounded by oil have their electrical double layer inside the particle and the double layer will have little effect on the external motion of the drop.

3.1.9 Particles in general linear two-dimensional flows

3.1.9.1 Description of a linear flow field

The motion of particles in a simple shear flow is a special example of a much wider class of motions of particles, in general two-dimensional flows given by

$$v_1 = 0;\ v_2 = v_2(x_2, x_3);\ v_3 = v_3(x_2, x_3) \tag{3.176}$$

relative to a fixed Cartesian coordinate system. In the vicinity of a colloidal particle, the flow field can be expressed as

$$v_i = A_{ij} x_j \tag{3.177}$$

i.e. it can be considered linear. By separating A_{ij} into a symmetric and an antisymmetric part, eq (3.177) can be rewritten as

$$\begin{pmatrix} v_1 \\ v_2 \\ v_3 \end{pmatrix} = \begin{pmatrix} 0 & 0 & 0 \\ 0 & a & b \\ 0 & b & a \end{pmatrix} \cdot \begin{pmatrix} x_1 \\ x_2 \\ x_3 \end{pmatrix} + \begin{pmatrix} 0 & 0 & 0 \\ 0 & 0 & -\frac{1}{2}\omega \\ 0 & \frac{1}{2}\omega & 0 \end{pmatrix} \cdot \begin{pmatrix} x_1 \\ x_2 \\ x_3 \end{pmatrix} \tag{3.178}$$

where a and b are constants and ω the local angular velocity about the X_1 axis. The rate of strain tensor can be diagonalized and a further rotation of $\pi/4$ about X_1 yields

$$\begin{pmatrix} v_1 \\ v_2 \\ v_3 \end{pmatrix} = \begin{pmatrix} 0 & 0 & 0 \\ 0 & 0 & c \\ 0 & c & 0 \end{pmatrix} \cdot \begin{pmatrix} x_1 \\ x_2 \\ x_3 \end{pmatrix} + \begin{pmatrix} 0 & 0 & 0 \\ 0 & 0 & -\frac{1}{2}\omega \\ 0 & \frac{1}{2}\omega & 0 \end{pmatrix} \cdot \begin{pmatrix} x_1 \\ x_2 \\ x_3 \end{pmatrix} \tag{3.179}$$

or

$$\begin{pmatrix} v_1 \\ v_2 \\ v_3 \end{pmatrix} = G \begin{pmatrix} 0 & 0 & 0 \\ 0 & 0 & \gamma \\ 0 & 1 & 0 \end{pmatrix} \cdot \begin{pmatrix} x_1 \\ x_2 \\ x_3 \end{pmatrix} \tag{3.180}$$

It follows that any linear two-dimensional flow field can be described by two parameters: G, describing the strength of the flow, and γ, defining the type of flow. Without loss of generality one can choose $-1 \leqslant \gamma \leqslant 1$. Flows with values of γ outside this range can be transformed into eq (3.180) with $-1 \leqslant \gamma \leqslant 1$ using a suitable transformation. When $\gamma = 0$, the flow is a simple shear flow and G is the rate of shear. When $\gamma = -1$, the flow is pure rotational and when $\gamma = 1$ there is pure shear flow (or hyperbolic extensional flow). The types of flow corresponding to various values of γ are shown in Fig. 3.34a and photographs of the flow field are shown in Fig. 3.34b.

It follows from eq (3.180) that the streamlines of such a flow field are

$$x_2^2 - \gamma x_3^2 = \text{constant} \tag{3.181}$$

which for $\gamma > 0$ represents a family of hyperbolas with asymptotes making an angle α with the X_2 axis, given by

$$\tan \alpha = \pm \gamma^{-1/2} \tag{3.182}$$

When $\gamma < 0$, eq (3.181) represents a family of ellipses with an axis ratio, r_e, given by

$$r_e = (-\gamma)^{-1/2} \tag{3.183}$$

3.1.9.2 A sphere in a linear flow field

The flow field around a sphere is governed by the creeping-flow and the continuity equations

$$\mu \nabla^2 v = \nabla p \tag{3.184}$$

$$\nabla \cdot v = 0 \tag{3.185}$$

with the usual no-slip boundary condition at the surface

$$v_1 = 0;\ v_2 = -\omega x_3;\ v_3 = \omega x_2 \text{ as } r \to 1 \tag{3.186}$$

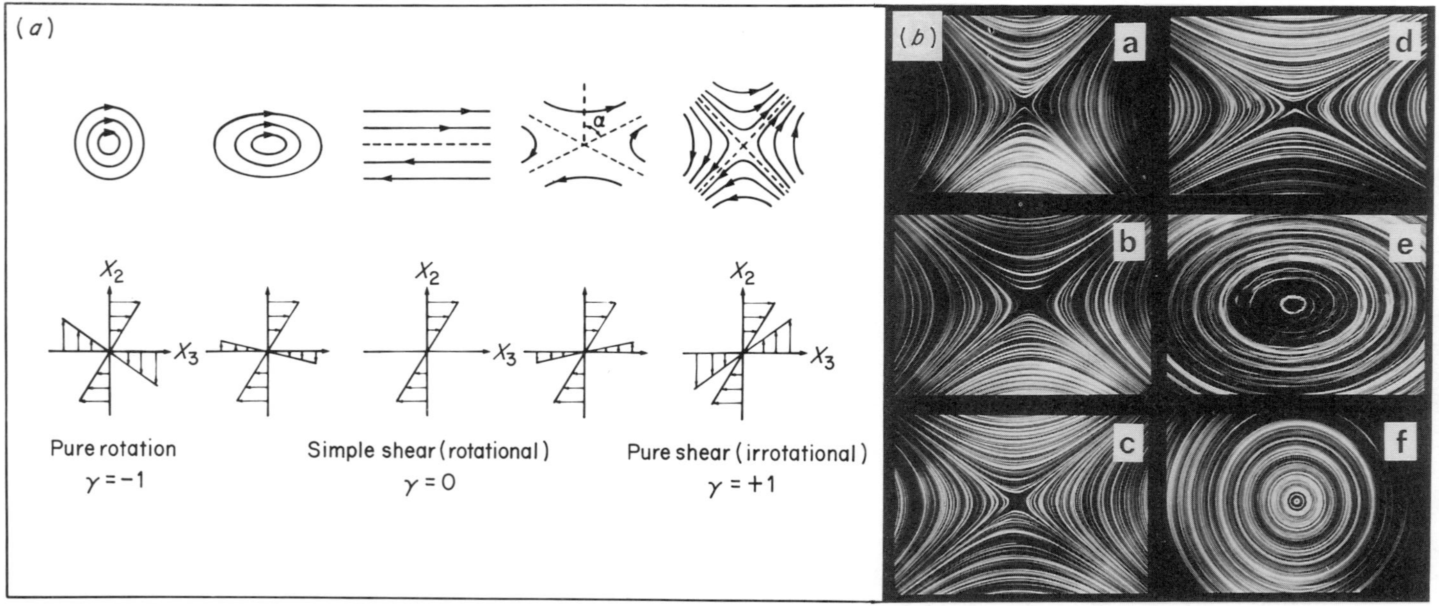

Fig. 3.34. (*a*) General two-dimensional linear flows characterized by the parameter γ which can vary from -1 (pure rotation) to $+1$ (pure shear). (*b*) Examples of streamlines for various values of γ varying from $+1$ in (a) to -1 in (f).

and the requirement of an undisturbed flow far from the sphere:

$$v_1 = 0;\ v_2 = G\gamma x_3;\ v_3 = Gx_2 \text{ as } r \to \infty \tag{3.187}$$

Here, ω is the angular velocity about the X_1 axis and r the distance from the centre, together with other distances rendered dimensionless by the sphere radius, a. The solution of eqs (3.184) and (3.185), subjected to eqs (3.186) and (3.187) has been given by Kao *et al.* (1977):

$$v_i = \epsilon_{ijk}\omega_j x_k + x_j S_{ij} - x_k S_{ik}\left[\frac{5}{2}\left(\frac{r^2-1}{r^7}\right) x_i x_j + \frac{\delta_{ij}}{r^5}\right] \tag{3.188}$$

where ϵ_{ijk}, δ_{ij}, S_{ij} and ω_i are the components of the alternating tensor, the Kronecker delta, the rate of strain tensor and the fluid angular velocity ($\omega = \frac{1}{2}\nabla \times v$), respectively. The solution represented by eq (3.188) can be written in spherical polar coordinates as

$$\begin{pmatrix} v_r \\ v_\theta \\ v_\phi \end{pmatrix} = \begin{pmatrix} \mathrm{d}r/\mathrm{d}t \\ r\,\mathrm{d}\theta/\mathrm{d}t \\ r\sin\theta\,\mathrm{d}\phi/\mathrm{d}t \end{pmatrix}$$

with

$$\frac{\mathrm{d}r}{\mathrm{d}t} = G(1+\gamma)\,A(r)\,\sin^2\theta\sin 2\phi \tag{3.189a}$$

$$\frac{\mathrm{d}\theta}{\mathrm{d}t} = \tfrac{1}{4}G(1+\gamma)\,B(r)\,\sin 2\theta\sin 2\phi \tag{3.189b}$$

$$\frac{\mathrm{d}\phi}{\mathrm{d}t} = \tfrac{1}{2}G[(1-\gamma) + B(r)(1+\gamma)\cos 2\phi] \tag{3.189c}$$

where the functions $A(r)$ and $B(r)$ are the same as those defined for the flow around a sphere in simple shear (cf. eqs (3.8) and (3.9)).

From the condition of zero torque it can be shown that

$$\omega = \tfrac{1}{2}G(1-\gamma) \tag{3.190}$$

When $\gamma = 0$, the results reduce to those for the flow around a sphere in simple shear. When $\gamma = -1$ (pure rotational flow), the angular velocity is twice that of a sphere in simple shear. When $\gamma = 1$ (pure strain), the angular velocity is zero.

As for the equations for simple shear flow, eqs (3.189) can be integrated to give

TABLE 3.3
Summary of streamline behaviour around a single sphere (after Kao *et al.*, 1977)

	$E < -\beta g(1)$	$-\beta g(1) < E < E_0$	$E_0 < E < 0$	$0 < E$
$\gamma = 0$	No streamlines are defined	(i) No open streamlines exist; (ii) as for $0 < \gamma < 1$, except that closed streamline region extends to infinity	$(E_0 = 0)$	As for $0 < \gamma < 1$
$0 < \gamma < 1$	Open streamline does not cut $\phi = \alpha$; r_{min} occurs at $\phi = \pi/2$	(i) Open streamline does not cut $\phi = \alpha$; r_{min} occurs at $\phi = \pi/2$; (ii) closed streamline cuts $\phi = \alpha$; r_{min} occurs at $\phi = 0$; r_{max} occurs at $\phi = \pi/2$; closed streamline region is finite	Open streamline cuts $\phi = \alpha$; r_{min} occurs at $\phi = 0$	Open streamline does not cut $\phi = \alpha$; r_{min} is at $\phi = 0$
$\gamma = 1$	As for $0 < \gamma < 1$	$(-\beta g(1) = 0)$		As for $0 < \gamma < 1$

$$x_1 = Crf(r) \tag{3.191}$$

$$x_2 = \pm r f(r)\left[E + \frac{\gamma f(r)^{-2}}{1+\gamma} + \frac{1-\gamma}{1+\gamma}\, g(r)\right]^{1/2} \tag{3.192}$$

with $f(r)$ and $g(r)$ defined as before (eqs (3.11) and (3.13)). The streamlines are, in general, three-dimensional, but when $C=0$ they are all confined to the X_2X_3 plane. It follows from eq (3.192) that E is restricted to values for which

$$0 \leqslant E + \frac{\gamma f(r)^{-2}}{1+\gamma} + \frac{1-\gamma}{1+\gamma}\, g(r) \leqslant f(r)^{-2} \tag{3.193}$$

It can be shown from this condition that when $-1 \leqslant \gamma < 0$, all streamlines are closed and no streamlines are defined when

$$E < -\frac{1-\gamma}{1+\gamma}\, g(1)$$

When $0 < \gamma \leqslant 1$, a variety of streamlines are possible. These possibilities are summarized in Table 3.3. In the table, E_0 is defined as

$$E_0 = -\frac{\gamma f(r_0)^{-2}}{1+\gamma} - \frac{1-\gamma}{1+\gamma}\, g(r_0)$$

with $r_0 = [(1+\gamma)/2\gamma]^{1/5}$; and β is defined as $\beta = (1-\gamma)/(1+\gamma)$. An example of streamlines for $\gamma = 1/3$ is given in Fig. 3.35.

3.1.9.3 An ellipsoidal particle in a linear flow field

The problem of the flow field around an ellipsoidal particle with semi-axes a_1, a_2 and a_3, arbitrarily oriented in a general two-dimensional flow, has been solved by Jeffery (1922). As can be seen from the special case of the flow field around a spheroid in simple shear (eq (3.84)), this flow field is rather complicated. For details the reader is referred to the original paper. From the condition of zero net torque on the ellipsoid, it can be shown that the particle rotates about its three symmetry axes with angular velocities, ω_i, given by eq (3.85) with (Zuck and van de Ven, 1989)

$$\frac{d\theta}{dt} = G(1+\gamma)\,[\tfrac{1}{4}B_3 \sin 2\theta \sin 2\phi$$

$$-\tfrac{1}{2}(B_3 - B_2)\sin\theta \sin\psi\, f(\theta,\phi,\psi)] \tag{3.194a}$$

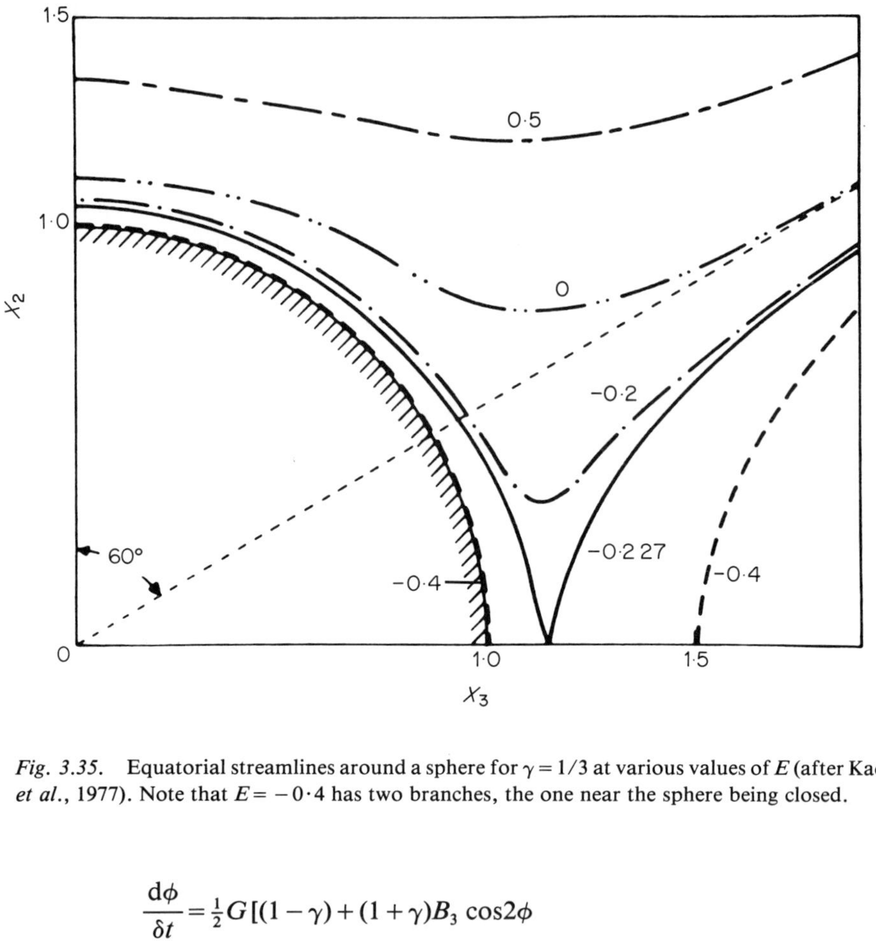

Fig. 3.35. Equatorial streamlines around a sphere for $\gamma = 1/3$ at various values of E (after Kao *et al.*, 1977). Note that $E = -0{\cdot}4$ has two branches, the one near the sphere being closed.

$$\frac{\mathrm{d}\phi}{\delta t} = \tfrac{1}{2}G[(1-\gamma) + (1+\gamma)B_3\cos 2\phi + (1+\gamma)(B_3 - B_2)\cos\psi\, f(\theta,\phi,\psi) \tag{3.194b}$$

$$\frac{\mathrm{d}\psi}{\mathrm{d}t} = -\tfrac{1}{2}G(1+\gamma)\,[B_3\cos\theta\cos 2\phi + B_1\{\tfrac{1}{2}\sin 2\phi\sin 2\psi \times (1+\cos^2\theta) - \cos\theta\cos 2\phi\cos\psi\} + (B_3 - B_2) \times \cos\theta\cos\psi\, f(\theta,\phi,\psi)] \tag{3.194c}$$

where $f(\theta,\phi,\psi)$ is given by eq (3.164d) and the B_i values are the shape factors defined by eq (3.163).

When $\gamma = 0$, these equations reduce to the equations of motion of an ellipsoidal particle in simple shear (eqs (3.164)). For spheroids, $a_1 = a$ and $a_2 = a_3 = b$ and, therefore, $B_1 = 0$ and $B_2 = B_3 = B = (r_e^2 - 1)/(r_e^2 + 1)$, where

$r_e = a/b$ is the axis ratio of the spheroid. In this case the set of eqs (3.194) reduces to

$$\frac{d\theta}{dt} = \tfrac{1}{4}BG\,(1+\gamma)\,\sin 2\theta \sin 2\phi \tag{3.195a}$$

$$\frac{d\phi}{dt} = \tfrac{1}{2}G\,[(1-\gamma)+(1+\gamma)\,B\cos 2\phi] \tag{3.195b}$$

$$\frac{d\psi}{dt} = -\tfrac{1}{2}BG\,(1+\gamma)\,\cos\theta \cos 2\phi \tag{3.195c}$$

When $\gamma = 0$, this set of equations, in turn, reduces to the classical Jeffery's equations for the rotary motion of an axisymmetric particle in simple shear (eqs (3.86)). The rotational behaviour of spheroids in two-dimensional flow depends on whether or not the flow parameter γ is larger or smaller than a critical value, γ_{crit}, given by

$$\gamma_{crit} = \frac{1-|B|}{1+|B|} \tag{3.196}$$

or

$$\gamma_{crit} = r_e^{-2} \qquad \text{if } r_e > 1$$

$$\gamma_{crit} = r_e^{2} \qquad \text{if } r_e < 1$$

For $\gamma < \gamma_{crit}$ (subcritical flows), the particle rotates periodically about the X_1 axis. The period can be found by integrating eqs (3.195) which leads to

$$\tan\phi = s\tan\tau \tag{3.197a}$$

$$\tan\theta = C(s^2\sin^2\tau + \cos^2\tau)^{1/2} \tag{3.197b}$$

where

$$s = \frac{r_e^2 - \gamma}{r_e\sqrt{(\gamma - r_e^2)(\gamma - r_e^{-2})}} \tag{3.198}$$

and $\tau = 2\pi t/T$, where T is the period of rotation given by

$$T = \frac{2\pi}{G}\,\frac{s+s^{-1}}{1-\gamma} \tag{3.199}$$

It can be seen that a spheroid in a subcritical linear flow field behaves as a spheroid in a simple shear flow of shear rate $(1-\gamma)G$ and of equivalent axis ratio, s. Indeed, if one defines $G^* = (1-\gamma)G$ and $B^* = (s^2-1)/(s^2+1)$, the set of eqs (3.195) reduces to (Brenner, 1974)

$$\frac{d\theta}{dt} = \tfrac{1}{4}B^*G^* \sin 2\theta \sin 2\phi \tag{3.200a}$$

$$\frac{d\phi}{dt} = \tfrac{1}{2}G^* \left(1 + B^* \cos 2\phi\right) \tag{3.200b}$$

$$\frac{d\psi}{dt} = -\tfrac{1}{2}B^*G^* \cos\theta \cos 2\phi \tag{3.200c}$$

For a simple shear, $-1 < B < 1$ and a spheroid rotates periodically. For a spheroid in a general two-dimensional flow, B^* can take any value and the spheroid rotates only for values in the range $-1 < B^* < 1$. This condition is equivalent to eq (3.196); as soon as $B^* = \pm 1$, then $\gamma = \gamma_{crit}$. For $\gamma > \gamma_{crit}$ (so-called supercritical fields), the particle attains a steady-state orientation with a relaxation time τ. Integrating eqs (3.195) for supercritical fields yields

$$\tan\phi = \frac{(1+B^*)\,(De^{t/\tau}+1)}{\sqrt{B^{*2}-1}\,(De^{t/\tau}-1)} \tag{3.201a}$$

$$\tan\theta = \frac{E}{(1+B^* \cos 2\phi)^{1/2}} \tag{3.201b}$$

where D and E are integration constants and

$$\tau = \frac{1}{G\sqrt{B^{*2}-1}} \tag{3.202}$$

As, $t \to \infty$, the particle attains the orientation

$$\theta_\infty = \pi/2$$

$$\cos 2\phi_\infty = \frac{\gamma-1}{B(\gamma+1)} \tag{3.203}$$

For disks ($r_e < 1$) $90° < \phi_\infty < 180°$, while for rod-like particles $0 < \phi_\infty < 90°$.

For ellipsoidal particles no analytical solutions of the equations of motion exist, except for the three orientations shown in Fig. 3.26. In these orientations the ellipsoids rotate periodically with a period given by

$$T=\frac{2\pi}{G}\frac{s_i+s_i^{-1}}{1-\gamma} \tag{3.204}$$

where s_i is given by eq (3.198) with r_e replaced by r_i. For other orientations the ellipsoid executes quasi-periodic rotations, similar to the motion in a simple shear, provided the flow field is subcritical. For ellipsoidal particles three critical values of the flow parameter, γ, can be defined:

$$\gamma_i=\frac{1-B_i}{1+B_i}$$

If the critical γ values are ordered in such a way that $\gamma_1 < \gamma_2 < \gamma_3$, it can be shown (Zuck and van de Ven, 1989) that for $\gamma<\gamma_1$ there are no possible steady-state orientations, for $\gamma_1 < \gamma < \gamma_2$ there are two such orientations, for $\gamma_2 < \gamma < \gamma_3$ there are four, and for $\gamma > \gamma_3$ there are six such orientations. It can be shown that of all these possible orientations only one is stable, the stable orientation given by eq (3.203) with γ given by γ_1 and B by the corresponding B_i. This means that the steady-state orientation of an ellipsoid is the same as that of a spheroid having the longest and the shortest axes of the ellipsoid as the major and minor axes.

3.1.9.4 A drop in a linear flow field

Analogous to drops in simple shear, the flow field inside and outside a drop in a general two-dimensional flow can be readily found from the creeping-flow equation and the relevant boundary conditions. The results for an undeformed drop are (Powell, 1983)

$$\frac{\mathrm{d}r}{\mathrm{d}t}=(1+\gamma)\; G\, A(r)\; \sin^2\theta\sin2\phi \tag{3.205a}$$

$$\frac{\mathrm{d}\theta}{\mathrm{d}t}=\tfrac{1}{4}(1+\gamma)\; G\, B(r)\; \sin2\theta\sin2\phi \tag{3.205b}$$

$$\frac{\mathrm{d}\phi}{\mathrm{d}t}=\tfrac{1}{2}G\; [(1-\gamma)+(1+\gamma)\; B(r)\; \cos2\phi] \tag{3.205c}$$

where for the flow field outside the drop $A(r)$ and $B(r)$ are given by eqs (3.174), while for the flow field inside the drop $A(r)$ and $B(r)$ are given by eqs (3.175).

Analogous to spheroids in linear flows (eqs (3.195)), we can define the functions

$$B(r)^* = B(r)\,\frac{1+\gamma}{1-\gamma} \quad \text{and} \quad A(r)^* = A(r)\,\frac{1+\gamma}{1-\gamma} \tag{3.206}$$

and an effective shear rate $G^* = (1-\gamma)G$, in which case eqs (3.205) reduce to

$$\frac{\mathrm{d}r}{\mathrm{d}t} = G^*\, A(r)^*\, \sin^2\theta \sin 2\phi \tag{3.207a}$$

$$\frac{\mathrm{d}\theta}{\mathrm{d}t} = \tfrac{1}{4} G^*\, B(r)^*\, \sin 2\theta \sin 2\phi \tag{3.207b}$$

$$\frac{\mathrm{d}\phi}{\mathrm{d}t} = \tfrac{1}{2} G^*\, [1 + B(r)^* \cos 2\phi] \tag{3.207c}$$

These sets of equations can be integrated in the usual way, resulting in eqs (3.191) and (3.192), but $f(r)$ and $g(r)$ are now defined as

$$f(r,p) = \left(r^3 - \frac{5p+2}{2(1+p)} + \frac{3p}{2(1+p)r^2}\right)^{-1/3} \tag{3.208}$$

and

$$g(r,p) = \frac{p}{1+p} \int_p^\infty y^{-3} f(y,p)\,\mathrm{d}y \tag{3.209}$$

As $p \to \infty$ these functions reduce to eqs (3.11) and (3.13) for a solid sphere.

As in the case of streamlines around a solid sphere in a linear flow field, the variety of streamlines in and around the drop shows a rich pattern of possibilities. An example of streamlines inside a drop is given in Fig. 3.36.

When the capillary number, Ca, is large enough for drop deformation to occur, the drop will, after the onset of the flow, undergo damped oscillations in shape, similar to a drop in shear flow, and attain a steady-state orientation given by (cf. eq (3.171))

$$\phi_{\mathrm{m}} = \frac{\pi}{4} + \tfrac{1}{2}\tan^{-1}\left[\frac{19}{20}\,(1-\gamma)\,pCa\right] \tag{3.210}$$

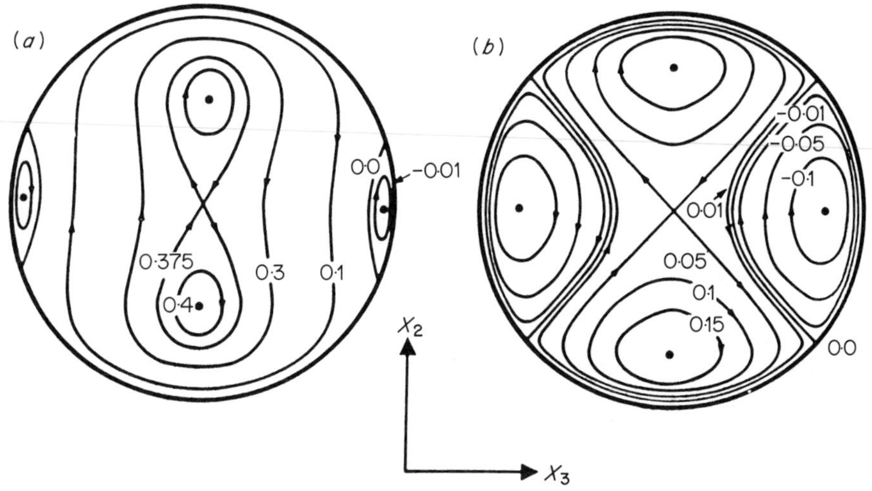

Fig. 3.36. Circulation inside a drop for $\gamma > 0$.

(*a*) $\lambda = p\left(\dfrac{1-\gamma}{1+\gamma}\right) - \dfrac{2\gamma}{1+\gamma} = -0\cdot 25$.

(*b*) $\lambda = -0\cdot 9$. The numbers inside the drops refer to values of the trajectory constant E (after Powell, 1983 with permission).

The extension or deformation of the drop in the steady-state equals (cf. eq (3.170))

$$E = \frac{5(19p+16)(1-\gamma)}{4(p+1)[(20/Ca)^2 + \{19(1-\gamma)p\}^2]^{1/2}} \tag{3.211}$$

It can be shown (Powell, 1983) that the steady-state deformation is reached with a relaxation time, τ, given by

$$\tau = \frac{p\,Ca}{G} \tag{3.212}$$

3.1.9.5 Diffusion in linear flows

The general problem of the diffusion of arbitrarily shaped particles in linear flows is complex. Some analytical results are available for the rotary motion of ellipsoidal particles, such as calculations of steady-state orientation distribution functions in simple shear (Workman and Hollingsworth, 1969). These could, in principle, be extended to general two-dimensional flows.

Another approach to the diffusion of ellipsoidal particles is to use dynamic simulation methods of the type described in Section 3.1.6.1(c). For ellipsoids, there are three rotary diffusion coefficients to consider for rotation about their three principal axes X_i'. Denoting the angles through which these axes rotate by α_i, we must have

$$\langle \Delta\alpha_i^2 \rangle = 2D_i \Delta t \tag{3.213}$$

For a given time interval Δt, random changes in the angles, $\Delta\alpha_i$, can be generated numerically using random-number generating techniques, which satisfy Gaussian statistics; i.e. the angles must be chosen in such a way that the distribution of the angles generated is Gaussian with a mean $\Delta\alpha_i$ of 0 and a standard deviation of $\sigma = (2D_i\Delta t)^{1/2}$. The randomly generated values of $\Delta\alpha_i$ can be transformed into random variations in the three Euler angles $\Delta\theta$, $\Delta\phi$ and $\Delta\psi$ using eq (3.85), with the result (Jayasuriya and van de Ven 1986b):

$$\begin{aligned}
\Delta\theta &= (\Delta\alpha_3 \sin\psi - \Delta\alpha_2 \cos\psi)/\sin\theta \\
\Delta\phi &= \Delta\alpha_3 \cos\psi + \Delta\alpha_2 \sin\psi \\
\Delta\psi &= \Delta\alpha_1 - \Delta\theta \cos\theta
\end{aligned} \tag{3.214}$$

These random displacements can be added to the deterministic displacements due to fluid flow

$$\begin{aligned}
\Delta\theta &= \dot{\theta}_0 \Delta t + (\Delta\theta)_r \\
\Delta\phi &= \dot{\phi}_0 \Delta t + (\Delta\phi)_r \\
\Delta\psi &= \dot{\psi}_0 \Delta t + (\Delta\psi)_r
\end{aligned} \tag{3.215}$$

where $\dot{\theta}_0$, $\dot{\phi}_0$ and $\dot{\psi}_0$ are the angular velocities given by eqs (3.194) and the subscript "r" refers to the random displacements given in eqs (3.214). The set of eqs (3.215) can be integrated numerically and stochastic trajectories in orientational space can be generated. By repeating this process many times the orientation distribution can be calculated as a function of time, from which all the relevant properties of the suspension can be calculated.

For spheroidal particles, the problem simplifies tremendously. Because the rotary motion of spheroids in linear flows is similar to that in simple shear (cf. eqs (3.200)), the rotary diffusion is identical to that of spheroids in simple shear, as discussed in Section 3.1.6. The only difference is that the shape parameter, B, must be replaced by B^* and G by G^*. This means that the rotary Péclet number must be redefined as

$$\mathrm{Pe}_r = \frac{(1-\gamma)G}{D_r} \tag{3.216}$$

All the analytical solutions discussed in Section 3.1.6 also hold for spheroids in linear flows. Numerical solutions of the rotary convective-diffusion equations (eqs (3.102) and (3.103)) are, unfortunately, only available for B values in the range $-1 < B < 1$.

Translational diffusion in linear flows has only been worked out for spheres (Foister and van de Ven, 1980), although in principle this approach could be extended to spheroids, similar to diffusion in simple shear. To describe the diffusion in linear flow fields one must solve the convective-diffusion equation

$$\frac{\partial n}{\partial t} = D\,\nabla^2 n - \nabla\cdot(\boldsymbol{u}n) \tag{3.217}$$

with the velocity $\boldsymbol{u}$ given by

$$u_1 = 0;\; u_2 = \gamma G x_3;\; u_3 = G x_2 \tag{3.218}$$

The general solution of eq (3.217) is

$$n(x_2, x_3, t) = \frac{\gamma}{2\pi}\left(\frac{2G}{Dg_1(t)}\right)^{1/2} \exp\left[-\frac{\gamma^{1/2}}{2g_1(t)}\left\{g_2(t)\,x_2^2 + g_3(t)x_3^2 + g_2(t)x_2x_3\right\}\right] \tag{3.219}$$

where

$$g_1(t) = DG^{-1}[(\gamma+1)^2\,\{\cosh(2\gamma^{1/2}Gt) - 1\} - 2\gamma(\gamma-1)^2\,G^2t^2]$$

$$g_2(t) = (\gamma+1)\,\sinh\,(2\gamma^{1/2}Gt) - 2\gamma^{1/2}(1-\gamma)Gt$$

$$g_3(t) = \gamma\,[(\gamma+1)\,\sinh\,(2\gamma^{1/2}Gt) + 2\gamma^{1/2}(1-\gamma)Gt]$$

$$g_4(t) = 2\gamma^{1/2}(\gamma+1)\,[\cosh(2\gamma^{1/2}Gt) - 1]$$

From eq (3.219), the first moments of the distribution can be found:

$$\langle x_i\, x_j\rangle = \int_{-\infty}^{\infty}\int_{-\infty}^{\infty} x_i x_j n\,(x_2, x_3, t)\,\mathrm{d}x_2\mathrm{d}x_3 \tag{3.220}$$

with the result

$$\langle x_i x_j \rangle = \begin{pmatrix} 2Dt & 0 & 0 \\ 0 & \langle x_2^2 \rangle & \langle x_2 x_3 \rangle \\ 0 & \langle x_2 x_3 \rangle & \langle x_3^2 \rangle \end{pmatrix} \tag{3.221}$$

with

$$\langle x_2^2 \rangle = \frac{\gamma+1}{2\gamma^{1/2}G} D \sinh(2\gamma^{1/2}Gt) - (\gamma-1) Dt \tag{3.222a}$$

$$\langle x_2 x_3 \rangle = \frac{\gamma+1}{2\gamma G} D [\cosh(2\gamma^{1/2}Gt) - 1] \tag{3.222b}$$

$$\langle x_3^2 \rangle = \frac{\gamma+1}{2\gamma^{3/2}G} D \sinh(2\gamma^{1/2}Gt) - \frac{\gamma-1}{\gamma} Dt \tag{3.222c}$$

It can readily be shown that when $\gamma = 0$, eqs (3.221) and (3.222) reduce to eq (3.54).

When $\gamma = -1$ (pure rotation), the equations reduce to

$$\langle x_2^2 \rangle = \langle x_3^2 \rangle = 2Dt; \ \langle x_2 x_3 \rangle = 0 \tag{3.223}$$

and the distribution eq (3.219) reduces to a simple Gaussian distribution. When $\gamma = 1$ (pure elongation), the equations reduce to

$$\langle x_2^2 \rangle = \langle x_3^2 \rangle = \frac{D}{G} \sinh(2Gt) \tag{3.224a}$$

and

$$\langle x_2 x_3 \rangle = \frac{D}{G} [\cosh(2Gt) - 1] \tag{3.224b}$$

As is well known, simple shear ($\gamma = 0$) is a superposition of pure rotational and pure elongational flow (see Chapter 1, Section 1.3.4.2), yet the diffusive convective coupling is not simply additive. In fact the rotational component in simple shear acts to retard convective enhancement of diffusion relative to pure elongation. This is illustrated graphically in polar plots of the probabilities $p(\phi)\mathrm{d}\phi$ of finding a particle between the angles ϕ and $\phi + \mathrm{d}\phi$, examples of which are given in Fig. 3.37 for $\gamma = 0$ and $\gamma = 1$. Transforming eq (3.219) into polar coordinates and performing the integration in eq (3.59) results in eq (3.60) for $\gamma = 0$, while for $\gamma = 1$ it results in

$$p(\phi) = \frac{[\cosh(2Gt) - 1]^{1/2}}{2^{1/2}\pi \sinh(2Gt) - [\cosh(2Gt) - 1] \sin 2\phi} \tag{3.225}$$

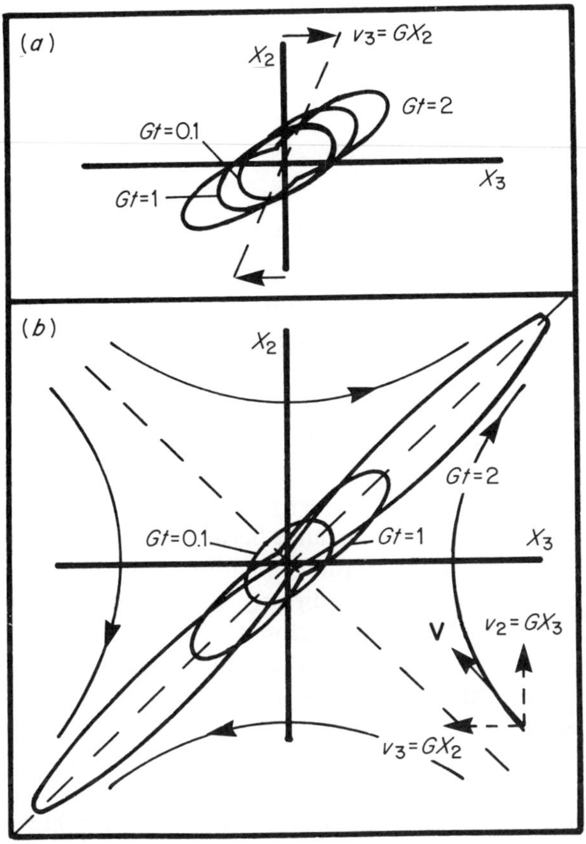

Fig. 3.37. Orientation distribution functions $p(\phi)$ (*a*) for simple shear and (*b*) pure shear or elongation for various dimensionless times Gt. For particles at the origin at $t=0$, the probability of finding the particle between ϕ and $\phi + d\phi$ at time t is $p(\phi)d\phi$. Note that in pure shear the orientation distribution is much more stretched than it is in simple shear and that its maximum always points in the same direction; this is due to the absence of a rotational component in such a flow (after Foister and van de Ven, 1980).

The results for the second moments are only valid in the diffusion limit $t \gg \beta^{-1}$ (see Section 2.1.3.1). The second moments which have been derived from a solution of the convective-diffusion equation, can also be obtained directly from the Langevin equation, which is valid for all values of t. The Langevin analysis also shows that for short times $t \ll \beta^{-1}$, there is coupling between the Brownian motion and the shear, as is evident from the second moments at short times, e.g. (Foister and van de Ven, 1980) when $t \ll \beta^{-1}$

$$\langle x_3^2 \rangle = \frac{kT}{m} t^2 + \frac{kT}{4m} G^2 t^4 \tag{3.226}$$

which clearly indicates that at short times the flow also contributes to the displacement.

3.1.10 Particles in Poiseuille flow

3.1.10.1 General considerations

For most colloidal dispersions subjected to Poiseuille flow, the diameter of the particles will be much smaller than the diameter of the tube or channel. In the vicinity of the particle, the flow resembles a simple shear flow with gradient

$$G = 2v_{\mathrm{m}} r / R^2 \tag{3.227}$$

where v_{m} is the maximum fluid velocity at the centre of the tube, R is the radius of the tube, and r is the distance between the particle centre and the tube axis. If follows that everything mentioned in the previous sections concerning simple shear applies directly to motion in Poiseuille flow, provided that the particle is not too close to the wall or to the centre of the tube. When the particle is close to the wall, the particle–wall interaction becomes important; this will be dealt with in Chapter 6. When a particle is close to the centre, the effects of the parabolic nature of the flow become important. If we consider the translational diffusion of a particle, then, no matter where it starts, after sufficient time the particle will reach either the wall or the centre of the tube and deviations from the behaviour in simple shear will occur. In principle, the behaviour of Brownian particles in Poiseuille flow is described by the convective-diffusion equation, but no general analytical solution of this equation presently exists. However, it is possible to calculate the moments of the distribution without actually solving the equation. This can be done by solving the appropriate Langevin equation, or by multiplying the convective-diffusion equation by the appropriate moment and integrating it over all space. Asymptotic results can also be obtained, which are valid either far from or close to the tube centre.

3.1.10.2 Solution of the Langevin equation

Consider a spherical particle of radius a in a Poiseuille flow given by

$$v = v_{\mathrm{m}} \left[1 - \left(\frac{r}{R} \right)^2 \right] \tag{3.228}$$

Let the particle initially (at time $t=0$) be at point (x_0, y_0, z_0) (see Fig. 3.38). It is convenient to define a coordinate system, x_i, that moves with the inital velocity, v_0, of the particle

$$v_0 = v_m \left(1 - \frac{x_0^2 + y_0^2}{R^2}\right) \tag{3.229}$$

and is related to the space-fixed coordinates x, y, z by

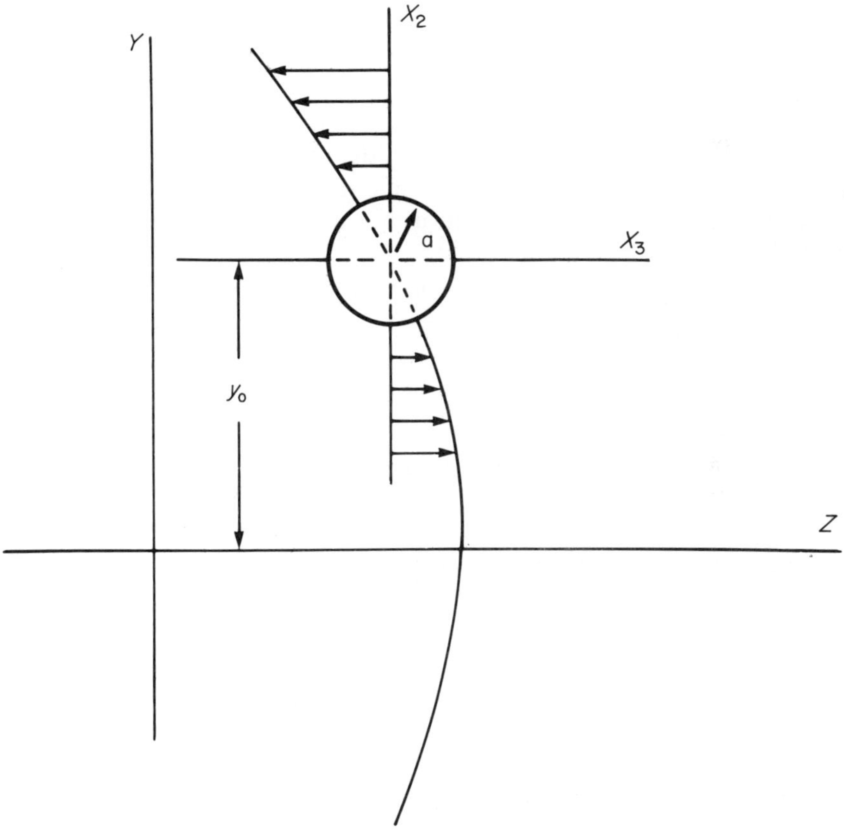

Fig. 3.38. Relation in Poiseuille flow between the initial velocity coordinates, x_i, of a sphere at time $t=0$ at a distance of $r_0\sqrt{x_0^2+y_0^2}$ from the centre and the space-fixed coordinates x, y, z. The X_1 and X axes (not shown) are directed out of the page.

$$x_1 = x - x_0$$
$$x_2 = y - y_0 \tag{3.230}$$
$$x_3 = z - (z_0 + v_0 t)$$

In this new-coordinate system, the velocity field becomes

$$v_1 = v_2 = 0;\ v_3 = \alpha[x_1^2 + x_2^2 + 2(x_0 x_1 + y_0 x_2)] \tag{3.231}$$

where $\alpha = v_m / R^2$.

With this choice of coordinates, the flow consists of a linear and a parabolic region for all x_0 and y_0, except at the centre where the flow is completely parabolic. Consequently, for particles with $r_0 \neq 0$ an enhancement of diffusion, similar to that calculated previously for simple shear (eq (3.55)) can be expected, which persists for as long as the particle has not had sufficient time to diffuse out of the linear region. This time will be of the order r_0^2/D. On the other hand, a qualitative difference may be expected between the enhancement of diffusion due to simple shear and Poiseuille flow for particles situated at or very near the tube centre where r_0 is small and the quadratic term becomes important.

The Langevin equation for a particle subjected to a flow field given by eq (3.231) can be written as (cf. eq (2.52))

$$\ddot{x}_1 + \beta\,\dot{x}_1 = A_1(t)$$
$$\ddot{x}_2 + \beta\,\dot{x}_2 = A_2(t) \tag{3.232}$$
$$\ddot{x}_3 + \beta\,\dot{x}_3 + \beta\alpha[x_1^2 + x_2^2 + 2(x_0 x_1 + y_0 x_2)] = A_3(t)$$

where $A_i(t)$ represents a random force per unit mass.

Equations (3.232) can be integrated formally to give

$$x_1(t) = \frac{x_0}{\beta}(1 - e^{-\beta t}) + \frac{1}{\beta}\int_0^t A_1(t)\,[1 - e^{-\beta(t-t')}]\,dt' \tag{3.233a}$$

$$x_2(t) = \frac{y_0}{\beta}(1 - e^{-\beta t}) + \frac{1}{\beta}\int_0^t A_2(t')\,[1 - e^{-\beta(t-t')}]\,dt' \tag{3.233b}$$

$$x_3(t) = \frac{z_0}{\beta}(1 - e^{-\beta t}) + \frac{1}{\beta}\int_0^t A_3(t')\,[1 - e^{-\beta(t-t')}]$$
$$-\beta\alpha\,[x_1^2(t') + x_2^2(t') + 2\{x_0 x_1(t') + y_0 y_1(t')\}]\,dt' \tag{3.233c}$$

From eqs (3.233) the second moments $\langle x_i x_j \rangle$ can be obtained by using the statistical properties of the forces $A_i(t)$ (see eqs (2.56) and (2.57)). The result of such a calculation is, in the diffusion limit $t \gg \beta^{-1}$:

$$\langle x_i x_j \rangle = \begin{pmatrix} \langle x_1^2 \rangle & 0 & \langle x_1 x_3 \rangle \\ 0 & \langle x_2^2 \rangle & \langle x_2 x_3 \rangle \\ \langle x_1 x_3 \rangle & \langle x_2 x_3 \rangle & \langle x_3^2 \rangle \end{pmatrix} \tag{3.234}$$

with

$$\langle x_1^2 \rangle = \langle x_2^2 \rangle = 2Dt \tag{3.235a}$$

$$\langle x_1 x_3 \rangle = -2\alpha\, x_0\, Dt^2 \tag{3.235b}$$

$$\langle x_2 x_3 \rangle = -2\alpha\, y_0\, Dt^2 \tag{3.235c}$$

$$\langle x_3^2 \rangle = 2Dt + \frac{8}{3}\,\alpha^2\, r_0^2\, Dt^3 + \frac{20}{3}\,\alpha^3 D^2 t^4 \tag{3.235d}$$

Because at time $t=0$ the local shear rate is

$$G_0 = 2\alpha\, r_0$$

the second term on the right-hand side of eq (3.235d) can be written as

$$\tfrac{2}{3} G_0^2\, Dt^3$$

which is identical to the shear-induced diffusion term in simple shear. The last term in eq (3.235d) represents the contribution of the quadratic term in the velocity profile. The t^4 dependence can be readily understood by realizing that $\langle r^2 \rangle \propto t$, $\mathrm{d}x_3/\mathrm{d}t \propto r^2 \propto t$ and, hence, $\langle x_3^2 \rangle \propto t^4$.

In the derivation of eq (3.235) it has been tacitly assumed that the particles are far away from the wall and, therefore, the results are rigorously correct only for particles in an infinite medium. Realistically, we expect eq (3.235) to hold for times, t, satisfying the condition

$$t \ll (R - r_0)^2 / D \tag{3.236}$$

This condition is often satisfied for colloidal particles. For a 100-nm particle in water the value for D is of the order of 10^{-12} m^2 s^{-1}. Hence, typically, a r.m.s. displacement of 10 μm takes about 1 min. For tubes with a radius $R \gg 10$ μm, the majority of the particles takes a time, t, of >1 min to reach the wall. It can be concluded that for colloidal dispersions, wall effects are important only in very thin tubes or after very long times. For molecular diffusion, D is several orders of magnitude larger so that condition (3.236) will be satisfied at much shorter times. For long times, satisfying the condition $t \gg (R-r_0)^2/D$, the molecules will have had time to sample all velocities in the parabolic profile, resulting in a different type of coupling between diffusion and convection, usually referred to as "Taylor diffusion". Although Taylor diffusion is less important for colloidal systems than the shear-induced diffusion discussed here, it has several important applications which will be discussed later (Sections 3.1.10.5 and 3.2.3).

We have seen that it is possible to obtain the moments of the distribution function without actually solving the convective-diffusion equation. It is clear that the solution of the convective-diffusion equation provides much more detail about the particle distribution than the moments. Although no complete solution exists, solutions for regions far from and near to the tube centre can be obtained.

3.1.10.3 Convective diffusion near the centre

Here we will consider convective diffusion in a parallel-plane channel. By introducing the dimensionless parameters

$$t^* = \frac{Dt}{y_0^2},\ x_i^* = \frac{x_i}{y_0},\quad \mathrm{Pe} = \frac{\alpha y_0^3}{D}$$

the convective-diffusion equation can be written as

$$\frac{\partial n}{\partial t^*} = \nabla^{*2} n + \mathrm{Pe}\,(2x_2^* + x_2^{*2})\,\frac{\partial n}{\partial x_3^*} \tag{3.237}$$

This equation can be solved by an expansion in Pe:

$$n = n_0 + \mathrm{Pe}\; n_1 + \mathrm{Pe}^2 n_2 + \ldots . \tag{3.238}$$

Note that when Pe is small the particle is close to the tube centre. Substituting the above expansion into eq (3.237), collecting terms of like order and solving the resulting equations yields (Foister and van de Ven, 1980):

$$n_i = f_i \exp\left[-\frac{x_1^{*2} + x_2^{*2} + x_3^{*2}}{4t^*}\right] \tag{3.239}$$

with

$$f_0=\frac{1}{8(\pi t^*)^{3/2}}$$

$$f_1=-\frac{x_3^*(t^*+3x_2^*+x_2^{*2})}{48(\pi t^*)^{3/2}}$$

$$f_2=(a_0+a_1x_2^*+a_2x_2^{*2}+a_3x_2^{*3}+a_4x_2^{*4}+a_5x_3^{*2}+a_6x_3^{*2}x_2^*$$
$$+a_7x_3^{*2}x_2^{*2}+a_8x_3^{*2}x_2^{*3}+a_9x_3^{*2}x_2^{*4})/48\pi^{3/2}$$

where

$$a_0=-\frac{3t^{*3/2}}{10}-t^{*1/2};\quad a_1=-\frac{t^{*1/2}}{2}$$

$$a_2=-\frac{3t^{*1/2}}{5}-\frac{3}{2t^{*1/2}};\quad a_3=-\frac{1}{t^{*1/2}}$$

$$a_4=-\frac{1}{6t^{*1/2}};\quad a_5=\frac{3t^{*1/2}}{20}+\frac{1}{2t^{*1/2}}$$

$$a_6=\frac{1}{t^{*1/2}};\quad a_7=\frac{3}{10t^{*1/2}}+\frac{3}{4t^{*3/2}}$$

$$a_8=\frac{1}{2t^{*3/2}};\quad a_9=\frac{1}{12t^{*3/2}}$$

From the solution of eq (3.239) the moments of the distribution (up to order Pe^2) can be obtained; for example

$$\langle r^{*2}\rangle=\iiint_{-\infty}^{\infty}(x_1^{*2}+x_2^{*2}+x_3^*)\,(n_0+\mathrm{Pe}\,n_1+\mathrm{Pe}^2\,n_2)\,dx_1^*\,dx_2^*\,dx_3^* \qquad (3.240)$$

Performing the integration and transforming the answer back to dimensional parameters results in

$$\langle r^2\rangle=6Dt+\frac{8}{3}\alpha^2y_0^2Dt^3+\frac{20}{3}\alpha^3D^2t^4 \qquad (3.241)$$

which is in agreement with eq (3.235). The occurrence of r_0 or y_0 in the second term on the right-hand side of eq (3.235d) and (3.230) arises from the fact that the first is for a plane channel and the second for a circular tube. Comparison with the results of the Langevin analysis shows that terms of order greater than Pe^2 do not contribute to the second moment.

3.1.10.4 Convective diffusion far from the centre

Let us again consider a plane channel and introduce the following dimensionless parameters

$$\tilde{x}_i = \frac{\alpha |y_0|^{1/2}}{D}, \quad \tilde{t} = \alpha\, y_0\, t$$

The convective-diffusion equation then reduces to

$$\frac{\partial n}{\partial \tilde{t}} = \tilde{\nabla}^2 n + \left(2\tilde{x}_2 + \frac{\tilde{x}_2^2}{\mathrm{Pe}^{1/2}}\right)\frac{\partial n}{\partial \tilde{x}_3} \tag{3.242}$$

This equation can be solved by a series expansion in $\mathrm{Pe}^{-1/2}$. When $\mathrm{Pe}^{-1/2}$ is small, the particle is far from the tube centre. The solution of eq (3.242) can be written as

$$n = \tilde{n}_0 + \mathrm{Pe}^{-1/2}\,\tilde{n}_1 + \mathrm{Pe}^{-1}\,\tilde{n}_2 + \ldots . \tag{3.243}$$

Substituting this expansion into eq (3.242) and solving for the various orders results in (Foister and van de Ven, 1980):

$$\tilde{n}_i = \tilde{f}_i \exp\left[-\left\{\frac{3(\tilde{t}\tilde{x}_2 + \tilde{x}_3)^2}{4\tilde{t}(\tilde{t}^2+3)} + \frac{\tilde{x}_2^2 + \tilde{x}_3^2}{4\tilde{t}}\right\}\right] \tag{3.244}$$

with

$$\tilde{f}_0 = \frac{1}{8(2\pi\,\tilde{t})^{3/2}}\left(\frac{3}{\tilde{t}^2+3}\right)^{1/2}$$

and

$$\tilde{f}_1 = \frac{1}{16}\left(\frac{3}{2\pi}\right)^{3/2}(b_1\tilde{x}_3 + b_2\tilde{x}_2 + b_3\tilde{x}_3^3 + b_4\tilde{x}_3^2\tilde{x}_2 + b_5\tilde{x}_3\tilde{x}_2^2 + b_6\tilde{x}_2^3)$$

where the b_i terms are functions of $\tilde{t}$ which can be found by solving a coupled set of equations.

It is possible to determine various moments of the distribution by using a similar method as that described before.

3.1.10.5 Taylor diffusion

Taylor diffusion is the diffusion of particles or molecules in Poiseuille flow for time scales that are large compared to the time it takes for a particle to sample all the radial positions in the flow. As we have seen, this time can be estimated as

$$t \gg R^2/D \tag{3.245}$$

In principle, one would like to solve the convective-diffusion equation

$$\frac{\partial n}{\partial t} = D\nabla^2 n - \nabla\cdot(un) \tag{3.246}$$

with

$$u = v_m\left[1 - \left(\frac{r}{R}\right)^2\right] \tag{3.247}$$

These equations are already an approximation, as the effects of the bounding walls are neglected. Near the wall the diffusion coefficient, D, is not a constant and the particle velocity, u, does not equal the fluid velocity, v. Even for the simplified eqs (3.246) and (3.247), no general solutions are available, but as we have seen, the moments of the distribution can be obtained by various methods. The solutions of eq (3.246) for long times is originally due to Taylor (1953). The main interest is in the distribution of particles, initially introduced into the tube, in the direction of flow (z direction) as a function of time. Taylor found that the average velocity of the particles is equal to the mean velocity in the tube and that the dispersion about the mean can be described by an effective diffusion constant, D^*, i.e.

$$\langle(z-\bar{z})^2\rangle = 2D^*t \tag{3.248}$$

where $\bar{z}$ is the average distance from the origin z_0:

$$\bar{z} = z_0 + \bar{v}t$$

$\bar{v}$ being the average velocity ($\bar{v} = \frac{1}{2}v_m$ for a tube, and $\bar{v} = 2v_m/3$ for a plane channel). The effective diffusion constant, D^*, is given by

$$D^* = D + \frac{a^2\bar{v}^2}{48D} \tag{3.249}$$

The second term on the right-hand side of eq (3.249) is an enhanced diffusion term which depends on the mean velocity, $\bar{v}$, and on D. Rather paradoxically, it appears that the smaller the value of D, the larger the value of D^*. However, a much longer time must elapse before the diffusion process

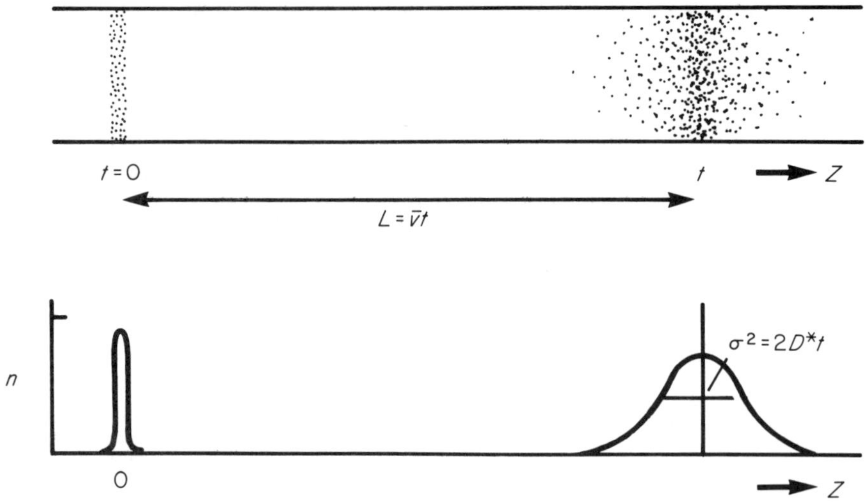

Fig. 3.39. The principle of Taylor diffusion: a narrow distribution of particles, being at $z=0$ at time $t=0$ in a capillary tube, finds itself, after sufficiently long times, at a position $z=\bar{v}t$ with a dispersion given by D^* (cf. eq (3.249)).

can be described by eq (3.249) (cf. eq (3.245)). The nature of Taylor diffusion in a tube is presented schematically in Fig. 3.39.

Thus the nature of convective diffusion changes from the short-time behaviour, given by eqs (3.235), to the long-time behaviour given by eq (3.248). For intermediate times, no solutions are available in the literature.

The phenomenon of Taylor diffusion can be generalized to a much larger class of problems. This has been done by Brenner and Dill (Brenner, 1980, 1982; Dill and Brenner, 1982a,b). In essence, many problems can be described by two sets of coordinates, local and global ones, and usually one is interested in the dispersion of particles in global coordinates. Generalized Taylor diffusion applies at times sufficiently long for the particles to have had the opportunity to sample the whole range of the local coordinates. Describing the tube flow by the cylindrical coordinates, r, ϕ and z, then in the classical Taylor diffusion problem discussed above, the local coordinates are r and ϕ and the global coordinate is z. Another example of Taylor diffusion is discussed in Section 3.2.3.

3.1.10.6 Hydrodynamic chromatography

Hydrodynamic chromatography is a technique in which colloidal particles can be separated according to size. The principle is based on the fact that

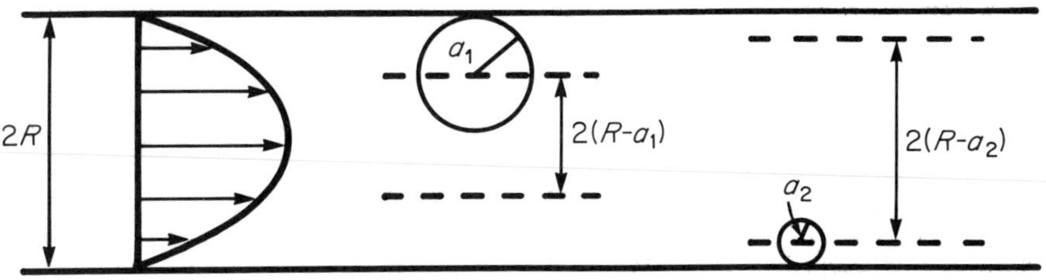

Fig. 3.40. The principle of hydrodynamic chromatography: a large sphere diffusing through a tube samples a smaller volume of the tube than does a small sphere. Because the velocity at the centre of the tube is larger than at the wall, the average velocity of the large sphere is larger than that of the small one.

smaller particles can approach the wall of a tube more closely than can larger particles. Therefore, in sampling all velocities in a tube, diffusing particles that are large will have a larger average velocity than small particles because they cannot sample the positions near the wall where the velocity is low. This is shown schematically in Fig. 3.40.

Neglecting shear-enhanced diffusion and particle–wall interactions, the average velocity of a particle of radius a, at times $t \gg R^2/D$ is given by

$$\bar{u} = \frac{1}{R-a} \int_0^{R-a} v_{\mathrm{m}} \left[1 - \left(\frac{r}{R} \right)^2 \right] \mathrm{d}r \tag{3.250a}$$

or

$$\bar{u} = \frac{1}{2} v_{\mathrm{m}} \left[1 - \frac{1}{2} \left(\frac{a}{R} \right)^2 \left(3 - \frac{a}{R} \right) \right] \frac{R}{R-a} \tag{3.250b}$$

This means that the average particle velocity, $\bar{u}$, increases monotonically from the value $\frac{1}{2}v_{\mathrm{m}}$ when $a=0$ to $\bar{u}=v_{\mathrm{m}}$ when $a=R$. Thus as a suspension containing particles of various sizes is pumped through a tube, particles with the largest dimensions will, on average, travel faster and exit first. The smallest particles will remain longest in the tube.

The boundary conditions on the wall will have a considerable effect on the mean velocity. Therefore, in most cases one must resort to calibration with samples of known sizes. It is clear that the separation process also depends crucially on the relative radius of the particle and the tube. Furthermore, the tube must be long enough to satisfy t $\gg R^2/D$, which becomes

$$\ell \gg \bar{u}R^2/D \tag{3.251}$$

where ℓ is the length of the tube.

The same principle applies to gel permeation chromatography. Here a column is packed with small (usually spherical) particles. The flow through the column is a flow through a porous medium. For the same reason as in hydrodynamic chromatography, the largest particles pass through the column the fastest.

3.1.11. Viscosity of dispersions of non-interacting particles

3.1.11.1 Dispersions of neutral spheres

The apparent viscosity of a liquid increases when small colloidal particles are added to it. The first person to calculate the increase in viscosity with volume fraction was Einstein (1906, 1911), who obtained the result for low-volume fractions $c \ll 1$:

$$\mu = \mu_0(1 + 2 \cdot 5c) \tag{3.252}$$

Here μ is the apparent viscosity of the dispersion and μ_0 that of the suspending medium. In his first paper Einstein gave the result $\mu = \mu_0(1 + c)$, which he subsequently corrected. So even someone of the calibre of Einstein can make mistakes in this subject. The result expressed by eq (3.252) was obtained by calculating the energy dissipation in a sphere of radius R around a reference sphere and letting $R \to \infty$. As noted by Batchelor and Green (1972), this leads to diverging integrals which can be evaluated by using special mathematical techniques. An alternative procedure is outlined in the book by Happel and Brenner (1983), who define the viscosity in an operational manner. The work done in maintaining the motion of the external boundaries of an apparatus used to measure the viscosity of a (homogeneous) liquid is directly proportional to the viscosity of the liquid. For a Newtonian liquid of viscosity μ_0, let the work done by the stresses acting on the boundaries of the apparatus be denoted by W_0. Similarly, for a liquid with suspended particles an apparent viscosity, μ, can be defined which is proportional to the work, W, done by the stresses over the boundaries

$$\frac{\mu}{\mu_0} = \frac{W}{W_0} \tag{3.253}$$

In linear experiments on homogeneous liquids the rate at which energy is dissipated in the apparatus equals the rate at which stresses acting over its surface are doing work. Hence, instead of work we can speak of energy dissipation and we can rewrite eq (3.253) as

$$\frac{\mu}{\mu_0}=\frac{\int_{S_0} \mathrm{d}\boldsymbol{S}\cdot\boldsymbol{\Pi}\cdot\boldsymbol{v}}{\int_{S_0} \mathrm{d}\boldsymbol{S}\cdot\boldsymbol{\Pi}_0\cdot\boldsymbol{v}_0} \tag{3.254}$$

where $\mathrm{d}\boldsymbol{S}$ is a vector normal to the infinitesimal surface element $\mathrm{d}S$, $\boldsymbol{\Pi}$ is the stress tensor, and $\boldsymbol{v}$ is the fluid velocity. The subscript "0" refers to the value in the absence of particles. S_0 is the total area of the bounding walls. Equations (3.254) can also be written as

$$\frac{\mu}{\mu_0}=1+\frac{E^*}{E_0} \tag{3.255}$$

where $E_0=W_0=\int_{S_0}\mathrm{d}\boldsymbol{S}\cdot\boldsymbol{\Pi}_0\cdot\boldsymbol{v}_0$ and E^* is the additional rate at which energy dissipation occurs within the measuring apparatus due to the presence of suspended particles. Happel and Brenner (1983) have shown that E^* can be expressed as a surface integral over the suspended particles, requiring only detailed knowledge of the flow field in the immediate neighbourhood of the particles. For a homogeneous suspension of identical particles of number concentration n:

$$E^*=n\int_S \mathrm{d}\boldsymbol{S}\cdot\boldsymbol{\Pi}\cdot(\boldsymbol{v}-\boldsymbol{v}_0) \tag{3.256}$$

where the integral is to be evaluated over the surface, S, of a suspended particle. From the solution of the flow field around an isolated sphere (eq (3.5)) one can determine $\boldsymbol{\Pi}$, while $\boldsymbol{v}_0$ is the undisturbed flow field. Evaluating E^* then results in

$$E^*=\frac{5}{2}\,c\,E_0 \tag{3.257}$$

Combining eqs (3.255) and (3.257), one recovers Einstein's result (eq (3.252)).

More generally, the rheological behaviour of a suspension can be described by an apparent macroscopic stress tensor, $\boldsymbol{P}$ (different from $\boldsymbol{\Pi}$ which describes the stresses in a fluid with particles). The viscosity corresponds to the P_{23} components of $\boldsymbol{P}$:

$$P_{23}=\mu G \tag{3.258}$$

To describe fully the behaviour of a dispersion, the stress tensor $\boldsymbol{P}$ must be evaluated. For dilute dispersions, $\boldsymbol{P}$ can be expanded in terms of the volume fraction c:

$$\boldsymbol{P} = \Pi_0 + \boldsymbol{A}c + \boldsymbol{B}c^2 + \ldots. \tag{3.259}$$

Here $\boldsymbol{A}$ is the first virial stress tensor, determined by the presence of non-interacting isolated particles, and $\boldsymbol{B}$ is the second virial stress tensor, determined by two-body interactions. Higher-order terms correspond to more-body interactions. Here we are interested in dilute dispersions without particle interactions whose behaviour is determined by the tensor $\boldsymbol{A}$. In general, for an arbitrarily shaped particle, the first virial stress tensor, $\boldsymbol{A}$, is determined by the leading term of the disturbance flow field far from the particle, which varies as r^{-2}. As the distance r from the particle approaches infinity, the disturbance field around a particle can be written as (Okagawa *et al.*, 1974):

$$v'_j = M'_{ik} \left[\frac{1}{r^3} (x'_i \delta_{jk} + x'_j \delta_{ik} - x'_k \delta_{ij}) - \frac{3}{r^5} x'_i x'_j x'_k \right] \tag{3.260}$$

where the prime denotes parameters in a particle-fixed coordinate system. M'_{ij} is a tensor determined by particle shape. Equation (3.260) represents the flow field produced by a point force doublet (or stresslet).

The first virial stress tensor $\boldsymbol{A}$ is related to M'_{ij} by

$$\boldsymbol{A} = \frac{8\pi\mu_0}{V} \langle \boldsymbol{M} \rangle \tag{3.261}$$

where V is the volume of a particle and $\langle \boldsymbol{M} \rangle$ the orientational average of $\boldsymbol{M}$, which is obtained by transforming $\langle \boldsymbol{M}' \rangle$ into space-fixed coordinates and averaging over the distribution.

For the case of a sphere, the leading term of the disturbance flow is given by (cf. eq (3.5)):

$$v_i = -\frac{5}{2} \frac{G}{r^5} x_i x_2 x_3 \tag{3.262}$$

This term can be compared with the last term of the right-hand side of eq (3.260). For torque-free particles, the first term $(x'_i \delta_{ik} + x'_j \delta_{ik} - x'_k \delta_{ij})$ can be omitted because the terms $x'_j \delta_{jk} - x'_k \delta_{ij}$ contribute an antisymmetric contribution to M'_{ij}, which can be shown to be symmetric for a torque-free particle. The remaining term only contributes to the isotropic pressure and does not affect rheological properties. Assigning an orientation to the sphere by considering the position of a dot on the sphere surface, one can choose a particle-fixed coordinate system such that the X_1' axis coincides with the X_1 axis. In this case, eq (3.262) represents the velocity field in particle-fixed coordinates and can be directly compared to the last term of eq (3.260)

$$v_i' = -\frac{3}{r^5}(M_{23}' + M_{32}')\, x_i' x_2' x_3' \tag{3.263}$$

Comparing eqs (3.262) and (3.263) and using $M_{ij}' = M_{ji}'$ and converting into dimensional parameters, results in

$$M_{23}' = M_{32}' = \frac{5}{12} G a^3 \tag{3.264}$$

other components being zero. Hence from eq (3.261):

$$A_{23} = \frac{5}{2} \mu_0 G \tag{3.265}$$

Combining this with eqs (3.258) and (3.259) again results in

$$\mu = \mu_0 \left(1 + \frac{5}{2} c\right) \tag{3.266}$$

This derivation shows clearly that all that is needed to calculate the intrinsic viscosity is the term of order r^{-2} in the disturbance flow. Terms which decay faster can be ignored.

3.1.11.2 Dispersions of electrically charged particles: primary electroviscous effect

Similar to a dispersion of neutral spheres, the viscosity of a dispersion of charged spheres is given by eq (3.255), but now with

$$E^* = n \int_S dS \cdot \Pi^* \cdot (v - v_0) \tag{3.267}$$

with

$$\Pi^* = \Pi + \Pi_{el} \tag{3.268}$$

Equation (3.267) follows directly from the derivation given by Happel and Brenner (1983) for neutral spheres by realizing that the electric field around a sphere makes an isotropic contribution to the local pressure (for low shear rates). The stress tensor $\mathbf{\Pi}$ can be evaluated from the flow field, v (eqs (3.34) and (1.30)), while the additional terms $\mathbf{\Pi}_{el}$ can be obtained from the leading terms of eqs (3.21) and (1.106). Therefore, all the functions appearing in eq (3.267) are known and it is possible to perform the integration to yield E^* and thus μ. Originally the problem was solved in a different way by Booth, who followed Einstein's method of calculating the energy dissipation in a sphere around a reference particle. His result for the case of symmetrical electrolytes can be written as

$$\mu = \mu_0 (1 + [\mu] c) \tag{3.269}$$

with the intrinsic viscosity $[\mu]$ given by

$$[\mu] = 2.5 + \bar{b}\Phi_s^2 Q(\tau) \tag{3.270}$$

where $\bar{b}$ is the average dimensionless ion mobility, Φ_s the reduced ζ potential, and $Q(\tau)$ is a function of the double-layer parameter $\tau = \kappa a$, κ^{-1} being the double-layer thickness (cf. eq (3.35c)).

The increase in viscosity due to the presence of electrical double layers around the colloidal particles is called the primary electroviscous effect.

The result expressed by eq (3.270) follows directly from the disturbance flow around a charged sphere, which at large r equals (cf.eq (3.34):

$$v_i = -\left(\frac{5}{2} + \bar{b}\Phi_s^2 Q\right) \frac{G}{r^5} x_i x_2 x_3 \tag{3.271}$$

Analogous to the derivation for a neutral sphere (eqs (3.262) to (3.266)), eq (3.270) follows immediately from eq (3.271).

As noted earlier, the function $Q(\tau)$ approaches infinity as $\tau \to 0$, i.e. as the double-layer thickness approaches infinity or, equivalently, as the ion concentration approaches zero. We concluded that as $\tau \to 0$, it is not allowed to make an expansion in Φ_s as this leads to infinite potentials and ion concentrations. Equation (3.270) predicts that as the ion concentration approaches zero, the intrinsic viscosity approaches infinity. Obviously the theory must break down before this happens. It can be readily shown from the governing equations that as $\tau \to 0$, the body force $\rho \nabla \psi \to 0$ and hence the flow field approaches that of a neutral sphere. It follows that as $\tau \to 0$, $[\mu] \to 2{\cdot}5$.

The intrinsic viscosity can be expressed in terms of the specific conductivity, χ, of the electrolyte, provided the mobilities (or friction coefficients) of all ions are the same. Using Ohm's law, the definition of κ, and the relation

$$F_{el} = zeE = fu_i \tag{3.272}$$

where F_{el} is the electrical force on an ion, z its valency, E the field strength, f the friction coefficient, and u_i its velocity, we can write

$$\chi \equiv \frac{i}{E} = \frac{\sum n_i z_i e u_i}{E} = \frac{\sum n_i z_i^2 e^2}{f} = \frac{\epsilon\, kT\, \kappa^2}{2f} \tag{3.273}$$

where i is the current per unit area.

Hence eq (3.270) can be written as

$$[\mu] = 2 \cdot 5 + \frac{\epsilon^2 \zeta^2}{\chi \mu_0 a^2}\, \tau^2 Q(\tau) \tag{3.274}$$

As $\tau \to \infty$, $Q \to 6/\tau^2$ and eq (3.274) reduces to

$$[\mu] = 2 \cdot 5 + \frac{6\epsilon^2 \zeta^2}{\chi \mu_0 a^2} \tag{3.275}$$

a result first derived by Krasny-Ergen (1936), who corrected an earlier result by Smoluchowski (1903, 1916). Thus, contrary to Booth's assertion (Booth, 1950), as $\kappa a \to \infty$, the exact result of eq (3.270) reduces to eq (3.275) when $b_1 = b_2$. Booth probably overlooked the fact that as $\tau \to \infty$, $\chi \to \infty$.

Because eq (3.270) is obtained by an expansion solution in Φ_s, it is obviously only valid when $\Phi_s \ll 1$. The problem of dilute dispersions of spheres with arbitrary Φ_s values in the limit $\kappa a \to \infty$ seems amenable to an analytical solution (Lever, 1979).

3.1.11.3 Dispersions of polymer-coated spheres

As we have seen in Section 2.3, the flow around a sphere coated with polymer depends on the polymer segment density distribution on the sphere surface. To obtain the viscosity of a dispersion of polymer-coated spheres one must solve the creeping-flow equation outside the polymer layer and the Brinkman equation inside it, with the no-slip boundary condition on the sphere surface, with continuity of stresses and velocities inside and outside the polymer layer and the solution must approach a simple shear flow at large distances from the sphere. The viscosity can then be found from the leading term of the disturbance flow far from the sphere. In principle this problem can be solved in a way analogous to the problem of a polymer-coated sphere in uniform flow.

As the result will depend on the particular choice of the segment density distribution, which is usually unknown, a more approximate procedure is often followed. An effective polymer-layer thickness, δ, is defined by

$$\mu = \mu_0(1 + [\mu_0]c) = \mu_0(1 + 2{\cdot}5c_{\text{eff}}) \tag{3.276}$$

where

$$c_{\text{eff}} = \frac{4}{3}\pi(a+\delta)^3 n \simeq c\left(1 + 3\frac{\delta}{a}\right) \tag{3.277}$$

n being the number concentration of particles.

It must be emphasized that the layer thickness, δ, obtained in this way differs from the hydrodynamic layer thickness determined from diffusion coefficients.

3.1.11.4 Emulsions

The fluid flow field around an undeformed drop subjected to a simple shear is given by eq (3.173). The viscosity is determined by the leading term of the disturbance flow (of order r^{-2}), which is given by the second term on the right-hand side of eq (3.174a). Comparing this term with the corresponding term for a solid sphere (eq (3.9a)), and noting that the intrinsic viscosity is (minus) twice the coefficient appearing in this term, it follows directly that for undeformed drops the intrinsic viscosity equals

$$[\mu] = \frac{5p+2}{2(1+p)} \tag{3.278}$$

As the viscosity ratio, p, approaches infinity, the intrinsic viscosity of a solid sphere is recovered; letting $p \to 0$, it is found that the intrinsic viscosity of a dispersion of bubbles is $[\mu] = 1$.

Similar to friction coefficients of emulsion droplets (Section 2.1.6), the intrinsic viscosity of emulsions of droplets covered with surfactant molecules is often equal to that of solid spheres.

When drops deform in shear, the flow field around a drop will be different from that given by eq (3.173). For small deformations it will have an additional term of order E (cf. eq (3.169)). In such cases, the intrinsic viscosity will show non-Newtonian transient behaviour prior to reaching a steady-state value. At the same time the emulsion will exhibit normal stress differences (absent for undeformed drops). In general, for deformed drops the macroscopic stress tensor, $\boldsymbol{P}$, will depend on the stress history of the suspension.

3.1.11.5 Dispersions of non-spherical particles

(a) General considerations

We have seen that for non-interacting particles the macroscopic stress tensor is given by

$$\boldsymbol{P} = \Pi_0 + \boldsymbol{A}c \tag{3.279}$$

with $\boldsymbol{A}$ given by eqs (3.261) and (3.260). The tensor $\boldsymbol{M}$ appearing in eq (3.261) can be determined from the leading terms of the disturbance flow around a particle in simple shear. For spheroidal particles the disturbance flow has been calculated by Jeffery (1922) and is given by eqs (3.84). By comparing this solution with eq (3.260), the tensor M' can be evaluated with the result (Okagawa *et al.*, 1974):

$$M'_{11} = \frac{G\,b^3 r_e(r_e^2-1)}{9[2r_e^2-(1-2r_e^2)A]}$$

$$M'_{22} = \frac{G\,b^3 r_e(r_e^2-1)}{27(2r_e^2-A)}\sin 2\phi\left[1+2\cos^2\theta-\frac{(4r_e^2-1)A-2r_e^2}{4[2r_e^2-(1+2r_e^2)A]}\sin^2\theta\right]$$

$$M'_{33} = -\frac{G\,b_e^3 r_e(r_e^2-1)}{27(2r_e^2-A)}\sin 2\phi\left[2+\cos^2\theta-\frac{(4r_e^2-1)A-2r_e^2}{4[2r_e^2-(1+2r_e^2)A]}\sin^2\theta\right] \tag{3.280}$$

$$M'_{12} = M'_{21} = \frac{G\,b^3 r_e(r_e^2-1)}{6(r^2+1)(3A-2)}\sin 2\theta\,\sin 2\phi$$

$$M'_{23} = M'_{32} = \frac{2G\,b^3 r_e(r_e^2-1)}{3(2r_e^2-3A)}\cos\theta\,\cos 2\phi$$

$$M'_{13} = M'_{31} = \frac{G\,b^3 r_e(r_e^2-1)}{3(r_e^2+1)(3A-2)}\sin\theta\,\cos 2\phi$$

where

$$A = \frac{r_e^2}{r_e^2 - 1} - \frac{r_e \cosh^{-1} r_e}{(r_e^2 - 1)^{3/2}} \qquad \text{for } r_e > 1$$

$$A = \frac{r_e \cos^{-1} r_e}{(1 - r_e^2)^{3/2}} - \frac{r_e^2}{1 - r_e^2} \qquad \text{for } r_e < 1 \tag{3.281}$$

$$A = \frac{2}{3} \qquad \text{for } r_e = 1$$

Transforming $\boldsymbol{M'}$ into $\boldsymbol{M}$, averaging over all orientations and substituting into eq (3.261) and subsequently into eq (3.279) yields for the components $P_{32} = P_{23}$:

$$P_{32} = \mu_0 G(1 + [\mu_0]c) \tag{3.282}$$

with the intrinsic viscosity $[\mu_0]$ given by

$$[\mu] = C_1\langle \sin^4\theta \sin^2 2\phi \rangle + C_2\langle \cos^2\theta \rangle + C_3\langle \sin^2\theta \rangle \tag{3.283}$$

with

$$C_1 = (r_e^2 - 1)\left[\frac{6r_e^2 - 4r_e^2 A - 5A}{2(2r_e^2 - 3A)\{2r_e^2 - (2r_e^2 + 1)A\}} - \frac{2}{(r_e^2 + 1)(3A - 2)}\right]$$

$$C_2 = \frac{4(r_e^2 - 1)}{2r_e^2 - 3A} \tag{3.284}$$

$$C_3 = \frac{2(r_e^2 - 1)}{(r_e^2 + 1)(3A - 2)}$$

When $r_e \to 1$, $C_1 \to 0$ and $C_2 = C_3 = 2.5$. Thus eq (3.283) reduces to eq (3.252): At the same time $\langle \sin^2\theta \rangle \to 2/3$ and $\langle \cos^2\theta \rangle \to 1/3$.

The other components of $\boldsymbol{P}$ can be evaluated. Similarly one finds

$$\frac{P_{12}}{\mu_0 Gc} = C_1\langle \sin^3\theta \cos\theta \sin\phi \cos^2\phi \rangle \tag{3.285}$$

$$\frac{P_{13}}{\mu_0 Gc} = 4C_1\langle \sin 3\theta \cos\theta \sin^2\phi \cos\phi \rangle + (C_3 - C_2)\langle \sin\theta \cos\theta \cos\phi \rangle \tag{3.286}$$

and for the intrinsic normal stress differences, defined by

$$[\xi_{ij}] = \frac{P_{ii} - P_{jj}}{\mu_0 Gc} \tag{3.287}$$

one finds

$$[\xi_{12}] = C_1[-\tfrac{1}{2}\langle \sin^4\theta \sin 4\phi\rangle - 3\langle \sin^4\theta \sin 2\phi\rangle + 2\langle \sin^2\theta \sin 2\phi\rangle] + (C_2 - C_3)\langle \sin^2\theta \sin 2\phi\rangle \quad (3.288)$$

$$[\xi_{23}] = C_1\langle \sin^4\theta \sin 4\phi\rangle \quad (3.289)$$

$$[\xi_{13}] = [\xi_{12}] + [\xi_{23}] \quad (3.290)$$

In the theory described above, the second virial stress tensor, $\boldsymbol{A}$, is solely determined by the disturbance flow around spheroidal particles and, therefore, only applies when the rotary Péclet number $\mathrm{Pe_r} = G/D_r = \infty$. When Brownian motion is present, the disturbance flow is altered and this results in an additional random torque exerted on the particle. It can be shown that, allowing for Brownian motion, the intrinsic viscosity and normal stress differences are given by (Brenner, 1974)

$$[\mu] = [\mu]_\infty + \frac{C_4}{\mathrm{Pe_r}} \langle \sin^2\theta \sin 2\phi\rangle \quad (3.291)$$

$$[\xi_{12}] = [\xi_{12}]_\infty + \frac{C_4}{\mathrm{Pe_r}} (2 - 3\langle \sin^2\theta\rangle - \langle \sin^2\theta \cos 2\phi\rangle) \quad (3.292)$$

$$[\xi_{23}] = [\xi_{23}]_\infty + \frac{C_4}{\mathrm{Pe_r}} (2 - 3\langle \sin^2\theta\rangle - \langle \sin^2\theta \cos 2\phi\rangle) \quad (3.293)$$

where the subscript "∞" refers to the values at $\mathrm{Pe_r} = \infty$, given by eqs (3.283), (3.288) and (3.289), and

$$C_4 = \frac{15}{2B} (3E_1 + 4E_2) \quad (3.294)$$

with

$$E_1 = \frac{2}{15A_4}\left(1 - \frac{A_4}{A_5}\right) \text{ and } E_2 = \frac{1}{5A_4}\left[\frac{r_e(A_1 + A_2)}{A_1 + r_e^2 A_2}\left(\frac{A_4}{A_3}\right) - 1\right]$$

where

$$A_1 = A = \frac{r_e^2}{r_e^2 - 1}(1 - \beta)$$

$$A_2 = \frac{2}{r_e^2 - 1}(r_e^2 \beta - 1)$$

$$A_3 = \frac{r_e^2}{(r_e^2-1)^2}\,(r_e^2+2-3r_e^2\beta)$$

$$A_4 = \frac{r_e^2}{4(r_e^2-1)^2}\,(3\beta+2r_e^2-5)$$

$$A_5 = \frac{r_e^2}{(r_e^2-1)^2}\,[(2r_e^2+1)\beta-3]$$

with

$$\beta = \frac{\cosh^{-1} r_e}{r_e(r^2-1)^{1/2}} \quad \text{if } r_e > 1$$

and

$$\beta = \frac{\cos^{-1} r_e}{r_e(1-r_e^2)^{1/2}} \quad \text{if } r_e < 1$$

Because of the relations (Brenner, 1974)

$$\langle \sin^4\theta \sin 2\phi \rangle = \langle \sin^2\theta \sin 2\phi \rangle - 2(B\mathrm{Pe})^{-1}\,(3\langle \sin^2\theta \rangle - 2) \tag{3.295}$$

$$\langle \sin^4\theta \sin^2 2\phi \rangle = \langle \sin^2\theta \rangle + B^{-1}\,\langle \sin^2\theta \cos 2\phi \rangle - 6(B\mathrm{Pe})^{-1}\langle \sin^2\theta \sin 2\phi \rangle\rangle \tag{3.296}$$

$$\langle \sin^4\theta \sin 4\phi \rangle = 2B^{-1}\langle \sin^2\theta \sin 2\phi \rangle - 12(B\mathrm{Pe})^{-1}\langle \sin^2\theta \cos 2\phi \rangle \tag{3.297}$$

the intrinsic viscosity and normal stress differences can be written in terms of the orientational averages $\langle \sin^2\theta \rangle$, $\langle \sin^2\theta \sin 2\phi \rangle$ and $\langle \sin^2\theta \cos 2\phi \rangle$ only.

The orientational average occurring in eqs (3.283) to (3.293) can be obtained from the orientation-distribution function $p(\theta, \phi, \mathrm{t})$ which is determined by the rotary convective-diffusion equation

$$\frac{\partial p}{\partial t} = D_r \nabla^2 p - \nabla \cdot (\omega p) \tag{3.298}$$

where the angular velocity, ω, is given by Jeffery's equations (cf. eqs (3.103) and (3.86)). As $p(\theta, \phi, \mathrm{t})$, in general, is a function of time, it follows that the viscosity and normal stress differences will also be functions of time. Hence a suspension, initially sheared at time $t=0$, will undergo transient behaviour before attaining a steady state.

The theory outlined above applies to all particles whose shape is a body of revolution, i.e. whose rotation can be described by an equivalent axis ratio, r_e. The theory has also been worked out for ellipsoidal particles or, in general, all triaxial particles. The flow field around an ellipsoidal particle has been given by Jeffery (1922) and from it the first virial stress tensor, $\boldsymbol{A}$, can be calculated. This was done by Haber and Brenner (1984). For particles with a more complex shape no results are available.

(b) Steady-state behaviour

For long times the orientation distribution attains a steady state $p(\theta, \phi, \infty)$, and the components of the macroscopic stress tensor attain steady-state values. The solution of the rotary convective-diffusion equation for spheroids was discussed in Section 3.1.6. The results of that section can be used to calculate the orientational averages $\langle \sin^2\theta \rangle$, $\langle \sin^2\theta \sin 2\phi \rangle$ and $\langle \sin^2\theta \cos 2\phi \rangle$ needed to calculate $\boldsymbol{P}$. Values of these orientational averages have been tabulated by Brenner (1974) for various values of the rotary Péclet number, Pe_r. As $Pe_r \to \infty$, the averages can be found from the procedure outlined in eqs (3.126) to (3.129).

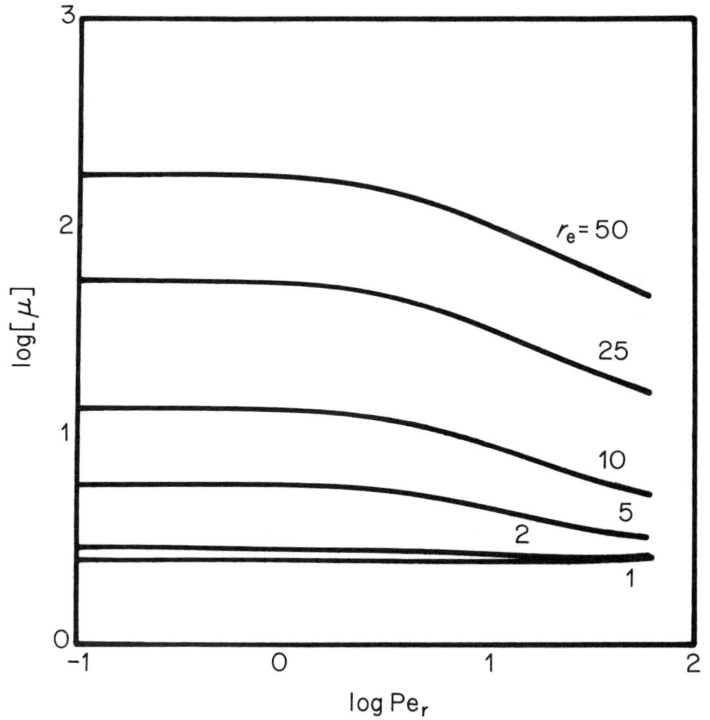

Fig. 3.41. Variation in the intrinsic viscosity, $[\mu]$, with shear rate for prolate spheroids of various aspect ratios suspended in a simple shear flow (after Brenner, 1974).

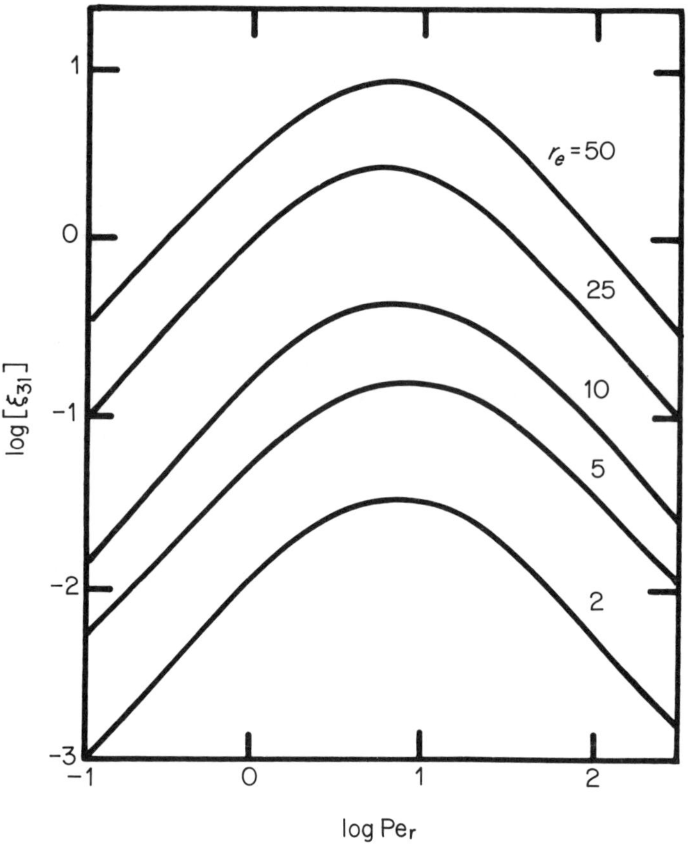

Fig. 3.42. Variation in the intrinsic stress difference, $[\xi_{31}]$, with shear rate for prolate spheroids of various aspect ratios subjected to a simple shear.

Results for the intrinsic viscosity, $[\mu]$, as a function of rotary Péclet number, Pe_r, are shown in Fig. 3.41 for various values of the axis ratio, r_e. It can be seen that dilute dispersions of spheroids are shear-thinning. The reason for the shear thinning is the change in the average orientation of particles as the system changes from being dominated by Brownian motion to being dominated by shear.

An example of the intrinsic first normal stress difference, $[\xi_{13}]$, is shown in Fig. 3.42. It can be seen that $[\xi_{13}]$ is zero when $\text{Pe}_r = 0$ or $\text{Pe}_r = \infty$ and is maximum when Pe_r is of the order of unity. Results given by Brenner (1974) and Chaffey and Porter (1984) show a monotonic increase in the first normal stress difference rather than a maximum because there the stresses are normalized by $\mu_0 D_r c$ rather than by $\mu_0 G c$. Other normal stresses vary in a similar way. The actual dimensional normal stress differences increases from zero to an asymptotic value as the shear rate varies from zero to infinity. The reason for stress differences at high shear is the non-uniform distribution of particle orientations.

The only explicit result for particles more complex than spheroids is the steady-state low shear limit ($\mathrm{Pe}_r \to 0$) for ellipsoidal particles (Haber and Brenner, 1984):

$$[\mu_0] = \frac{4}{15}\left(\frac{\alpha_1'' + \alpha_2'' + \alpha_3''}{\alpha_1''\alpha_2'' + \alpha_2''\alpha_3'' + \alpha_3''\alpha_1''}\right) + \frac{1}{5}\left[\frac{a_2 a_3(\alpha_2+\alpha_3)}{\alpha_1'(a_2^2\alpha_2 + a_3^2\alpha_3)} + \frac{a_3 a_1(\alpha_3+\alpha_1)}{\alpha_2'(a_3^2\alpha_3 + a_1^2\alpha_1)} + \frac{a_1 a_2(\alpha_1+\alpha_2)}{\alpha_3'(a_1^2\alpha_1 + a_2^2\alpha_2)}\right] - \frac{1}{5}\frac{\left(\dfrac{a_1^2 - a_2^2}{a_1^2\alpha_1 + a_2^2\alpha_2} + \dfrac{a_2^2 - a_3^2}{a_2^2\alpha_2 + a_3^2\alpha_3} + \dfrac{a_3^2 - a_1^2}{a_3^2\alpha_3 + a_1^2\alpha_1}\right)^2}{\left(\dfrac{a_1^2 + a_2^2}{a_1^2\alpha_1 + a_2^2\alpha_2} + \dfrac{a_2^2 + a_3^2}{a_2^2\alpha_2 + a_3^2\alpha_3} + \dfrac{a_3^2 + a_1^2}{a_3^2\alpha_3 + a_1^2\alpha_1}\right)} \tag{3.299}$$

where α_i and α_i' are defined below eqs (3.84).

The same analysis can be made for spheroids and ellipsoids in a general two-dimensional linear flow. For example, in extensional flow $\boldsymbol{P}$ is similar to simple shear, with the exception that suspensions of prolate spheroids are shear thickening for particles with $r_e < \sim 15$ and when $r_e > \sim 15$ they are shear thickening at low Pe_r but shear thinning at high Pe_r (Brenner, 1974).

It would also be of interest to know what the effect of electrical double layers surrounding colloidal particles would be on the intrinsic viscosity. Because of the geometrical complexity of the problem, no results are available for non-spherical particles. From the general nature of the governing equations, it can be anticipated that for low surface potentials the intrinsic viscosity will be of the form

$$[\mu] = [\mu]_0 + \bar{b}\,\Phi_s^2 f(\kappa a, r_e) \tag{3.300}$$

where $[\mu]_0$ is the intrinsic viscosity in the absence of a double layer, $\bar{b}$ is the normalized average ion size, Φ_s is the reduced ζ potential, and $f(\kappa a, r_e)$ an as yet unknown function of the double-layer parameter, κa, and the axis ratio, r_e.

(c) Transient behaviour

A spheroidal particle subjected to a simple shear flow will execute a periodic motion. For a dispersion of identical particles, all rotating according to

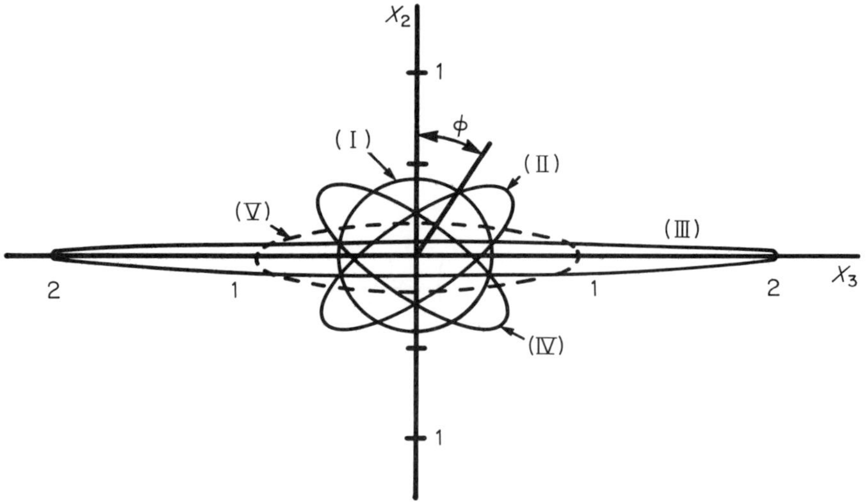

Fig. 3.43. Orientation distribution function for a suspension of initially randomly oriented rods of $r_e = 5$. The plot is of $\sqrt{p(\phi)}$ for various times t/T: (I) $t/T = 0$, (II) 0·03, (III) 0·25, (IV) 0·47, and (V) steady state (after Okagawa *et al.*, 1973).

Jeffery's equations, the orientation distribution will oscillate with frequency $2/T$. (The factor 2 arises because of fore–aft symmetry.) An example of the orientation distribution, $p(\phi, t)$, for an initially random orientation distribution is shown in Fig. 3.43. It can be seen that even dispersions that have initially randomly oriented particles show oscillations. A random distribution (at time $t = 0$) will be random again at time $t = T/2$ when each particle has rotated through 180°. The distribution is non-random at intermediate times. This is a direct consequence of the non-uniformity of the rotational motion. Rod-shaped particles spend more time aligned with the flow than perpendicular to it and, therefore, the probability of finding a particle oriented with the flow is higher than finding one normal to it. The oscillations in the orientation distribution are reflected in a variety of macroscopic suspension properties. For example, the viscosity and normal stress differences will also oscillate with a period $T/2$. An example of the intrinsic viscosity of a suspension of initially randomly oriented rods is shown in Fig. 3.44. In real suspensions such oscillatory behaviour is never observed. Due to a variety of causes, which will be discussed later, these oscillations are damped and after some time the suspension reaches a steady state. The steady-state viscosities and normal stresses were discussed earlier.

Particles with more complex geometries, such as ellipsoids or other triaxial particles, do not rotate with a constant period (see Section 3.1.7). As a result

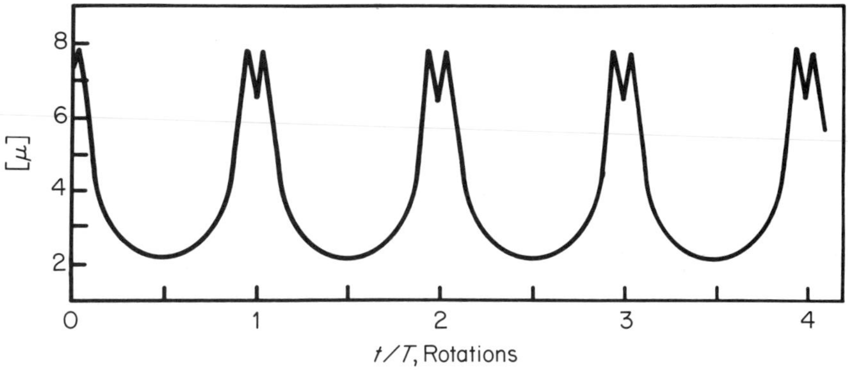

Fig. 3.44. Intrinsic viscosity of an initially randomly oriented suspension of prolate spheroids of axis ratio $r_e = 10$ (after Okagawa *et al.*, 1973).

the viscosity and normal stresses of such suspensions are not periodic. Analogous to the intrinsic viscosity of a suspension of spheroids, the viscosity can be found from their orientation distribution which, for ellipsoids, can be calculated numerically (Zuck and van de Ven, 1989). An example of the viscosity of an initially random suspension of ellipsoids is shown in Fig. 3.45 (for $\mathrm{Pe_r} = \infty$). The results suggest in such suspensions no true steady state exists as the viscosity continues to show fluctuations at all times.

The discussion above applies to dispersions in which Brownian motion is absent ($\mathrm{Pe_r} = \infty$). Allowing for Brownian motion causes a damping of the transient oscillations. As usual, the transient intrinsic viscosity (for spheroids) can be determined from

$$[\mu] = \int_0^{\pi} \int_0^{2\pi} [\mu(\theta, \phi)]\, p(\theta, \phi; t) \sin\theta \mathrm{d}\theta \mathrm{d}\phi \tag{3.301}$$

with p determined by eq (3.298). An example of a transient intrinsic viscosity for a suspension of rods is shown schematically in Fig. 3.46 for $\mathrm{Pe_r} > 1$. One can define a relaxation time, τ_B, as the time at which the amplitude of the damped oscillations has decayed to $1/e$ of its initial value. For a monodisperse system, this relaxation time is approximately given by (Okagawa and Mason, 1977)

$$\tau_B = 1/6Dr \tag{3.302}$$

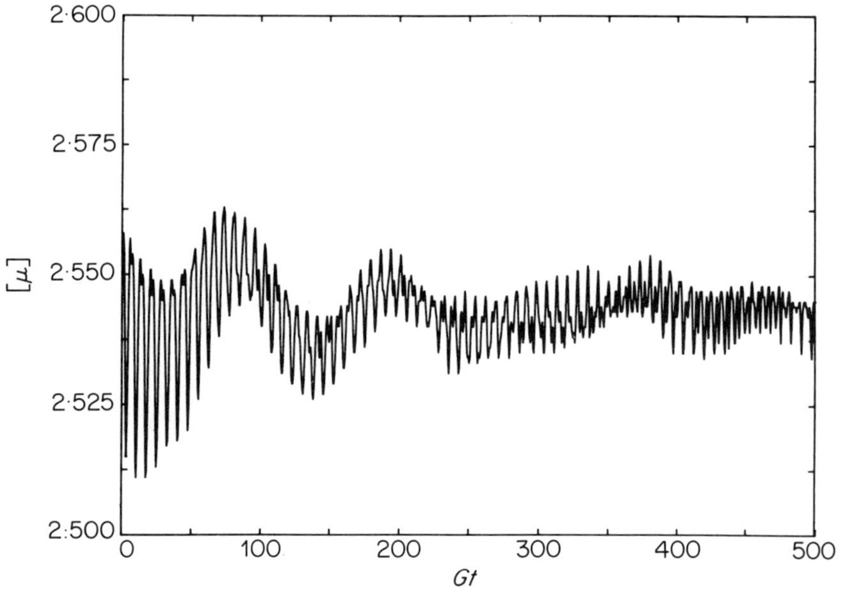

Fig. 3.45. Intrinsic viscosity of a suspension of ellipsoidal particles as a function of time. $a_1/a_2 = 1\cdot42$, $a_1/a_3 = 1\cdot88$, $Pe_r = \infty$.

It can be shown that, in reality, the relaxation is governed by more than one relaxation time. In the limit $Pe_r \to \infty$, the decay is governed by two relaxation times: $\tau_1 = 1/6Dr$ and $\tau_2 = 1/20Dr$ (Jayasuriya and van de Ven, 1985).

When $Pe_r \ll 1$, the rotary convective-diffusion equation can be solved analytically (Section 3.1.6.1.(b)) and the intrinsic viscosity, $[\mu]$, and the intrinsic stress differences, $[\xi_{ij}]$, can be evaluated (Zuzovski *et al.*, 1980b). Examples are shown in Fig. 3.47. It can be seen that in this case $[\mu]$ varies monotonically from its initial to its steady-state value. $[\xi_{21}]$ becomes negative after passing through a positive transient regime and $[\xi_{31}]$ is positive and reaches its steady state monotonically.

(d) Effects of polydispersity

As no real suspension consists of identical particles, it is of interest to consider the effects of polydispersity in particle size and shape on the properties of dispersions. Although the intrinsic viscosity and normalized normal stress differences depend only on the shape of the particle (when $Pe_r = \infty$), the variations in particle volume must also be considered, because they determine the volume fraction, c. For axisymmetric particles of constant volume it is only necessary to consider variations in the equivalent axis ratio, r_e.

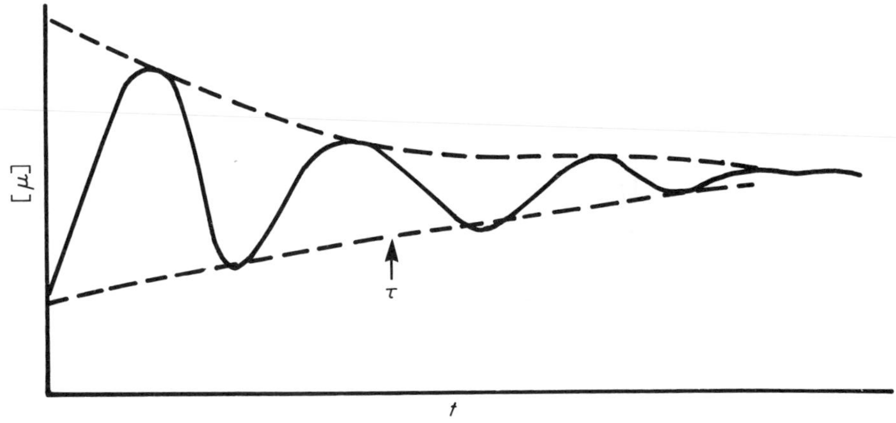

Fig. 3.46. Variation in the intrinsic viscosity in a suspension of rods (schematic) subjected to shear and Brownian motion. The damping in the oscillation can be quantified by a relaxation time, τ, defined as the time where the amplitude of the damped oscillation has decayed to $1/e$ of its initial value.

Consider a dispersion of axisymmetric particles, all having the same volume, with a normalized shape distribution $g(r_e)$. The intrinsic viscosity is given by

$$[\mu] = \int_0^\infty [\mu(r_e)]\, g(r_e)\, dr_e \tag{3.303}$$

with similar expressions for the normal stress differences.

For a dispersion of polydisperse spheroids, not all particles rotate with the same period and, as a result, the viscosity is no longer periodic. Instead polydisperse dispersions exhibit damped oscillations similar to those for monodisperse Brownian spheroids as shown in Fig. 3.46. An example of an experimentally observed transient intrinsic viscosity of a polydisperse suspension of rods is shown in Fig. 3.48 for comparison.

As in the case of dispersions of Brownian spheroids, a relaxation time, τ_σ, can be defined for the damping in polydisperse systems. When the distribution in particle axis ratio, r_e, is Gaussian, τ_σ can be readily calculated. The relaxation occurs in all properties, S, that are orientation-dependent:

$$\langle S \rangle = \int_0^\pi \int_0^{2\pi} S(\theta, \phi)\, p(\theta, \phi; t) \sin\theta d\theta d\phi \tag{3.304}$$

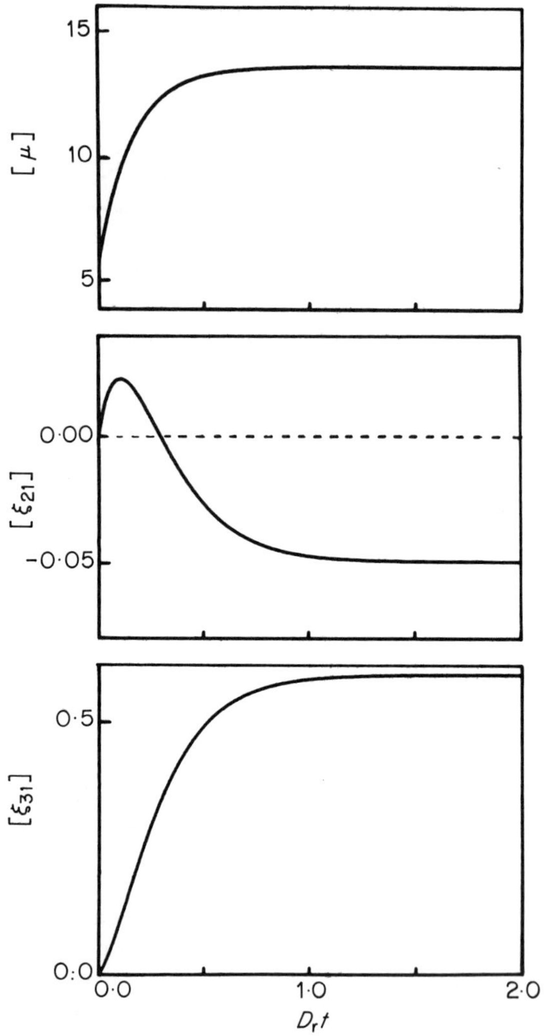

Fig. 3.47. Intrinsic viscosity, $[\mu]$, and normal stress differences, $[\xi_{21}]$, and $[\xi_{31}]$ versus $D_r t$ for $r_e = 10$ and $Pe_r = 0 \cdot 25$. While both normal stresses initially increase with time, $[\xi_{21}]$ becomes negative in the steady state (after Zuzovski *et al.*, 1980b).

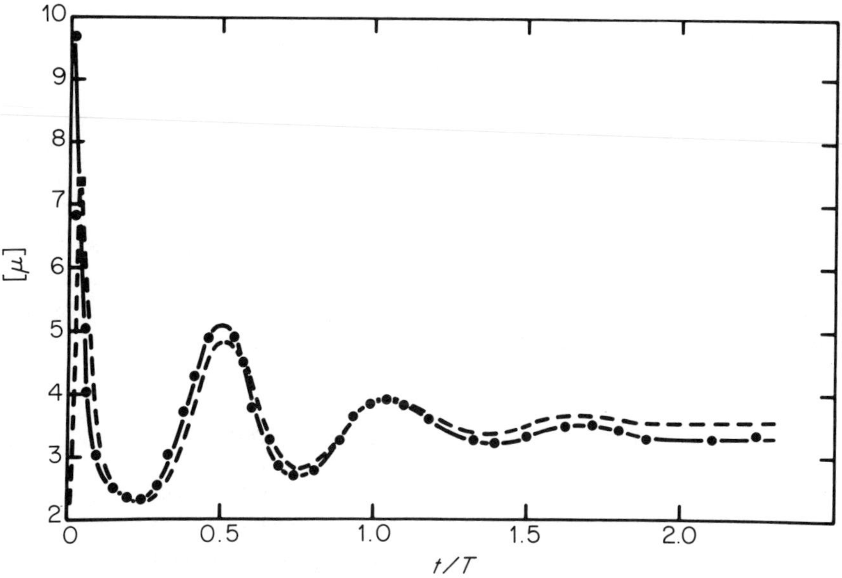

Fig. 3.48. Comparison of experimental data (———) for $[\mu]$ with theoretical predictions (----) calculated using eq (3.305); $\bar{r}_e = 5\cdot2$, standard deviation $\sigma = 0\cdot9$, $G = 2\cdot51\ \mathrm{s}^{-1}$; initially aligned at $\theta = \pi/2$ and $\phi = 0$ (after Ivanov *et al.*, 1982).

For a polydisperse system

$$\overline{\langle S\rangle} = \int_0^\infty \langle S\rangle\, g(r_e)\, \mathrm{d}r_e \tag{3.305}$$

When $g(r_e)$ is Gaussian,

$$g(r_e) = \frac{1}{(2\pi\sigma^2)^{1/2}}\, \mathrm{e}^{-(r_e - \bar{r}_e)/2\sigma^2} \tag{3.306}$$

where $\bar{r}_e$ is the average axis ratio and σ the spread in the distribution.

The integral occurring in eq (3.305) can be evaluated by expanding $\langle S\rangle$ in a Fourier cosine series (Okagawa *et al.*, 1973; Cerda *et al.*, 1981):

$$\langle S\rangle = \tfrac{1}{2}A_0 + \sum_{n=1}^{\infty} A_n(r_e)\cos\alpha_n t \tag{3.307}$$

where $\alpha_n = 4\pi n/T$, T being the period of rotation (eq (3.91)) and $A_n(r_e)$ are the Fourier coefficients

$$A_n(r_e) = \frac{2}{T}\int_0^T \langle S\rangle \cos\alpha_n t \, dt \tag{3.308}$$

For small spreads σ, the coefficients A_n can be expanded in powers of σ and can be found from eq (3.308) with S being a function of r_e. Next, the integral in eq (3.305) can be integrated term by term after substitution of eq (3.307). When $\bar{r}_e > 1 + 0(\sigma)$, this results in:

$$\overline{\langle S\rangle} = \tfrac{1}{2}A_0 + \frac{1}{(2\pi)^{1/2}}\sum_{n=1}^{\infty} A_n(\bar{r}_e)\cos\bar{\alpha}_n t \, e^{-\lambda_n t^2} \tag{3.309}$$

where

$$\lambda_n = \frac{2n^2G^2\sigma^2(\bar{r}_e - 1)^2}{(\bar{r}_e + 1)^4} \tag{3.310}$$

It can be seen that the relaxation time of the nth harmonic can be defined as

$$\tau_n = \frac{1}{G\sigma}\,\frac{(\bar{r}_e^2 + 1)^2}{2^{1/2}n\,|\bar{r}_e - 1|} \tag{3.311}$$

When $(\bar{r}_e - 1) = 0(\sigma)$, the relaxation times are

$$\tau_n = \frac{4 \cdot 401}{nG\sigma^2} \tag{3.312}$$

Analogous to relaxation due to Brownian motion, relaxation due to polydispersity is governed by more than one relaxation time. When both Brownian motion and polydispersity are important, we have the approximate relation

$$\frac{1}{\tau} \simeq \frac{1}{\tau_B} + \frac{1}{\tau_\sigma} \tag{3.313}$$

τ_B and τ_σ being leading relaxation times ($n = 1$).

It follows from the above analysis that the average axis ratio, $\bar{r}_e$, and the spread, $\bar{\sigma}$, from the frequency and the damping of the transient oscillations can be determined.

Owing to the linearity of the creeping-flow equation, all particle motions are reversible in the absence of Brownian motion. As a consequence, the damped oscillations in the intrinsic viscosity (or any other property) can be retraced by reversing the flow at a given time. For polydisperse dispersions in the absence of rotary Brownian motion, the reversal is perfect and the suspension is said to have a perfect memory. Allowing for perturbing effects (rotary Brownian motion, fluctuations in the flow field, irreversible collisions between particles, etc.) results in a fading memory. Because the effects of reversibility of flow can be more readily studied with optical methods, they will be considered in more detail at the end of the following section.

3.1.12 Rheo-optics of dilute dispersions

3.1.12.1 General considerations

When a suspension containing particles is subjected to a fluid flow, the particles will execute rotations, which are periodic for axisymmetric particles and non-periodic for particles with less symmetry. When a light beam passes through a flowing suspension, the amount of light scattered by each particle depends on its orientation. As particle orientations are continuously changing, the intensity of the light scattered by each single particle changes with a frequency corresponding to the period of rotation of the particle. Because the period depends on particle shape, it follows that the shape can be deduced from observing the variations in the intensity of scattered light. These variations are similar to the transient variations in rheological properties discussed in the previous section and show up in a variety of optical properties such as light scattering, light transmittance, streaming birefringence, linear dichroism, etc. Several aspects of the way in which optical measurements can be used to determine the properties of suspended particles are discussed below.

3.1.12.2 Angular light scattering

Here we will discuss the light-scattering theory for particles whose maximum dimension, ℓ, is small compared to the wavelength, λ, of the incident light or, more precisely, for particles for which

$$\frac{2\pi n_0 \ell}{\lambda}(m^2-1) \ll 1 \qquad (3.314)$$

where n_0 is the refractive index of the medium and m is the ratio of the refractive index, n_p, of the particle to n_0. When a light beam of wavelength λ and intensity I_0 impinges on a particle of volume V, it will scatter light in all directions. According to the Rayleigh–Debye–Gans theory, for a perpendicularly polarized incident light beam, the intensity of the scattered light in a direction (θ_s, ϕ_s) at a distance r from the particle is given by:

$$I_2(\theta_s, \phi) = I_0 \frac{9\pi^2 V^2 n_0^2}{r^2 \lambda^4} \left(\frac{m^2-1)}{m^2+1}\right)^2 \left[(1-\sin^2\theta_s \sin^2\phi_s) P(\theta_s, \phi_s; \theta, \phi, \psi)\right] \quad (3.315)$$

where P is a form factor which depends on the scattering geometry and the particle shape.

Many different scattering geometries can be considered. Okagawa and Mason (1977) considered two classes which are readily accessible experimentally: one in which the bisectrix (i.e. the direction which bisects the angle between I_0 and I_s) and another in which the incident light coincides with one of the fixed coordinates X_i. For such geometries $\phi_s = 0$ and the only scattering angle to be considered is θ_s. In this case

$$\frac{I_s}{I_0} = K\, P(\theta_s; \theta, \phi, \psi) \quad (3.316)$$

K being a constant which is independent of scattering angle. For ellipsoids the form factor, P, is given by (Paterson, 1939)

$$P(\theta_s; \theta, \phi, \psi) = \left[\frac{3}{u^3} (\sin u - u \cos u)\right]^2 \quad (3.317)$$

where

$$u = \frac{4\pi n_0}{\lambda} (a_1^2 \cos^2\alpha + a_2^2 \cos^2\beta + a_3^2 \cos^2\gamma) \sin\left(\frac{\theta_s}{2}\right) \quad (3.318)$$

Cosα, cosβ and cosγ are the direction cosines of the semi-axes a_1, a_2 and a_3, respectively with the bisectrix. They can be found from the scattering geometry. For light incident along the X_2 axis and scattering in the X_2X_3 plane (Jayasuriya and van de Ven, 1986a):

$$\begin{aligned} \cos\alpha &= \alpha_{13} \cos(\theta_s/2) - \alpha_{12} \sin(\theta_s/2) \\ \cos\beta &= \alpha_{23} \cos(\theta_s/2) - \alpha_{22} \sin(\theta_s/2) \\ \cos\gamma &= \alpha_{33} \cos(\theta_s/2) - \alpha_{32} \sin(\theta_s/2) \end{aligned} \quad (3.319)$$

where the components α_{ij} are given by eq (3.79). For spherical particles $P=1$.

A convenient way to represent light-scattering data is to employ terms of the Rayleigh ratio, $R(\theta_s)$, defined as

$$R(\theta_s) = \frac{I_s(\theta_s)}{I_0} r^2 \tag{3.320}$$

When an ellipsoid is subjected to shear, its orientation varies with time. This is reflected in variations in the components α_{ij} which depend on the particle orientation (θ, ϕ, ψ). To obtain the Rayleigh ratio for a suspension, we must average over all orientations:

$$\langle R(\theta_s)\rangle = \int_0^{\pi}\int_0^{2\pi}\int_0^{2\pi} R(\theta_s; \theta, \phi, \psi)\, p(\theta, \phi, \psi) \sin\theta \, d\phi \, d\psi \tag{3.321}$$

In general, these integrals must be evaluated numerically.

The Rayleigh ratio of a spheroidal particle ($a_2 = a_3$) subjected to a simple shear (or a flow with $\gamma < \gamma_c$) in the absence of Brownian motion is, for many

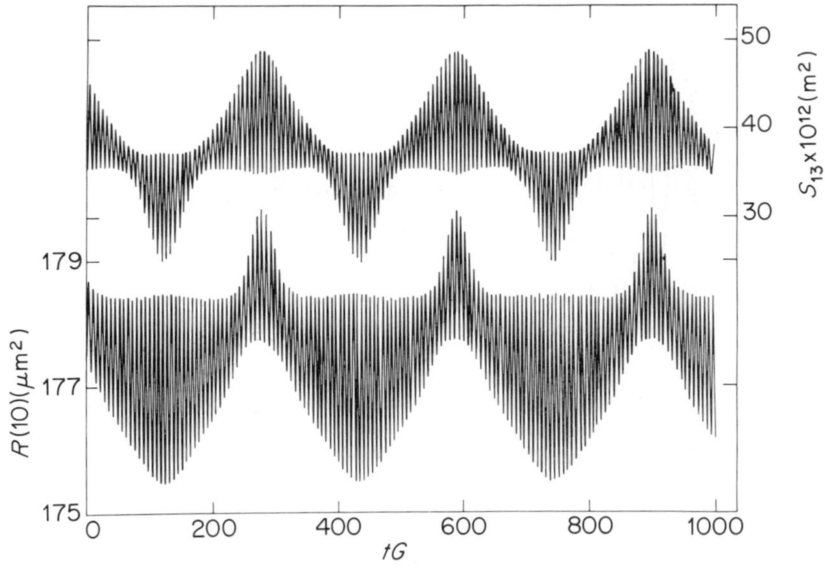

Fig. 3.49. Variation in the Rayleigh ratio, $R(10)$, and projected area, S_{13}, with time, Gt, for a single ellipsoid in simple shear. $a_1 = 473$ nm, $a_2 = 334$ nm, $a_3 = 248$ nm, $n_p = 1{\cdot}59$, $n_0 = 1{\cdot}46$, $\lambda = 633$ nm. Initial orientation: $\theta = 30°$, $\phi = 40°$, $\psi = 50°$ (after Jayasuriya and van de Ven, 1986a).

experimentally convenient geometries, periodic with period $T/2$, where T is given by eq (3.91) (or a modification thereof). The factor $\frac{1}{2}$ arises because after half a rotation the particle finds itself at an orientation with similar symmetry. Ellipsoidal particles in simple shear do not rotate with a constant period and the Rayleigh ratio will be non-periodic. An example is shown in Fig. 3.49. Although the variations are non-periodic, Fig. 3.49 shows that quasi-periodic patterns repeat themselves at regular intervals. Also shown in the same figure is the projected area, S_{13}, of the ellipsoid on the X_1X_3 plane, given by

$$S_{13} = \pi\, a_1 a_2 a_3 \left[\left(\frac{1}{a_1^2} - \frac{1}{a_3^2} \right) \alpha_{21}^2 + \left(\frac{1}{a_2^2} - \frac{1}{a_3^2} \right) \alpha_{22}^2 + \frac{1}{a_3^2} \right]^{1/2} \quad (3.322a)$$

For spheroids ($a_2 = a_3$), this expression reduces to

$$S_{13} = \pi\, a_1 a_2 \left[1 - \left(1 - \frac{a_2^2}{a_1^2} \right) \sin^2\theta \cos^2\phi \right]^{1/2} \quad (3.322b)$$

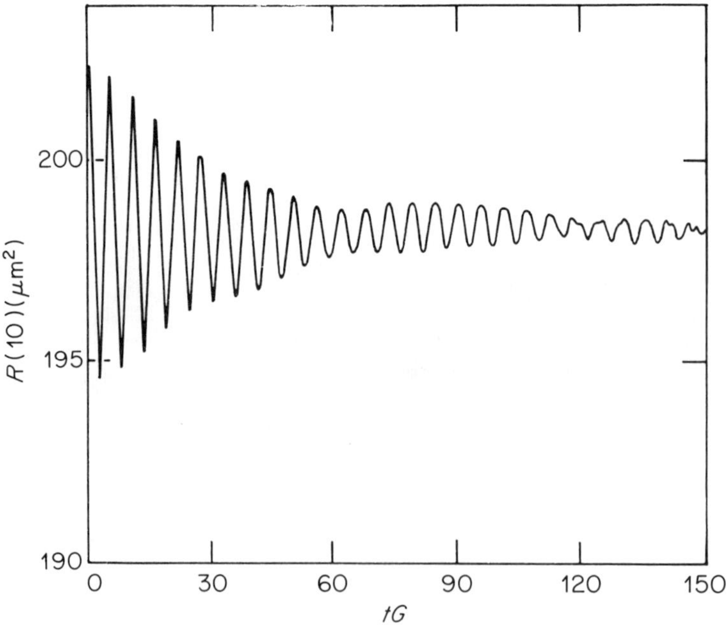

Fig. 3.50. Rayleigh ratio, $R(10)$, as a function of time for a suspension of initially randomly oriented ellipsoids. For the conditions see Fig. 3.49 (after Jayasuriya and van de Ven, 1986b).

Both $R(\theta_s)$ and S_{13} depend on the particle orientation.

An example of $R(\theta_s)$ for a suspension of ellipsoidal particles initially randomly oriented is shown in Fig. 3.50.

In suspensions of spheroidal particles, Brownian motion and polydispersity cause an attenuation in the periodic oscillations in $R(\theta_s)$. For a distribution $g(a_1, a_2)$ in particle axes:

$$R(\theta_s) = \langle \overline{R(\theta_s)} \rangle = \int_0^\infty \int_0^\infty R(\theta) g(a_1, a_2)\, da_1 da_2 \tag{3.323}$$

Variations in a_1 and a_2 affect both the volume, V, of the particles (for ellipsoids $V = 4\pi a_1 a_2 a_3/3$) and the form factor, P. An example of the Rayleigh ratio, $R(\theta_s)$, of a suspension of oblate spheroids is given in Fig. 3.51. The damping of the oscillations can be characterized by a relaxation time τ, at which the amplitude of the oscillations decays to $1/e$ of its initial

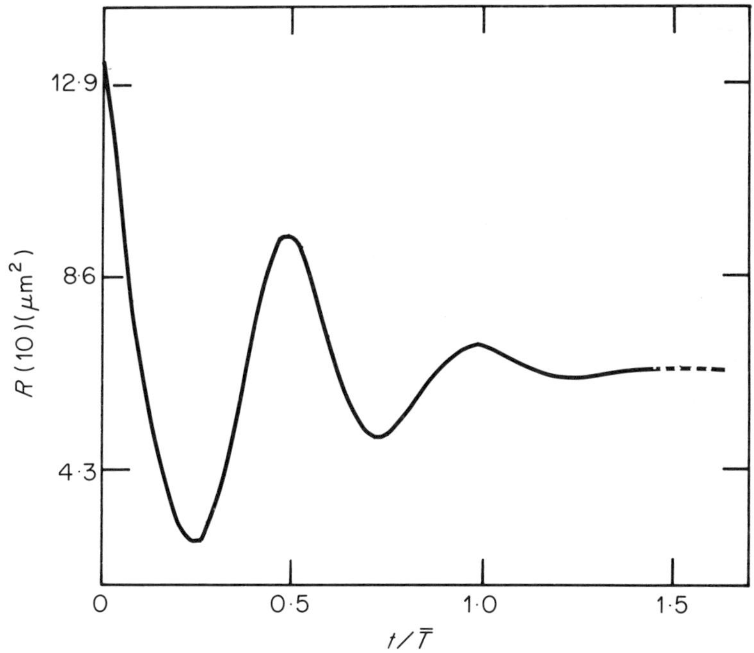

Fig. 3.51. Rayleigh ratio, $R(10)$, as a function of time for a suspension of initially randomly oriented oblate spheroids. $\bar{r}_e = 0{\cdot}34 \pm 0{\cdot}07$ (characteristic of human erythrocytes). $\bar{T}$ is the period corresponding to $\bar{r}_e$ (after Cerda *et al.*, 1981).

value. For a suspension of spheroidal particles of constant volume, but with variations in r_e, eq (3.323) becomes:

$$\overline{R(\theta_s)} = \int_0^\infty R(\theta_s) g(r_e)\, dr_e \qquad (3.324)$$

In the absence of diffusion, $R(\theta_s)$ can be expanded in a Fourier series which can be integrated term by term. When $g(r_e)$ is a Gaussian distribution of mean r_e and standard deviation σ (and when $r_e - 1 \neq 0(\sigma)$) (Cerda *et al.*, 1981) (cf. eq (3.309))

$$R(\theta_s) = R_\infty(\theta_s) + \sum_{n=1}^{\infty} f_n(t/T)\, e^{-\lambda_n t^2} \qquad (3.325)$$

where $R_\infty(\theta_s)$ is the Rayleigh ratio at time $t = \infty$, and f_n are known periodic functions of time. The relaxation time of the nth harmonic is given by eq (3.310). When the polydispersity is large, eq (3.324) must be integrated numerically.

Often it is easier to obtain information from the Fourier transform of $R(\theta_s)$, especially for polydisperse or multimodal particle shape distributions. The Fourier transform is defined as

$$\Delta F(\Omega) = \int_0^\infty \cos(\Omega t)\, \Delta R(\theta_s, \mathrm{t})\, dt \qquad (3.326)$$

where $\Delta R(\theta_s, t) = R(\theta_s, t) - R(\theta_s, \infty)$.

Performing the integration, using the expansion described by eq (3.325), results in

$$\Delta F(\Omega) = \frac{1}{8} \sum_{n=1}^{\infty} \frac{A_n(\bar{r}_e)}{f_n} \left(e^{-(\Omega + \bar{\alpha}_n)^2/2f_n^2} + e^{-(\Omega - \bar{\alpha}_n)^2/2f_n^2} \right) \qquad (3.327)$$

where

$$A_n(\bar{r}_e) = \frac{2}{\bar{T}} \int_0^{\bar{T}} R(\theta_s, t) \cos\bar{\alpha}_n t\, dt \qquad (3.328)$$

with $\bar{\alpha}_n = 4\pi n/\bar{T}$, $\bar{T}$ determined by $\bar{r}_e$, and

$$f_n = \frac{2n\,G\sigma\,r_e^{-2} - 1}{(r_e + r_e^{-1})^2} \tag{3.329}$$

It can be seen from eq (3.327) that the frequency spectrum consists of Gaussian pulses (the width of which is determined by σ, see eq (3.329)), centred at the fundamental mean frequency $\bar{\alpha}_1$ and integer multiples thereof. Similarly it can be shown (Cerda *et al.*, 1981) that for multimodal dispersions consisting of several fractions i of mean $\bar{r}_{ei}$ and spread σ_i, the frequency spectrum consists of Gaussian pulses centred at $\bar{\alpha}_{1i}$ (and integer multiples), where $\bar{\alpha}_{1i}$ is determined by the mean period of each fraction i.

An example of an experimentally observed transient of a bimodal dispersion, together with its frequency spectrum is shown in Fig. 3.52. In these experiments the higher order integer multiples of the two main frequencies do not show up. The two main frequencies correspond to the periods of the two fractions ($\nu_{1,2} = 2/\bar{T}_{1,2}$).

For ellipsoidal particles the behaviour is more complex. An example of

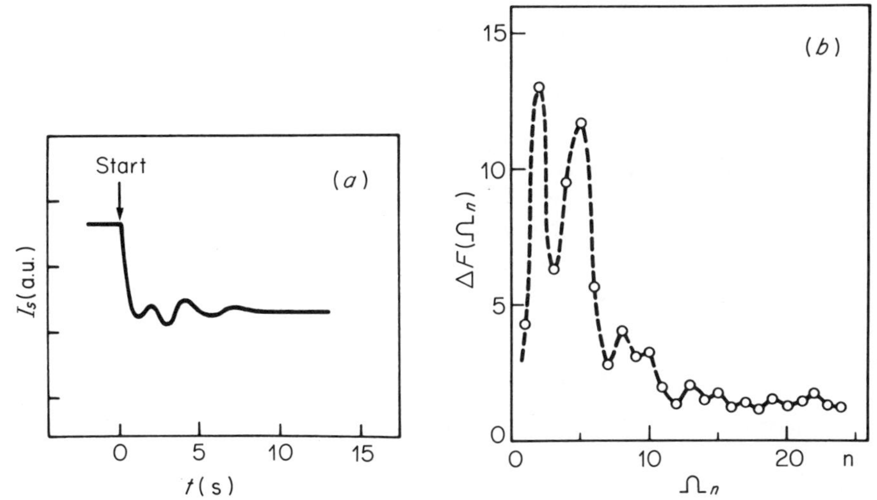

Fig. 3.52. (*a*) Intensity, I_s, of scattered light at $\theta_s = 10°$ of a bimodal suspension of hardened frog and human erythrocytes as a function of time, t, subjected to a simple shear of $G = 5\ s^{-1}$. (*b*) Corresponding discrete Fourier transform showing two peaks corresponding to the rotational frequencies, Ω_n, of both types of cells (after Cerda *et al.*, 1981).

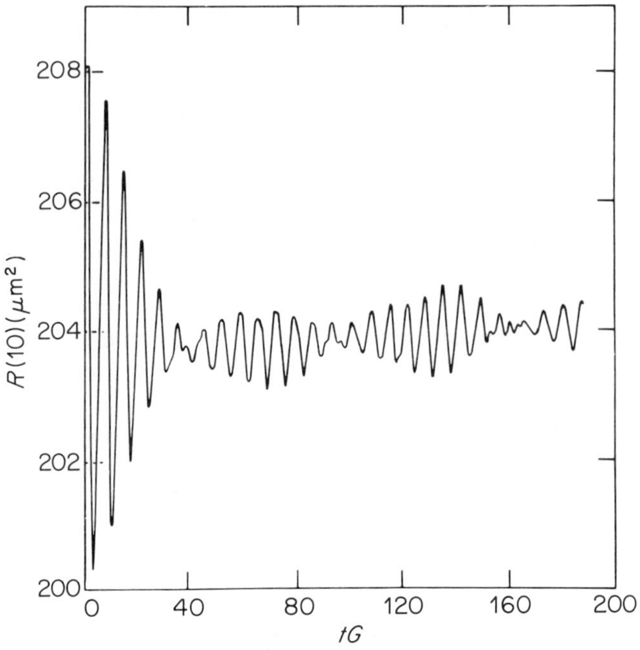

Fig. 3.53. Variation in the Rayleigh ratio, $R(10)$, with time for a suspension of initially random ellipsoids: $a_1 = 470 \pm 25$ nm, $a_2 = 330 \pm 33$ nm, and $a_3 = 250 \pm 33$ nm. Other parameters as in Fig. 3.49 (after Jayasuriya and van de Ven, 1986b).

$R(\theta_s)$ for a polydisperse system of initially randomly oriented particles is given in Fig. 3.53. It can be seen that for such systems no steady state is reached, but instead fluctuations in $R(\theta_s)$ occur which resemble acoustic "beats". Brownian motion causes damping and distorts the beat pattern. An example is shown in Fig. 3.54.

3.1.12.3 Small-angle light scattering

The theory outlined in Section 3.1.12.2 applies to particles that are small compared to the wavelength of light. For larger particles, light absorption and reflection at the surface can occur and the light-scattering pattern can be very different. Except for spheres, very few theoretical calculations for larger particles have been performed. However, the theory outlined above still applies at small scattering angles, θ_s. This is because $P(\theta_s) \to 1$ as $\theta_s \to 0$. By directing a laser beam through a suspension subjected to a simple shear and focusing the forward-scattered light onto a photo-diode matrix (eliminating the directly transmitted light with a beam stop), it is possible to

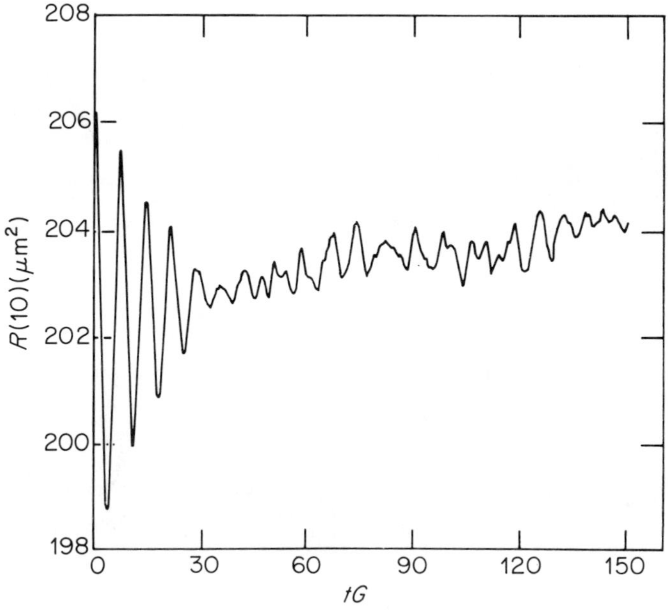

Fig. 3.54. Rayleigh ratio for a suspension of ellipsoidal particles subjected to shear and Brownian motion. $\mathrm{Pe}_r = G/D_r \approx 250$. Other parameters are as in Fig. 3.53 (after Jayasuriya and van de Ven, 1986b).

measure the spatial variation in the forward-scattered light. Such measurements have been performed by Salem and Fuller (1985) on suspensions of hardened human erythrocytes. Examples of such measurements of suspensions subjected to a simple shear at various times are shown in Fig. 3.55. Like $R(\theta_s)$, the system shows damped oscillations before attaining the steady state.

3.1.12.4 Light transmittance

For non-absorbing particles, the intensity of the transmitted light, I_t, is the difference between the intensity of the incident light and the sum of the intensities of the light scattered in all directions. The transmittance is defined as

$$T = I_t / I_0 \tag{3.330}$$

and is given by the Beer–Lambert law:

$$T = e^{-\nu d} \tag{3.331}$$

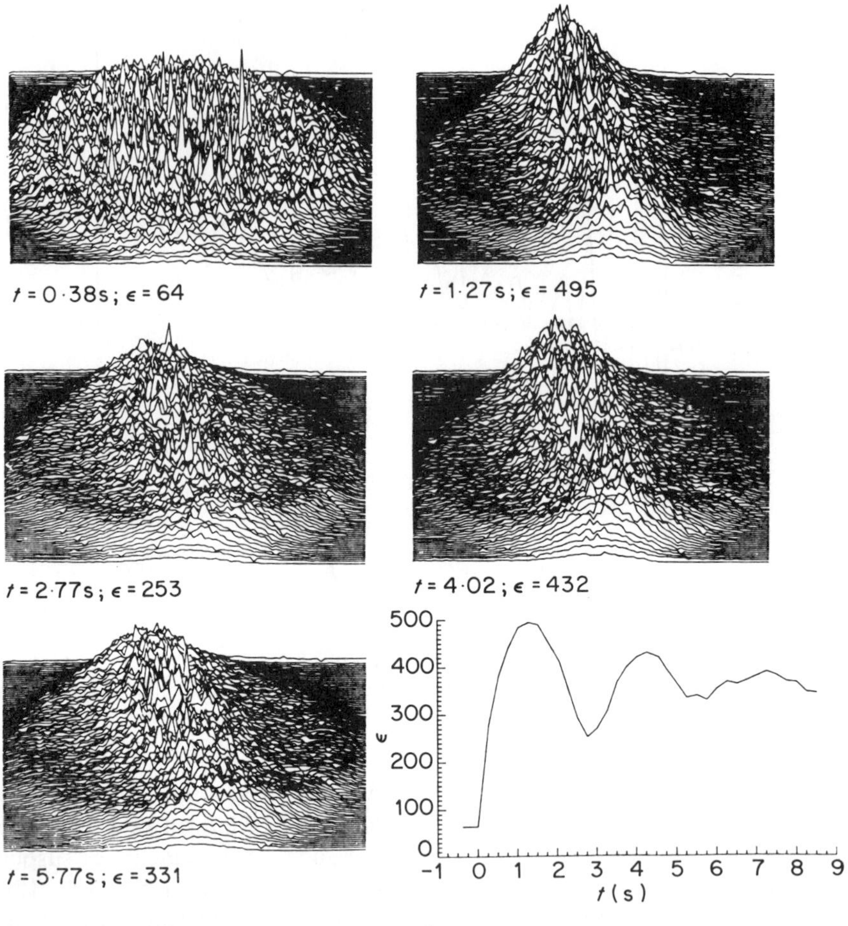

Fig. 3.55. Experimental observations of small angle light scattering from a suspension of hardened human erythrocytes subjected to simple shear. Epsilon is a measure of the anisotropy (after Salem and Fuller, 1985 with permission).

where ν in the turbidity of the suspension and d the path length, i.e. the length of the light beam traversing the suspension. Neglecting secondary scattering and particle interactions, which is allowed for dilute suspensions, eq (3.331) reduces to

$$T = 1 - \nu d \tag{3.332}$$

In this case the turbidity is simply proportional to the number concentration, n, of particles and the amount of light scattered by each particle, which can be expressed in terms of the scattering cross-section C_{sca} as:

$$\nu = n\langle C_{sca}\rangle \tag{3.333}$$

where

$$C_{sca} = 2\pi \int_0^{\pi} R(\theta_s) \sin\theta_s \, d\theta_s$$

and

$$\langle C_{sca}\rangle = \int_0^{\pi}\int_0^{2\pi}\int_0^{2\pi} C_{sca}(\theta, \phi, \psi)\, p(\theta, \phi, \psi; t) \sin\theta \, d\theta d\phi d\psi \tag{3.334}$$

with $R(\theta_s)$ given by eq (3.320) and p by eq (3.102).

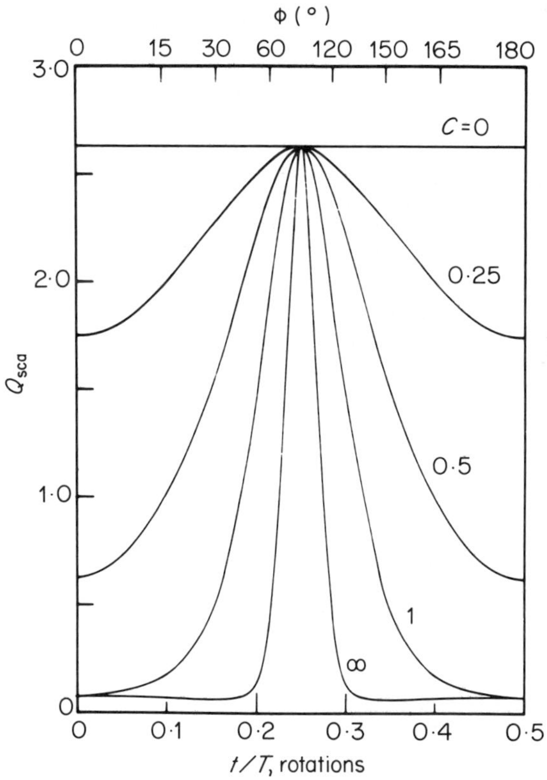

Fig. 3.56. Scattering efficiency of a single oblate spheroid subjected to simple shear for various orbit constants C. $r_e = 0{\cdot}4$, $n_0 = 1{\cdot}44$, $n_p = 1{\cdot}40$, $\lambda = 633$ nm, $V = 91$ μm^3 (after Okagawa and Mason, 1977).

Let us again consider scattering with an incident beam along the X_2 axis, then we can define a scattering efficiency Q_{sca} as

$$Q_{sca} = C_{sca}/S_{13} \tag{3.335}$$

where S_{13} is given by eq (3.322). The parameter S_{13} is of interest in its own right as, for certain particles which absorb or reflect light, the turbidity is given by (Okagawa and Mason, 1977):

$$\nu = n\langle S_{13}\rangle \tag{3.336}$$

For observations in the X_1X_2 plane of identical spheroids subjected to simple shear in the absence of Brownian motion, $R(\theta_s)$, Q_{sca}, C_{sca}, S_{13}, T and

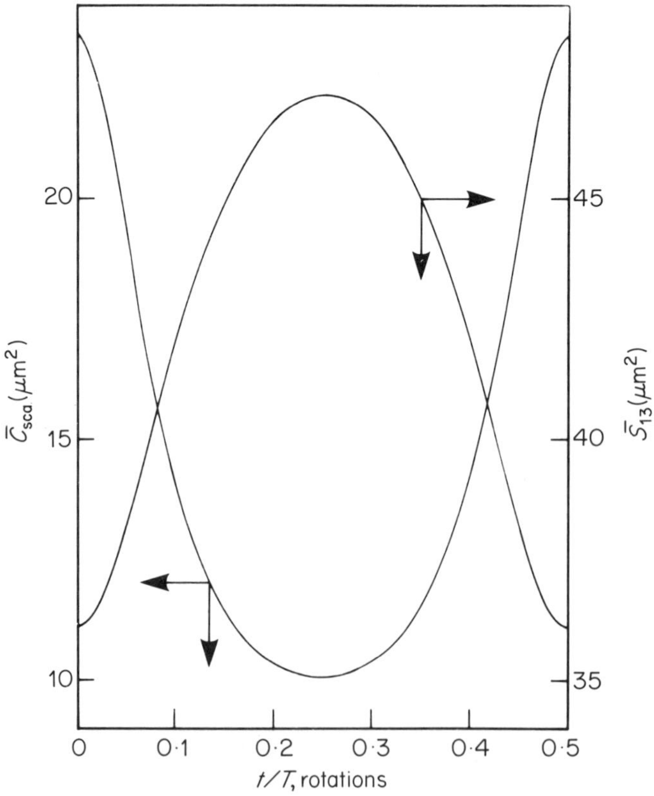

Fig. 3.57. Mean scattering cross-section, $\bar{C}_{sca}$, and mean projected area, $\bar{S}_{13}$, for initially isotropic dilute suspensions of oblate spheroids ($r_e = 0 \cdot 4$) (conditions are as those for Fig. 3.56) (after Okagawa and Mason, 1977).

ν are all periodic with a period $T/2$. The scattering efficiency of a single oblate spheroid is shown in Fig. 3.56 for various orbit constants, C. When $C=0$, the spheroid axis of revolution is aligned with the vorticity axis, X_1, and the light beam impinges the spheroid edge-on at all times. When $C=\infty$, the spheroid axis rotates in the X_2X_3 plane and the light beam "sees" the particle face-on when $\phi=0°$ and edge-on when $\phi=90°$. Somewhat surprisingly, it can be seen that oblate spheroids scatter more light when oriented edge-on than when oriented face-on. This is confirmed experimentally. If particles are made absorbing by silvering, then the scattering becomes a maximum when $\phi=0°$ (Ennis *et al.*, 1978).

An example of the scattering and geometrical cross-sections of an ensemble of identical spheroids is shown in Fig. 3.57. It can be seen that when C_{sca} is maximum, S_{13} is minimum and vice versa. Here C_{sca} is minimum at $t=\frac{1}{4}T$, because at this time more disks are oriented face-on (minimum scattering) than at time $t=0$ when the orientation distribution is random.

Polydispersity results in damped oscillations. An example is given in Fig. 3.58.

An example of the transmittance as a function of time for monodisperse ellipsoids is given in Fig. 3.59, which also includes the results for the Rayleigh ratio at 10°. It can be seen that the transmittance contains the same information as $R(\theta_s)$. The data can be considered to be complementary: as T

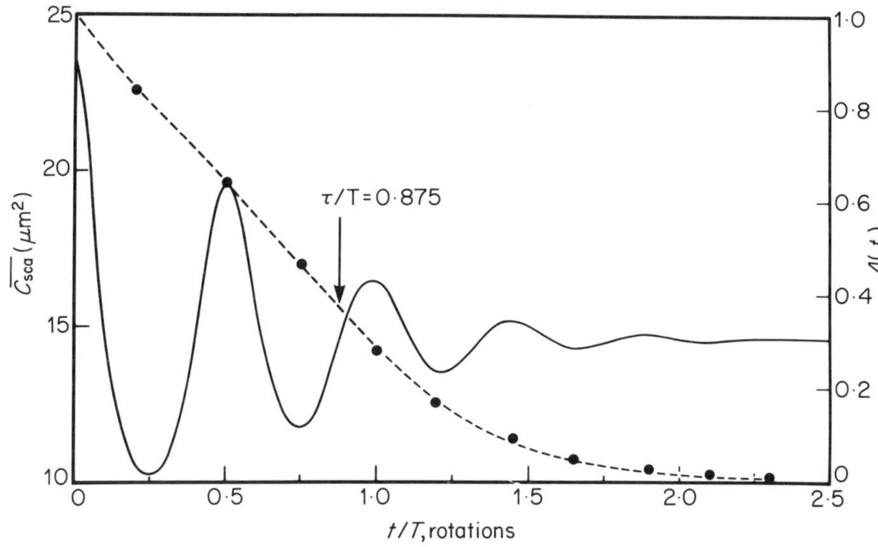

Fig. 3.58. Mean scattering cross-section calculated for an initially isotropic suspension of polydisperse oblate spheroids ($r_e=0{\cdot}40\pm0{\cdot}07$) subjected to a simple shear. The damping of the relative amplitude, $A(t)$, is also shown (after Okagawa and Mason, 1977).

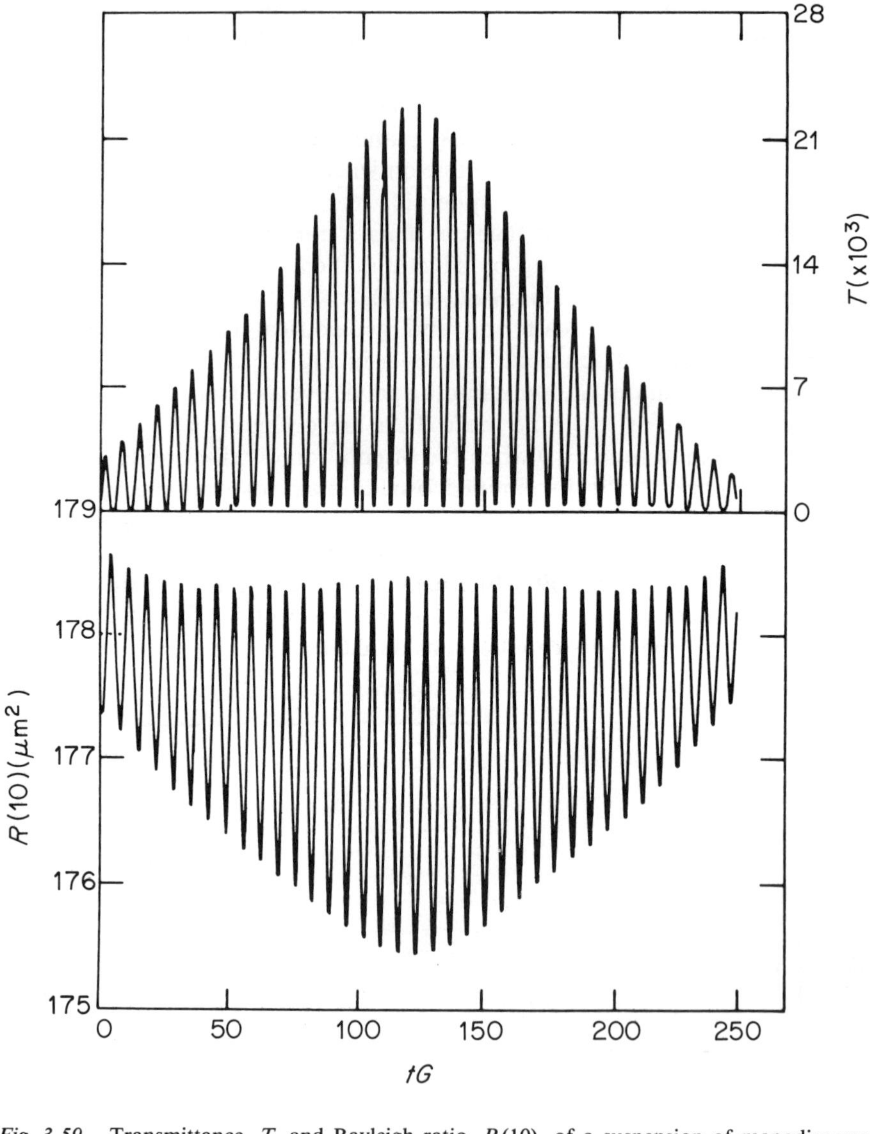

Fig. 3.59. Transmittance, *T*, and Rayleigh ratio, *R*(10), of a suspension of monodisperse ellipsoids subjected to simple shear. Conditions are as those in Fig. 3.49 (after Jayasuriya and van de Ven, 1986a).

increases, $R(\theta_s)$ decreases and vice versa. An example of an experimental transient of a suspension of hardened erythrocytes is shown in Fig. 3.60. Also shown is the relaxation after switching off the shear when the orientation distribution relaxes to a random distribution of orientations.

3.1.12.5 Streaming birefringence and linear dichroism

When a suspension of non-spherical particles is subjected to shear, the orientation distribution is no longer isotropic. As a result, the refractive index of the suspension is no longer a scalar, but becomes a tensor. In general this tensor is complex:

$$\boldsymbol{n} = \boldsymbol{n}' - \mathrm{i}\boldsymbol{n}'' \qquad (3.337)$$

where $\boldsymbol{n}'$ is related to the refracted light, and $\boldsymbol{n}''$ to the amount of light absorbed or attenuated by scattering.

For Rayleigh–Debye–Gans scatterers subjected to a shear in the X_2X_3 plane (cf. Fig. 3.1), $\Delta n'$ and $\Delta n''$ can be defined as the differences in the principal axes of $\boldsymbol{n}'$ and $\boldsymbol{n}''$. $\Delta n'$ is called linear birefringence and $\Delta n''$ linear dichroism. For spheroids these quantities can be expressed in terms of orientational averages as (Johnson *et al.*, 1985):

$$\Delta n' = \frac{n_0 n}{2\epsilon_0}(\alpha_1 - \alpha_2)(\langle \sin^2\theta \sin 2\phi \rangle^2 + \langle \sin^2\theta \cos 2\phi \rangle^2)^{1/2} \qquad (3.338)$$

and

$$\Delta n'' = \frac{n_0 n\, k^3}{12\pi\, \epsilon_0}(\alpha_1^2 - \alpha_2^2)(\langle \sin^2\theta \sin 2\phi \rangle^2 + \langle \sin^2\theta \cos 2\phi \rangle^2)^{1/2} \qquad (3.339)$$

where n_0 is the refractive index of the medium, n the particle number concentration, ϵ_0 the permittivity of free space, α_1 and α_2 the principal values of the polarizability tensor, and k the wave number ($k = 2\pi/\lambda$). For such suspensions, $\Delta n'$ and $\Delta n''$ are coaxial and the angle χ between the direction of $\Delta n'$ (or $\Delta n''$) and the X_3 axis is given by

$$\tan 2\chi = \frac{\langle \sin^2\theta \sin 2\phi \rangle}{\langle \sin^2\theta \cos 2\phi \rangle} \qquad (3.340)$$

It can be seen that for suspensions of spheroids, $\Delta n'$ and $\Delta n''$ contain the same information. As for all other optical properties discussed above, $\Delta n'$, $\Delta n''$ and χ are periodic functions for identical particles in the absence of

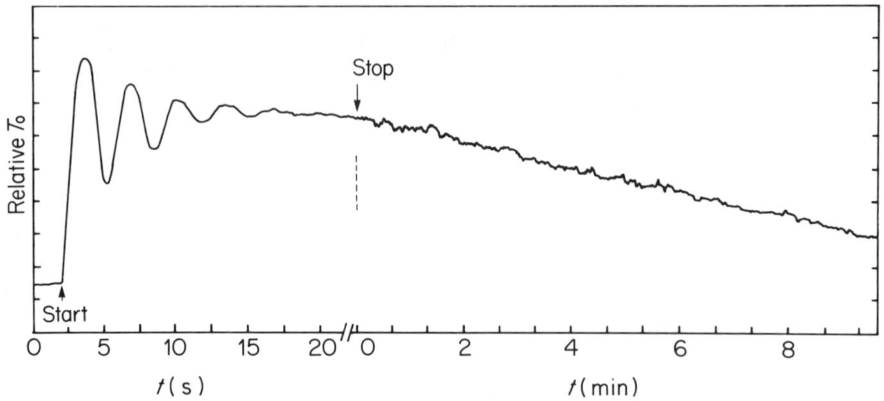

Fig. 3.60. Experimentally observed transient transmittance of a suspension of hardened human erythrocytes ($N = 4 \times 10^8$ particles ml^{-1}, $G = 1{\cdot}6\ s^{-1}$). At the time indicated by "stop", the shear was switched off and the system was allowed to relax towards a random distribution of orientations (after Okagawa and Mason, 1977).

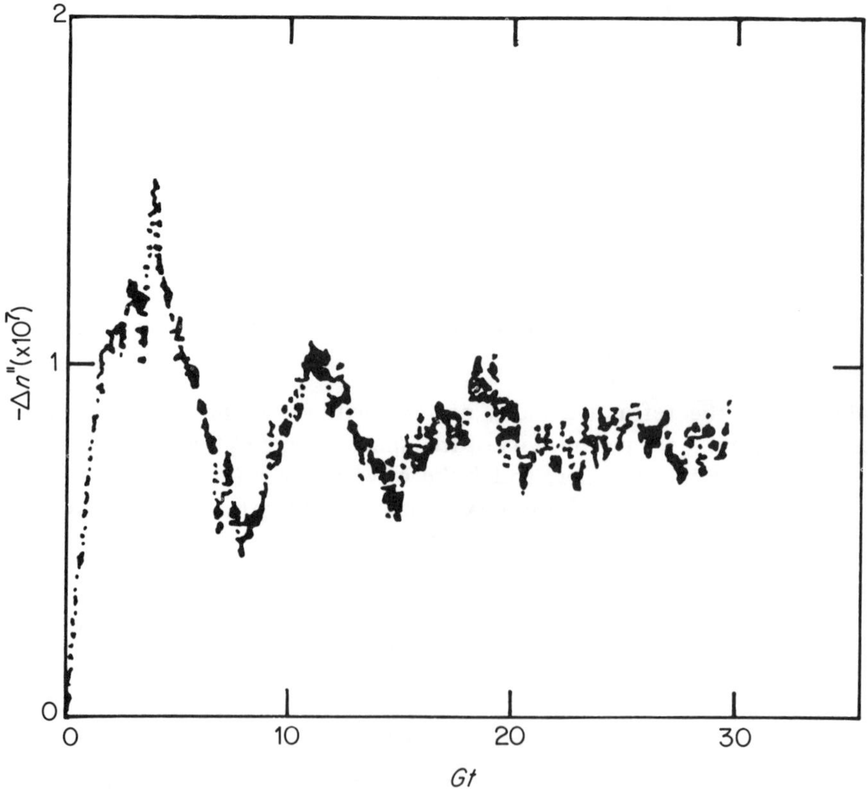

Fig. 3.61. An example of the observed linear dichroism as a function of time for suspensions of bentonite in 90/10% glycerol/water (after Frattini and Fuller, 1984).

Brownian motion. Brownian motion and polydispersity introduce damping in the oscillations with a relaxation time, approximately given by

$$\frac{1}{\tau}=\frac{1}{\tau_B}+\frac{1}{\tau_\sigma} \tag{3.341}$$

where τ_B and τ_σ are the leading relaxation times for Brownian motion and polydispersity.

It can be shown (see next section) that

$$\tau_B = 1/6D_r \tag{3.342}$$

D_r being the rotary diffusion coefficient; τ_σ is given by τ_1 (see eq (3.310)).

Examples of transients of $\Delta n''$ and χ are shown in Figs. 3.61 and 3.62.

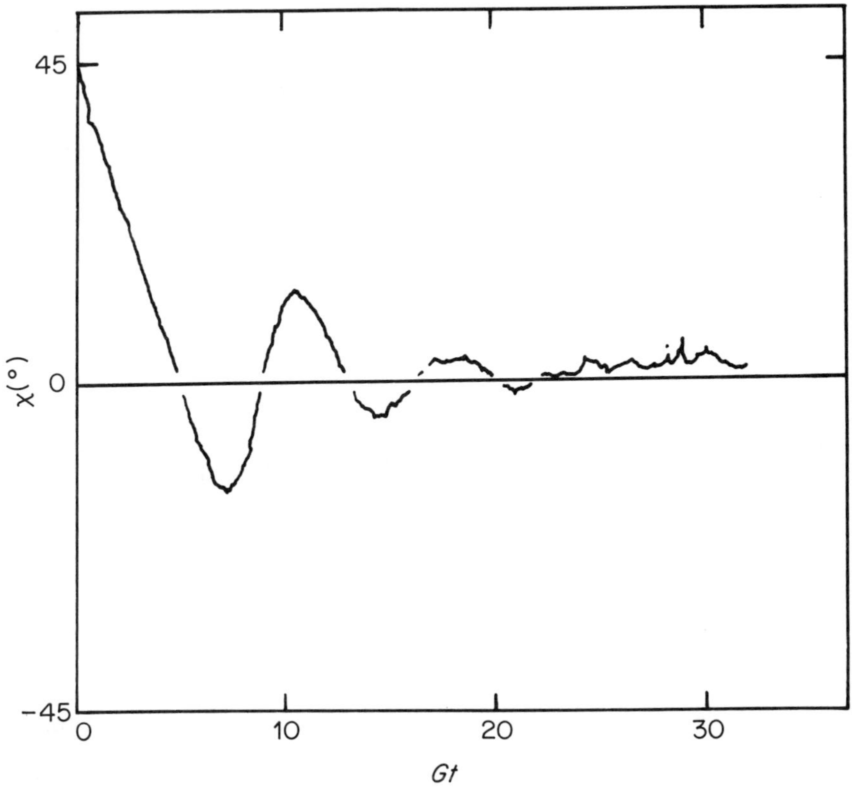

Fig. 3.62. An example of the observed orientation angle, χ, as a function of time for a bentonite suspension (after Frattini and Fuller, 1984).

3.1.12.6 Relaxation in optical properties

Useful information can be obtained from measurements of the optical properties after switching off the shear. An example has already been shown in Fig. 3.60. The relaxation towards the random orientation distribution contains information about the rotary diffusion coefficient and, for more concentrated systems, about particle interactions. For polydisperse systems it provides information about the polydispersity. Rather interestingly, the relaxation behaviour depends on what optical property is being measured (Jayasuriya and van de Ven, 1985). In general the relaxation behaviour of a property S can be obtained as

$$\langle S \rangle = \int_0^{\pi} \int_0^{2\pi} \int_0^{2\pi} S\, p \sin\theta \; d\theta d\phi d\psi \tag{3.343}$$

with p governed by

$$\frac{\partial p}{\partial t} = \nabla \cdot (\boldsymbol{D}_r \cdot \nabla p) \tag{3.344}$$

For spheroids, only one component of the rotary diffusion tensor $\boldsymbol{D}_r$ (the rotary diffusion coefficient D_r) need be considered. For ellipsoids not too far from axisymmetry, one can define an effective rotary diffusion coefficient by

$$D_r^{\text{eff}} = \frac{D_2 + D_3}{2} \tag{3.345}$$

where D_2 and D_3 are the two smallest components of the three principal values of $\boldsymbol{D}_r$.

For spheroids, the solution of eq (3.344) can be readily found and the integration in eq (3.343) can be performed for a variety of properties S. For the angular light scattering $I_s(\theta_s)$, the Rayleigh ratio $R(\theta_s)$, the scattering cross-section C_{sca}, and the transmittance T, one finds

$$R(\theta_s) = R_r(\theta_s) + \sum_{n=1}^{\infty} a_n e^{-2n(2n+1)D_r t} \tag{3.346}$$

with similar results for the other properties. Here $R_r(\theta_s)$ is the value for the random distribution at $t = \infty$. The coefficients a_n depend on the initial orientation distribution at time $t = 0$ (i.e. the moment the shear is switched off), on particle shape, on the wavelength λ, and (except for T and C_{sca}) on θ_s. At large times the leading relaxation time is $\tau = 1/6D_r$.

For the birefringence and dichroism the relaxation of the orientational averages is given by:

$$\langle \sin^2\theta \sin 2\phi \rangle = A\, e^{-6D_r t} \tag{3.347a}$$

$$\langle \sin^2\theta \cos 2\phi \rangle = B\, e^{-6D_r t} \tag{3.347b}$$

where A and B depend on the initial orientation and the shape. It follows from eqs (3.338), (3.339) and (3.347) that

$$\Delta n = (\Delta n)_0\, e^{-6D_r t} \tag{3.348}$$

where Δn stands for either $\Delta n'$ or $\Delta n''$.

Rather surprisingly, the orientation angle, χ, does not decay:

$$\tan 2\chi = A/B = \text{constant} \tag{3.349}$$

The above results can readily be extended to polydisperse particles. For spheroids, the intensity of the scattered light depends on the particle volume and shape and hence

$$\overline{R(\theta_s)} = \int_0^\infty \int_0^\infty R(\theta_s)\, g(a_1, a_2)\, da_1 da_2 \tag{3.350}$$

with similar expressions for T and C_{sca}. An experimentally observed relaxation in the transmittance of a suspension of non-spherical latex particles is shown in Fig. 3.63.

For the birefringence and dichroism one has a relation of the type

$$\Delta n = K(\bar{S}_1^{-2} + \bar{S}_2^{-2})^{1/2} \tag{3.351a}$$

K being a constant and

$$\bar{S}_i = \int S_i g(r_e)\, dr_e \tag{3.351b}$$

where S_i is one of the orientational averages occurring in eq (3.338) or (3.339).

It can be seen that, in general, the decay of the above properties is no longer a simple exponential in time. Allowing for polydispersity, it is now possible for the angle χ to relax:

$$\tan 2\chi = S_1/S_2 \tag{3.352}$$

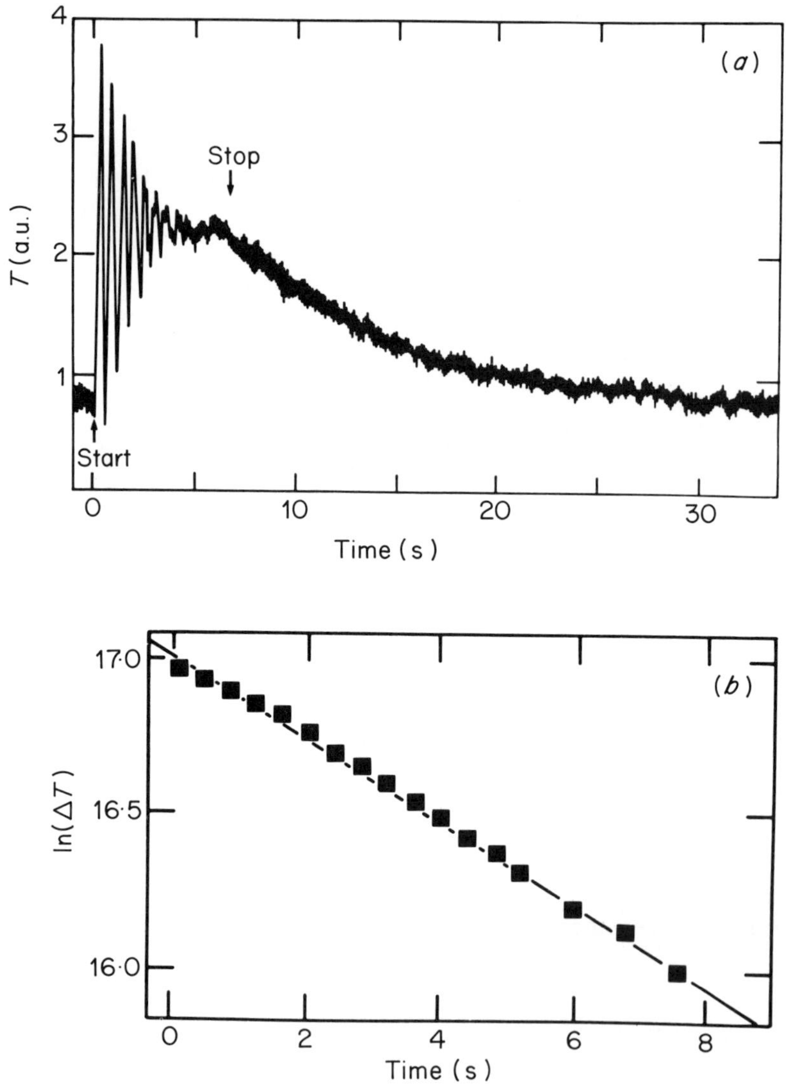

Fig. 3.63. (*a*) Relaxation (after transient regime) for a suspension of non-spherical latex particles due to rotary Brownian motion.
(*b*) The first few seconds of the relaxation behaviour on a semi-log plot, showing that the leading relaxation time is obtained after about 2 s (after Jayasuriya and van de Ven, 1985).

Because the coefficients A and B in eq (3.347) are functions of r_e, S_1/S_2 will, in general, be a function of time.

It is of interest to note that the integral appearing in eq (3.351b) is the so-called Fredholm integral. It is possible to obtain the size distribution, $g(r_e)$, from measured values, $\bar{S}_i(t)$, by inverting the integral. This can be done by converting the Fredholm integral to a set of linear algebraic equations. In general, such a set has multiple solutions for the unknown distribution $g(r_e)$, but by minimizing a modified weighted sum of the squared residuals, together with the constraint that $g(r_e) > 0$, it is possible to find a unique regularized solution for $g(r_e)$ (Provencher, 1979). The function S can be any function, e.g. a relaxation or transient in whatever optical property one is interested in.

3.1.12.7 Effects of double layers on rheo-optical properties

By analogy with known solutions of spherical particles we may assume that for electrically charged axisymmetric particles surrounded by a diffuse ionic layer, the period of rotation will be given by (cf. eq (3.98)):

$$T = T_0\,[1 + \bar{b}\,\Phi_s^2 f(\kappa a_1, r_e)] \qquad (3.353)$$

where T_0 is the period for a neutral particle, $\bar{b}$ the average scaled friction coefficient of the ions surrounding the particle, Φ_s a reduced ζ potential, and $f(\kappa a_1, r_e)$ an unknown function (of order unity) of κa_1 and r_e, κ^{-1} being the double-layer thickness. Equation (3.353) applies to symmetrical electrolytes and low ζ potentials.

It follows that the frequency of the damped rheo-optical oscillations will depend on the properties of the double layer surrounding the particles. The same applies to the relaxation. For spheroids surrounded by double layers it may be speculated that

$$D_r = D_r^0\,[1 - \bar{b}\,\Phi_s^2\, g(\kappa a_1, r_e)] \qquad (3.354)$$

where $g(\kappa a_1, r_e)$ is an unknown function of the double-layer thickness and the particle size and shape.

3.1.12.8 Memory of suspensions

In the absence of Brownian motion in systems obeying the Newtonian creeping-flow equation, all particle motions are reversible. This is indicated in Fig. 3.64a which shows the mean projection of the length (on the X_2X_3 plane) of a suspension of polydisperse rods subjected to simple shear.

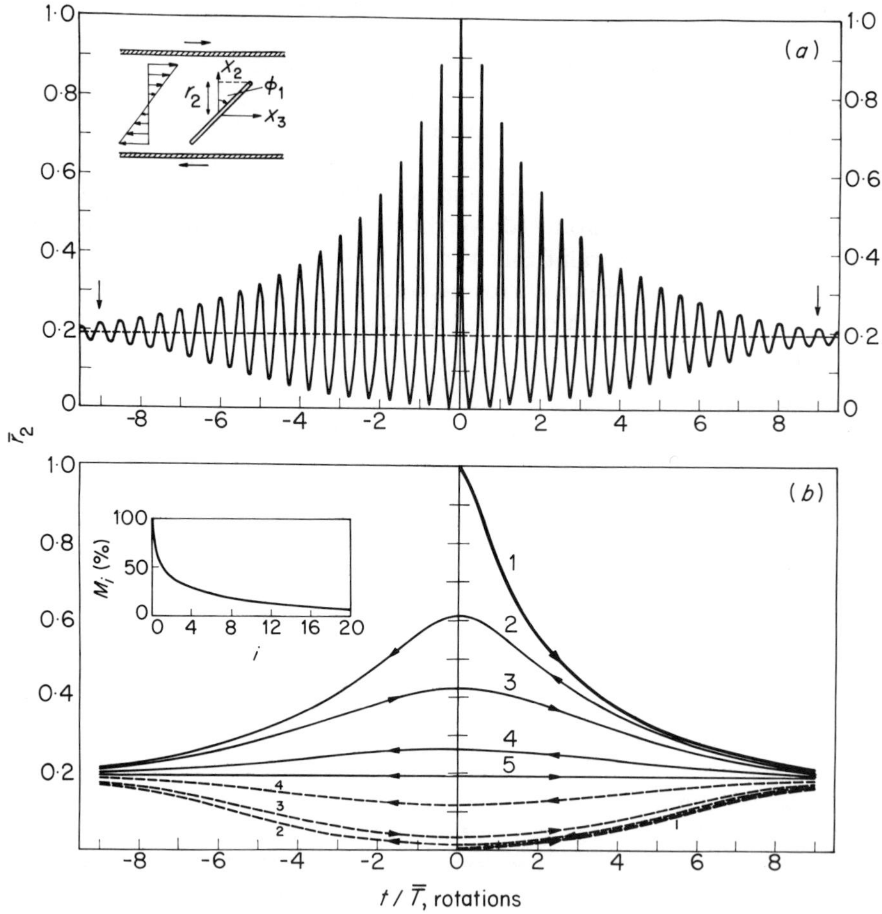

Fig. 3.64. (*a*) Calculated changes in mean projections $\bar{r}_2$ showing damped oscillations resulting from the spread in axis ratios of the rods ($\bar{r}_e = 10$, $\sigma = 0{\cdot}2$) in a system with perfect memory. The flow is reversed after nine rotations as indicated by the arrows.
(*b*) Loss of memory when the system is perturbed at each flow reversal. Curves are envelopes of damped oscillations. The inset shows the loss of memory as a function of flow reversals (after Okagawa and Mason, 1973).

Figure 3.64a shows the case of a perfect memory where all the translational and rotational motions are retraced exactly. In this particular calculation, the flow is reversed after nine rotations (indicated by the arrow). For the first nine rotations after flow is reversed, the system retraces exactly the damped oscillations and for the next nine rotations it shows the same damped oscillations (in negative time, i.e. going backwards). For systems with perfect memory this reversibility can be repeated indefinitely. For real systems, however, there are a variety of factors which cause loss of memory, e.g. Brownian motion, irreversible particle interactions, variations in the flow field, etc. Hence real suspensions can be said to have a fading memory (Okagawa and Mason, 1973). An artificial example is shown in Fig. 3.64b, which shows the behaviour of a system which at each reversal is perturbed by some external disturbance. Now the system is no longer perfectly reversible as shown by the envelopes of the damped oscillations. One way of defining the memory for such systems is

$$M = \frac{\bar{r}_{2i} - \bar{r}_{2\infty}}{\bar{r}_{20} - \bar{r}_{2\infty}} \tag{3.355}$$

where $\bar{r}_{2i}$ is the value of the mean projection $\bar{r}_e$ at time $t = 0$ after i reversals. The memory as defined in this way is shown in the inset of Fig. 3.64 and changes from 1 to 0 as i goes from 0 to ∞.

Another extreme example is a perfectly monodisperse system in which the loss of memory is due only to rotary Brownian motion. In such systems one also observes damped oscillations with a characteristic damping time of $\tau = 1/6D_r$. These oscillations are non-reversible. In such systems a convenient way to define the memory is

$$M = A(t)/A_0 \tag{3.356}$$

where $A(t)$ is the amplitude of the envelope of the damped oscillations, and A_0 is the initial value at time $t = 0$.

Calculations of M have been made by Zuzovski *et al.* (1980a) by using dynamic simulation calculations, for rotary Péclet numbers, Pe_r, in the range 100–2000 and by van de Ven *et al.* (1981b) by obtaining exact solutions of the rotary convective diffusion equation for $\mathrm{Pe}_r > \sim 2000$ for spheroids and for all Pe_r for spheres. In the range $100 < \mathrm{Pe}_r < 2000$, the results for damping in $\langle \sin^2\phi \rangle$ can be summarized as

$$M = \mathrm{e}^{-\beta(D_r t)^n} \tag{3.357}$$

where $\beta = 8{\cdot}25(r_e + r_e^{-1} - 2)^{0{\cdot}24}$ and $n = 2/(r_e + r_e^{-1})^{1/2}$. Note that in this regime M is independent of Pe_r and only depends on r_e and D_r. Results for various r_e are shown in Fig. 3.65. For $\mathrm{Pe}_r > 2000$, an expansion yields (for $r_e > 1$)

$$M = 1 - \frac{3r_e^4 + 2r_e^2 + 3}{4}\left(1 + r_e^{-4} + \frac{1 + r_e^{-6}}{C^2}\right) D_r t - \ldots \tag{3.358}$$

where C is the orbit constant.

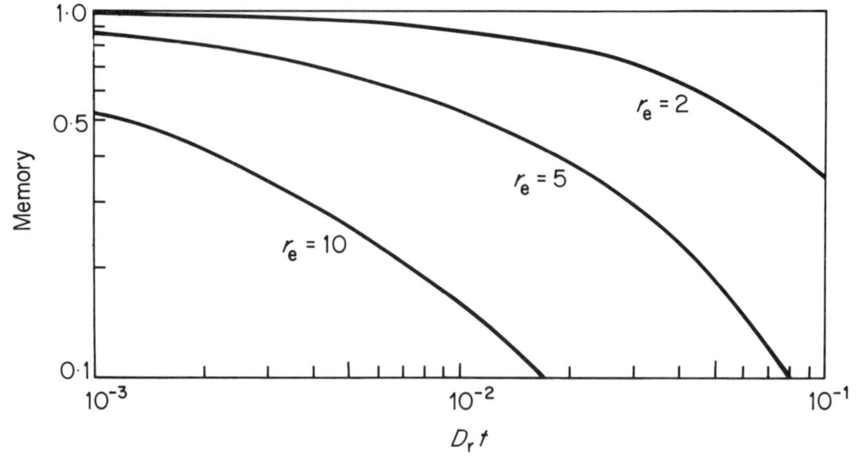

Fig. 3.65. Loss of memory for a suspension of monodisperse prolate spheroids due to rotary Brownian motion for various axis ratios, r_e (after Zuzovski *et al.*, 1980a).

In general, damping is due to both rotary Brownian motion and polydispersity. In such a case there is no obvious way to define a memory. It is neither determined by the amount of damping, nor by the flow reversals because Brownian motion continues to be present after reversing the flow. An experimental example is shown in Fig. 3.66. It shows a flow reversal experiment with a suspension of hardened human red blood cells. As can be

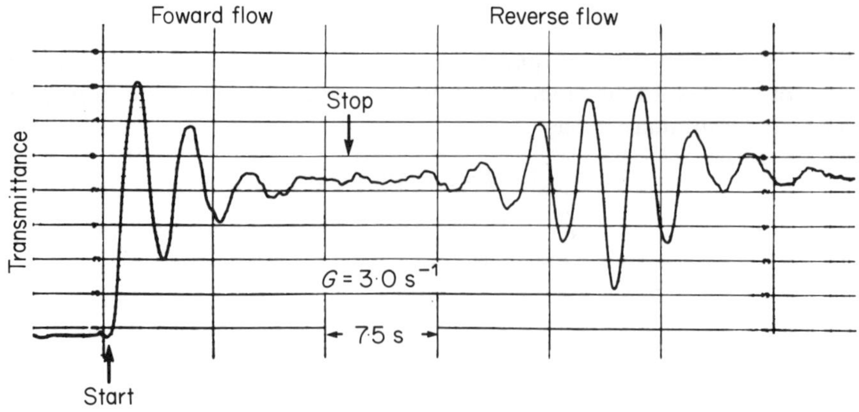

Fig. 3.66. Flow reversal for a suspension of hardened human red blood cells in which damping is due to both polydispersity and rotary Brownian motion. The damped oscillations after flow reversal differ slightly from the initial ones.

Fig. 3.67. Sequence of photographs of letters of dye, placed in the gap of two concentric cylinders. Rotating one of the cylinders causes the letters to smear out (1–5). By rotating the cylinder in the opposite direction (4–9), the original letters reappear (after Mason, 1973). (PPRIC stands for Pulp and Paper Research Institute of Canada)

seen, due to the loss of memory the oscillations after flow reversal are slightly different from the initial ones.

A striking demonstration of the fact that suspensions subjected to a low Reynolds number flow have a memory is shown in Fig. 3.67, which shows the dispersion of letters made by placing dye in the gap between two concentric cylinders filled with silicone oil. By rotating one of the cylinders, the dye is smeared out in streaks. Rotation in the opposite direction results in the reappearance of the letters.

3.2 Colloidal particles subjected to gravity or centrifugal fields

3.2.1 Sedimentation velocity

When the density of a suspended colloidal particle is different from that of the fluid, it will experience a buoyancy force and, as a result, it will either settle to the bottom or rise to the top of the mixture. The velocity with which it sediments depends on the density difference between the particle and the fluid, on the viscosity of the fluid and on the size and shape of the particle. The sedimentation velocity can be found from a balance of forces acting on the particle.

For an isolated spherical particle of volume V, mass m and friction coefficient f, suspended in a fluid medium and sedimenting in the negative X_3 direction, a force balance yields

$$m\ddot{x}_3 = \rho_p Vg - \rho_f Vg - f\dot{x}_3 \tag{3.359}$$

where ρ_p and ρ_f are the densities of the particle and fluid, respectively, and g is the acceleration due to gravity. As usual, the dots denote differentation with respect to time t. The three terms on the right-hand side of eq (3.359) represent the gravity force, the buoyancy force and the drag force. Integrating eq (3.359) gives for the velocity $u = \dot{x}_3$:

$$u = u_0 e^{-\beta t} - \frac{g\Delta\rho V}{f}(1 - e^{-\beta t}) \tag{3.360}$$

where $\beta = f/m$ and $\Delta\rho = \rho_p - \rho_f$.

As $t \to \infty (t \gg \beta^{-1})$, the velocity attains a steady-state value given by

$$u_{sed} = -\frac{g\Delta\rho V}{f} = -\frac{2g\Delta\rho a^2}{9\mu} \tag{3.361}$$

μ being the viscosity of the suspending medium.

From eq (3.361) it can be seen that either the size or the density of the particle can be determined by measuring its sedimentation velocity. Instead of sedimenting in a gravity field, a particle can be subjected to a centrifugal field. This provides a wide range for the parameter g.

For non-spherical particles, the situation is slightly more complex. In general they do not settle in the direction of gravity. As an example let us consider the sedimentation of an axisymmetric particle. Analogous to the case of a sphere, we can write

$$\boldsymbol{u}_{\text{sed}} = -\Delta\rho V \boldsymbol{f}^{-1} \cdot \boldsymbol{g} \tag{3.362}$$

where $\boldsymbol{u}_{\text{sed}}$ is the sedimentation velocity vector (three components), $\boldsymbol{f}$ the resistance tensor (3×3 components) and $\boldsymbol{g}$ the acceleration vector (one non-zero component, of magnitude g, in the direction of gravity). For an axisymmetric particle, $\boldsymbol{f}$ (in body-fixed coordinates) is of the form

$$\boldsymbol{f} = \begin{pmatrix} f_1 & 0 & 0 \\ 0 & f_2 & 0 \\ 0 & 0 & f_2 \end{pmatrix} \tag{3.363}$$

An axisymmetric particle will settle in a plane formed by the direction of gravity and the axis of revolution of the particle. A space-fixed coordinate system can be chosen such that the particle settles in the X_1X_3 plane (see Fig. 3.68). Expressing $\boldsymbol{f}$ in space-fixed coordinates and substituting into eq (3.362) yields

$$\begin{aligned} u_1 &= -g\Delta\rho V\left(\frac{1}{f_1} - \frac{1}{f_2}\right)\sin\theta\cos\theta \\ u_2 &= 0 \\ u_3 &= -g\Delta\rho V\left[\left(\frac{1}{f_1} - \frac{1}{f_2}\right)\cos\theta + \frac{1}{f_2}\right] \end{aligned} \tag{3.364}$$

where θ is the angle between the axis of revolution X_1' of the particle and the axis X_1.

From the set of eqs (3.364) it follows that the particle settles in a direction making an angle β with the direction of gravity (see Fig. 3.68), given by

$$\tan\beta = \frac{\sin 2\theta}{\lambda + \cos 2\theta} \tag{3.365}$$

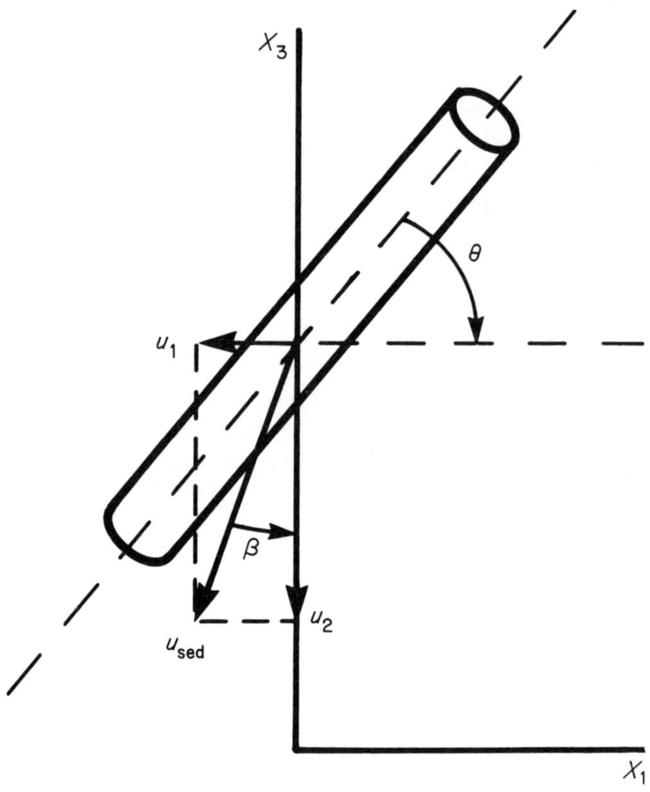

Fig. 3.68. Orientation of sedimenting axisymmetric particle. The particle settles in the X_1X_3 plane, X_3 being the direction of gravity. The particle settles in a direction that makes an angle β with the direction of gravity.

where

$$\lambda = \frac{q+1}{q-1} \text{ with } q = \frac{f_2}{f_1} = \frac{D_1}{D_2} \tag{3.366}$$

It can be seen that when $\theta = 0$ or $90°$, $u_1 = 0$. When $\theta = 0°$, $u_3 = -g\Delta\rho V/f_1$ and when $\theta = 90°$, $u_3 = -g\Delta\rho V/f_2$; in both cases the sedimentation is determined by only one component of the friction tensor, as is obvious from geometry.

The angle β is maximum when

$$\cos 2\theta_m = -\lambda^{-1} \tag{3.367}$$

For thin rods ($r_e = \infty$, $q = 2$ and $\lambda = 3$) this results in $\theta_m = 54{\cdot}8°$ and $\beta_m = 19{\cdot}5°$. For thin disks ($r_e = 0$, $q = 2/3$ and $\lambda = -5$) $\theta_m = 39{\cdot}0°$ and $\beta_m = -11{\cdot}5°$.

3.2.2 Sedimentation equilibrium

In the preceding section we discussed the settling of particles in the absence of Brownian motion. The sedimentation of particles in a suspension with boundaries creates concentration gradients. As a result diffusion will occur, trying to equalize the concentration. In a steady state, an equilibrium is established between the gravitational flux and the diffusional flux. This sedimentation equilibrium is governed by the steady-state convective-diffusion equation

$$\nabla \cdot (\boldsymbol{D} \cdot \nabla n) = \nabla \cdot (\boldsymbol{u}_{\text{sed}} n) \tag{3.368}$$

For spherical particles, variations in concentration only occur in the direction of gravity (the negative X_3 direction) and, upon integrating the above equation once, we obtain

$$D \frac{\mathrm{d}n}{\mathrm{d}x_3} = -ng \frac{\Delta m}{f} + K \tag{3.369}$$

When Δm or $\Delta \rho = 0$, no sedimentation occurs and hence $\mathrm{d}n/\mathrm{d}x_3 = 0$. It follows that the integration constant, K, is zero. Integrating once more and using the Stokes–Einstein equation $Df = kT$, results in

$$n = n_0 \mathrm{e}^{-\frac{\Delta m\, g x_3}{kT}} \tag{3.370}$$

which is an example of a Boltzmann distribution. Equation (3.370) shows that the mass of a colloidal particle can be obtained by measuring the concentration profile of the suspended particles in a gravitational (or centrifugal) field. If the density of the particles is known, their volume can be obtained.

The sedimentation of non-spherical Brownian particles is complicated by the fact that the translational and rotary motions are coupled. As we have seen, a non-spherical particle settles in a direction that depends on its orientation. Rotary Brownian motion will change the orientation of the particles and hence their sedimentation speeds and directions. After some time a steady state will be reached in which each particle is equally likely to move to one or the other side, and thus, on average, the particles will move in the direction of gravity only. For homogeneous particles (i.e. in the absence of gravitational torques), $\langle\cos^2\theta\rangle = 1/3$ and the average sedimentation velocity equals (cf. eq (3.364)):

$$\bar{u}_{\text{sed}} = -\frac{g\Delta\rho V}{3}\left(\frac{1}{f_1}+\frac{2}{f_2}\right) \tag{3.371}$$

Similarly, the diffusion constant can be replaced by an average value

$$\bar{D} = \frac{1}{3}(D_1 + 2D_2) = \frac{kT}{3}\left(\frac{1}{f_1}+\frac{2}{f_2}\right) \tag{3.372}$$

Substituting the values for $\bar{u}_{\text{sed}}$ and $\bar{D}$ into eq (3.368) and integrating, gives the same result as that obtained for spheres (eq (3.370)). The same is true for ellipsoidal particles. It follows that for such particles their mass can be determined by measuring the concentration profile of the sedimentation equilibrium, independently of particle shape.

3.2.3 Taylor diffusion of non-spherical sedimenting particles

It is of interest to consider what happens to an ensemble of non-spherical particles, initially introduced at some position in a liquid and subsequently allowed to sediment (see Fig. 3.69). The dispersion of the ensemble of

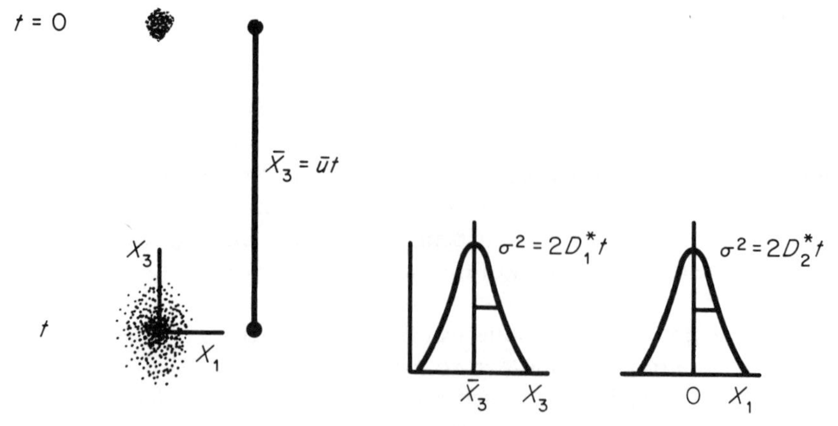

Fig. 3.69. Generalized Taylor diffusion of an ensemble of non-spherical particles, initially at the origin of a Cartesian coordinate system X_i. At sufficiently large times t, the average of the ensemble is at position $(0, 0, \bar{x}_3)$, with $\bar{x}_3 = \bar{u}t$. The position dispersion in the gravity direction is determined by D_1^* and in the direction perpendicular to gravity by D_2^*.

particles can be described by the generalized Taylor dispersion theory developed by Dill and Brenner (1982a,b). In this approach the coordinates are divided into local and global coordinates and one is only interested in the convective diffusion expressed in global coordinates after a proper averaging over the local coordinates. For axisymmetric sedimenting particles, the local coordinates are the orientational angles θ and ϕ and the global coordinates are the space coordinates. For times sufficiently long for the particles to have sampled all possible orientations, i.e. for times $t \gg D_r^{-1}$, D_r being the rotary diffusion coefficient, the convective-diffusion equation in global coordinates reduces to (Brenner, 1979):

$$\frac{\partial n}{\partial t} = D_2^* \left(\frac{\partial^2 n}{\partial x_1^2} + \frac{\partial^2 n}{\partial x_2^2} \right) + D_1^* \frac{\partial n^2}{\partial x_3^2} - \bar{u}^* \frac{\partial n}{\partial x_3} \tag{3.373}$$

where $\bar{u}^*$ is the mean sedimentation velocity in the direction of gravity, given by eq (3.371). It follows that, on average, the ensemble of particles settles with velocity $\bar{u}^*$.

D_1^* and D_2^* are the effective diffusion coefficients for diffusion parallel with and perpendicular to the direction of gravity, respectively. They are given by

$$D_1^* = \bar{D} + \frac{2\sigma\bar{u}^{*2}}{135\, D_r} \tag{3.374a}$$

$$D_2^* = \bar{D} + \frac{\sigma\bar{u}^{*2}}{90\, D_r} \tag{3.374b}$$

where $\sigma = (D_1 - D_2)^2/\bar{D}$, with $\bar{D}$ given by eq (3.372).

Note that by measuring the sedimentation concentration profile of an ensemble of particles, the values $\bar{D}$ and σ/D_r can be obtained which, in turn, yield information regarding the particle size and shape.

3.2.4 Sedimentation of charged particles

In Chapter 2 (Section 2.2.2) we discussed the problem of the friction coefficient, f, of a charged sphere. It was found that for symmetrical electrolytes with an average normalized friction coefficient $\bar{b}$ and for particles with a reduced ζ potential, Φ_s ($\Phi_s \ll 1$)

$$f = f_0[1 + \bar{b}\, \Phi_s^2 \tilde{f}(\kappa a)] \tag{3.375}$$

where f_0 is the friction coefficient of a neutral sphere and $\tilde{f}(\kappa a)$ is a known function of the double-layer parameter, κa, given by eq (2.173).

It follows directly from the definition of the friction coefficient that

$$u_{\text{sed}} = u_{\text{sed}}^0 [1 - \bar{b}\, \Phi_s^2 \tilde{f}(\kappa a)] \tag{3.376}$$

where u_{sed}^0 is the sedimentation velocity of a neutral sphere (eq (3.361)). For more general conditions the friction coefficient, f, and the sedimentation velocity, u_{sed}, can be found from a numerical solution of the governing equations (Ohshima *et al.*, 1984).

When a number of particles are sedimenting with a velocity given by eq (3.376), the combined electric fields surrounding each particle give rise to a macroscopic field. The potential associated with this macroscopic field is called the sedimentation potential. This potential is defined as

$$E_{\text{sed}} = -\frac{1}{V}\int \nabla\psi(r)\, \mathrm{d}V \tag{3.377}$$

where ψ is the electrical potential around a particle and V is the suspension volume. In the absence of a flow, $\nabla\psi$ is zero and the first contribution to E_{sed} comes from the first term of an expansion in the Péclet number, Pe (see eq (2.154)):

$$\psi = \psi_0 + \text{Pe}\, \psi_1 \tag{3.378}$$

Hence

$$E_{\text{sed}} = -\frac{\text{Pe}}{V}\int \nabla\psi_1\, \mathrm{d}V \tag{3.379}$$

To calculate E_{sed}, it is required that the net current in the suspension must be zero, i.e.

$$\langle i \rangle = \frac{1}{V}\int i(r)\, \mathrm{d}V = 0 \tag{3.380}$$

with the current $i(r)$ given by

$$i(r) = \sum_i z_i e n_i u_i \tag{3.381}$$

where u_i is obtained by solving eqs (2.164) and (2.165) and n_i by solving eqs (2.166) and (2.169). Evaluating the integral in eq (3.380) results in (Ohshima *et al.*, 1984)

$$\langle i \rangle = \chi_\infty \left[E_{\text{sed}} + \frac{\epsilon_r \epsilon_0 g \Delta\rho \zeta h(\kappa a) c}{\mu \chi_\infty} \right] \tag{3.382}$$

where χ_∞ is the electric conductivity of the medium, given by

$$\chi_\infty = \sum_i z_i^2 e n_i^0 / f_i \tag{3.383}$$

ϵ_r and ϵ_0 are the dielectric constant of the medium and the permittivity of free space, ζ is the ζ potential, c the particle volume fraction, and $h(\kappa a)$ is given by

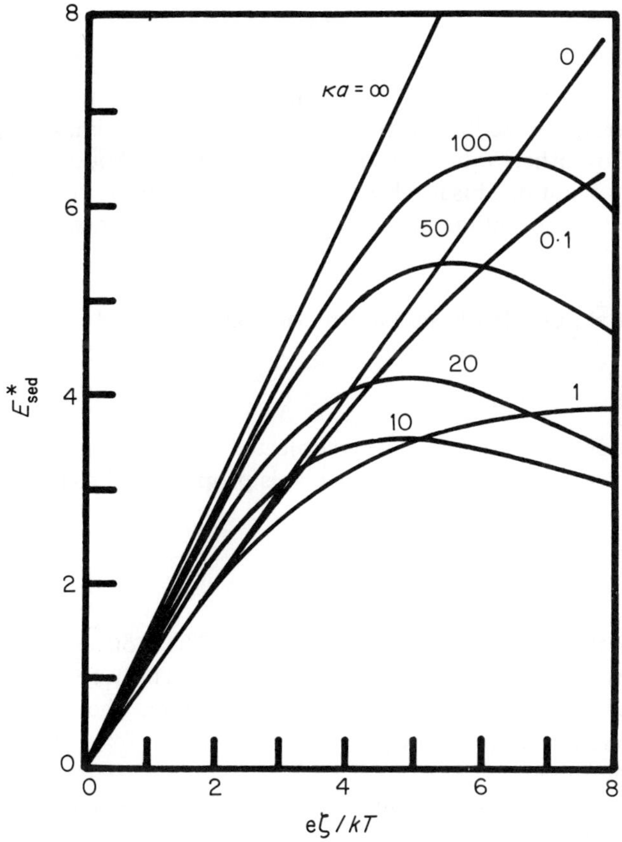

Fig. 3.70. Sedimentation field, E^*_{sed}, in a dilute suspension of spherical particles in a KCl solution at 25°C as a function of ζ and κa (shown on the curves) (after Ohshima *et al.*, 1984).

$$h(\kappa a) = 1 + 2e^{\kappa a} E_5(\kappa a) - 5e^{\kappa a} E_7(\kappa a) \tag{3.384}$$

with $E_i(\kappa a)$ as defined beneath eq (2.173).

Since $\langle i \rangle = 0$, it follows from eq (3.382) that

$$E_{\text{sed}} = -\frac{\epsilon_r \epsilon_0 g \Delta\rho \zeta h(\kappa a) c}{\mu \chi_\infty} \tag{3.385}$$

For higher potentials, results can be calculated numerically. Typical results are given in Fig. 3.70 where the sedimentation potential is represented in non-dimensional units as:

$$E^*_{\text{sed}} = \frac{3\mu e \chi_\infty}{2\epsilon_r \epsilon_0 kT \Delta\rho c} \frac{|E_{\text{sed}}|}{g}$$

It can be seen from Fig. 3.70 that, when $\kappa a = 0$ or ∞ (infinitely thick or thin double layers), the sedimentation potential increases linearly with the ζ potential. For other values of the double-layer thickness, the curves go through a maximum. This implies that one cannot always determine ζ potentials uniquely from sedimentation-potential measurements.

3.3 Colloidal particles in an electric or magnetic field

3.3.1 Electrophoresis

When a charged colloidal particle is subjected to an electric field it will move with a velocity u which, for weak fields, is proportional to the electric field strength E:

$$u = mE \tag{3.386}$$

where m is called the electrophoretic mobility. The motion of the particle is governed by the same set of equations (eqs (2.137) to (2.141)) which describes the friction coefficient of a charged sphere. The only difference is the boundary condition for the potential ψ, which far from the particle must approach the value $-\boldsymbol{E}\cdot\boldsymbol{r}$.

In the absence of a double layer, a spherical charged particle would simply move according to

$$F = qE = 6\pi\mu a u \tag{3.387}$$

where q is the charge on the particle. For a particle with a surface charge density σ (cf. eq (1.104)) and a zeta-potential ζ:

$$q = 4\pi a^2\sigma = 4\pi a(1 + \kappa a)\epsilon\zeta \tag{3.388}$$

Combining eqs (3.387) and (3.388) leads, when $\kappa a \ll 1$ (i.e. in the limit of zero electrolyte concentration), to

$$m = \frac{2}{3}\frac{\epsilon\zeta}{\mu} \tag{3.389}$$

Defining a dimensionless electrophoretic mobility E_m as

$$E_m = \frac{3\mu ze}{2\epsilon kT} m$$

we can write when $\kappa a \ll 1$:

$$E_m = \Phi_s \tag{3.390}$$

where $\Phi_s = ze\zeta/kT$. Equation (3.390) gives the electrophoretic mobility for low surface potentials and low ion concentrations.

It is worthwhile pointing out what precisely is meant by a zeta potential. The set of eqs (2.137) to (2.141) contains as parameters the viscosity, μ, and the dielectric constant, ϵ_r ($\epsilon = \epsilon_r\epsilon_0$), of the medium. In integrating these equations it is implicitly assumed that μ and ϵ_r are constant everywhere. In reality, μ and ϵ_r are functions of the distance r from the surface of the particle. Near the surface is a strong electric field as well as an accumulation and depletion of counter- and co-ions, respectively, which affect these parameters. Other assumptions are that the hydrodynamic velocity of the ions is the same as the fluid velocity and that their diffusion constant is not affected by the proximity of a surface. If the relations $\mu(r)$ and $\epsilon_r(r)$, as well as the precise motions of the ions, were known, the governing equations could, in principle, be solved exactly, in which case the boundary condition at the surface of the particle would be $\psi \to \psi_0$ as $r \to 0$, ψ_0 being the surface potential. The exact solution would lead to a relation of the type

$$m = f_1(\psi_0)$$

where the mobility m depends on, besides ψ_0, a variety of other parameters. In taking μ and ϵ_r as contants, making the above-mentioned assumptions about ions and again requiring that $\psi \to \psi_0$ as $r \to 0$, a different relation is obtained:

$$m = f_2(\psi_0)$$

Obviously this equation is only an approximation due to the simplifying assumptions. Inserting the true surface potential into the above relation will lead to an incorrect value of the mobility. The zeta potential is defined as

$$m = f_2(\zeta)$$

i.e. the potential in the relation $m = f_2(\psi_0)$ is adjusted in such a way that the correct mobility m is obtained. It follows that the relation between the surface and the zeta potential is

$$f_1(\psi_0) = f_2(\zeta)$$

Since f_2 can be calculated but f_1 is as yet unknown, it is clear that at present it is not possible to obtain ψ_0 from ζ. It also follows that the zeta potential depends on the way in which it is measured. For example, it has been shown above that it is possible to determine ζ from the diffusion constant of small colloidal particles. Equation (2.174) is of the type $D = g_2(\zeta)$, while the exact result would lead to $D = g_1(\psi_0)$. Thus ζ is defined by $g_1(\psi_0) = g_2(\zeta)$, which is different from the above equation.

In the literature the zeta potential is often defined as the potential a distance of, say, δ away from the surface of the particle. It is assumed that the liquid within this distance sticks to the surface and the plane $r = \delta$ is referred to as the "plane of slip" (or plane of shear). This picture of the zeta potential is highly idealized and incorrect. Even if for distances $r > \delta$, μ and ϵ_r can be considered constant, the no-slip boundary condition must be applied at $r = 0$.

When the condition $\kappa a \ll 1$ is not met, the complete set of governing equations must be solved. This problem is simplified by realizing that the solution can be written as the sum of the solutions of two simpler problems: (i) a particle held fixed in a flow field $\boldsymbol{u}$ with no electric field. This problem has been discussed previously with the result that

$$F_h = -fu \tag{3.391}$$

where f is given by eq (3.375); (ii) a particle held fixed in an electric field in the absence of a flow field (i.e. $v \to 0$ as $r \to \infty$). The solution of this problem yields

$$F_e = -\beta E \tag{3.392}$$

When a colloidal particle moves in an electric field both forces are acting and a force balance leads to

$$u = \frac{\beta}{f} E \tag{3.393a}$$

or

$$m = \beta / f \tag{3.393b}$$

The coefficient β can be found in a manner similar to f. In fact this problem is simpler because only the first three equations of the set (eqs (2.137) to (2.141)) are needed (Pe = 0). The set of the remaining three equations can be solved by making an expansion in Φ_s. Going through this procedure eventually leads to (Henry, 1931)

$$E_m = \frac{3}{2} \Phi_s h(\kappa a) \tag{3.394}$$

where $h(\kappa a)$ is as given by eq (3.384).

Note the reciprocal relationship between the sedimentation potential and the electrophoretic mobility. This is a direct consequence of an Onsager relation between sedimentation and electrophoresis. In electrophoresis, an electric field is applied and as a result the suspended particles move. In sedimentation, the particles move which results in an electric field.

From eqs (3.384) and (3.394) it can be seen that as $\kappa a \to 0$, $h(\kappa a) \to 2/3$ and eq (3.390) is recovered. As $\kappa \to \infty$, $h(\kappa a) \to 1$ and

$$E_m = 3/2\, \Phi_s \tag{3.395}$$

Besides making an expansion in Φ_s, it is also possible to make an expansion in $(\kappa a)^{-1}$ (Ohshima *et al.*, 1983). This leads to a rather complicated equation which is valid when $\kappa > 50$. From this equation Ohshima *et al.* derived an equation which is valid (within 1%) for $\kappa a > 10$:

$$\begin{aligned} E_m = {} & \frac{3}{2} \Phi_s - \frac{3F}{1+F} G_- + \frac{1}{\kappa a}\Bigg[-18\left(y + \frac{y^3}{9}\right) K + \frac{15\,F}{1+F}\left(y + \frac{7y^2}{20} + \frac{y^3}{9}\right) \\ & - 6(1+3b_+)\left(1 - e^{-\Phi_s/2}\right) G_+ + \frac{12\,F}{(1+F)^2} G_- + \frac{9\Phi_s}{1+F}(b_+G_+ + b_-G_-) \\ & - \frac{36\,F}{1+F}\left(b_+G_+^2 + \frac{b_-}{1+F} G_-^2\right)\Bigg] \end{aligned} \tag{3.396}$$

where $G_{\pm}$ is as defined beneath eq (2.175); $y = \tanh(\Phi_s/4)$;

$$F = \frac{2}{\kappa a}(1 + 3b_-)\left(e^{\Phi_s/2} - 1\right)$$

and

$$K = 1 - \frac{25}{3(\kappa a + 10)} \exp\left[-\frac{\kappa a}{6(\kappa a - 6)}\Phi_s\right]$$

Because the condition $\kappa a > 10$ is often satisfied for colloidal particles, the above equation is an extremely useful expression for determining ζ potentials

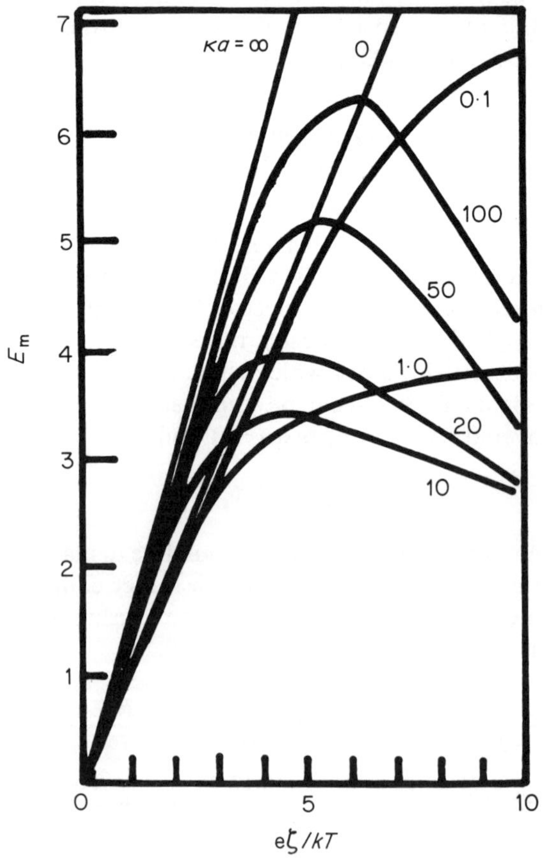

Fig. 3.71. Reduced mobility, E_m, of a spherical particle in a KCl solution as a function of the ζ potential for various values of κa (after O'Brien and White, 1978).

from electrophoretic mobilities. For conditions outside the range of validity of eqs (3.394) and (3.396), the electrophoretic mobility must be calculated numerically. Numerical solutions by O'Brien and White (1978) are shown in Fig. 3.71. Notice the similarity with the calculations for the sedimentation potential (Fig. 3.70). From Fig. 3.71 it can be seen that as $\kappa a \to 0$ or $\kappa a \to \infty$, the limiting results, eqs (3.390) and (3.395), are recovered (straight lines in Fig. 3.71). For other values of κa the electrophoretic mobility goes through a maximum, indicating that the retarding force due to the double-layer deformation increases faster than does the driving force. As a result, interpreting mobility data in terms of ζ potentials can be ambiguous. Often two different ζ potentials can give rise to the same mobility.

Recently, ζ potentials of model latex particles have been observed (Chow and Takamura, 1988b) which are larger than the maximum values shown in Fig. 3.71 (for fixed κa values). This seems to imply that the theory needs some revision. One possible weakness of the theory could be the choice of boundary conditions on the sphere surface in the presence of an electric field.

One property of a colloidal dispersion related to the electrophoretic mobility of particles is its conductivity, χ. The presence of charged particles surrounded by double layers increases the conductivity beyond the value of the conductivity of the suspending medium, χ_0:

$$\chi = \chi_0(1 + [\chi]c) \tag{3.397}$$

The intrinsic conductivity, $[\chi]$, can be found from the same set of equations as used in calculating f or m. An expansion in Φ_s yields (O'Brien, 1981):

$$[\chi] = -\frac{3}{2} - 3\,\frac{b_1 - b_2}{b_1 + b_2}\,\Phi_s f(\kappa a) \tag{3.398}$$

where

$$f(\kappa a) = \frac{1}{\kappa a} + \frac{1}{(\kappa a^2)}\,\frac{e^{\kappa a}}{2}\int_1^\infty \frac{e^{-\kappa a x}}{x^5}\,dx$$

An explicit expression for the term $0(\Phi_s^2)$ is also given by O'Brien. The first term on the right-hand side of eq (3.398) represents the intrinsic conductivity of a suspension of uncharged spheres, while the second term is the first-order contribution of the electric double layer. It is also possible to obtain an expansion in $(\kappa a)^{-1}$ (Ohshima *et al.*, 1983).

3.3.2 Non-spherical particles in electric or magnetic fields

When a non-spherical particle is subjected to an electric or magnetic field it will align itself in the direction of the field, provided the rotary Brownian motion is not too strong to cause a random distribution of orientations. The changes in direction of the particles manifest themselves in changes in a number of properties such as birefringence, light scattering, turbidity, dielectric constant, etc. This is shown schematically in Fig. 3.72. Here we are concerned with the approach to the steady state after switching on the field and the characteristic time over which changes occur. The relaxation after switching off the field was discussed earlier (Sections 2.4.4 and 3.1.12.6). The electrically or magnetically induced changes in orientation provide information about particle size and shape, dielectric or magnetic properties of particles and medium and about the surface potential of the particles and the nature of the surrounding double layer. In what follows, discussion is restricted to particles in an electric field. The theory can be readily modified to describe the behaviour of particles in a magnetic field. For example, for particles without a permanent magnetic dipole, exactly the same theory applies as described below, but with the electric field $\boldsymbol{E}$ replaced by the magnetic field $\boldsymbol{H}$ and the dielectric permittivity $\epsilon_i\epsilon_0$ by the magnetic permeability $\mu_i\mu_0$.

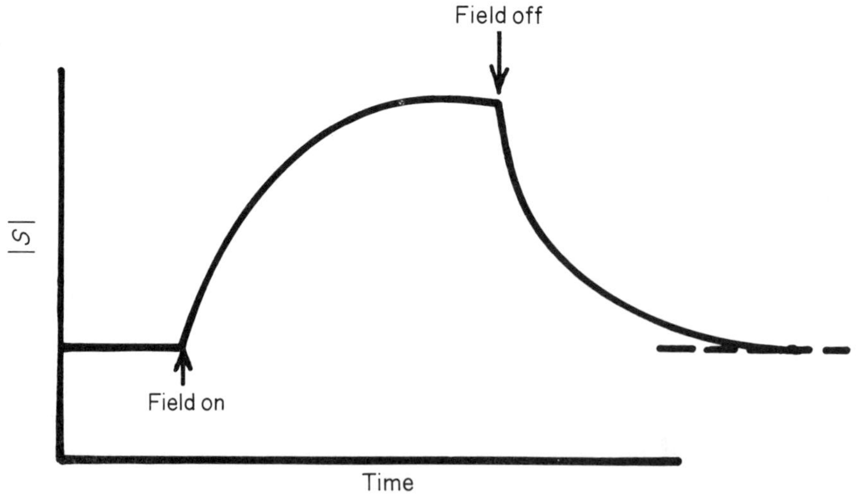

Fig. 3.72. Changes in suspension properties, $|S|$, as a function of time after switching an electric field on and off.

As an example, consider an ellipsoidal particle suspended in a homogeneous fluid and subjected to a uniform electric field. Such a particle will experience an electrostatic torque, $\boldsymbol{L}$, and hence rotate according to

$$\boldsymbol{L} = \boldsymbol{f}_r \cdot \boldsymbol{\omega} \tag{3.399}$$

where $\boldsymbol{f}_r$ is the friction tensor and $\boldsymbol{\omega}$ the angular velocity. It is convenient to express all vectors and tensors in terms of a body-fixed coordinate system. The electrostatic torque exerted on the ellipsoid is

$$\boldsymbol{L} = \boldsymbol{P} \times \boldsymbol{E} \tag{3.400}$$

where $\boldsymbol{P}$ is the dipole moment of the particle (above that which would exist if the ellipsoid were replaced by the suspending fluid) and $\boldsymbol{E}$ the strength of the external electric field. The dipole moment $\boldsymbol{P}$ consists of two parts, a permanent dipole moment and an induced dipole moment, so that

$$\boldsymbol{L} = \boldsymbol{P}_1 \times \boldsymbol{E} + \boldsymbol{P}_2 \times \boldsymbol{E} \tag{3.401}$$

For a homogeneous particle

$$\boldsymbol{P}_1 = \frac{4}{3} \pi a_1 a_2 a_3 \boldsymbol{p}_1 \tag{3.402}$$

where $\boldsymbol{p}_1$ is the dipole moment per unit volume.

For an electrically anisotropic particle the dielectric constant must be replaced by a tensor, ϵ_{ij}, which is symmetric ($\epsilon_{ij} = \epsilon_{ji}$) and thus has six independent components. Here we consider particles that have only diagonal components, ϵ_{11}, ϵ_{22} and ϵ_{33}, the off-diagonal components being zero. For such a particle the induced polarization per unit volume (over that of the medium) is given by

$$p_{2i} = \epsilon_0 \epsilon_2 (q_i - 1) E_i^* \tag{3.403}$$

where $q_i = \epsilon_{ii}/\epsilon_2$, ϵ_2 being the dielectric constant of the medium, ϵ_0 the permittivity of free space, and E_i^* the electric field inside the particle given by (Jones, 1945):

$$E_i^* = \frac{E_i}{1 + \frac{1}{2} a_1 a_2 a_3 (q_i - 1)\, \alpha_i} \tag{3.404}$$

where E_i are the components of E, a_i the semi-axes of the ellipsoids, and α_i the elliptic integrals as defined beneath eq (2.104).

Combining eqs (3.401) to (3.404) shows that the electrostatic torque on an ellipsoidal particle can be expressed as

$$L_i = V\,\epsilon_{ijk}\left(p_{ij}E_k + \frac{4\pi\epsilon_2\epsilon_0(q_j-1)\,E_jE_k}{1+\frac{1}{2}a_1a_2a_3(q_j-1)\,\alpha_j}\right) \tag{3.405}$$

where V is the volume of the particle.

From eqs (3.405) and (3.399) the components of ω in body-fixed coordinates can be obtained, which can then be transformed into the angular velocities in space-fixed coordinates. This leads to expressions of the variations of the three Euler angles θ, ϕ and ψ with time (Okagawa *et al.*, 1973).

These expressions simplify considerably for isotropic spheroids with or without a permanent dipole. In this case $a_1 = a$, $a_2 = a_3 = b$, $\epsilon_{11} = \epsilon_{22} = \epsilon_{33} = \epsilon_1$ and $p_{11} = m$, $p_{12} = p_{13} = 0$. Hence we consider particles with a permanent dipole moment parallel to the axis of revolution. Without loss of generality we can choose the electric field parallel to the X_2 axis. In this case the above analysis leads to (Chaffey and Mason, 1964)

$$\frac{\mathrm{d}\theta}{\mathrm{d}t} = \frac{mQ(r_e)}{\mu}E\cos\theta\cos\phi - \frac{\epsilon_2\epsilon_0 P(q,r_e)E^2}{\mu}\sin2\theta\cos^2\phi \tag{3.406a}$$

$$\frac{\mathrm{d}\phi}{\mathrm{d}t} = -\frac{mQ(r_e)}{\mu}E\frac{\sin\phi}{\sin\theta} + \frac{\epsilon_2\epsilon_0 P(q,r_e)}{\mu}E^2\sin2\phi \tag{3.406b}$$

where

$$Q(r_e) = \frac{2r_e^2 + (1-2r_e^2)A}{4(r_e^2+1)} \tag{3.407}$$

and

$$P(q,r_e) = \frac{(3A-2)(q-1)^2\,Q(r_e)}{2[2+(q-1)A][(q-1)A-q]} \tag{3.408}$$

where A is as given by eq (3.281) and $q = \epsilon_1/\epsilon_2$. The function $Q(r_e)$ is positive for all r_e. It is maximum (0·169) at $r_e = 0{\cdot}842$ and approaches zero as $r_e \to 0$ or ∞. The function $P(q,r_e)$ is positive when $r_e < 1$, negative when $r_e > 1$ and zero when $r_e = 1$ or $q = 1$. It also vanishes as $r_e \to 0$ or ∞, except when $q = 0$ or ∞, in which case

$$P(0,r_e) = \frac{(3A-2)\,Q(r_e)}{2A(A-2)} \tag{3.409}$$

and

$$P(\infty, r_e) = \frac{(3A-2)\,Q(r_e)}{2A(A-1)} \tag{3.410}$$

The last case ($q=\infty$) corresponds to conductive particles in a dielectric medium. $P(\infty, r_e)$ varies from $0{\cdot}25$ when $r_e=0$ to a maximum of $0{\cdot}256$ at $r_e=0{\cdot}043$, a minimum of $-0{\cdot}294$ at $r_e=10{\cdot}4$ and approaches $-0{\cdot}25$ as $r_e\to\infty$.

Equations (3.406) can be simplified considerably by realizing that motion only occurs in a plane made by the X_2 axis and the axis of revolution, X_1' of the spheroid (see Fig. 3.73). Let us denote the angle between these axes by θ_2, then from simple geometry it follows that θ_2 is related to the angles (θ, ϕ) by

$$\cos\theta_2 = \sin\theta\cos\phi \tag{3.411}$$

and eq (3.406) reduces to

$$\frac{d\theta_2}{dt} = -\frac{\sin\theta_2}{\tau_1} - \frac{\sin\theta_2\cos\theta_2}{\tau_2} \tag{3.412}$$

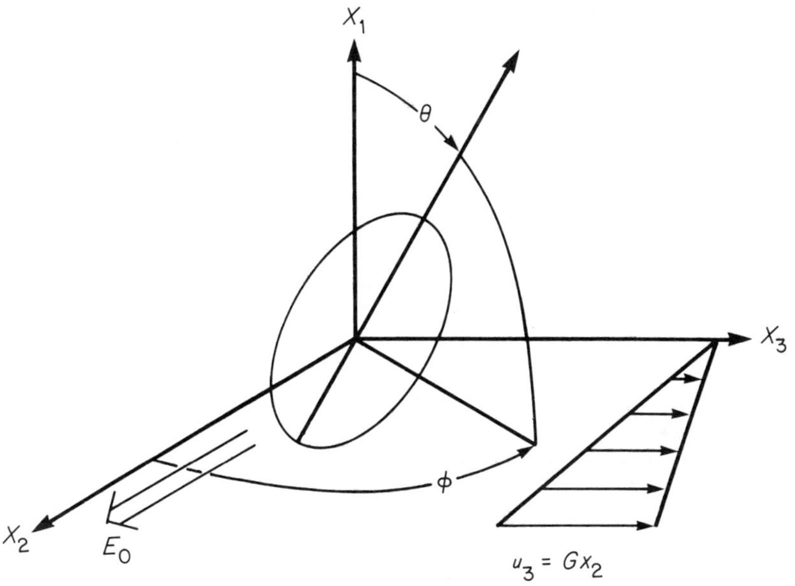

Fig. 3.73. Coordinate system for a non-spherical particle subjected to an electric field. The angle θ_2 is the angle between the axis of revolution of the particle and the direction of the electric field. θ and ϕ are the polar coordinates describing the orientation with respect to the coordinates X_i. Also shown is the superposition of a simple shear flow, discussed in Section 3.4.

where

$$\tau_1 = \frac{\mu}{mQ(r_e)E} \tag{3.413}$$

and

$$\tau_2 = -\frac{\mu}{2\epsilon_0\epsilon_2 P(q, r_e)E^2} \tag{3.414}$$

τ_1 and τ_2 are the relaxation times for alignment of the permanent and induced dipole moments, respectively. When $\tau_2 = \infty$ (i.e. $P = 0$), integration of eq (3.412) gives

$$\tan(\tfrac{1}{2}\theta_2) = \tan(\tfrac{1}{2}\theta_{20})\, e^{-t/\tau_1} \tag{3.415}$$

and when $\tau_1 = \infty$ ($m = 0$):

$$\tan\theta_2 = (\tan\theta_{20})\, e^{-t/\tau_2} \tag{3.416}$$

θ_{20} being the value of θ_2 at time $t = 0$. As $t \rightarrow \infty$, in the absence of an induced dipole, $\theta_2 \rightarrow 0°(\tau_1 > 0)$. In the absence of a permanent dipole, $\theta_2 \rightarrow 0$ for rods ($\tau_2 > 0$) and $\theta_2 \rightarrow 90°$ for disks ($\tau_2 < 0$).

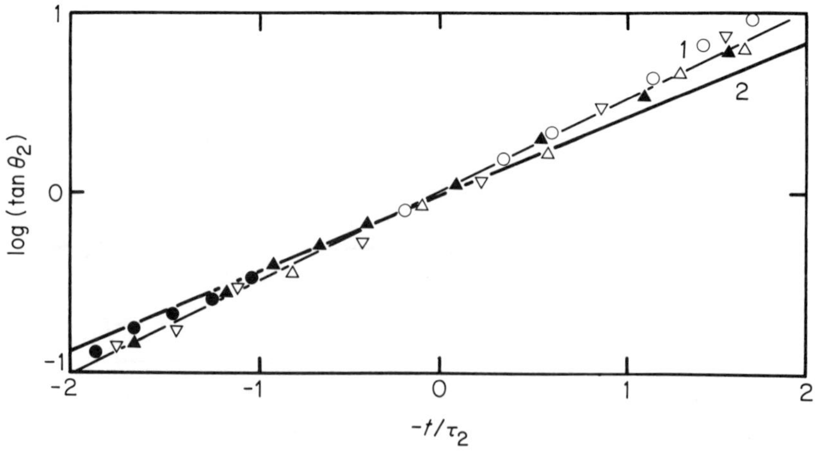

Fig. 3.74. Rotation of a disk with $r_e = 0{\cdot}216$ in an electric field. The points are experimental: ●, $E_0 = 1{\cdot}7$ kV cm^{-1}; ○, $E_0 = 3{\cdot}3$ kV cm^{-1}; ▲, $E_0 = 4.5$ kV cm^{-1}; △, $E_0 = 5{\cdot}3$ kV cm^{-1}; ▽, $E_0 = 6{\cdot}3$ kV cm^{-1}. Curve 1: best fit through experimental points. Curve 2: theoretical curve assuming the disk is an oblate spheroid of axis ratio 0·216 (after Chaffey and Mason, 1964).

An example of an observation of the alignment of a conducting disk without a permanent dipole moment ($q=\infty$, $m=0$) is given in Fig. 3.74, which shows values of log (tan θ_2) as a function of time. It can be seen that the data fall on a straight line, in agreement with eq (3.416). The slope of the line is slightly different from the slope calculated from eq (3.416), by making the (approximate) assumption that the disk is a spheroid of axis ratio $r_e=0{\cdot}216$.

From eq (3.412) it can be seen that $d\theta_2/dt=0$ when either

$$\sin\theta_2=0 \quad \text{or} \quad \cos\theta_2=-\tau_2/\tau_1$$

It can be readily shown that, for rods, the angle $\theta_2=\cos^{-1}(-\tau_2/\tau_1)$ is unstable, while $\theta_2=0$ is stable. For disks the angle $\theta_2=0$ is unstable and the equilibrium orientation is given by

$$\theta_{2\infty}=\cos^{-1}(-\tau_2/\tau_1) \tag{3.417}$$

if $|\tau_2| < \tau_1$, or $\theta_{2\infty}=90°$ if $|\tau_2| \geqslant \tau_1$.

It is possible to integrate eq (3.412) for arbitrary values of τ_1 and τ_2 with the result ($\tau_1 \neq \tau_2$):

$$\left(\frac{\tau_2+\tau_1\cos\theta_2}{\tau_1\sin^2\theta_2}\right)^{\tau_1/\tau_2}\tan\left(\frac{\theta_2}{2}\right)=\text{const}\cdot\exp\left[-\frac{(\tau_2^2-\tau_1^2)}{\tau_1\tau_2^2}\,t\right] \tag{3.418a}$$

As τ_1 or $\tau_2 \to \infty$, eq (3.418a) reduces to eq (3.416) or eq (3.415), respectively. When $\tau_1=\tau_2$, we have

$$e^{(1+\cos\theta_2)^{-1}}\tan(\theta_2/2)=e^{-2t/\tau_1} \tag{3.418b}$$

It can be seen from eqs (3.418) that the steady-state orientation is reached with a complicated time dependence.

It can be shown that eq (3.412) also holds when E is a function of time, e.g. an oscillatory field $E=E_0\cos\omega t$. If the frequency, ω, is much smaller than τ_1^{-1}, then the permanent dipole can follow the field and the particle will flip over periodically with frequency ω. When $\omega \gg \tau_1^{-1}$ the particle is unable to follow the field and behaves as if $m=0$ ($\tau_1=\infty$). When $\omega \gg (\tau_2{}^{-1}, \tau_1{}^{-1})$, the particle behaves as a spheroid without a permanent dipole in a field of strength equal to the r.m.s. value of E.

The above theory is valid for strong electric fields or large particles, in which case rotary Brownian motion can be neglected. For small colloidal particles Brownian motion must be taken into account. Let us for the moment consider particles with $m=0$ which, in the absence of Brownian motion, align according to

$$\frac{d\theta_2}{dt} = -\frac{\sin\theta_2\cos\theta_2}{\tau_2} \tag{3.419}$$

with τ_2 as given by eq (3.414). Properties of suspensions of spheroidal Brownian particles are governed by the rotary convective-diffusion equation

$$\frac{\partial p}{\partial t} = D_r\nabla^2 p - \nabla(\omega p) \tag{3.420}$$

with $\boldsymbol{\omega}$ having one non-zero component, given by eq (3.419). Writing the above equation explicitly yields:

$$\frac{\partial f}{\partial t} = D_r\left(\frac{\partial^2 f}{\partial\theta_2^2} + \frac{\cos\theta_2}{\sin\theta_2}\frac{\partial f}{\partial\theta_2}\right) + \frac{1}{\tau_2\sin\theta_2}\frac{\partial}{\partial\theta_2}(f\sin^2\theta_2\cos\theta_2) \tag{3.421}$$

where $f(\theta_2, t) = 2\pi p(\theta_2, \phi_2, t)$. The factor 2π arises because of the normalization of the function p; θ_2, ϕ_2 are the polar angles with respect to the axis X_2. Defining a normalized time $t^* = D_r t$ and a dimensionless number

$$H = \frac{1}{D_r\tau_2} \tag{3.422}$$

which denotes the ratio of electrical to Brownian torques acting on the particle, eq (3.421) can be rewritten as

$$\frac{\partial f}{\partial t^*} = \frac{\partial^2 f}{\partial\theta_2^2} + \frac{\cos\theta_2}{\sin\theta_2}\frac{\partial f}{\partial\theta_2} + \frac{H}{\sin\theta_2}\frac{\partial}{\partial\theta_2}(f\sin^2\theta_2\cos\theta_2) \tag{3.423}$$

When $H \to 0$, the orientation distribution is diffusion controlled, while when $H \to \infty$ it is controlled by the external electric field. It is possible to solve eq (3.423) as an expansion in H:

$$f = f_0 + Hf_1 + H^2 f_2 + \ldots\ldots \tag{3.424}$$

For an initially random orientation $f_0 = \frac{1}{2}$ and the governing equation for f_1 becomes:

$$\frac{\partial f_1}{\partial t^*} = \frac{\partial^2 f_1}{\partial\theta_2^2} + \frac{\cos\theta_2}{\sin\theta_2}\frac{\partial f_1}{\partial\theta_2} + \tfrac{1}{2}(3\cos^2\theta_2 - 1) \tag{3.425}$$

the normalized solution of which is (cf.eq (3.140))

$$f_1 = \frac{1}{8}\left(\cos 2\theta_2 + \frac{1}{3}\right) + \sum_n A_n \mathrm{P}_n^0(\cos\theta_2)\, \mathrm{e}^{-n(n+1)D_r t} \tag{3.426}$$

As $t \to \infty$, the steady-state solution up to order H becomes

$$f = \frac{1}{2} + \frac{H}{8}\left(\cos 2\theta_2 + \frac{1}{3}\right) = \frac{1}{2} + \frac{H}{12}\,(2 - 3\sin^2\theta_s) \tag{3.427}$$

which is maximum at $\theta_2 = 0$ and minimum at $\theta_2 = 90°$. This is as would be expected because more particles will be aligned with the field ($\theta_2 = 0$) than are perpendicular to it ($\theta_2 = 90°$). Higher-order terms can also be found in a similar way.

The change in the orientation distribution of suspended colloidal particles subjected to an electric field manifests itself in changes in a variety of properties. Benoit (1951) solved eq (3.420) in an analogous manner and calculated the electrically induced birefringence, Δn, of the dispersion (the so-called Kerr effect). Expressing eq (3.338) in terms of the angles (θ_2, ϕ_2) and averaging over ϕ_2 yields

$$\Delta n = \tfrac{1}{2} K \langle 2 - 3\sin^2\theta_2 \rangle \tag{3.428}$$

where K is the constant in front of the orientational averages in eq (3.338).

Note that for random orientations, $\Delta n = 0$. With the aid of eq (3.426), the orientational average in eq (3.428) can be evaluated with the result

$$\Delta n = \Delta n_\infty\,(1 - \mathrm{e}^{-6D_r t}) \tag{3.429}$$

with the steady-state birefringence, Δn_∞, given by

$$\Delta n_\infty = HK/15 \tag{3.430}$$

It can be readily shown that all the f_i terms in the expansion of f attain the steady state in a similar manner as f_1. Hence eq (3.429) holds when $H < 1$. When $H \gg 1$, particles align with a relaxation time given by eq (3.414). Hence the alignment time will, in general, show the behaviour illustrated schematically in Fig. 3.75. For low electric-field strengths, the time required to reach the steady-state birefringence is independent of E and depends only on D_r. For large fields the time becomes inversely proportional to E^2.

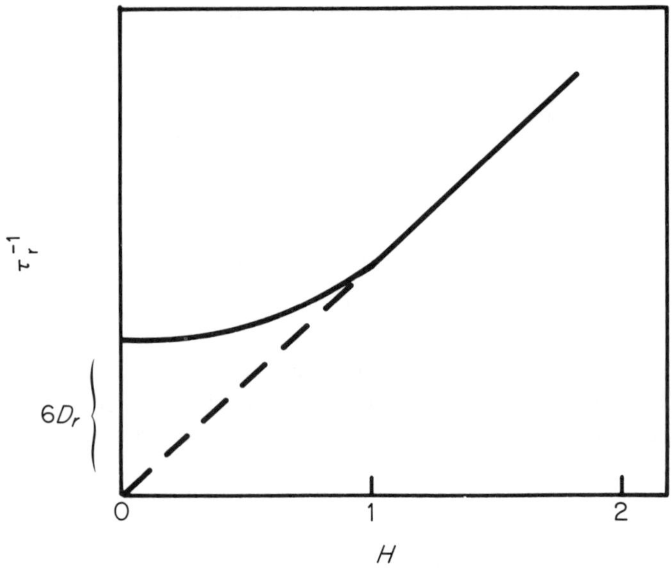

Fig. 3.75. Response time, τ_r, for the alignment of a non-spherical particle in an electric field. For low values of the ratio, H, of electrical to diffusion torques, τ_r is determined by the rotary diffusion constant, D_r. For large values of H, τ_r^{-1} becomes proportional to H. At very high electric fields (very high H) saturation will occur (not shown), manifesting itself in a slower increase in τ_r^{-1}.

The same analysis as above can be made for particles with a permanent dipole. This results in an additional term in eq (3.426) of magnitude

$$\tfrac{1}{2}(\tau_2/\tau_1)\cos\theta_2 \tag{3.431}$$

Hence, when $D_r\tau_1 \gg 1$, more particles have their dipoles parallel with the field ($\theta_2=0$) than antiparallel ($\theta_2=180°$). It is of interest to note that the above term does not contribute to Δn; hence up to order $H_1=(D_r\tau_1)^{-1}$, $\Delta n=0$. The first appearance of birefringence is of the order of H_1^2. If both permanent and induced dipoles are present, then eq (3.431) contributes to the way in which the steady state is reached. Using eq (3.431) to evaluate the orientational average in eq (3.428), gives the terms (cf. eq (3.426)):

$$\int_0^\pi \cos\theta\, P_n^0(\cos\theta)\sin\theta d\theta e^{-n(n+1)D_r t} = \sum_{m=0}^{n} a_{mn}\int_0^\pi \sin 2\theta\cos(n-2m)\theta d\theta\; e^{-n(n+1)D_r t}$$

which are non-zero if n is odd. Hence, in the presence of permanent dipoles, the steady state is reached with a leading relaxation time of $1/2D_r$.

Besides the birefringence, other properties of the suspension can also be found from the orientation distribution. Of importance is the intensity, I, of scattered or transmitted light. It can be readily shown that if $H < 1$:

$$I = I_\infty \left(1 - \sum_n a_n e^{-n(n+1)D_r t}\right) \tag{3.432}$$

where for spheroids without a permanent dipole n is even, for spheroids without induced dipoles n is odd, and if both types of dipole are present the sum in eq (3.432) is over all positive integers. For $H \gg 1$, the relaxation time for alignment is again given by eq (3.414).

For colloidal particles surrounded by electrical double layers, the behaviour is more complicated. The external electric field polarizes the double layer, thus creating a dipole moment in the double layer which, in general, will have a different orientation than the external field. For the double-layer dipole to align, it must drag the (non-spherical) particle along. Therefore, in the presence of a double layer the alignment of the particles is expected to be is faster. In the limit $\kappa a \to \infty$ or 0, i.e. very thin or very thick double layers, effects of the double layer on alignment are expected to be absent. The decrease in the alignment time at intermediate double-layer thickness has been observed experimentally (Baloch and van de Ven, 1989). As yet no theory for the alignment of non-spherical particles surrounded by electrical double layers is available.

3.3.3 Dielectric constants of suspensions

When colloidal particles are dispersed in a fluid of dielectric constant ϵ°, the dispersion will behave as having an apparent dielectric constant, ϵ, given by (Brugmann, 1935):

$$\epsilon = \epsilon^\circ\ (1 + [\epsilon]c) \tag{3.433}$$

where c is the volume fraction of particles and $[\epsilon]$ is the intrinsic dielectric constant. Equation (3.433) applies when particle interactions are absent ($c \ll 1$). For electrically conducting spheres $[\epsilon] = 3$, a result confirmed experimentally by Voet (1947). For non-spherical particles the dielectric constant must be replaced by a tensor, which can be found from

$$\boldsymbol{D}_s = \epsilon_0 \epsilon^\circ \boldsymbol{E} + \mathrm{n}\bar{\boldsymbol{P}} = \epsilon_0 \boldsymbol{\epsilon} \cdot \boldsymbol{E} \tag{3.434}$$

where $\boldsymbol{D}_s$ is the macroscopic electric displacement of the dispersion, $\boldsymbol{E}$ is the electric field strength, ϵ_0 the permittivity of free space, n the number concentration of particles, and $\boldsymbol{P}$ the mean polarization of the particle. For spheroidal particles without a permanent polarization (Jones, 1945; Chaffey and Mason, 1964):

$$\boldsymbol{P} = \epsilon_0 \epsilon^\circ\,(\alpha_1 \boldsymbol{dd} + \alpha_2 \boldsymbol{I}) \cdot \boldsymbol{E} \tag{3.435}$$

where the axis of revolution is given by the vector $\boldsymbol{d}$ and $\boldsymbol{I}$ is the idemfactor; α_1 and α_2 are functions of $q = \epsilon_1/\epsilon_2$, where ϵ_1 is the dielectric constant of the particles and $\epsilon_2 = \epsilon^\circ$ is that of the medium. Combining eq (3.435) with eq (3.434) and averaging, gives the dielectric tensor

$$\epsilon = \epsilon^\circ \boldsymbol{I} + \epsilon^\circ n(\alpha_1\, \boldsymbol{dd} + \alpha_2 I) \tag{3.436}$$

Let us define a Cartesian coordinate system in such a way that the electric field coincides with the X_2 axis (see Fig. 3.65) in which case

$$\boldsymbol{d} = \begin{pmatrix} \cos\theta \\ \sin\theta\cos\phi \\ \sin\theta\sin\phi \end{pmatrix} \tag{3.437}$$

and (Okagawa *et al.*, 1978):

$$\alpha_1 = \frac{4\pi ab}{3}\,\beta_1 \quad \text{and} \quad \alpha_2 = \frac{4\pi ab}{3}\,\beta_2 \tag{3.438}$$

with

$$\beta_1 = \frac{(q-1)^2(3A-2)}{[q-(q-1)A][2+(q-1)A]}$$

and

$$\beta_2 = \frac{2(q-1)}{2+(q-1)A}$$

where A is given by eq (3.281).

Substituting eqs (3.437) and (3.438) into eq (3.436) gives

$$\epsilon = \epsilon^\circ\,(\boldsymbol{I} + [\epsilon]c) \tag{3.439}$$

where

$$[\epsilon] = \beta_2 \boldsymbol{I} + \beta_1 \begin{pmatrix} \langle\cos^2\theta\rangle & \langle\sin\theta\cos\theta\cos\phi\rangle & \langle\sin\theta\cos\theta\sin\phi\rangle \\ \langle\sin\theta\cos\theta\cos\phi\rangle & \langle\sin^2\theta\cos^2\phi\rangle & \langle\sin^2\theta\sin\phi\cos\phi\rangle \\ \langle\sin\theta\cos\theta\sin\phi\rangle & \langle\sin^2\theta\sin\phi\cos\phi\rangle & \langle\sin^2\theta\sin^2\phi\rangle \end{pmatrix} \tag{3.440}$$

It can be seen that $[\epsilon]$ depends on the orientation distribution of the particles. For random orientations, eq (3.440) reduces to

$$[\epsilon] = \left(\beta_2 + \frac{1}{3}\beta_1\right) \boldsymbol{I} \tag{3.441}$$

i.e. the dispersion behaves as a homogeneous isotropic medium with an intrinsic dielectric constant of $\beta_2 + \beta_1/3$. For conducting spheres, $\beta_1 = 0$ and $\beta_2 = 3$ and we recover the classical result $[\epsilon] = 3$.

The dielectric tensor, ϵ, can be changed by applying an external field. If the suspension is subjected to a simple shear flow, the orientational averages can be found from Section 3.1.6. It can be concluded that the components of ϵ will go through a transient regime characterized by damped oscillations, prior to reaching a steady state. When the particles are subjected to an electric field the dielectric components vary as shown schematically in Fig. 3.72.

3.3.4 Fluid drops in an electric field

Fluid drops in an electric field behave qualitatively differently from solid particles. Usually, fluid drops deform into prolate or oblate spheroids, except in very weak fields. Furthermore, they exhibit internal circulation and at high field strengths the drops may burst. Similar to solid particles, the drops will have an induced dipole moment, but, depending on whether $\lambda q < 1$ or $\lambda q > 1$, the dipole will be parallel or antiparallel to the external field. Here $\lambda = \chi_2/\chi_1$ is the ratio of the conductivity of the medium to that of the particle and $q = \epsilon_1/\epsilon_2$ is the dielectric constant ratio. The induced dipole moment can be found from the surface charge density, Q_s, which is related to the normal component of the discontinuity in the dielectric displacement at the fluid/liquid interface. For an external electric field $\boldsymbol{E} = \boldsymbol{E}_0 \cos\omega t$, Q_s is given by (Torza *et al.*, 1971a)

$$Q_s = \frac{3\epsilon_0\epsilon_2 E_0(1 - \lambda q)\cos\theta}{[(2\lambda + 1)^2 + \omega^2\tau^2]^{1/2}} \cos(\omega t - \delta) \tag{3.442}$$

where the polar angle, θ, is defined in Fig. 3.76a;

$$\tau = (q + 2)\, \epsilon_0\epsilon_2/\chi_1$$

and the phase lag, δ, is

$$\delta = \tan^{-1}\left(\frac{\omega\tau}{2\lambda + 1}\right)$$

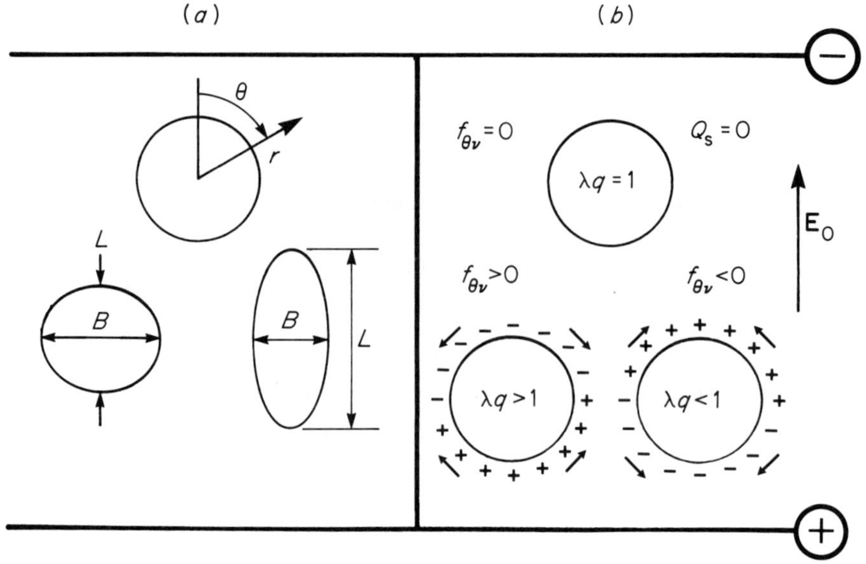

Fig. 3.76. (*a*) Polar coordinates (r,θ) of a drop in an electric field, E_0. The drop deforms into an oblate or prolate shape depending on whether the drop deformation, D, is larger or smaller than zero.
(*b*) Schematic representation of charge distribution, Q_s, and the direction of flow (indicated by the tangential stresses, $f_{\theta\nu}$) at the interface for various λq (after Arp *et al.*, 1980).

with $0 < \delta < \pi/2$.

The charge distribution is shown schematically in Fig. 3.76b. When $\lambda q=1$, $Q_s=0$ and the two phases are electrically indistinguishable.

The deformation of the drop, D, consists of a steady component, D_s, and an oscillating component, D_o:

$$D=D_s+D_o \tag{3.443}$$

If $D>0$ the drop assumes a prolate form, whereas if $D<0$ the drop is deformed into an oblate spheroid. To obtain the deformation, one must solve the Laplace equation

$$\Delta p=\sigma\left(\frac{1}{R_1}+\frac{1}{R_2}\right) \tag{3.444}$$

where Δp is the sum of the steady and time-dependent normal stresses acting at the interface, σ is the interfacial tension, and R_1 and R_2 are the two principal radii of curvature of the drop surface. This problem was solved by

Torza *et al.* (1971a) and Sozou (1972, 1973). The result for the steady deformation is

$$D_s = \frac{9\epsilon_0\epsilon_2\bar{E}_0^2 a}{16\sigma} f(\lambda, p, q, \omega\tau) \quad (3.445)$$

where a is the radius of the undeformed drop, p the viscosity ratio μ_1/μ_2 and

$$f(\lambda, p, q, \omega\tau) = 1 - \lambda \left(\frac{a_1 + a_2\lambda}{a_3[(2\lambda+1)^2 + \omega^2\tau^2]} \right) \quad (3.446)$$

where

$$a_1 = 11p + 14$$

$$a_2 = [15(p+1) + q(19p+16) + 15\omega^2\tau^2(1+p)(1+2q)/(q+2)^2]$$

$$a_3 = 5(1+p)$$

From eqs (3.445) and (3.446) it can be seen that a frequency, ω_c, exists for which the drop deformation is zero, given by

$$\omega_c = \frac{\{\lambda^2[q(19p+16) - 5(1+p)] - 3\lambda(3p+2) - 5(1+p)\}^{1/2}(q+2)}{|q-1|\ [5(1+p)]^{1/2}\tau} \quad (3.447)$$

Consequently, ω_c exists whenever

$$\lambda q > (\lambda q)_c;\ q \neq 1$$

where

$$(\lambda q)_c = 1 + \frac{5(1+p)(\lambda-1)^2}{\lambda(19p+16)} \quad (3.448)$$

From the above analysis it follows that one can distinguish three different classes of drop deformation in an electric field, which are

Class A: $\lambda q \leqslant 1,\ \partial D_s/\partial\omega \leqslant 0,\ D_s \geqslant 0$

Class B: $1 < q < (\lambda q)_c,\ \partial D_s/\partial\omega < 0,\ D_s > 0$

Class C: $\lambda q > (\lambda q)_c,\ \partial D_s/\partial\omega > 0 \begin{cases} D_s < 0 \text{ if } \omega < \omega_c \\ D_s > 0 \text{ if } \omega > \omega_c \end{cases}$

For systems in Class A, the drop deformation decreases with increasing frequency, systems belonging to Class B show an increase in deformation with increasing frequency, while Class C is characterized by systems that deform to oblate spheroids at flow frequencies and to prolate spheroids at high frequencies. An observation of the deformation of a Class-C drop is shown in Fig. 3.77.

The theory described above is valid for small deformations. As the field strength increases, the drops will continue to deform until they finally rupture or burst. Figure 3.78a shows deformation and burst in a steady field for a system belonging to Class A. Figure 3.78b shows the deformation of a Class-C system.

The time scale for reaching the steady-state deformation is (in the absence of rotary Brownian motion) given by

$$D(t) = D_\infty(1 - e^{-t/\tau}) \tag{3.449}$$

while after switching off the electric field

$$D(t) = D_\infty e^{-t/\tau} \tag{3.450}$$

where $\tau = (\mu_1 + \mu_2)\, a/\sigma$.

Other restrictions of the above theory are that it applies to drops with no net surface charge, no rotary Brownian motion and no surfactants adsorbed at the interface. An electric charge will cause the drop to migrate; rotary Brownian motion causes changes in the orientation of the deformed drop, which will lead to additional time-dependent deformations; the presence of surfactants will result in changed boundary conditions at the interface; and the internal flow will cause variations in the surfactant concentration at the surface, resulting in Marangoni effects.

The deformation of drops in a gravitational field is similar to that in an electric field. A well-known example on a different scale is the earth in the gravitational field of the moon. The earth, surrounded by oceans, can be thought of as a large drop that deforms into a spheroid in a gravitational field. Combined with the rotation of the earth around its axis, this deformation is responsible for the two high and low tides observed each day.

3.4 Electro- and magneto-rheology of dilute dispersions

Here we will discuss some aspects of the behaviour of dispersions subjected to both shear and electric fields. Because the theory of the motion of suspended particles in a magnetic field is analogous to that of the motion in an electric field, all results from an electric field also apply, *mutatis mutandis*, to a magnetic field. A summary of various electrohydrodynamic effects in fluid dispersions is given by Arp *et al.* (1980). Because the subject

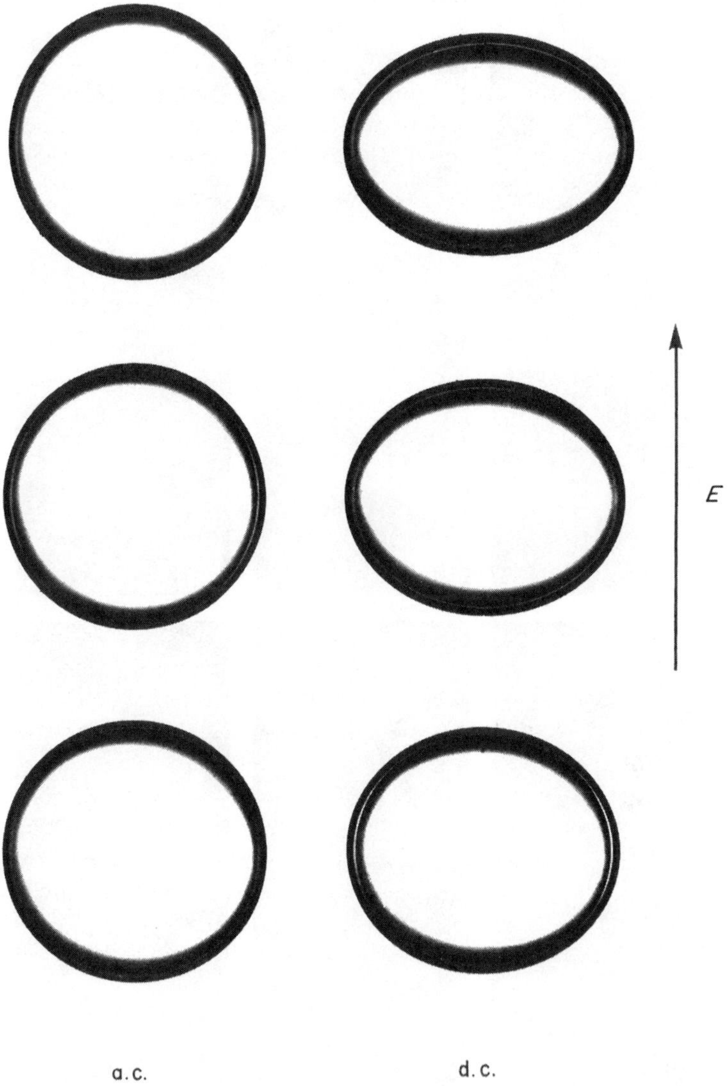

Fig. 3.77. Photographs of a silicone oil drop ($a = 0{\cdot}6$ mm) in castor oil. The drop is prolate in an a.c. field and oblate in a d.c. field, indicating that the system belongs to Class C. The field strengths are (from top to bottom) $3{\cdot}5$, $2{\cdot}5$ and $1{\cdot}5$ kV cm^{-1} (after Torza *et al.*, 1971a).

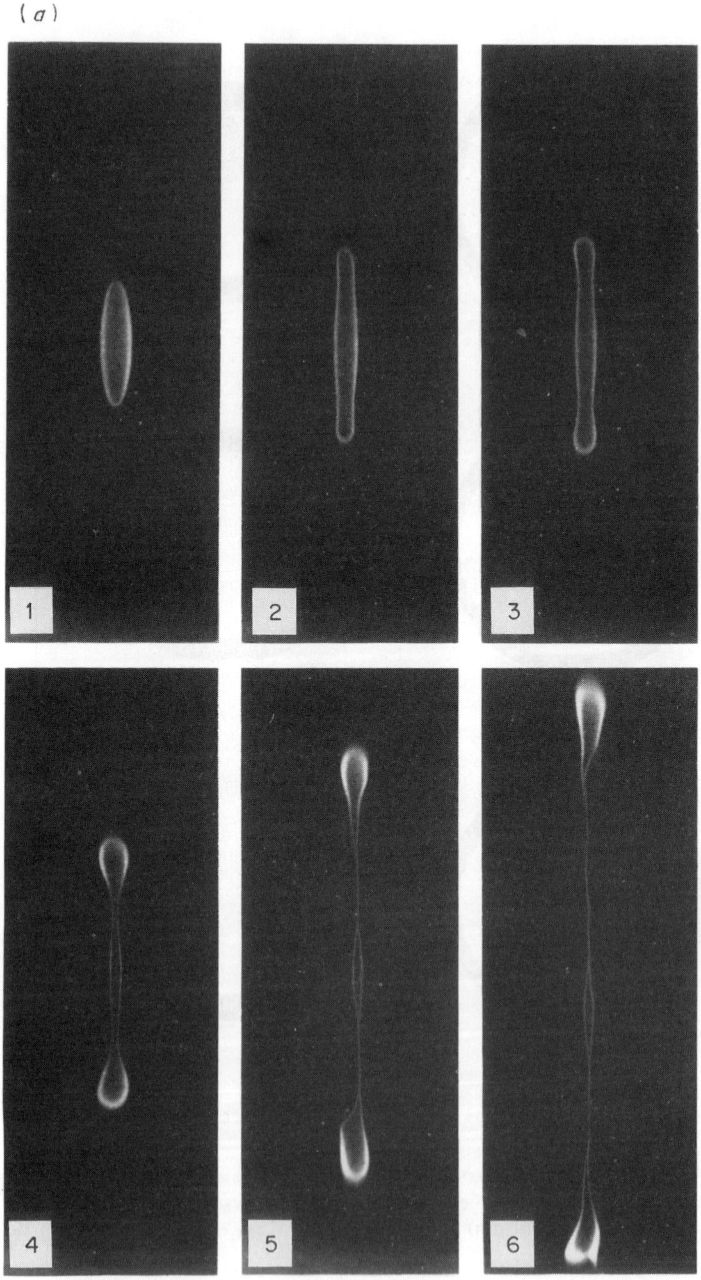

Fig. 3.78. (*a*) Photographs of the electric burst of a drop of Class A ($a = 0{\cdot}3$ mm). Time elapsed from frame 1 to 6 is about $0{\cdot}1$s. Frequency $\nu = 0$.

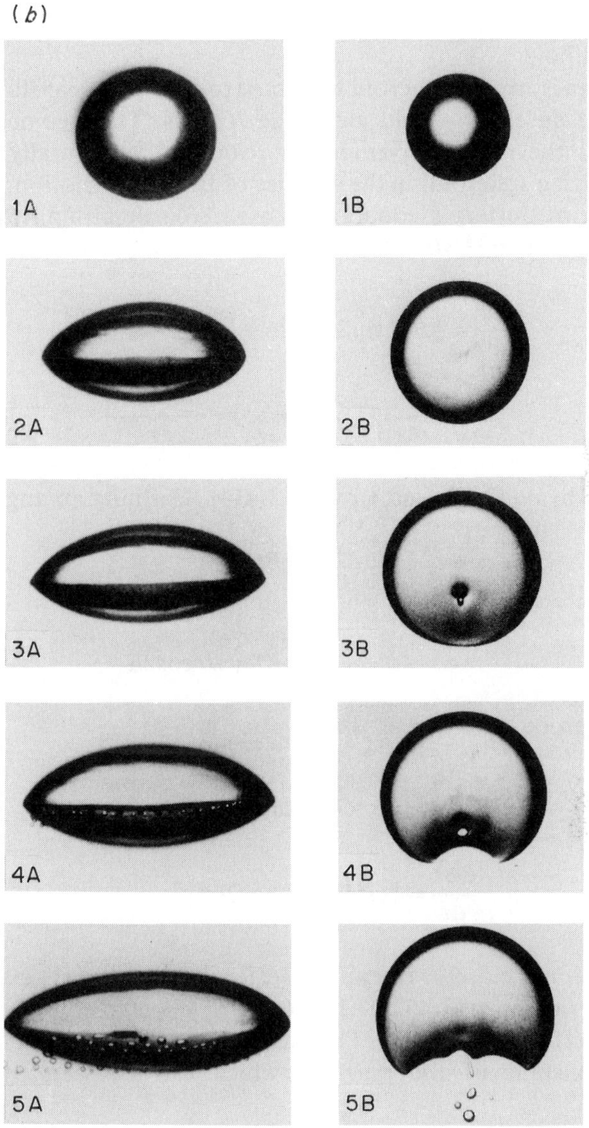

(b) Photographs of the burst of two drops from Class C at $\nu = 1$ Hz. Drop A: $a = 0{\cdot}4$ mm; $E_0 = 4{\cdot}5$ kV cm^{-1}; Drop B: $a = 0{\cdot}6$ mm (different magnification), $E_0 = 4{\cdot}0$ kV cm^{-1} viewed at an angle perpendicular to the viewing direction of drop A. Burst occurs after the formation of a biconcave lens (after Torza *et al.*, 1971a).

area of electrohydrodynamics is a large one, the discussion here will be restricted to the important case of a dispersion of isotropic spheroids with no permanent dipoles.

Consider an isotropic spheroid subjected to a simple shear flow ($v_1 = v_2 = 0$; $v_3 = Gx_2$) and an electric field along the X_2 axis. This geometry is chosen because it is the most convenient to realize experimentally. The rotary motion of such a spheroid, in the absence of Brownian motion, is simply the combination of Jeffery's equations (eqs (3.86)) describing the motion in simple shear, and eq (3.406) describing the motion in an electric field. Thus

$$\frac{\mathrm{d}\theta}{\mathrm{d}t} = \tfrac{1}{4}BG\,\sin 2\theta \sin 2\phi + \frac{\sin\theta\cos\theta\cos^2\phi}{\tau} \tag{3.451a}$$

$$\frac{\mathrm{d}\phi}{\mathrm{d}t} = \tfrac{1}{2}\mathrm{G}\,(1 + B\cos 2\phi) - \frac{\sin\phi\cos\phi}{\tau} \tag{3.451b}$$

with B given by eq (3.87) and τ by eq (3.414). Defining an angle α by

$$\tan 2\alpha = (BG\tau)^{-1}$$

and defining

$$\phi^* = \phi + \alpha \text{ and } B^* = B/\cos 2\alpha$$

eqs (3.451) can be rewritten as

$$\frac{\mathrm{d}\theta}{\mathrm{d}t} = \tfrac{1}{4}B^*G\,\sin 2\theta \sin 2\phi^* + \frac{\sin 2\theta}{4\tau} \tag{3.452a}$$

$$\frac{\mathrm{d}\phi^*}{\mathrm{d}t} = \tfrac{1}{2}G[1 + B^*\cos 2\phi^*] \tag{3.452b}$$

These equations closely resemble Jeffery's equations, except for an additional term in eq (3.452a). For $\theta = 0$ or $\theta = 90°$, these equations are identical to Jeffery's equations. The extra term in eq (3.452a) causes a drift in the orbit constant of the particle, which can be seen by integrating eqs (3.452):

$$\tan\theta = \frac{Cr_e e^{t/\tau}}{(r_e^2\cos^2\phi + \sin^2\phi - fr_e\sin 2\phi)^{1/2}} \tag{3.453}$$

Because for prolate spheroids $\tau > 0$, it follows that $\theta \to 90°$ as $t \to \infty$. Similarly, for oblate spheroids $\tau < 0$ and $\theta \to 0$, as $t \to \infty$. In eq (3.453) C is an integration constant and f is defined as

$$f = \frac{r_e + r_e^{-1}}{2G\tau} \tag{3.454}$$

The parameter f is related to the Mason number, $Mn = \mu G/\epsilon E^2$, which is the ratio of hydrodynamic to electric forces by

$$f = -(r_e + r_e^{-1})\, P(q, r_e)/Mn$$

As $\tau \to \infty$, $f \to 0$ and eq (3.453) reduces to eq (3.88), C being the orbit constant. It follows from eq (3.453) that as $t \to \infty$, the spheroid rotates according to Jeffery's equations with an orbit constant $C = 0$ for disks and $C = \infty$ for rods. The difference between a particle in simple shear and one in a combination of shear and an electric field is that in the latter case B^* can have any value. In this respect the motion of a particle subjected to shear and an electric field is very similar to the motion of a spheroid in a two-dimensional linear flow field (cf. eq (3.207)). From eq (3.452b) it can be seen that, when $-1 < B^* < 1$, the spheroid executes complete rotations through the angle ϕ. Expressed in terms of f, this occurs when

$$f^2 < 1 \tag{3.455}$$

Electric fields satisfying eq (3.455) are called subcritical fields. Integrating eq (3.452b) shows that the period of rotation is given by

$$T = \frac{2\pi}{G}(s + s^{-1}) = \frac{2\pi}{G}\,\frac{r_e + r_e^{-1}}{(1 - f^2)^{1/2}} \tag{3.456}$$

where s is defined by

$$B^* = (s^2 - 1)/(s^2 + 1) \tag{3.457}$$

when $f^2 > 1$, the electric field is supercritical and the particles attain a steady state orientation (cf. eq (3.203)).

From the flow field around a spheroid the macroscopic stress tensor $\boldsymbol{P}$ (cf. eq (3.259)) and thus the intrinsic viscosity and normal stress differences can be calculated. This leads to (Okagawa *et al.*, 1974):

$$[\mu] = [\mu]_0 + fC_5[(r_e^2 - 1)\langle \sin^4\theta \sin\phi \cos^3\phi\rangle + \langle \sin^2\theta \sin\phi \cos\phi\rangle] \tag{3.458a}$$

$$[\xi_{12}] = [\xi_{12}]_0 - fC_5(r_e^2 - 1)(\tfrac{1}{2}\langle \sin^2 2\theta \cos^2\phi\rangle + \tfrac{1}{4}\langle \sin^4\theta \sin^2 2\phi\rangle) \tag{3.458b}$$

$$[\xi_{23}] = [\xi_{23}]_0 + fC_5(r_e^2 - 1)(\tfrac{1}{2}\langle \sin^2 2\theta \cos^2\phi\rangle + \tfrac{1}{4}\langle \sin^4\theta \sin^2 2\phi\rangle) \tag{3.458c}$$

where

$$C_5 = \frac{8r_e}{(r_e^2+1)[2r_e^2+(1-2r_e^2)A]}$$

where A is as given by eq (3.281). The subscripts "0" refer to the values in the absence of an electric field (cf. eqs (3.283), (3.288) and (3.289)). It should be noted that, in the presence of an electric field, $\boldsymbol{P}$ is no longer symmetric.

The orientational averages in eqs (3.458) depend on the rotary Péclet number, $\mathrm{Pe_r}$, and on f and can be evaluated in the usual way. Results for the steady-state values of $[\mu]$ and $[\xi_{23}]$ for $\mathrm{Pe_r}=\infty$ were calculated by Okagawa *et al.* (1974) and are shown in Fig. (3.79). For spheres ($r_e=1$), $[\mu]$ increases from 5/2 when $f=0$ and the spheres rotate freely, to a value of 4 when $f \geqslant 1$ and the spheres do not rotate. The intrinsic normal stress difference, $[\xi_{23}]$, is zero in the limits $f \to 0$ and $f \to \infty$ and passes through a minimum at some value of f depending on r_e; it has a discontinuity in slope at $f=1$. The intrinsic viscosity approaches a constant in the limits $f \to 0$ and $f \to \infty$ and is minimum at $f=1$ (except for spheres), where it also exhibits a discontinuity in slope.

Besides the viscosity and normal stresses in a simple shear flow, the behaviour of a dispersion in an oscillatory shear flow is also of interest. In steady shear the macroscopic stress tensor is proportional to the rate of shear G. In oscillatory shear, each component of the stress tensor is the sum of a viscous and an elastic contribution. Concentrating on the component P_{32} we have

$$P_{32} = \mu'\dot{\gamma} + G'\gamma \tag{3.459}$$

where $\dot{\gamma} = G$ is the rate of shear (or rate of strain) and γ is the strain; μ' is called the dynamic viscosity and G' the dynamic rigidity (or storage modulus).

If a small-amplitude oscillating strain of amplitude ϵ and angular velocity ω is applied:

$$\gamma = \epsilon \sin \omega t \tag{3.460}$$

then

$$\mathrm{G} = \dot{\gamma} = \epsilon\omega \cos \omega t \tag{3.461}$$

and the response can be described by a complex dynamic viscosity

$$\mu^* = \mu' - \frac{iG'}{\omega} \tag{3.462}$$

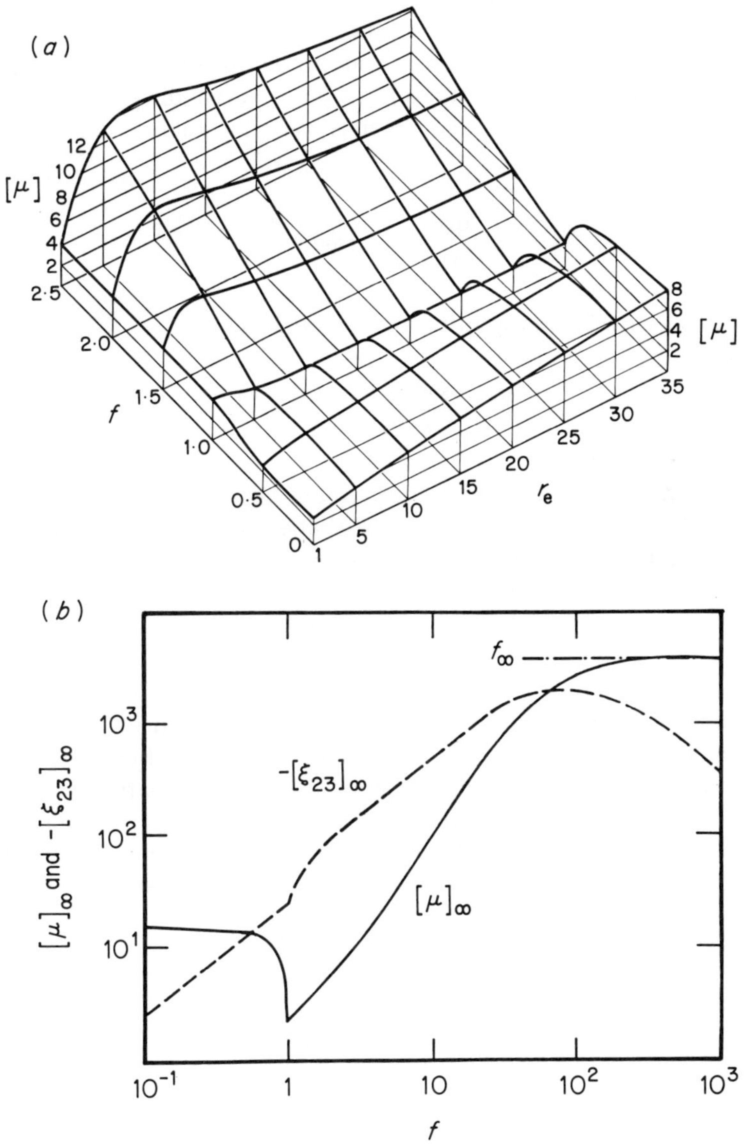

Fig. 3.79. (*a*) Three-dimensional plot of the equilibrium intrinsic viscosity, $[\mu]_\infty$, for prolate spheroids as functions of r_e and f.
(*b*) Plots of $[\mu]_\infty$ and $[\zeta_{23}]_\infty$ for $r_e = 100$ (after Okagawa *et al.*, 1974).

The rotary motion of a spheroid subjected to an oscillatory shear and an electric field is given by eq (3.451), but the shear rate, G, is now a function of time given by eq (3.461). For small-amplitude oscillations the angle ϕ will never be far from the orientation $\phi = 0$, corresponding to the angle of alignment in an electric field. Hence we can replace eq (3.451b) by (neglecting terms of order ϕ^2):

$$\frac{d\phi}{dt} = \epsilon k \cos\omega t - \frac{\phi}{\tau} \qquad (3.463)$$

where $k = (1 + B)/2$. The solution of eq (3.463) is

$$\phi = \frac{\epsilon k \omega\tau(\cos\omega t + \omega\tau\sin\omega t)}{1 + \omega^2\tau^2} - \frac{\epsilon k \omega\tau e^{-t/\tau}}{1 + \omega^2\tau^2} \qquad (3.464)$$

Disk-shaped particles ($\tau < 0$) do not remain near the orientation $\phi = 0$ but drift away from that orientation. Rod-shaped particles ($\tau > 0$) execute, at times $t \gg \tau$, an oscillation of frequency ω described by

$$\phi = \frac{\epsilon k \omega\tau(\cos\omega t + \omega\tau\sin\omega t)}{1 + \omega^2\tau^2} \qquad (3.465)$$

and as a result all rheological properties oscillate with an angular frequency ω. If all the particles are first aligned in an electric field and a small-amplitude oscillatory strain is subsequently applied, all particles oscillate in phase with a ϕ angle given by eq (3.465) and $\theta = 90°$. From the orientation of the particle axes, the components of the stress tensor $\boldsymbol{P}$ can be evaluated. For the component P_{32}, eq (3.459) becomes (van de Ven, 1984)

$$\mu' = \mu_0 (1 + [\mu']c) \qquad (3.466)$$

where

$$[\mu'] = C_3 + \frac{S}{1 + \omega^2\tau^2} \qquad (3.467)$$

and

$$G' = \mu_0 S c \frac{\omega^2\tau}{1 + \omega^2\tau^2} \qquad (3.468)$$

where C_3 and S are functions of particle shape, with C_3 as given by eq (3.284) and S by

$$S = \frac{2r_e^2(1+B)}{2r_e^2 + (1-2r_e^2)A} \tag{3.469}$$

Equations (3.467) and (3.468) show that dispersions of rods subjected to an electric field exhibit viscoelastic behaviour. For a dispersion of disks

$$[\mu'] = C_2 \text{ and } G' = 0 \tag{3.470}$$

Thus, in contrast to dispersions of rods, dispersions of disks are Newtonian. This conclusion is true only for the special geometry considered here; with an electric field along the X_1 axis, all disks are oriented with their axis of revolution perpendicular to the direction of the electric field and no electric torques are acting on the disks. For other directions of the electric field, dispersions of disks will also exhibit electro-viscoelastic behaviour.

Some interesting phenomena are predicted when a suspension of rods is subjected to both an electric field and a small-amplitude oscillatory shear superimposed on a steady shear. In this case the shear rate is given by

$$G = G_0 + \epsilon\omega\cos\omega t \tag{3.471}$$

For supercritical fields ($f^2 > 1$), the particles attain a steady-state orientation ϕ_∞ (cf. eq (3.457)) and $\theta = 90°$; with an oscillating field superimposed, for times $t \gg \tau$, the particle will oscillate about ϕ_∞ according to (van de Ven, 1984):

$$\phi = \phi_\infty + \frac{\epsilon k_2 \omega \tilde{\tau}}{1+\omega^2\tilde{\tau}^2}(\cos\omega t + \omega\tilde{\tau}\sin\omega t) \tag{3.472}$$

where $k_2 = (1 + B\cos 2\phi_\infty)/2$ and

$$\tilde{\tau} = \tau\,\frac{1 + B\cos 2\phi_\infty}{B + \cos 2\phi_\infty} \tag{3.473}$$

In this case

$$P_{32} = P_{32}^{\circ} + P_{32}' \tag{3.474}$$

where P_{32}° is the stress in the absence of the oscillation (steady shear plus electric field only) and P_{32}' is due to the oscillatory shear. Evaluating the orientational averages occurring in P_{32} using eq (3.458a) and comparing P_{32}' with eq (3.459) gives

$$[\mu'] = C_1\sin^2 2\phi_\infty + C_3 + \frac{\tilde{S}}{1+\omega^2\tilde{\tau}^2} \tag{3.475}$$

and

$$G' = \mu_0 \tilde{S} c \, \frac{\omega^2 \tilde{\tau}}{1 + \omega^2 \tilde{\tau}^2} \tag{3.476}$$

where

$$\tilde{S}(f, r_e) = \frac{(r_e^2 + 1)(1 + B\cos 2\phi_\infty)}{2} \left[\frac{C_1 \sin 4\phi_\infty}{f r_e} + \frac{2(\cos 2\phi_\infty + B\cos 4\phi_\infty)}{2r_e^2 + (1 - 2r_e^2)A} \right] \tag{3.477}$$

As $f \to \infty$, $\phi_\infty \to 0$ and eqs (3.475) to (3.477) reduce to eqs (3.467) to (3.469). The function $\tilde{S}$ is shown in Fig. 3.80. It can be seen that $\tilde{S}$ is positive when f is near one and for large values of f. For intermediate f-values $\tilde{S}$ is negative, implying the existence of negative dynamic rigidities G'. From eq (3.475) it can be shown that the intrinsic viscosity $[\mu']$ can also become negative for intermediate f values.

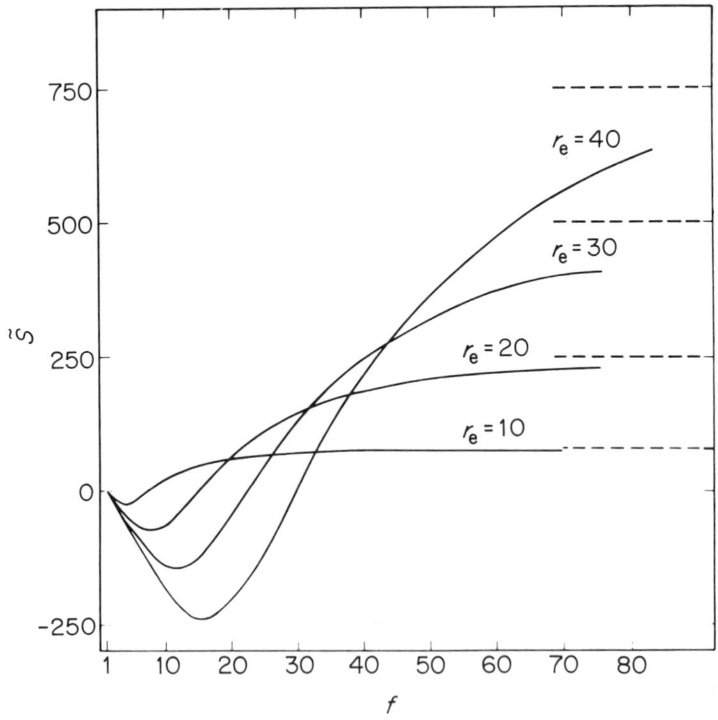

Fig. 3.80. Function $\tilde{S}$, describing the viscoelastic behaviour of a suspension of hard rods, as a function of the ratio, f, of electric to hydrodynamic torques, for various axis ratios, r_e (after van de Ven, 1984).

For subcritical fields ($f^2 < 1$), the situation is more complex because all particles undergo quasi-periodic rotations. It can be shown (Zuck *et al.*, 1989) that when $\nu T = 2$, where $\nu = \omega/2\pi$ and T the period of rotation given by eq (3.456), resonance occurs. At resonance all particles become aligned in phase while rotating and the orientation distribution becomes a δ function rotating with frequency $2/T$. An example of resonance is given in Fig. 3.81, which shows variations in the intrinsic viscosity, reflecting variations in the phase angle κ, defined as an integration constant in the integration of eq (3.451b):

$$\tan\phi = r_e(1-f^2)^{1/2} \tan\left(\frac{2\pi t}{T} + \kappa\right) + fr_e \tag{3.478}$$

Different values of κ reflect different values of the orientation ϕ at $t = 0$.

For Brownian particles, the orientation distribution is governed by the rotary convective-diffusion equation with the angular velocity given by

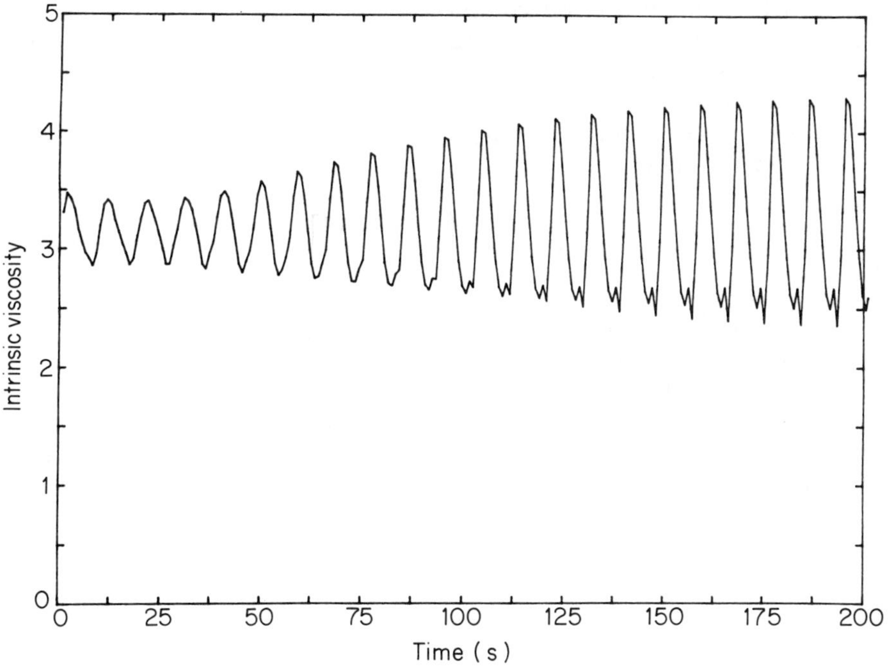

Fig. 3.81. Intrinsic viscosity of a suspension of spheroids of axis ratio $r_e = 2$, initially randomly oriented, subjected to a superposition of a steady shear and an oscillatory shear at resonance ($\nu T = 2, f = 0{\cdot}5$, Pe $= \infty$). Calculated from dynamic simulation calculations.

eq (3.451). For particles surrounded by electrical double layers, the period of rotation, T, and the alignment time, τ, will be affected by the surface potential of the particles and the nature and concentration of the electrolyte in solution.

3.5 Experimental techniques

3.5.1 Travelling microtube

A very powerful and versatile technique for observing, filming and analysing the motions of colloidal particles subjected to shear is the travelling microtube (Vadas *et al.*, 1973). The technique consists of a precision-bore glass tube, typically with an inner diameter in the range 50–200 μm, through which a colloidal dispersion is either pumped by a hydraulic drive (see Fig. 3.82a) or, even better, allowed to flow by a gravity feed (Fig. 3.82b). The microtube is mounted on a moveable platform. When a particle suspended in the dispersion is flowing down the tube with a velocity $-u$, the platform on which the particle is mounted is moved upwards with velocity u, thus keeping the particle at the same position in laboratory-fixed coordinates. In this way the particle can be observed directly through an optical microscope to which either a cine film or a video camera can be attached.

When the flow through the tube is laminar the flow profile is parabolic (cf.eq (1.51)), so that

$$u_3(R) = u_m \left[1 - \left(\frac{R}{R_0}\right)^2\right] \tag{3.479}$$

where $u_3(R)$ is the velocity in the X_3 direction at a radial distance R from the tube centre, at which the velocity u_m is maximum; R_0 is the inner radius of the tube (cf. Fig. 3.83).

Because in most cases one wants to obtain as large a magnification as possible, usually only part of the diameter of the tube can be seen in the field of view. In such a case it is advantageous to express eq (3.479) in terms of the distance, d, from the tube wall:

$$\frac{u_3(d)}{d} = \frac{2u_m}{R_0}\left(1 - \frac{d}{2R_0}\right) \tag{3.480}$$

The shear rate, G, expressed in terms of d becomes

$$G = \frac{2u_m}{R_0}\left|1 - \frac{d}{R_0}\right| \tag{3.481}$$

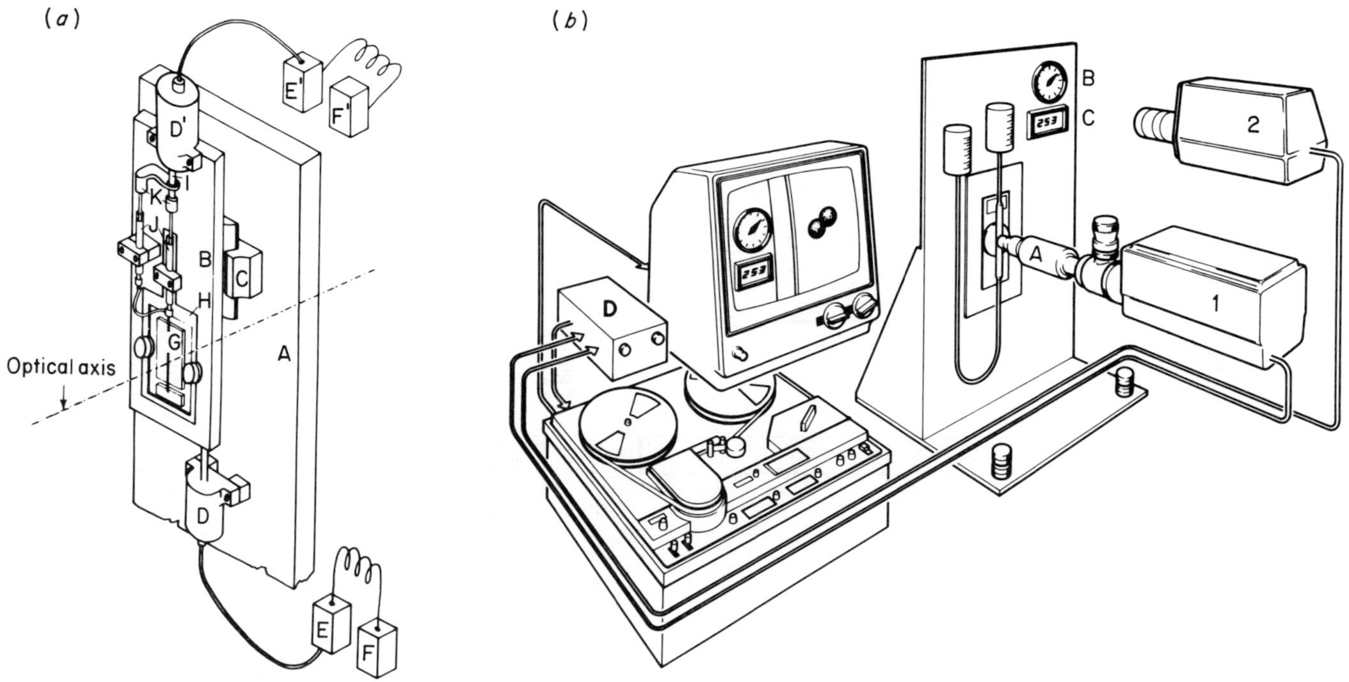

Fig. 3.82. Diagram of the travelling microtube apparatus.
(*a*) A microtube (G) is mounted on a platform (B) that can be moved up by means of hydraulic drive (D). A suspension is pumped through the tube by moving the piston of a syringe, connected to the microtube with hydraulic drive (D′). A particle moving down the tube can be kept on the optical axis by moving the tube upwards with the same velocity.
(*b*) Instead of using a pump to flow the suspension through the tube, a gravity feed can be used. To measure the velocity of a moving particle, a clock and a micrometer can be filmed at the same time in order to indicate the position of the platform (after Vadas *et al.*, 1973; Takamura *et al.*, 1979).

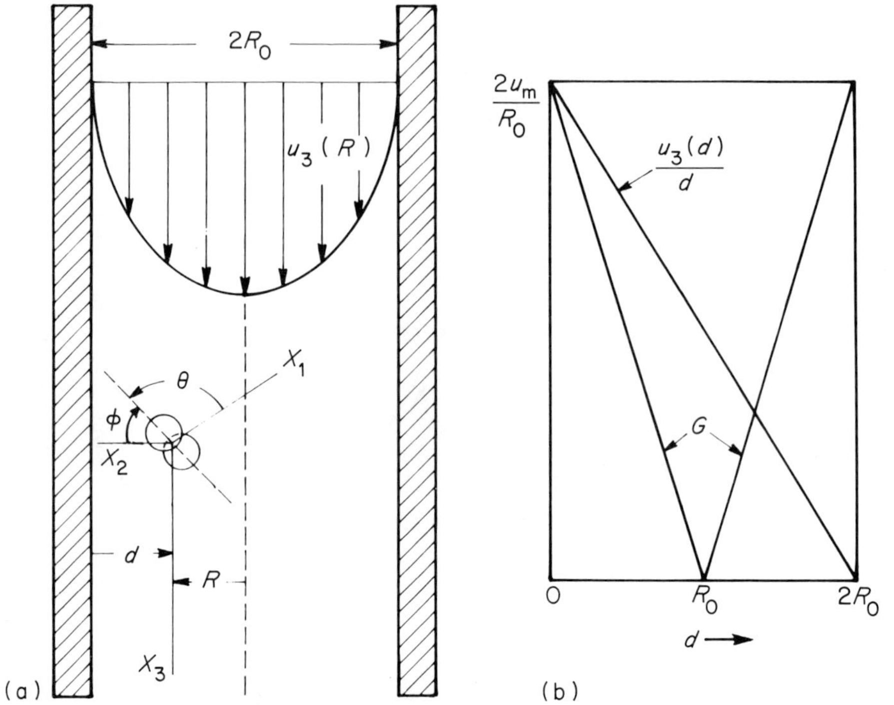

Fig. 3.83. Velocity distribution in a microtube (schematic).
(*a*) Cross-section of tube showing the parabolic flow profile, a doublet in the median plane at R from the tube centre and its orientation with respect to the Cartesian coordinates X_i, the X_1 axis (the vorticity axis of the shear) being directed towards the viewer. A particle moving down (or up) the tube with velocity $u_3(R)$ is observed in a fixed microscope by moving the tube at a measured $-u_3(R)$.
(*b*) Velocity, u_3, and shear rate, G, as functions of the distance, d, from the wall (after van de Ven and Mason, 1976a).

To calibrate the apparatus for a particular experiment, one first measures the velocity, u_3, of the suspended particles in the median plane of the tube over a range of d values. According to eq (3.480), a plot of u_3/d versus d yields a straight line from which R_0 and $G_{max} = 2u_m/R_0$ can be calculated. As an additional check, R_0 can be measured independently by photographing the tube at known magnification. When a particle is subsequently observed in the median plane of the tube at a distance d from the wall, the shear rate to which it is subjected can be found from eq (3.481). Alternatively, the velocity of the particle can be determined by measuring the velocity with which the platform is pushed upwards (e.g. with a velocity transducer), from which G

can be calculated from eqs (3.480) and (3.481), if u_m and R_0 are known from calibration. The pressure difference, Δp, over the tube can also be measured, from which u_m can be either calculated or found by calibration.

With the travelling-microtube apparatus, individual particles, collisions between particles, break-up of aggregates, etc., can be observed directly while the particles are subjected to flow. Because of limits in optical resolution, this technique is restricted to particles larger than about $0 \cdot 5\,\mu$m.

3.5.2. Couette apparatus

Another way of studying particles subjected to shear is to observe them in a Couette apparatus. Such a device consists of two concentric cylinders which can be rotated in opposite directions and in which a liquid containing the particles is placed in the gap between the two cylinders. When the cylinders are rotating in opposite directions there exists a stationary layer between the cylinders in which the fluid velocity and thus the particle velocity is zero (see Fig. 3.84), thus allowing direct observation of the particles.

If the angular velocities of the inner and outer cylinders are denoted by Ω_1 and by Ω_2, respectively, and if the outer radius of the inner cylinder is R_1 and the inner radius of the outer cylinder is R_2, the angular velocity of the fluid in the gap at a distance R from the axis of rotation is given by (Trevelyan and Mason, 1951)

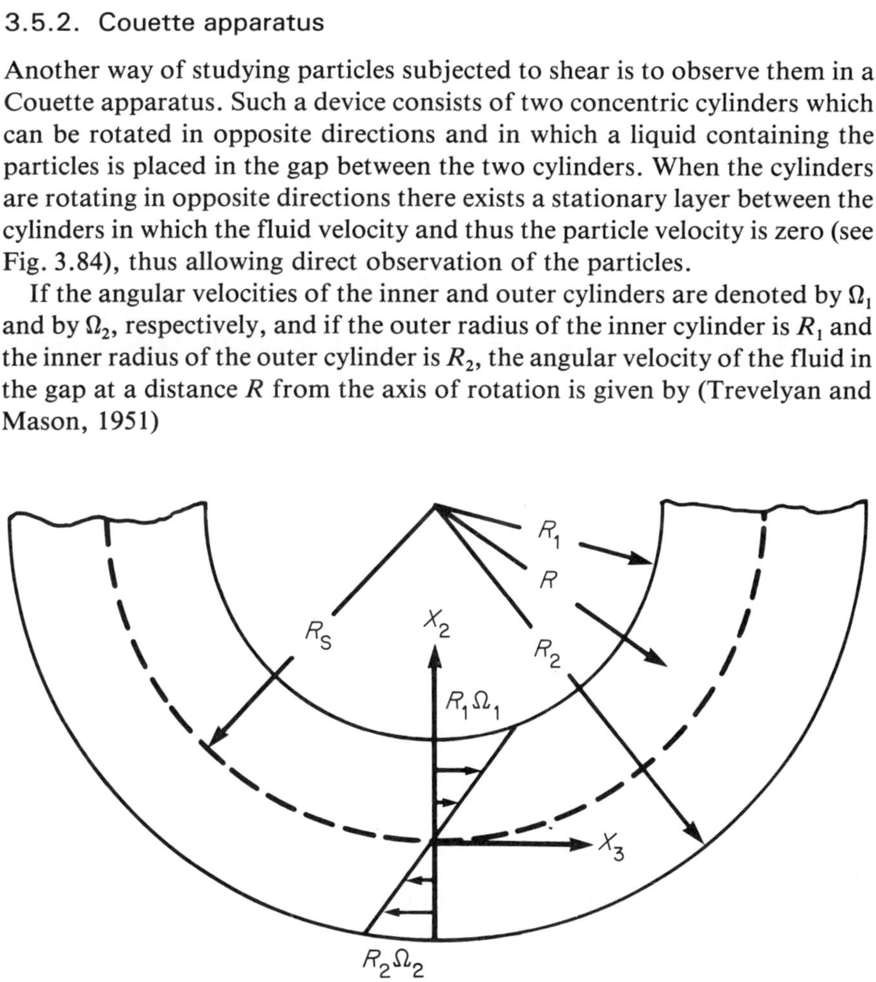

Fig. 3.84. A Couette apparatus consists of two concentric cylinders, with fluid in the gap between them. If the cylinders are rotated in opposite directions, a stationary layer (at a distance R_s from the centre) exists in which the particles can be easily observed.

$$\Omega(R)=\Omega_1+(\Omega_2-\Omega_1)\frac{R^2-R_1^2}{R_2^2-R_1^2}\frac{R_2^2}{R^2} \tag{3.482}$$

i.e. it varies smoothly from $\Omega=\Omega_1$ at $R=R_1$ to $\Omega=\Omega_2$ at $R=R_2$; it follows that the shear rate equals

$$G(R)=\frac{R\mathrm{d}\Omega}{\mathrm{d}R}=\frac{2(\Omega_2-\Omega_1)}{R_2^2-R_1^2}\frac{R_1^2 R_2^2}{R^2} \tag{3.483}$$

At the stationary plane $\Omega(R_s)=0$ and from eq (3.482)

$$R_s^2=\frac{R_1^2R_2^2(\Omega_2-\Omega_1)}{R_2^2\Omega_2-R_1^2\Omega_1} \tag{3.484}$$

and the shear rate

$$G(R_s)=\frac{2(R_2^2\Omega_2-R_1^2\Omega_1)}{R_2^2-R_1^2} \tag{3.485}$$

When R_1 and R_2 are large and $h=R_2-R_1$ is small, the velocity profile approaches a simple shear flow with a shear rate

$$G=\frac{R(\Omega_2-\Omega_1)}{h} \tag{3.486}$$

which equals the relative velocity between the cylinders divided by the gap width.

By manipulating the angular velocities of the cylinders it is possible to keep a particle in the stationary layer and observe it through a microscope. The shear rate to which it is subjected can be found from the measured values of Ω_i and the known dimensions of the cylinders by means of eq (3.485). To avoid wall effects the particle size must be much smaller than the gap width and the distance to the wall.

3.5.3 Four-roller apparatus

With the four-roller apparatus one can observe particles subjected to any type of linear two-dimensional flow (cf. Section 3.1.9.1). The device consists of four rollers or cylinders which are mounted in a square arrangement and are placed in a chamber containing a fluid with suspended particles (Fig. 3.85). By manipulating the angular velocity of the cylinders, any desired two-dimensional flow field can be created in the centre of the apparatus.

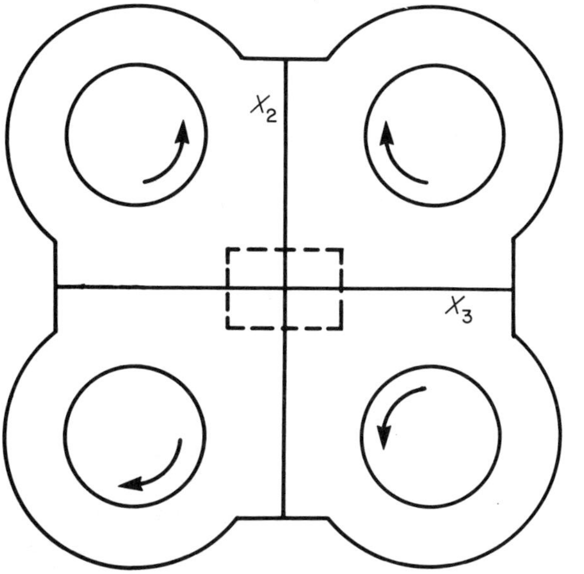

Fig. 3.85. Principle of the four-roller apparatus. Four cylinders are placed in a container which holds a fluid with particles. By rotating the cylinders a flow is created within the container which at the centre region (indicated by the dashed lines) closely resembles a linear flow. Depending on the relative velocities and the directions of rotation of the respective cylinders, flow fields of various γ can be created. When the four cylinders are rotated in the directions indicated in the figure with equal velocities, $\gamma = 1$ and the flow in the centre is an extensional flow.

The device can be calibrated by analysing the trajectories of tracer particles such as those shown in Fig. 3.34b. When $-1 \leqslant \gamma < 0$, γ can be obtained from the axis ratio of the elliptical path traced out by a tracer particle (cf. eq (3.183)). When $0 < \gamma \leqslant 1$, γ can be obtained from the asymptotes to which the tracer particles approach (cf.eq (3.182)). The shear rate, G, can be obtained either by taking photographs of tracer particles with stroboscopic illumination, or by measuring the period of rotation of suspended particles of known equivalent axis ratio (cf. eq (3.199)).

The velocity at the centre of the flow field is zero and thus, in principle, particles can be observed there. However, the symmetry point is unstable and all particles at the centre will slowly drift away. To observe a particle for a long time the angular velocity of the cylinders must be constantly adjusted, in essence moving the position of the stagnation point along with the motion of the particle (Powell and Mason, 1982). The tracking of the particle can be automated by monitoring the particle with a video system (Bentley and Leal, 1986).

3.5.4 Rheo-optical measurements

3.5.4.1 Light scattering and turbidity

The smaller the colloidal particles suspended in a liquid are, the more difficult it is to observe them directly by microscopy. Instead the motion of the particles can be deduced from light-scattering or light-transmittance (or turbidity) measurements. In essence the method consists of sending a light beam through a dispersion subjected to a well-defined flow field and measuring the intensity of the scattered or transmitted light. The simplest method is to subject the dispersion to a simple shear flow because, in this case, all the particles are subjected to the same rate of shear and the results can be readily interpreted. A simple shear flow can be created in either a Couette apparatus (described in Section 3.5.2) or between a cone and a plate (cf. Fig. 1.36). As all that is required in rheo-optical measurements is that all the particles in the light beam are subjected to the same shear rate, the simplest method is to place the dispersion between two parallel circular plates one of which rotates. This creates a flow field which has a zero rate of shear at the plate axis and a maximum rate at the edge of the plates. When a laser beam traverses the sample at some radial distance R from the axis, the rate of shear in the beam can be taken as constant when $d/R \ll 1$, d being the beam diameter. An example of a parallel-plate device, referred to as the microCouette apparatus, is shown schematically in Fig. 3.86. The rate of shear can be determined from measurements of the radial distance, R, the gap width, h, and the angular velocity, ω, of the moving plate:

$$G = \frac{\omega R}{h} \tag{3.487}$$

For the flow between the plates to closely resemble a simple shear, the Reynolds number based on the gap width must be small:

$$\mathrm{Re} = \frac{\omega R h}{\nu} \ll 1 \tag{3.488}$$

Fig. 3.86. Principle of the microCouette apparatus.
(*a*) A laser beam traverses a polarizer (P1), a pin hole (P2) and after reflection at mirror M, traverses the microCouette (MC). The scattered or transmitted beam falls on a photodiode (PD), is amplified in a photomultiplier (PM) and the signal is recorded on an oscilloscope (OS) and a recorder.
(*b*) Details of the microCouette, consisting of two parallel plates, one of which can be rotated with angular velocity ω.
(*c*) Flow field and coordinate system between plates (after Cerda *et al.*, 1981).

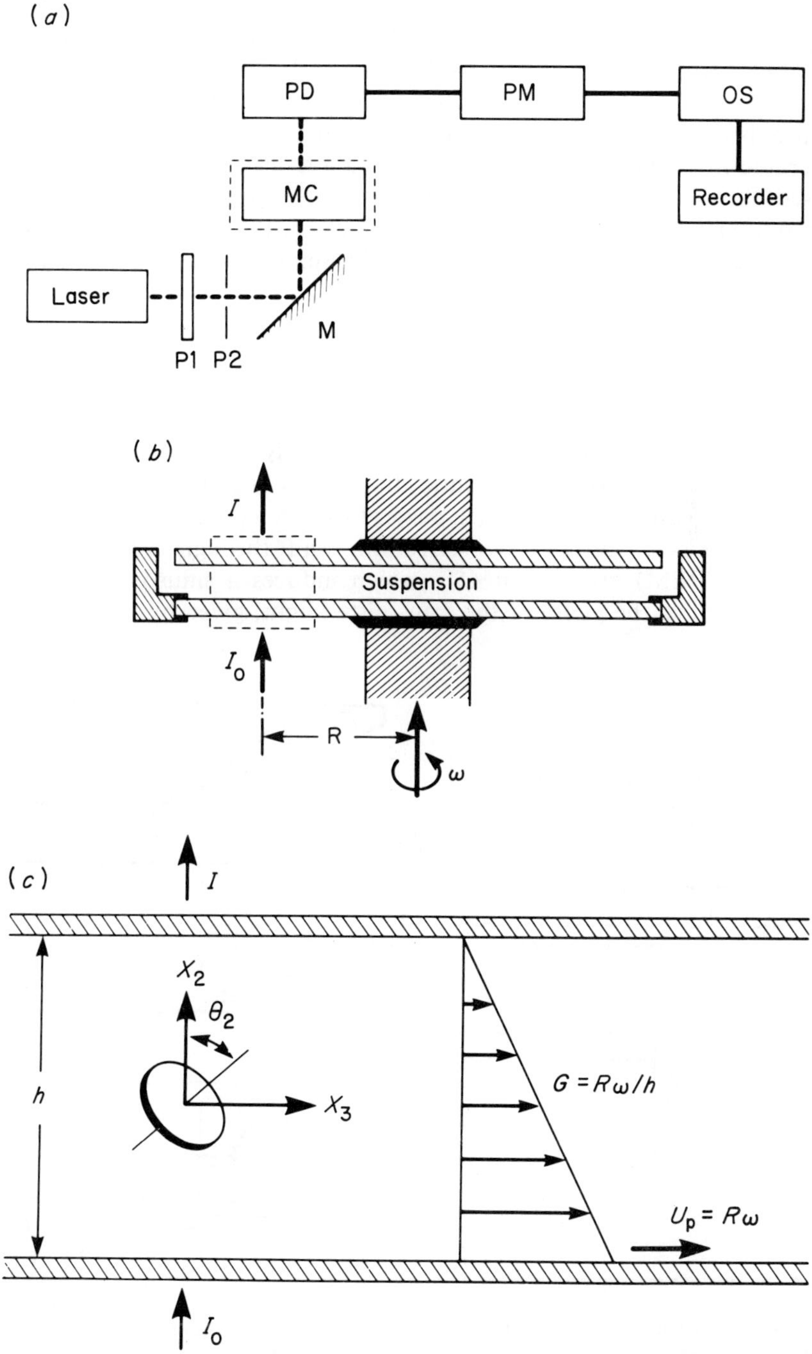
(a)
PD
PM
OS
MC
Recorder
Laser
M
P1 P2
(b)
I
Suspension
I_0
R
ω
(c)
I
X_2
θ_2
h
X_3
$G = R\omega/h$
$U_p = R\omega$
I_0

ν being the kinematic viscosity of the dispersion. If condition (3.488) does not hold, secondary flows arise which destroy the linear flow field.

It is possible to obtain rheo-optical measurements for flows other than simple shear. Other flows that can be exploited are two-dimensional flows generated in a four-roller apparatus. In such a device a laser beam can be passed through the stagnation point and the intensity of scattered and transmitted light measured (Zuck and van de Ven, 1989). It is also possible to perform experiments in Poiseuille flow by sending a laser beam along a streamline for which G is constant.

3.5.4.2 Birefringence and dichroism

An experimental apparatus designed to measure birefringence, linear dichroism and orientation angles simultaneously has been described by Johnson *et al.* (1985). A schematic diagram is shown in Fig. 3.87. The apparatus consists of a laser which passes through a polarizer (P1), oriented 90° with respect to the flow direction, and through a photoelectric modulator (PEM) oriented at 45°, which induces a sinusoidally varying retardance, δ, to the phase of the transmitted light, given by

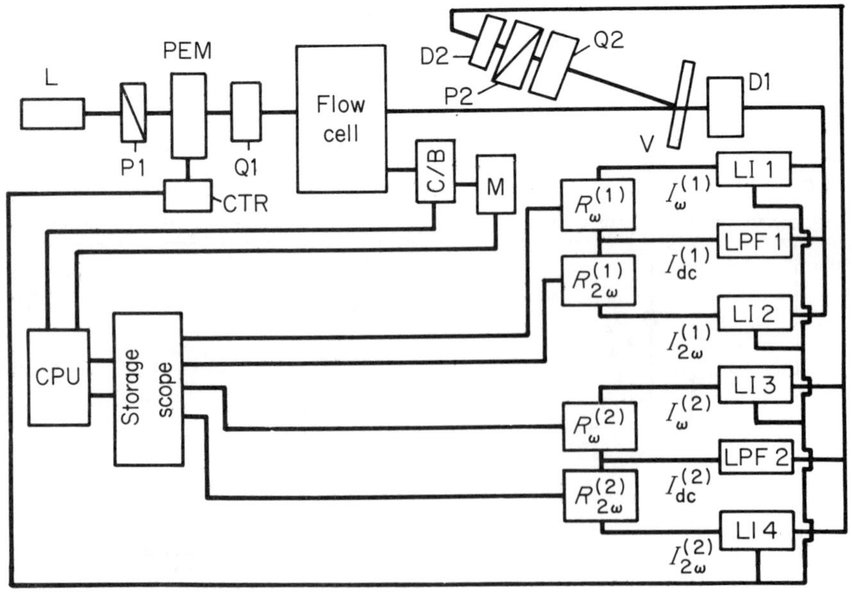

Fig. 3.87. Schematic diagram of the experimental apparatus for simultaneously measuring birefringence, linear dichroism and orientation angles (after Johnson *et al.*, 1985).

$$\delta = A \sin\omega t \tag{3.489}$$

where A is an adjustable experimental parameter and ω a fixed frequency (ca. 50 kHz). The beam passes through a quarter wave retarder (Q1), oriented at 0°, prior to passing through the flow cell. The flow cell is a Couette device controlled by a central processing unit (CPU) and coupled to a clutch/brake mechanism (C/B). After passing through the flow cell, the beam is split by a variable attenuator (V) with a small angle of incidence in order to not affect the polarization. The intensity of the transmitted beam is measured with a photodiode (D1) and the reflected beam is passed through a quarter wave retarder (Q2) (at 0°) and a polarizer (P2) (at 45°) and its intensity is measured by a diode (D2). The measured signals are passed through lock-in amplifiers (LI) and low-pass filters (LPF) which, in essence, decompose the signals into a Fourier series. The information that is extracted is the mean intensity, I_{dc}, and the first and second harmonics, I_ω and $I_{2\omega}$, for each signal. From these one can obtain the ratios

$$R_\omega^{(1)} = \frac{I_\omega^{(1)}}{I_{dc}^{(1)}} \; ; \; R_\omega^{(2)} = \frac{I_\omega^{(2)}}{I_{dc}^{(2)}} \; ; \; R_{2\omega}^{(1)} = \frac{I_{2\omega}^{(1)}}{I_{dc}^{(1)}} \; ; \; R_{2\omega}^{(2)} \frac{I_{2\omega}^{(2)}}{I_{dc}^{(2)}} \tag{3.490}$$

These four parameters are related to $\Delta n'$, $\Delta n''$, χ and ψ, where $\Delta n'$ is the birefringence, $\Delta n''$ the linear dichroism, χ the orientation angle of the principal axis of $\Delta n'$ and $\chi + \psi$ the orientation angle of $\Delta n''$. Thus after measuring the four parameters defined in eq (3.490), one has four equations with four unknowns from which $\Delta n'$, $\Delta n''$, χ and ψ can be calculated (Johnson *et al.*, 1985). As noted before (cf. Section 3.1.12.5) for suspensions of spheroids subjected to simple shear $\psi = 0$ and $\Delta n'$, $\Delta n''$ and χ are given by eqs (3.338) to (3.340).

3.5.5 Rheometry

The rheological properties of suspensions discussed in Sections 3.1.11.1 and 3.4 can be measured using a variety of rheometric techniques. For a summary the reader is referred to the excellent book by Walters (1975). Here we will discuss, as a representative example, the bob-and-cup rheometer, which can be used to measure either viscosities or viscoelastic properties. It is also ideally suited for electrorheological measurements because conditions can be created (large cylinders and small gap widths) for which the electric field is approximately uniform. The principle of the method is shown schematically in Fig. 3.88. The outer cylinder, referred to as the cup, is either fixed or allowed to rotate with a constant angular velocity. The inner

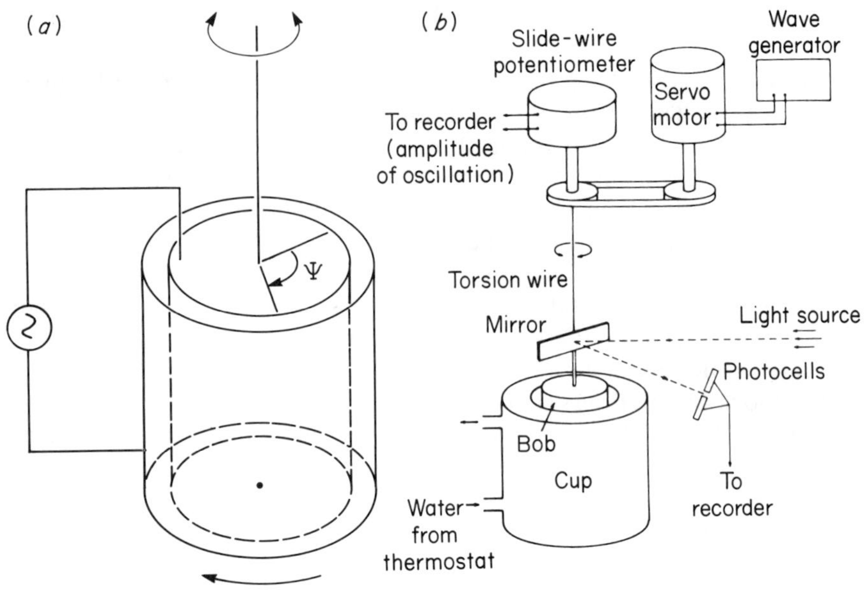

Fig. 3.88. Principle of the bob-and-cup viscometer.
(*a*) Scheme to show the geometry. The sample is placed in the gap between the two concentric cylinders. This geometry allows the application of an approximately uniform electric field. To measure viscosities, the outer cylinder is rotated and the torque on the inner cylinder is measured, e.g. by measuring the twist of the torsion wire.
(*b*) Dynamic viscometer for measuring the complex viscosity. Here the torsion wire is oscillated at the top and the response of the bob is measured (after van de Ven and Hunter, 1977).

cylinder, referred to as the bob, is freely suspended from a torsion wire. If the bob and cup are made of metal, an electric field can readily be applied over the suspension in the gap. The viscosity can be determined by rotating the outer cylinder with constant angular velocity and measuring the angle ψ through which the torsion wire is twisted. Alternatively, the bob can be kept fixed and the torque needed to keep the position fixed measured using a null balancing technique. The torque, L, required to twist a torsion wire of length ℓ, radius R and shear modulus G_w by an angle ψ is (van de Ven and Firth, 1978)

$$L = K\psi = \frac{\pi R^4}{2\ell} G_w \psi \tag{3.491}$$

where K is the torsional constant of the wire. This torque equals the torque exerted by the dispersion:

$$L = 2\pi h R_1^2 \mu G \tag{3.492}$$

where h is the height of the immersed part of the bob in contact with the dispersion and R_1 the bob radius; G is the rate of shear to be found from eq (3.486). From a measurement of ψ, the viscosity, μ, can be found from eqs (3.491) and (3.492).

The same geometery (Fig. 3.88b) can be used to measure the viscoelastic properties of dispersions, which are determined by the dynamic viscosity, μ', and the dynamic rigidity, G'. This can be achieved by oscillating the top of the torsion wire at a frequency ω and measuring the resulting oscillation of the bob. The bob will oscillate with the same frequency, ω, but with a different amplitude and phase. From the phase lag, δ, and the amplitude ratio, $Q = A_1/A_2$ (A_1 being the amplitude of the bob and A_2 of the top of the wire), the dynamic viscosity, μ', and the dynamic rigidity, G', can be determined (van de Ven and Firth, 1978):

$$\mu' = \frac{K \sin\delta}{SQ\omega} \tag{3.493}$$

$$G' = \frac{1}{S}\left[\frac{K\cos\delta}{Q} - (K - I\omega^2)\right] \tag{3.494}$$

where K is the torsional constant of the wire, I the moment of inertia of the bob, and S a geometrical factor given by

$$S = 2\pi h \frac{R_2^2}{R_2 - R_1} \tag{3.495}$$

R_2 being the inner radius of the cup.

3.5.6 Photon correlation spectroscopy

Similar to photon correlation spectroscopy (PCS) experiments on quiescent samples, discussed in Section 2, PCS experiments can be performed on samples subjected to flow. The only analysis in the literature is for dispersions of spherical particles subjected to simple shear (Ackerson and Clark, 1981). A schematic diagram of the experimental set-up is given in Fig. 3.89a for both homo- and heterodyne experiments. Heterodyne spectra give information about the velocity of the particle (or fluid), while a homodyne spectrum mainly provides information about diffusion constants and shear rates, although it usually also depends on the particle velocity. With the homodyne method, the autocorrelation function for a dispersion of spheres

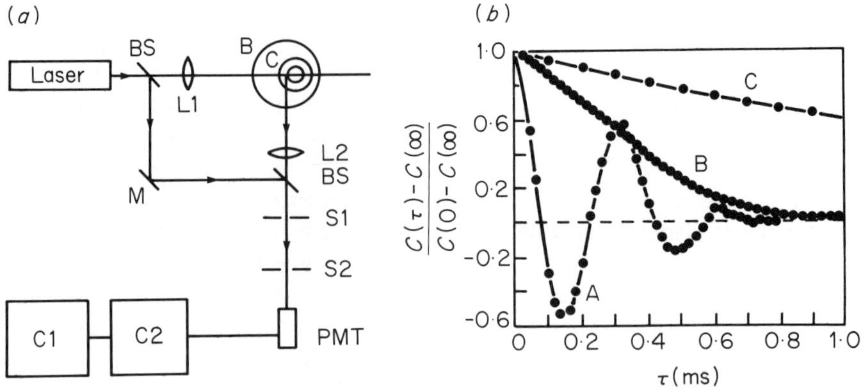

Fig. 3.89. Photon-correlation spectroscopy experiments on dispersions subjected to flow. (*a*) Scheme to show the experimental technique. The sample is between the gap of a Couette apparatus (C) kept at constant temperature with waterbath (B). L1 and L2 are lenses. S1 is a stop and S2 a pin hole. C1 is a computer and C2 a digital correlator. In a heterodyne experiment, the light is split by a beam splitter (BS) and via a mirror (M) and a second beam splitter combined with the scattered beam. For homodyne spectra this beam is blocked.
(*b*) Experimentally determined normalized autocorrelation function for dispersions of monodisperse latex particles. $C(\tau)$ is the time-dependent part of the autocorrelation function. Curve A is a heterodyne spectrum and B and C are homodyne spectra. For A and B, the shear rate is $G = 1{\cdot}8\ \mathrm{s}^{-1}$, and for curve C, $G = 0$ (after Ackerson and Clark, 1981).

moving in the X_3 direction and sheared in the X_2X_3 plane with shear rate, G equals (cf. eqs (2.216) to (2.223)):

$$\langle i(t)\, i(t+\tau)\rangle = I_0^2(1+\gamma)\, F(q,\tau) \tag{3.496}$$

where I_0 is the average intensity at the detector and

$$F(q,\tau) = \exp\left(-Dq^2\tau + GDq_2q_3\tau^2 - \frac{1}{3}\, G^2Dq_3^2\tau^3\right) \tag{3.497}$$

where q_i are the components of the wave vector $\boldsymbol{q}$ in the X_i direction. Analogously to the derivation of eq (2.228), eq (3.497) follows directly from the solution of the convective-diffusion equation for Brownian particles in simple shear (cf. eq (3.54)). The three terms on the right-hand side of eq (3.497) are due to pure diffusion, coupling in the X_2X_3 plane and enhanced diffusion in the X_3 direction. The signal-to-background factor, γ, depends on τ and on the velocity field $v_0(\boldsymbol{r}_0)$, which is assumed to be constant and determined by the initial position $\boldsymbol{r}_0$. An expression for γ can be found in

Ackerson and Clark (1981). Furthermore, γ depends on the scattering volume and the detector area. Examples of experimentally determined autocorrelations are shown in Fig. 3.89b.

When the incident and scattered light are such that $\boldsymbol{q}$ is perpendicular to the fluid velocity, then γ is independent of τ. At the same time, $q_3=0$ and, hence, in this arrangement eq (3.496) reduces to the result for a quiescent dispersion of spheres.

No theoretical or experimental data are available for dispersions of non-spherical particles subjected to shear. Here we will derive effective diffusion coefficients for the simple case of ellipsoids subjected to uniform flow. The results also apply to simple shear, provided that the variations in velocity are small in the scattering volume and that the rate of shear is small, such that the coupling terms in eq (3.497) can be neglected. The results are generally valid when $\boldsymbol{q} \perp v$ and the coupling terms vanish accordingly.

Consider an ellipsoidal particle with orientation (θ,ϕ,ψ) in a uniform flow of velocity v (see Fig. 3.90). In the absence of Brownian motion, a particle at the origin at time $t=0$ will be at position $(0,0,L)$ at time Δt, where $L=v\Delta t$. With Brownian motion, the particle will end up in position (x_1,x_2,x_3+L). When a light beam of frequency ν hits the particle, there will be a Doppler shift due to the fact that the particle is moving with respect to the beam. Since the particle velocity, u, is much smaller than the speed of light, c, the Doppler shift is given by

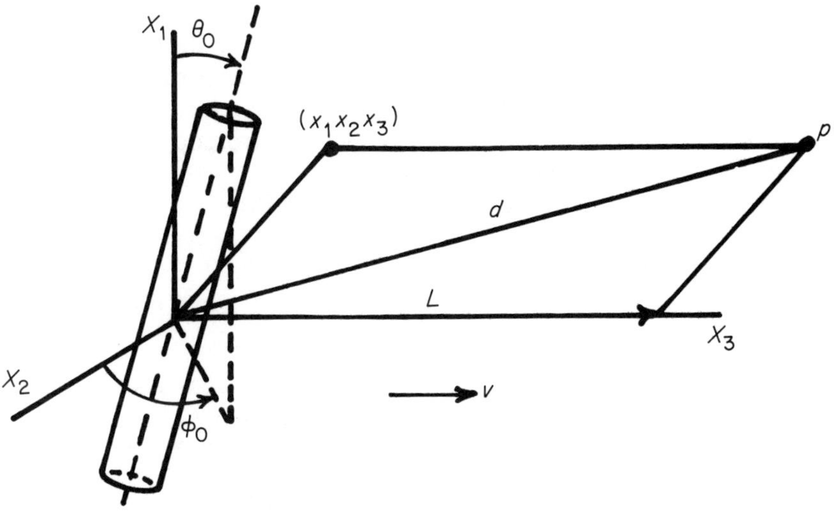

Fig. 3.90. Scheme of a rod-shaped particle in a uniform flow of velocity v.

$$\Delta\nu = \frac{\nu}{c}\, u \cos\gamma \tag{3.498}$$

where γ is the angle between the moving particle and the observer. Notice that $\Delta\nu$ in eq (3.498) does not equal the shift in the power spectrum, which depends on the direction of the scattering vector. To find the effective diffusion constant of non-spherical particles, the actual Doppler shift for non-spherical particles is compared with that of spheres.

Indicating the direction in which the particle moves by the angles (θ_0, ϕ_0), it follows from geometry that

$$\cos\gamma = \sin\theta_0 \sin(\phi_0 + \theta_s) \tag{3.499}$$

θ_s being the scattering angle. Inserting eq (3.499) into eq (3.498) and transforming to space-fixed coordinates yields

$$\Delta\nu = \frac{\nu}{c\Delta t}\, [x_2 \sin\theta_s + (x_3 + L) \cos\theta_s] \tag{3.500}$$

For a dispersion of colloidal particles, the mean frequency shift is found by averaging over all positions:

$$\overline{\Delta\nu} = \frac{\nu}{c}\, v \cos\theta_s \tag{3.501}$$

Hence a dispersion flowing with velocity v causes a shift in the average frequency of the scattered light. Measuring the frequency of the incident and scattered light provides a method for measuring the velocity, v, of the fluid. The heterodyne method and laser Doppler spectroscopy discussed later (Section 3.5.7.2) are based on this principle.

The broadening of the frequency spectrum by Brownian motion is given by

$$\overline{\Delta\nu^2} \propto \langle (u\cos\gamma - \overline{u\cos\gamma})^2 \rangle \propto \langle (x_2 \sin\theta_s + x_3 \cos\theta_s)^2 \rangle \tag{3.502}$$

For spherical particles

$$\overline{\Delta\nu^2} \propto \langle x_2^2 \rangle \sin^2\theta_s + \langle x_3^2 \rangle \cos^2\theta_s = \langle x_{2,3}^2 \rangle = 2D\Delta t \tag{3.503}$$

is independent of the scattering angle. For non-spherical particles the above is no longer true, but:

$$\overline{\Delta\nu^2} \propto \langle x_2^2 \rangle \sin^2\theta_s + \langle x_3^2 \rangle \cos^2\theta_s + \langle x_2 x_3 \rangle \sin 2\theta_s \tag{3.504}$$

Comparing eqs (3.503) and (3.504) shows that non-spherical particles behave as spherical particles with an effective diffusion constant, D_{eff}, given by

$$D_{eff} = \overline{D_{22}} \sin^2\theta_s + \overline{D_{33}} \cos^2\theta_s + \overline{D_{23}} \sin 2\theta_s \tag{3.505}$$

with

$$D_{ij} = \alpha_{im} \alpha_{jn} D'_{mn} \tag{3.506}$$

where α_{ij} is the transformation matrix (eq (3.79)) and D'_{mn} is the translational diffusion tensor in body-fixed coordinates. For ellipsoidal particles with the diagonal components D_1, D_2 and D_3, we have

$$\overline{D_{22}} = \langle \alpha_{21}^2 \rangle D_1 + \langle \alpha_{22}^2 \rangle D_2 + \langle \alpha_{23}^2 \rangle D_3$$

$$\overline{D_{23}} = \langle \alpha_{21} \alpha_{31} \rangle D_1 + \langle \alpha_{22} \alpha_{32} \rangle D_2 + \langle \alpha_{23} \alpha_{33} \rangle D_3$$

$$\overline{D_{33}} = \langle \alpha_{31}^2 \rangle D_1 + \langle \alpha_{32}^2 \rangle D_2 + \langle \alpha_{33}^2 \rangle D_3$$

For spheroids where $D_2 = D_3$, the expression for D_{eff} reduces to

$$D_{eff} = D_2 + (D_1 - D_2)(\langle \sin^2\theta \cos^2\phi \rangle \sin^2\theta_s + \langle \sin^2\theta \sin^2\phi \rangle \cos^2\theta_s + \langle \sin^2\theta \sin\phi \cos\phi \rangle \sin 2\theta_s) \tag{3.507}$$

The orientational averages depend on the rotary Péclet number, Pe_r, and the equivalent axis ratio, r_e.

For geometries other than $\boldsymbol{q} \perp \boldsymbol{v}$, the coupling terms in eq (3.497) must also be considered, as must the signal-to-background factor, γ. However, γ is independent of particle shape and size. At present, no general theory is available for describing dynamic light scattering from non-spherical particles subjected to flow.

3.5.7 Electrophoresis

3.5.7.1 Microscopic methods

Traditionally, electrophoretic mobility experiments are made by subjecting the suspended particles to an electric field, observing the motion under a microscope and measuring their velocity and the magnitude of the field.

From the mobility (ratio of velocity and field strength) the ζ potential of the particles can be found from eq (3.396). Nowadays it is possible to automate the process, but the measurements are still based on the same principle. Usually the motion is observed in closed cylindrical or parallel-plate channels, which must be well thermostated to avoid convection currents. Because in a closed vessel the electro-osmotic flow near the wall causes a hydrodynamic back flow, care must be taken to observe particles in the correct position in the channel. In a cylindrical channel, the velocity profile is given by

$$v = v_{eo} + k(r^2 - R^2) \qquad (3.508)$$

where v_{eo} is the electro-osmotic velocity and k a constant which can be determined from the condition that the net flow through a cross-section of the tube is zero:

$$\int_0^R 2\pi r v \mathrm{d}r = 0$$

from which it follows that $k = 2v_{eo}/R^2$. Inserting this value for k in eq (3.408) shows that a stationary layer exists in the capillary for which $v = 0$, given by $r = \frac{1}{2}R\sqrt{2}$. It is in this layer that observations must be made. For flat channels a similar analysis results in $r = \frac{1}{3}R\sqrt{3}$.

3.5.7.2 Laser Doppler spectroscopy

We have seen that, in principle, the velocity of a particle can be determined by measuring the frequency of the incident and scattered light (cf. eq (3.501)). The same applies when the velocity is caused by an external electric field. The problem in making such measurements is that the frequency shift $\Delta\nu$ (of the order of 10 Hz) is very small compared to the frequency of light $\nu = c/\lambda \simeq 10^{15}$ Hz. The small frequency shift can, however, be readily measured by mixing the scattered beam with, for example, the transmitted or incident beam. The mixing of the two beams with slightly different frequencies causes beats which can be detected. From the beats the shift $\Delta\nu$ can be found. The mixing of two beams can be achieved experimentally by splitting a laser beam in two beams and directing them both onto the same sample from different directions (see Fig. 3.91). Usually the intensity of the transmitted (weakest) beam, mixed with the scattered light of the strongest beam, is measured with a photomultiplier. The signal from the photomultiplier is amplified and analysed using a spectrum analyser. The

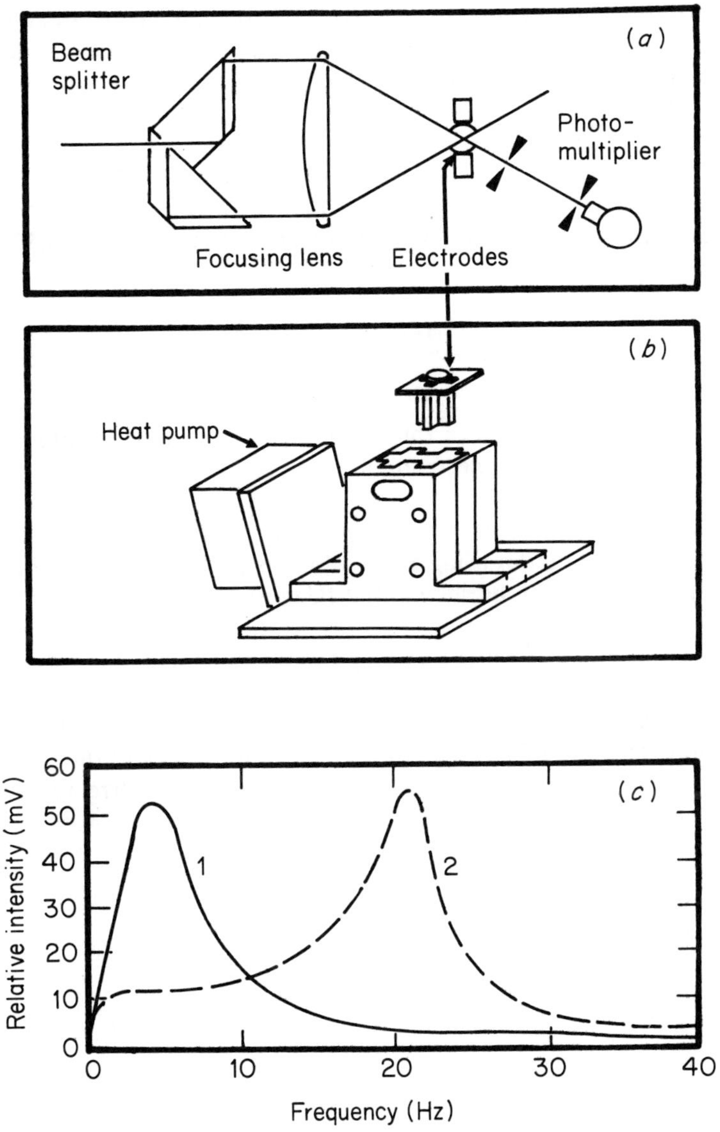

Fig. 3.91. The principle of laser Doppler spectroscopy.
(*a*) A laser beam is split into two beams, one strong (ca. 99·5%) and one weak (ca. 0·5%). The two beams are focused onto the sample, placed between two electrodes, by using a lens.
(*b*) Details of the electrodes and the heat pump for monitoring the sample at constant temperature.
(*c*) Examples of frequency spectra measured on bitumen emulsions (after Ross and Takamura 1988a).

output of the analyser is a laser Doppler spectrum which shows a peak at a characteristic frequency shift from which the electrophoretic mobility can be determined. The shift in angular frequency in a power spectrum is given by (Ware, 1974):

$$\Delta\omega = \boldsymbol{q}\cdot\boldsymbol{u} \tag{3.509}$$

$\boldsymbol{u}$ being the particle velocity. Using eqs (2.215) and (3.386), eq (3.509) can be written

$$\Delta\omega = \frac{4\pi nmE}{\lambda}\sin\left(\frac{\theta}{2}\right)\cos\psi \tag{3.510}$$

where ψ is the angle between the scattering vector and the applied electric field. From eq (3.510) the electrophoretic mobility, m, can be found from the measurement of $\Delta\omega$.

An alternative way of obtaining m is to measure the heterodyne autocorrelation function using PCS, which is the Fourier transform of the power spectrum. It can be shown that (Uzgiris, 1981)

$$\langle i(t)i(t+\tau)\rangle = I_0^{\,2}\,\mathrm{e}^{-Dq^2\tau}\cos(\tau\Delta\omega) \tag{3.511}$$

The autocorrelation given by eq (3.511) has the form of a damped oscillation. From the frequency of the oscillation, $\Delta\omega$ and thus the particle velocity can be determined. An example of such an autocorrelation is shown in Fig. 3.89b (curve A).

Chapter 4

Brownian Motion of Interacting Pairs of Particles

4.1 Brownian motion of two hard spheres

4.1.1 Diffusion tensor of a two-sphere system

When two particles suspended in a liquid come close together they interact hydrodynamically. In general their friction coefficients will change and, as a result, the Brownian motion of either particle is affected by the presence of the other. Here we will be looking at the relatively simple problem of the Brownian motion of two hard spheres. We define hard spheres as rigid spheres on which only hydrodynamic forces are acting. The spheres undergo no other interaction forces except for an infinitely strong repulsion on contact (to prevent interpenetration). In contrast, we define soft spheres as spheres subject to interactions, other than hydrodynamic ones, felt at finite distances. One can think of such spheres as having a surface boundary layer, the thickness of which is determined by a characteristic interaction length. For electrostatically stabilized dispersions this thickness can be identified with the double-layer thickness, for van der Waals interactions with the thickness where the interaction energy becomes comparable to the thermal energy, and for sterically stabilized colloids with the length of the polymer tails adsorbed or anchored on the surface of the suspended particles. A sphere with a surface boundary layer can be considered soft because this layer can be deformed and penetrated during the approach of two particles.

We have seen that the diffusion tensor of a particle can be found from the resistance tensor (cf. Section 2.1.5). For a single particle the resistance tensor has 6×6 components because each particle has six degrees of freedom for particle motion (three translational velocities and three rotational velocities). For a two-particle system the resistance tensor has 12×12 components because there are 12 degrees of freedom, six for each particle. Fortunately, many of the components of the resistance matrix of two spheres are zero and for most problems knowledge of only one or two components is needed. The 12×12 matrix can be reduced to an 8×8 matrix because two of the three translational and rotational components are equivalent. This is shown in Fig. 4.1, where it can be seen that translational motion in the 2 direction is equivalent to translational motion in the 3 direction. Similarly, rotary motion about the 2 axis is equivalent to motion about the 3 axis. Therefore, only the translational velocities, u_1 and u_2, and the angular velocities, ω_1 and ω_3 (causing motion in the 1–2 plane) need be considered.

The forces, $\boldsymbol{F}$, and torques, $\boldsymbol{L}$, on each sphere can be written as a linear combination of the velocity components, e.g.

$$F_1^{(1)} = au_1^{(1)} + bu_2^{(1)} + cu_1^{(2)} + du_2^{(2)} + e\omega_1^{(1)} + f\omega_3^{(1)} + g\omega_1^{(2)} + h\omega_3^{(2)} \tag{4.1}$$

The superscripts (1) and (2) refer to spheres 1 and 2, respectively. Similar equations hold for the other forces and torques, $F_j^{(i)}$, $L_j^{(i)}$. This leads to eight expressions which can be combined in tensor notation as:

$$\tilde{\boldsymbol{F}} = -\mu\, \boldsymbol{R} \cdot \boldsymbol{U} \tag{4.2a}$$

or

$$\tilde{\boldsymbol{F}} = -\mu \begin{matrix} R_1 & 0 & R_2 & 0 & 0 & 0 & 0 & 0 \\ 0 & R_3 & 0 & R_4 & 0 & R_5 & 0 & R_6 \\ R_2 & 0 & R_7 & 0 & 0 & 0 & 0 & 0 \\ 0 & R_4 & 0 & R_8 & 0 & R_9 & 0 & R_{10} \\ 0 & 0 & 0 & 0 & R_{11} & 0 & R_{12} & 0 \\ 0 & R_5 & 0 & R_9 & 0 & R_{13} & 0 & R_{14} \\ 0 & 0 & 0 & 0 & R_{12} & 0 & R_{15} & 0 \\ 0 & R_6 & 0 & R_{10} & 0 & R_{14} & 0 & R_{16} \end{matrix} \cdot \boldsymbol{U} \tag{4.2b}$$

where $\tilde{\boldsymbol{F}}$ is the generalized force–torque vector containing the two forces and torques on each sphere

$$\tilde{\boldsymbol{F}} = (F_1^{(1)}, F_2^{(1)}, F_1^{(2)}, F_2^{(2)}, L_1^{(1)}, L_2^{(1)}, L_1^{(2)}, L_2^{(2)})$$

and $\boldsymbol{U}$ is the generalized velocity vector

$$\boldsymbol{U} = (u_1^{(1)}, u_2^{(1)}, u_1^{(2)}, u_2^{(2)}, \omega_1^{(1)}, \omega_3^{(1)}, \omega_1^{(2)}, \omega_3^{(2)})$$

The zeros appearing in the resistance matrix, $\boldsymbol{R}$, can be readily found by looking at the geometry of the problem. For example, applying a force in the 1 direction on sphere 1 results in the motions $u_1^{(1)}$ and $u_2^{(1)}$, other velocity components being zero. Hence the only non-zero components in eq (4.1) are a and c, which are denoted by R_1 and R_2 in eq (4.2). Similarly, applying a

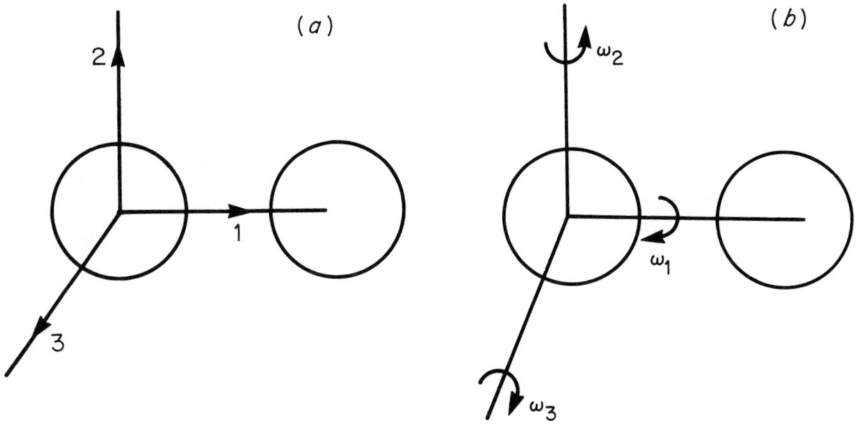

Fig. 4.1. The relative motion of two spheres.
(*a*) The translational motion in the 2 and 3 directions is equivalent.
(*b*) The rotational motion about the 2 and 3 axes is equivalent.

force on sphere 1 in the 2 direction results in the velocities $u_2^{(1)}$, $u_2^{(2)}$, $\omega_3^{(1)}$ and $\omega_3^{(2)}$, the others being zero. The coefficients corresponding to these velocities are R_3 to R_6. All the non-zero components of the resistance matrix, $\boldsymbol{R}$, can be found in a similar way. This procedure yields the 24 non-zero components in $\boldsymbol{R}$ in eq (4.2). Because $\boldsymbol{R}$ is symmetrical, it is sufficient to know just 16 components. However, not all these components are independent. Sphere 1 can be interchanged with sphere 2, resulting in the transformation

$$q \rightarrow 1/q$$

where q is the radius ratio a_2/a_1. This results in six identities

$$R_7(q) = R_1(q^{-1}) \qquad R_{10}(q) = -R_6(q^{-1})$$
$$R_8(q) = R_3(q^{-1}) \qquad R_{15}(q) = R_{11}(q^{-1})$$
$$R_9(q) = -R_5(q^{-1}) \qquad R_{16}(q) = R_{13}(q^{-1})$$

and thus we are left with 10 independent components. The components R_i are functions of the radius ratio, q, and the non-dimensional separation, r, between the spheres, where r is defined as

$$r = \frac{2r^*}{a_1 + a_2} \tag{4.3}$$

r^* being the distance between the sphere centres. Asymptotic expressions for small and large values of r are given by Jeffrey and Onishi (1984). From these expressions values of R_i can be calculated for every distance r by taking a sufficient number of terms. As the distance between the spheres approaches infinity, all the off-diagonal components approach zero, as do R_3, R_8, R_{13} and R_{16}. The four remaining terms R_1, R_7, R_{11} and R_{16} correspond to the translational- and rotational-friction coefficients of each of the spheres. As the distance between the spheres approaches zero, the coefficients associated with translational motion along the line of the centres R_1, R_2 and R_7 behave as

$$R_i = \frac{A_i}{h} + B_i \ln h + C_i + \ldots; \tag{4.4}$$

those associated with a spin about the line of centres (R_{11}, R_{12}, R_{15}) approach a constant D_i

$$R_i = D_i + \ldots \tag{4.5}$$

and the remaining terms approach infinity with a leading term

$$R_i = E_i \ln h + \ldots \tag{4.6}$$

where h is the dimensionless gap width ($h = r - 2$). The constants $A_i - E_i$ are all functions of the radius ratio, q.

The diffusion tensor can be found by inverting the resistance tensor

$$\boldsymbol{D} = \boldsymbol{R}^{-1} kT \tag{4.7}$$

$\boldsymbol{D}$, like $\boldsymbol{R}$, contains 24 non-zero terms, 10 of which are independent.

Although the validity of the Stokes–Einstein relation of an isolated particle is a well-established fact, there is some doubt of its validity in a two-particle system. In the derivation it is assumed that the random forces exerted on colloidal particles cause the particles to move and to rotate and only the flow field caused by this motion is responsible for the hydrodynamic drag. However, the fluid molecules, which collide with the particles and cause their motion, transfer momentum to the particle which must cause a disturbance flow in the wake of the particle. This disturbance flow could conceivably affect the motion of a second particle located near the first, an effect not taken into consideration in the derivation of the Stokes–Einstein equation for a multi-particle system. However, in what follows it will be assumed that the Stokes–Einstein relation remains valid for a two-particle system.

4.1.2 Relative diffusion between two spheres

It is often more important to know the relative motion between two spheres than their absolute motion. The relative motion can be split into two components, one along the line of centres and one perpendicular to it. The translational relative motion is the most important one because it determines the rate at which particles coagulate within a dispersion.

The component along the line of centres can be found by considering the forces acting along the line of centres

$$\begin{aligned} F_1^{(1)} &= f_{11} u_1^{(1)} + f_{12} u_1^{(2)} \\ F_1^{(2)} &= f_{21} u_1^{(1)} + f_{22} u_1^{(2)} \end{aligned} \tag{4.8}$$

where f_{ij} are the components corresponding to R_1, R_2 and R_7 in eq (4.2). Equation (4.8) can be rewritten as

$$\begin{pmatrix} F_1^{(1)} \\ F_1^{(2)} \end{pmatrix} = \begin{pmatrix} f_{11} & f_{12} \\ f_{21} & f_{22} \end{pmatrix} \cdot \begin{pmatrix} u_1^{(1)} \\ u_1^{(2)} \end{pmatrix} \tag{4.9}$$

and thus the corresponding diffusion tensor is

$$\boldsymbol{D} = \frac{kT}{f_{11}f_{22} - f_{12}f_{21}} \begin{pmatrix} f_{22} & -f_{21} \\ -f_{12} & f_{11} \end{pmatrix} = \begin{pmatrix} D_{11} & D_{12} \\ D_{21} & D_{22} \end{pmatrix} \tag{4.10}$$

The relative displacement in the X direction (the 1 direction) is

$$\langle (x_{(1)} - x_{(2)})^2 \rangle = \langle x_{(1)}^2 - 2x_{(1)}x_{(2)} + x_{(2)}^2 \rangle \tag{4.11}$$

Therefore the relative diffusion along the line of centres is given by

$$D_{\parallel}(r, q) = D_{11} - 2D_{12} + D_{22} \tag{4.12}$$

It is clear from eq (4.12) that there exists a coupling between the Brownian motion of the two spheres, i.e. the coupling term $\langle x_{(1)}x_{(2)} \rangle$ is non-zero. This can also be seen by considering the Langevin equations for the translational motion long the line of sphere centres (cf. eq (2.51)):

$$\begin{aligned} m_1 \ddot{x}_1 + f_{11}\dot{x}_1 + f_{12}\dot{x}_2 &= R_1(t) \\ m_2 \ddot{x}_2 + f_{21}\dot{x}_1 + f_{22}\dot{x}_2 &= R_2(t) \end{aligned} \tag{4.13}$$

where the subscripts 1 and 2 now refer to sphere 1 and sphere 2, respectively. Dividing by m_1 and m_2 gives:

$$\begin{aligned}\ddot{x}_1 + \beta_{11}\dot{x}_1 + \beta_{12}\dot{x}_2 &= A_1(t)\\ \ddot{x}_2 + \beta_{21}\dot{x}_1 + \beta_{22}\dot{x}_2 &= A_2(t)\end{aligned} \tag{4.14}$$

Again it is clear from eq (4.14) that the motion between the spheres is coupled. The coupling can be partly undone by the transformation

$$\begin{aligned}\dot{s}_1 &= a\dot{x}_1 + \dot{x}_2\\ \dot{s}_2 &= \dot{x}_1 - a\dot{x}_2\end{aligned} \tag{4.15}$$

with a a root of

$$a\beta_{11} - a^2\beta_{21} + \beta_{12} - a\beta_{22} = 0 \tag{4.16}$$

which leads to

$$\begin{aligned}\ddot{s}_1 + \gamma_1\dot{s}_1 &= aA_1 - A_2\\ \ddot{s}_2 + \gamma_2\dot{s}_2 &= A_1 - aA_2\end{aligned} \tag{4.17}$$

with

$$\gamma_1 = (a\beta_{11} + \beta_{12})/a \tag{4.18a}$$

$$\gamma_2 = (a\beta_{22} - \beta_{12})/a \tag{4.18b}$$

Equation (4.17) has the same form as the equation for a single sphere (eq (2.52)). However, in eq (4.17), γ_i is a function of both x_1 and x_2. For equal spheres (equal-sized spheres with equal mass), $a = 1$ and

$$\gamma_1 = (\beta_{11} + \beta_{12})\ ;\ \gamma_2 = (\beta_{11} - \beta_{12})$$

Assuming that in a short time interval Δt, γ_i does not change very much and can be considered constant, eq (4.17) can be solved in an analogous way as eq (2.52). Squaring and averaging the solution yields, for times $t \gg \gamma_i^{-1}$

$$\langle (x_1 - x_2)^2 \rangle = \frac{2K\Delta t}{(\beta_{11} - \beta_{12})^2} \tag{4.19}$$

where $K = \langle A^2 \rangle$. For a single sphere K is given by eq (2.62). In the case of two spheres, the motion is coupled, and comparison of eq (4.19) with eqs (4.11) and (4.12) gives

$$K = 2(f_{11} - f_{12})\,\frac{kT}{m^2} \tag{4.20}$$

When the distance between the two spheres approaches infinity, the coupling coefficient $f_{12} \to 0$ and the result for a single sphere is recovered. Approximate solutions of the Langevin equation for unequal spheres can be found in a similar way.

Let us now consider the relative translational motion perpendicular to the line of sphere centres. Since perpendicular motion gives rise to torques, one must consider the forces and torques which result in such motions. These are the forces, $F_2^{(1)}$ and $F_2^{(2)}$, and the torques, $L_2^{(1)}$ and $L_2^{(2)}$, involving the components occurring in the second, fourth, sixth and eighth row of the resistance matrix in eq (4.2). For torque-free spheres

$$L_1^{(1)} = L_2^{(2)} = 0 \tag{4.21}$$

from which the two angular-velocity components can be expressed in terms of the two translational components. Substituting these components into the expressions for $F_2^{(1)}$ and $F_2^{(2)}$ yields

$$\begin{pmatrix} F_2^{(1)} \\ F_2^{(2)} \end{pmatrix} = \begin{pmatrix} h_{11} & h_{12} \\ h_{21} & h_{22} \end{pmatrix} \cdot \begin{pmatrix} u^{(1)} \\ u_2^{(2)} \end{pmatrix} \tag{4.22}$$

with the corresponding diffusion tensor

$$\boldsymbol{D}^* = \begin{pmatrix} D_{11}^* & D_{12}^* \\ D_{21}^* & D_{22}^* \end{pmatrix} = \frac{kT}{h_{11}h_{22} - h_{12}h_{21}} \begin{pmatrix} h_{22} & -h_{21} \\ -h_{12} & h_{11} \end{pmatrix} \tag{4.23}$$

Notice that $\boldsymbol{D}^*$, in contrast to $\boldsymbol{D}$ given by eq (4.10), is no longer symmetrical.

From eq (4.23) it follows that the relative diffusion coefficient perpendicular to the line of centres is given by

$$D_\perp(r,q) = D_{11}^* - D_{12}^* - D_{21}^* + D_{22}^* \tag{4.24}$$

Equation (4.24) together with eq (4.12) determine the relative diffusion between two spheres, which can be described by the diffusion equation

$$\frac{\partial n}{\partial t} = \nabla \cdot (\boldsymbol{D} \cdot \nabla n) \tag{4.25}$$

where

$$\boldsymbol{D} = \begin{pmatrix} D_\parallel(r,q) & 0 \\ 0 & D_\perp(r,q) \end{pmatrix} \tag{4.26a}$$

or

$$\boldsymbol{D} = (D_A + D_B)\begin{pmatrix} d_{\parallel}(r,q) & 0 \\ 0 & d_{\perp}(r,q) \end{pmatrix} \tag{4.26b}$$

where D_A and D_B are the diffusion constants for spheres A and B, respectively (isolated spheres in an infinitely dilute medium).

It can be shown from eqs (4.24) and (4.12) that both $d_{\parallel}$ and $d_{\perp}$ approach zero as the gap width $h \to 0$. From eq (4.4) it follows that for small h, $d_{\parallel} \propto h$, the proportionality constant depending on q (varying from 2 for equal spheres to 1 for a sphere near a flat wall), while $d_{\perp} \propto (\ln h)^{-1}$ (cf. eq (4.6)). As the distance between the spheres approaches infinity, $d_{\parallel}$ and $d_{\perp}$ approach 1. The functions $d_{\parallel}$ and $d_{\perp}$ are shown in Fig. 4.2 for the case of equal-sized spheres.

The functions $d_{\parallel}$ and $d_{\perp}$ are given by Batchelor (1976) as

$$d_{\parallel} = G(r,q) = \frac{q A_{11} + A_{22}}{1+q} - \frac{4q A_{12}}{(1+q)^2} \tag{4.27a}$$

$$d_{\perp} = H(r,q) = \frac{q B_{11} + B_{22}}{1+q} - \frac{4q B_{12}}{(1+q)^2} \tag{4.27b}$$

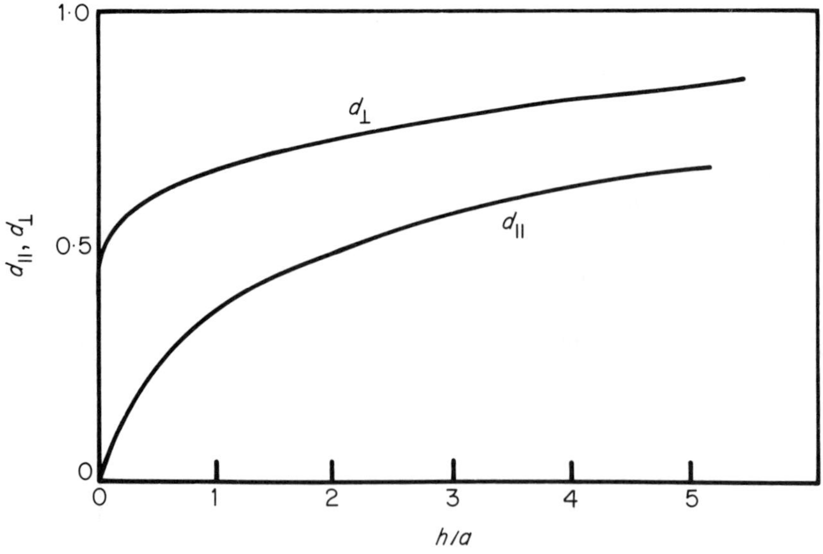

Fig. 4.2. Variation of $d_{\parallel}$ and $d_{\perp}$ with gap width, h, between two equal-sized spheres.

together with asymptotic expressions for the functions A_{ij} and B_{ij}. The function $d_\parallel$ can be expressed as an infinite series. For equal-sized spheres

$$d_\parallel^{-1} = \frac{4}{3} \sinh\xi \sum_{n=1}^{\infty} \frac{n(n+1)}{2\Delta n} \left[\frac{2n+1}{2(2n-1)} e^{2\xi} + \frac{2n+1}{2(2n+3)} e^{-2\xi} + \frac{4e^{-(2n+1)\xi}}{(2n-1)(2n+3)} - 1 \right] \quad (4.28a)$$

where $\Delta n = 2\sinh(2n+1)\xi - (2n+1)\sinh 2\xi$ and $\xi = \cosh^{-1}(r/2)$.

For arbitrary axis ratios, expressions for $d_\parallel$ and experimental verifications can be found in Adamczyk *et al.* (1983a).

For equal-sized spheres, a useful approximation (with an error of less than 1%) is (Honig *et al.*, 1971):

$$d_\parallel \simeq \frac{6h^2 + 4h}{6h^2 + 13h + 2} \quad (4.28b)$$

4.1.3 Brownian motion of pairs of spheres

Instead of considering the relative motion between two spheres, it is often useful to consider two spheres as a single entity (a doublet) to which rotary- and translational-diffusion coefficients can be ascribed. For a doublet consisting of two spheres moving with velocity u along the line of centres, forces $F_A = F_1^{(1)}$ and $F_B = F_1^{(2)}$ act on the spheres (cf. eq (4.8))

$$F_A = (f_{11} + f_{22})\, u$$
$$F_B = (f_{21} + f_{22})\, u \quad (4.29)$$

The total force on the doublet is

$$F = F_A + F_B = (f_{11} + f_{12} + f_{21} + f_{22})\, u \quad (4.30)$$

and thus the diffusion constant for motion along the line of sphere centres equals

$$D_\parallel = \frac{kT}{f_{11} + f_{12} + f_{21} + f_{22}} \quad (4.31)$$

Similarly, from eq (4.22) the diffusion coefficient for motion perpendicular to the line of sphere centres (for torque-free spheres) is found to be:

$$D_{\perp} = \frac{kT}{h_{11} + h_{12} + h_{21} + h_{22}} \tag{4.32}$$

From eqs (4.4) and (4.6) it might appear that $D_{\parallel,\perp} \to 0$ as $h \to 0$. However, in eqs (4.31) and (4.32) the terms approaching infinity cancel and $D_{\parallel}$ and $D_{\perp}$ approach a constant value as $h \to 0$. Examples of $D_{\parallel}$ and $D_{\perp}$ for a pair of torque-free equal-sized spheres are given in Fig. 4.3.

Besides doublets of torque-free spheres, we can consider two spheres which are rigidly connected by a rod of vanishing dimensions. A doublet of such spheres is usually referred to as a dumbbell. In a dumbbell each sphere is subjected to a torque and the condition of zero-torque must be replaced by the condition of zero relative translational and angular velocity. Because for a doublet, translating with velocity u along the line of centres, the angular velocity is zero, eq (4.31) also applies to rigid dumbbells. However, eq (4.32) needs modification. In this case, for motion perpendicular to the line of sphere centres

$$\begin{aligned} F_A &= g_{11} u_1 + g_{12} u_2 \\ F_B &= g_{21} u_1 + g_{22} u_2 \end{aligned} \tag{4.33}$$

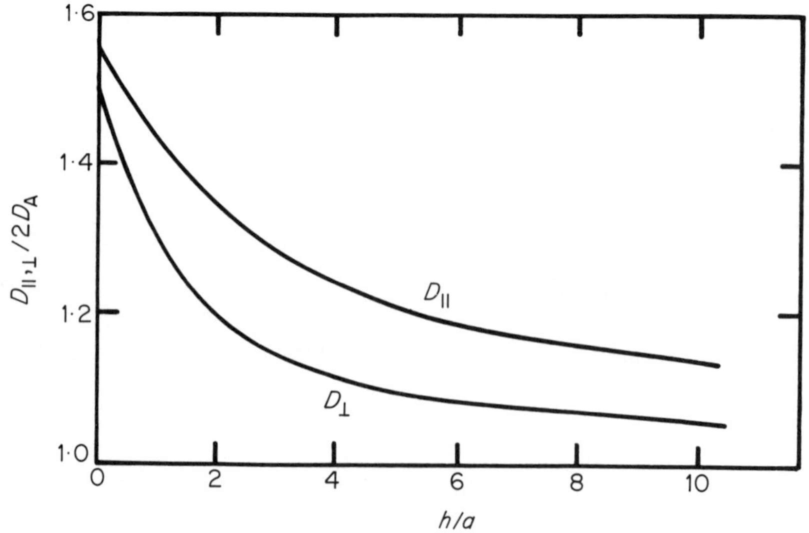

Fig. 4.3. Translational-diffusion coefficients for a pair of equal-sized torque-free spheres, for motion along the line of centres ($D_{\parallel}$) and perpendicular to it ($D_{\perp}$) (adapted from Happel and Brenner, 1983).

where g_{ij} are the friction coefficients, corresponding to the components R_3, R_4 and R_8 of eq (4.2b). Hence, for rigid dumbbells, the diffusion constant for perpendicular motion is

$$D_{\perp} = \frac{kT}{g_{11}+g_{12}+g_{21}+g_{22}} \tag{4.34}$$

The rotary Brownian motion of a pair of spheres can be described by the rotary diffusion constant, D_r, for motion about an axis perpendicular to the line joining the sphere centres, and by the spin diffusion coefficient, D_s, corresponding to motion about the line of centres.

If ϕ is the angle between the projection of the line of centres on the X_2X_3 plane and the X_2 axis, then translations of spheres A and B in the X_2 direction of magnitude Δx_A and Δx_B results in

$$\Delta\phi = \frac{\Delta x_A - \Delta x_B}{r\sin\theta} \tag{4.35}$$

where r is the dimensionless distance between sphere centres and θ is the other polar coordinate. The mean-square angular displacement of an axisymmetric body is given by (Brenner, 1967)

$$\langle(\sin\theta\Delta\phi)^2\rangle = 2D_r\Delta t \tag{4.36}$$

Using eq (4.35),

$$\langle(\sin\theta\Delta\phi)^2\rangle = \frac{\langle(\Delta x_A - \Delta x_B)^2\rangle}{r^2} = \frac{\langle\Delta x_A^2\rangle - 2\langle\Delta x_A x_B\rangle + \langle\Delta x_B^2]}{r^2} \tag{4.37}$$

From the relations (cf. eq (4.23))

$$\langle\Delta x_A^2\rangle = 2D_{11}^* t\ ;\ \langle\Delta x_B^2\rangle = 2D_{22}^* t\ ;\ \langle\Delta x_A \Delta x_B\rangle = (D_{12}^* + D_{21}^*)t \tag{4.38}$$

it follows, by comparing eq (4.37) with eq (4.36), that

$$D_r = (D_{11}^* + D_{22}^* - D_{12}^* - D_{21}^*)/r^2 \tag{4.39}$$

From the coefficients given by Arp and Mason (1977a) it can be shown that for equal small gap widths

$$\frac{D_r}{D_A} = \frac{1}{(1+\frac{1}{2}h)^2}\left(\frac{0\cdot2673 - 0\cdot724/\ln h}{1 - 1\cdot350/\ln h}\right) \tag{4.40}$$

From the asymptotic expressions given by Batchelor (1976) it can be shown that for large gap widths:

$$\frac{D_r}{D_A} = \frac{16}{3(1+q)^2 r^2}\left[\frac{1+q}{q} - \frac{3}{(1+q)r} - \frac{4(1+q^2)}{(1+q)^3 r^3}\right] \tag{4.41}$$

It should be noted that this expression is not in agreement with that derived from asymptotic expressions for the force–torque coefficients of Arp and Mason (1977a), seemingly because the last term in the expansion of the coefficient denoted by them as a is incorrect. Values of D_r for doublets of torque-free equal-sized spheres are given in Fig. 4.4, together with the asymptotic expression, eqs (4.40) and (4.41). It can be seen that the asymptotic expressions are extremely good, even having a region of overlap near $h = 0.5$. Also included in Fig. 4.4 are the rotary-diffusion constants for rigid dumbbells (curve 2). These can be found by replacing h_{ij} in eq (4.22) by g_{ij}, defined in eq (4.33). For equal-sized spheres at small gap widths:

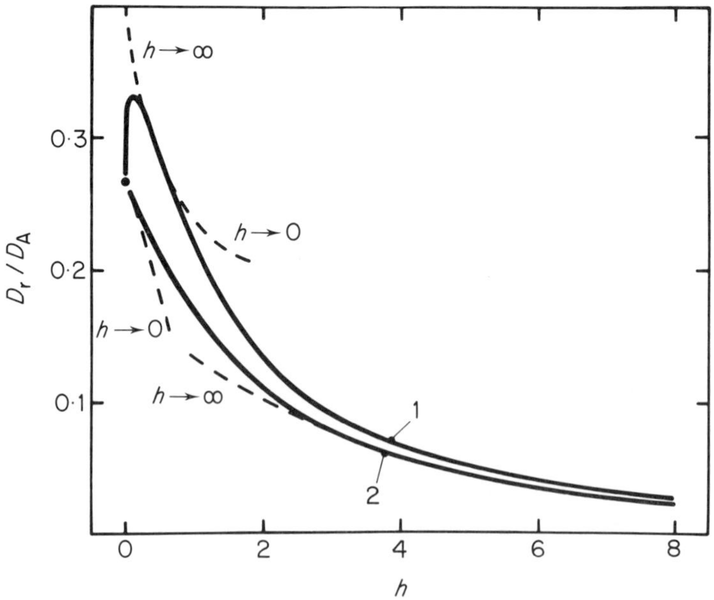

Fig. 4.4. Rotary-diffusion coefficient, D_r, of a pair of equal-sized spheres (divided by the rotary-diffusion coefficient, D_A, of a single sphere) as a function of distance between the spheres. Curves 1 and 2 correspond to a pair of torque-free spheres and a rigid dumbbell, respectively. Included are the asymptotic results for the two cases (eqs (4.40) to (4.43)). When $h = 0$ both curves approach the same value (•). Note the large difference between the two cases at short distances.

$$\frac{D_r^{rd}}{D_A} = 0\cdot2673 - 0\cdot153h - 0\cdot0341h^2 + 0\cdot0089h^2 \ln h + \ldots \tag{4.42}$$

and for large separations:

$$\frac{D_r^{rd}}{D_A} = \frac{8}{3r^2}\left(1 - \frac{3}{4r} - \frac{4}{r^2} + \frac{4}{r^3}\right) \tag{4.43}$$

Equations (4.42) and (4.43) are also shown in Fig. 4.4 for comparison. It can be seen from Fig. 4.4 that the rotary-diffusion constant of a torque-free pair

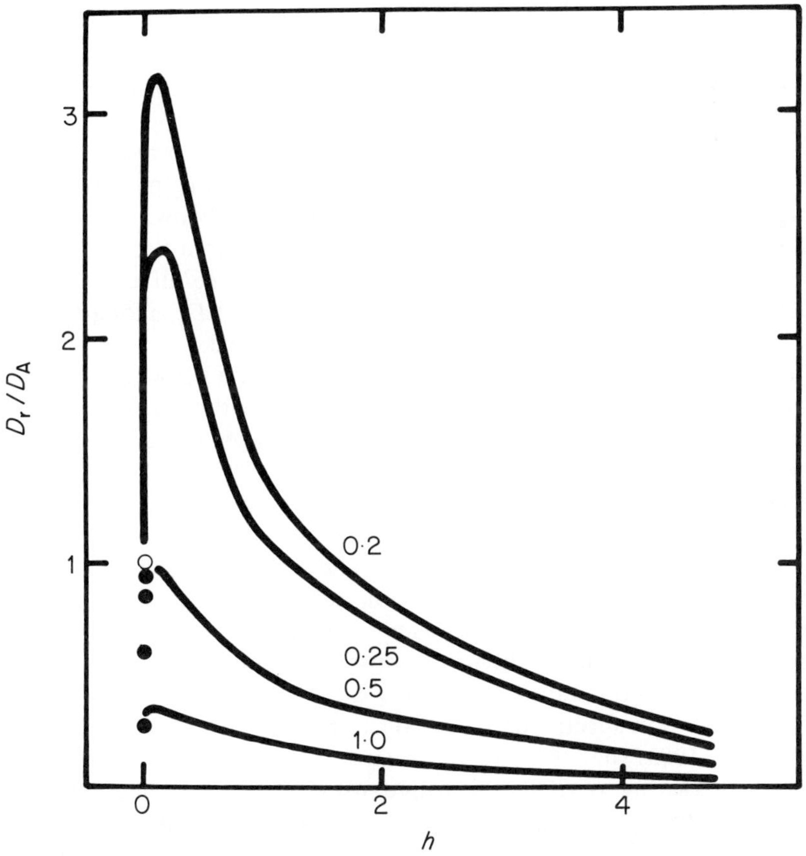

Fig. 4.5. Rotary-diffusion coefficient, D_r of a pair of torque-free spheres (divided by the coefficient of the larger sphere) as a function of distance, h, for various values of the radius ratio, q, indicated in the figure. (●) Values at $h=0$; (○) value when $h=0$ and $q=0$. Note that as q decreases, the relative increase in D_r increases rapidly at short separations.

is considerably larger than that of a rigid dumbbell. The relative increase in D_r is even larger for the case of unequal spheres, as can be seen from Fig. 4.5.

The spin diffusion coefficient, D_s, can be found by considering a pair of spheres which rotate about the line of centres with angular velocity ω. The torque, L, is given by

$$L = L_A + L_B = -\mu(R_{11} + 2R_{12} + R_{15})\omega = -f_s\omega \tag{4.44}$$

Since the rotation of the spheres does not cause a translation, D_s is the same for rigid dumbbells as for unconnected spheres. From eq (4.44):

$$D_s = kT/f_s \tag{4.45}$$

4.2 Concentration dependence of diffusion constants of hard spheres

When a colloidal particle undergoes Brownian motion in the presence of other particles, its motion will be affected by the neighbouring particles. For a single isolated spherical particle the translational Brownian motion is characterized by a single constant $D_0 = kT/6\pi\mu a$. When a Brownian sphere interacts with other particles, one can distinguish three different types of diffusion constant: a short-time self-diffusion constant D_0^s, a long-time self-diffusion constant D_∞^s, and a mutual-diffusion constant D_m.

The self-diffusion constants describe the motion of a Brownian particle in a swarm of other particles. Observed for a short time, the particle diffuses in a "cage" of other particles, while if the particle is observed for a long time it is seen to move through the swarm of particles. The diffusion constants associated with these time scales are denoted here by D_0^s and D_∞^s. As a result of the Brownian motion of each particle, local fluctuations in particle concentration occur in the dispersion. This in turn results in gradient diffusion: particles at higher concentrations diffuse towards regions of lower concentrations. The diffusion constant associated with this collective process is called the mutual-diffusion constant, denoted by D_m.

The most common way of measuring diffusion coefficients is by using photon-correlation spectroscopy (PCS) (see Section 2.4.3). It can be shown (Dickenson, 1983) that the effective-diffusion coefficient, D_{eff}, for monodisperse interacting particles measured using PCS is a function of the scattering vector, q, or, alternatively, of the scattering angle, θ_s. It has been shown (Russel and Glendinning, 1981) that at small scattering angles D_{eff} equals the mutual-diffusion constant D_m, and at large angles it equals the self-diffusion constant D_0^s. In the range $0{\cdot}2 < qa < 4$, a transition occurs from D_m to D_0^s. The diffusion coefficient, D_m, can also be found by boundaryspreading experiments (cf. Section 2.4.2), while D_0^s and D_∞^s can be obtained from microscopic observations (cf. Section 2.4.1).

The coefficient D_m is the one needed to describe convective diffusion. For interacting particles the convective-diffusion equation can be written as

$$\frac{\partial n}{\partial t} = \nabla \cdot (D_m \nabla n) - \nabla \cdot (un) \tag{4.46}$$

The magnitude of D_m can be found from the theory of irreversible thermodynamics (Fitts, 1962):

$$D_m = \frac{(1-c)n}{N_A f(n)} \left(\frac{\partial \mu^*}{\partial n} \right)_{P,T} \tag{4.47}$$

where c is the hydrodynamic volume fraction, N_A is Avogadro's number, μ^* is the chemical potential, $f(n)$ is the friction coefficient which depends on the number density n, P is the pressure, and T is the temperature. The variations in chemical potential can be expressed in terms of the osmotic compressibility $\partial\pi/\partial n$ (Kops-Werkhoven and Fijnaut, 1981):

$$\left(\frac{\partial \mu^*}{\partial n} \right)_{P,T} = \frac{(1-c_t) N_A}{n} \left(\frac{\partial \pi}{\partial n} \right)_{P,\mu_0^*} \tag{4.48}$$

where c_t is the thermodynamic volume fraction (not necessarily equal to c), and μ_0^* the chemical potential of the solvent. An extension of van 't Hoff's law (eq (2.72)) which allows for particle interactions gives

$$\frac{\pi}{kT} = n + B_2 n^2 + \dots \tag{4.49}$$

where B_2 is the second virial coefficient. For hard spheres

$$B_2 = \frac{16}{3} \pi a^3 \tag{4.50}$$

Combining eqs (4.47) to (4.50) and noting that for hard spheres $c_t = c$, results in

$$D_m = \frac{kT}{f(c)} (1 + 8c) \tag{4.51}$$

The above equation clearly shows the two main effects that determine D_m: (i) friction, which slows down diffusion; and (ii) an osmotic contribution which speeds up diffusion. The dependence of f on volume fraction c is related to the sedimentation velocity u of a sphere by

$$\frac{f(0)}{f(c)} = \frac{u(c)}{u(0)} \tag{4.52}$$

This velocity was determined by Batchelor (1972) who found

$$u(c) = u(0)\,[1 + (k_1 + k_2 + k_3 + k_4)c] \tag{4.53}$$

where $k_1 = -1$, $k_2 = -9/2$, $k_3 = 0{\cdot}5$ and $k_4 = -1{\cdot}55$. The coefficient k_1 is due to the fact that a downward flux of particles is accompanied by an equal flux of fluid going upward. The coefficient k_2 arises from the upward flow of fluid in the region accessible to the particle centre, corresponding to fluid dragged down in the excluded overlap region around each particle. The coefficient k_3 takes into account the particle response to the net average stress acting on its surface and k_4 is the reflection contribution due to the boundaries of the spheres. Combining all these effects leads to

$$u = u_0(1 - 6{\cdot}55c) \tag{4.54}$$

Combining eqs (4.51), (4.52) and (4.53) leads to

$$D_m = D_0\,(1 + k_m\,c) \tag{4.55}$$

with $k_m = 1{\cdot}45$. Other authors have made various estimates of k_m, a summary of which is shown in Table 4.1. It can be seen that the estimates vary from -6 to $+3$. The best available experimental measurements (on silica dispersions from PCS data) give (Kops-Werkhoven and Fijnaut, 1981) $k_m = 1{\cdot}3 \pm 0{\cdot}2$, which is in agreement with the calculations of Burgers (1941), Batchelor (1976) and Felderhof (1978). It should be noted that more recently Carter and Phillies (1985) used the same method as Felderhof (a cluster expansion) but concluded that Felderhof neglected the factor $k_1 = -1$ as well as a divergence term of $-1{\cdot}457$, which combined, leads to their value of $-0{\cdot}898$. It appears that the close agreement between the result of Felderhof (1978) and Batchelor (1976) is accidental. Carter and Phillies (1985) explain the discrepancy between their result and the experimental data as being due to the assumption that the silica particles used for the experiments were not truly hard spheres. We can conclude that not only do the experts disagree about the value of k_m, but they do not even agree on its sign. Batchelor (1976)

TABLE 4.1
Comparison of the virial coefficients, k_m, for hard spheres reported by various authors

k_m	Reference
1·12	Burgers (1941)
0·84	Pyun and Fixman (1964)
2·0	Altenberger and Deutch (1973)
3·0	Harris (1976)
−1·83	Anderson and Reed (1976)
1·45	Batchelor (1976)
1·0	Hess and Klein (1976)
−6·0	Ackerson (1978)
1·56	Felderhof (1978)
2·50	Altenberger (1979)
−2·0	Allison *et al.* (1979)
−0·875	Phillies (1982)
−0·898	Carter and Phillies (1985)

is the only one who used in his calculation the exact form of the hydrodynamic interactions between spheres and his result seems to be the most reliable.

Similarly, the concentration dependence of self-diffusion constants can be expressed as

$$D_0^s = D_0\,(1 + k_0^s c) \tag{4.56}$$

and

$$D_\infty^s = D_0\,(1 + k_\infty^s c) \tag{4.57}$$

where according to Batchelor (1976) (cf. Kops-Werkhoven and Fijnaut, 1981) for hard spheres $k_0^s = -1.83$ and $k_\infty^s = -2{\cdot}68$. Other quoted values are $k_0^s = 1{\cdot}73$ (Felderhof, 1978; Muthukumar and Freed, 1983); $k_\infty^s = -1{\cdot}89$ (Felderhof, 1977); $k_0^s = -2$ (Yoshida, 1983); and $k_\infty^s = -2{\cdot}5$ (Dickenson, 1983), obtained from the classical result of Einstein (1906). It can be seen that self diffusion over short time intervals is faster than over longer time intervals.

In a similar way, a concentration dependence is expected for rotary-diffusion coefficients. The only theoretical results are for rotary self-diffusion coefficients. Muthukumar and Freed (1983) found from a cluster expansion that

$$D_r = D_r^0\left(1 - \frac{5}{16}c\right) \tag{4.58}$$

while Masliyah and van de Ven (1987) found, from a cell model, that

$$D_r = D_r^0(1 - c) \tag{4.59}$$

Because of the assumptions made in deriving eqs (4.58) and (4.59), eq (4.58) corresponds to the short-time diffusion coefficient and eq (4.59) to the long-time one.

4.3 Brownian motion of soft spheres

We defined soft spheres as those on which colloidal forces act. In the absence of Brownian motion, two soft spheres will either approach one another or separate, depending on whether the colloidal force is attractive or repulsive. When a force, $\boldsymbol{F}_{\text{coll}}$, is acting on the spheres they will move with a relative velocity $\boldsymbol{u}_{\text{coll}}$ given by:

$$\boldsymbol{F}_{\text{coll}} = \boldsymbol{f}\cdot\boldsymbol{u}_{\text{coll}} \tag{4.60}$$

where

$$\boldsymbol{f} = \boldsymbol{D}^{-1}kT \tag{4.61}$$

$\boldsymbol{D}$ being the diffusion tensor for relative motion between the spheres. In eq (4.60), $\boldsymbol{u}_{\text{coll}}$ is defined in such a way that when a positive (repulsive) force is acting between the spheres, the distance between them increases with time and $\boldsymbol{u}$ is positive.

When only van der Waals forces are operating between the spheres the relative diffusion tensor is given by eq (4.26). When double-layer forces are operative, $\boldsymbol{D}$ must be modified. As yet the correction to $\boldsymbol{D}$ due to double-layer interactions has not been calculated. As we have seen, electrostatic and hydrodynamic interactions lead to coupling forces and torques (van de Ven, 1988b), which affect diffusion. In principle, the coupled system of differential equations discussed in Section 2 must be solved (convective-diffusion, Poisson, creeping-flow and continuity equations). The general conclusions derived in Section 2 are also valid for a two-sphere system, the only difference being the boundary conditions. It is now necessary to impose the requirement that the potential on both spheres is equal to the respective surface potentials. We have seen that the diffusion constant of a charged sphere differs from that of a hard one, due to an electroviscous contribution to the drag coefficient of the order of $\bar{b}\Phi_s^2\,\tilde{f}(\kappa a)$. A similar result can be expected for the relative friction coefficient between two charged spheres, in which case the correction also depends on the radius ratio, q, and on the ratio of surface potentials, β, of both spheres. Hence for two charged spheres, at low potentials, both components of the relative diffusion tensor need corrections of the form

$$\bar{b}\Phi_s^2\,f(\kappa a, q, \beta, r)$$

Clearly much more research is needed to clarify the precise nature of $\boldsymbol{D}$ for spheres with double layers.

The diffusion of soft spheres is described by the convective-diffusion equation which takes the form:

$$\frac{\partial n}{\partial t} = \nabla\cdot(\boldsymbol{D}\cdot\nabla n) - \nabla\cdot\left(\frac{\boldsymbol{D}\cdot\boldsymbol{F}_{\text{coll}}\,n}{kT}\right) \tag{4.62}$$

Usually the colloidal force only acts along the line of sphere centres and for most problems the distribution of particles is radially symmetric. In such cases the above relation reduces to

$$\frac{\partial n}{\partial t} = (D_{\mathrm{A}} + D_{\mathrm{B}})\left[\frac{\partial}{\partial r}\left\{r^2 d_{\parallel}(r)\left(\frac{\partial n}{\partial r} + \frac{n}{kT}\frac{\partial V_{\text{int}}}{\partial r}\right)\right\}\right] \tag{4.63}$$

where $d_{\parallel}(r)$, besides being a function of the dimensionless separation r, also depends on q, $\bar{b}$, κa, Φ_s and β. D_{A} and D_{B} are diffusion constants of isolated spheres given by eq (2.174). V_{int} is the colloidal interaction energy.

At long times the distribution of particles around a reference sphere located at the origin approaches a steady state, governed by

$$\frac{\mathrm{d}}{\mathrm{d}r}\left[r^2 d_{\parallel}(r)\left(\frac{\mathrm{d}n}{\mathrm{d}r} + \frac{n}{kT}\frac{\partial V_{\text{int}}}{\partial r}\right)\right] = 0 \tag{4.64}$$

the solution of which, satisfying the boundary conditions $n(\infty) = n_0$ and the no-penetration boundary condition at the surface, is:

$$n = n_0 \mathrm{e}^{-V_{\text{int}}/kT} \tag{4.65}$$

where n is the concentration of particles of radius a_2 around a reference sphere of radius a_1. The concentration profiles for various interaction functions are shown schematically in Fig. 4.6a.

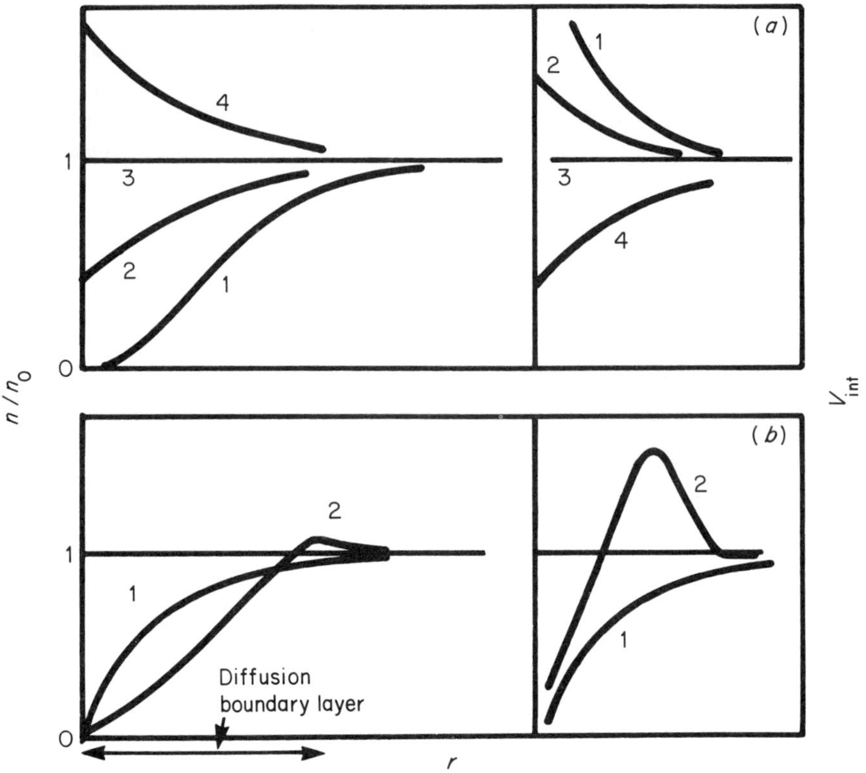

Fig. 4.6. Concentration profiles around a reference sphere (schematic).
(*a*) Under no-penetration boundary conditions for the various interaction energies shown on the right-hand side.
(*b*) Under the perfect-sink boundary condition.

When there is a large attractive force between the spheres, eq (4.65) predicts that n becomes very large. In such a case it takes a long time for the steady-state distribution to be reached (typically of the order of the time needed to escape out of the minimum). For systems with a deep primary minimum it is of more interest to know the rate at which particles accumulate in the primary minimum region. A sufficiently deep minimum can be approximated by an infinitely deep minimum and the boundary condition at the surface changed to the so-called perfect-sink boundary condition. This condition can be written as $n(2) = 0$ and can be interpreted to mean that every particle coming into contact with the reference sphere "disappears" from the system, i.e. it no longer contributes to the convective-diffusion process. Justifications of the perfect-sink boundary condition and comparisons with the no-penetration boundary condition are given in Chapter 6.

With the perfect-sink boundary condition, the solution of eq (4.64) (van de Ven and Mason, 1977b) is

$$n = n_0 e^{-V_{int}/kT} \left[\alpha \int_{\infty}^{r} \frac{e^{V_{int}/kT}}{d_{\parallel}(r) r^2} dr + 1 \right] \tag{4.66}$$

where

$$\alpha = 1 \Big/ \int_{2}^{\infty} \frac{e^{-V_{int}/kT}}{d_{\parallel}(r) r^2} dr \tag{4.67}$$

When $\alpha = 0$, eq (4.66) reduces to eq (4.65). When $\alpha \neq 0$, the above equation can be written as

$$n = \alpha n_0 e^{-V_{int}/kT} \int_{2}^{r} \frac{e^{-V_{int}/kT}}{d_{\parallel}(r) r^2} dr \tag{4.68}$$

The concentration profiles for some interaction energies are shown in Fig. 4.6b. It should be noted that when $\alpha = 0$, no particle ever makes contact with the reference particle, i.e. no coagulation occurs. Ideal smooth particles can only come into contact with the reference particle when an infinite attractive force exists at zero gap width because the relative diffusion constant is zero at contact. In Fig. 4.6b, the region where $n < n_0$, and where a concentration gradient exists, is called the diffusion boundary layer.

4.4 Perikinetic coagulation

From the solution of the convective-diffusion equation we can calculate the number of particles that are captured by the reference sphere per unit time. Since the formation of doublets is the first step in the coagulation process, knowing the rate at which doublets form under various conditions is of enormous importance. The frequency of doublet formation can be found by considering the flux of particles crossing the collision sphere, which is defined as a sphere of radius $a_1 + a_2$ around a reference sphere of radius a_1 (see Fig. 4.7). Each sphere of radius a_2, whose centre enters the collision sphere, is captured by the reference particle. According to Fick's law (eq (2.71)), the flux particles crossing the surface of the collision sphere equals

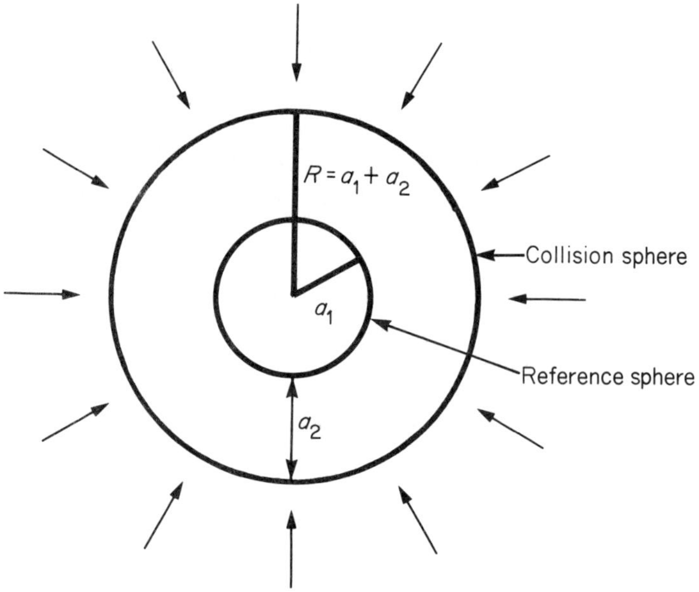

Fig. 4.7. Definition of the collision sphere of radius $a_1 + a_2$ around a reference particle of radius a_1. All particles of radius a_2, whose centres enter the collision sphere, are captured by the reference particle.

$$J = -A\left(D\,\frac{\partial n}{\partial r}\right)_{r=2} \tag{4.69}$$

where A is the area of the collision sphere

$$A = 4\pi(a_1 + a_2)^2 \tag{4.70}$$

and D the relative diffusion constant

$$D = (D_A + D_B)\, d_{\parallel}(r, q) \tag{4.71}$$

Evaluating the derivative in eq (4.69), using eq (4.68) and substituting eqs (4.70) and (4.71) into eq (4.69) yields for the flux

$$J = -4\pi\alpha_p n_0\,(a_1 + a_2)(D_A + D_B) \tag{4.72}$$

where $\alpha_p = \alpha/2$.

When only van der Waals forces are acting, $d_{\|}(r,q)$, which enters into the definition of α_p, is given by eq (4.27a), and D_A and D_B are the diffusion constants for isolated hard spheres. When double-layer forces are present, $d_{\|}$, D_A and D_B are no longer given by the hard-sphere expressions. D_A and D_B can be found from the theory outlined in Section 2.2.2, while for $d_{\|}(r, q)$ no theory is yet available. Usually the known results for hard spheres are used. It can be seen that double-layer interactions affect the flux in several ways: by affecting α_p, via the functions V_{int} and $d_{\|}$, and by affecting the diffusion constants D_A and D_B.

Traditionally, coagulation of soft spheres is compared to the hypothetical coagulation of hard spheres with no hydrodynamic interactions (for which D is independent of the distance between the spheres) and which stick on contact (Smoluchowski, 1917). In such a case the convective-diffusion equation reduces to the diffusion equation

$$\frac{\mathrm{d}}{\mathrm{d}r}\left(r^2\frac{\mathrm{d}n}{\mathrm{d}r}\right)=0 \tag{4.73}$$

the solution of which, with the boundary conditions $n(2)=0$ and $n(\infty)=n_0$, is

$$n=n_0\left(1-\frac{2}{r}\right) \tag{4.74}$$

Inserting eq (4.74) into eq (4.69) yields for the flux:

$$J_s=-4\pi\, n_0\,(a_1+a_2)(D_A+D_B) \tag{4.75}$$

This is the flux first calculated by Smoluchowski (1917) and is usually taken as a reference flux. Comparing eqs (4.72) and (4.75), it can be seen that α_p can be interpreted as a capture efficiency:

$$J=\alpha_p J_s \tag{4.76}$$

α_p is referred to as the *perikinetic* doublet-formation efficiency. Coagulation is perikinetic when it is caused solely by Brownian motion. When caused solely by deterministic forces it is called *orthokinetic* coagulation.

Notice that α_p can have any positive value, i.e. $0 \leqslant \alpha_p < \infty$. Sometimes α_p is referred to as the collision efficiency. However, in a hydrodynamic context, the word collision has no precise meaning. In the absence of strong attractive forces, two particles will never come into contact and, in a strict sense, they never collide.

For two equal spheres, eq (4.72) reduces to

$$J = -16\pi\alpha_p n_0 Da \tag{4.77}$$

To obtain the doublet-formation frequency, f, i.e. the rate at which doublets are formed in a unit volume of a dispersion, the flux must be multiplied by half the number concentration of particles:

$$f = \tfrac{1}{2} n_0 J \tag{4.78}$$

The factor of $\frac{1}{2}$ is included so that each collision is not counted twice.

The efficiency α_p is largest when there are strong attractive forces between the particles. In such a case the coagulation is called fast. For spheres with a large energy barrier in the interaction energy V_{int}, the coagulation is slow; the higher the barrier, the slower the coagulation. The ratio of fast to slow coagulation rates is usually called the stability ratio, W (Reerink and Overbeek, 1954). For electrostatically stabilized dispersions, the fastest coagulation rate is achieved when the double layer is completely compressed, i.e. at high electrolyte concentrations. In this case

$$(\alpha_p^{-1})_f = 2 \int_2^\infty \frac{e^{V_{vdW}/kT}}{d_\parallel(r) r^2}\, dr \tag{4.79}$$

where V_{vdW} is the van der Waals dispersion energy (cf. Section 1.4.2). For slow coagulation

$$(\alpha_p^{-1})_s = 2 \int_2^\infty \frac{e^{(V_{rep} + V_{vdW})/kT}}{d_\parallel(r) r^2}\, dr \tag{4.80}$$

where V_{rep} is the repulsive electrostatic interaction (cf. Section 1.4.3). Hence

$$W = \frac{(\alpha_p)_f}{(\alpha_p)_s} \tag{4.81}$$

It is often assumed that for fast coagulation $\alpha_p = 1$ and, therefore, the slow: fast ratio is approximately

$$W \simeq \frac{1}{\alpha_p} \tag{4.82}$$

Exact calculations of α_p for fast coagulation show that, depending on the value of the Hamaker constant and the particle radius, α_p is in the range 0·4–1·0 (Honig *et al.*, 1971).

For slow coagulation, characterized by a relatively high barrier in the energy of interaction curve, α_p can be evaluated by assuming that the mean contribution of the integral in eq (4.80) comes from the region near the maximum. Using a Taylor expansion near the maximum, it can be shown that ln α_p varies approximately linearly with the logarithm of the electrolyte concentration (Reerink and Overbeek, 1954). This linear relationship breaks down when the electrolyte concentration reaches the critical coagulation concentration (CCC), beyond which α_p is constant. If the stability ratio with neglect of the effect of double layers on the hydrodynamic friction is denoted by W_0 and with the incorporation of these electroviscous effects by W, it follows readily from eqs (4.79) to (4.82) that when the radial relative diffusion coefficient is given by

$$D_{\|} = D_{\|}^{0}[1 - \bar{b}\Phi_s^2 h(\kappa a, r)] \tag{4.83}$$

we have (van de Ven, 1988a)

$$W \simeq W_0[1 + \bar{b}\Phi^2 h(\kappa a, r_m)] \tag{4.84}$$

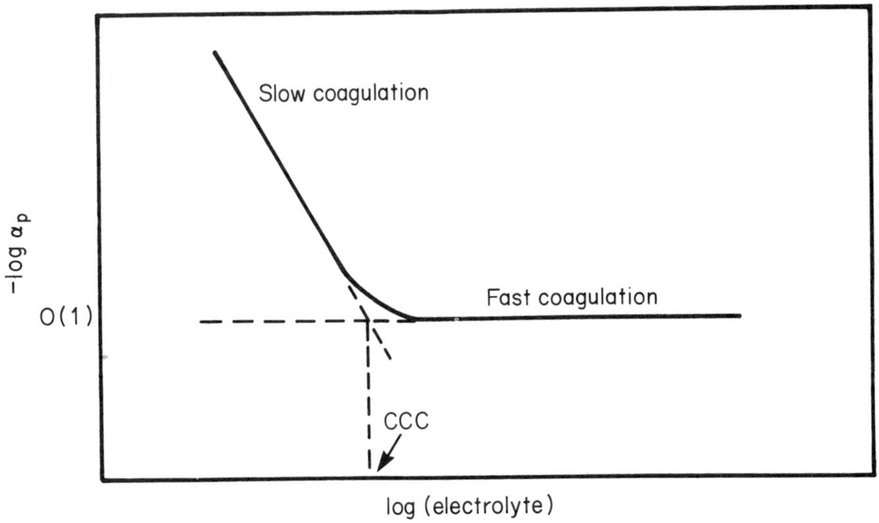

Fig. 4.8. Scheme to show the variation in the perikinetic-capture efficiency with electrolyte concentration. The critical coagulation concentration (CCC) denotes the onset of fast coagulation.

where r_m is the distance where V_{int} is maximum. More generally,

$$W \simeq W_0(1 + b_1 h_1 + b_2 h_2) \tag{4.85}$$

where h_1 and h_2 are functions of Φ_s, κa and V_{int}. From eq (4.85) it follows that the CCC is given by

$$\mathrm{CCC} = \mathrm{CCC}_0(1 + b_1 p_1 + b_2 p_2) \tag{4.86}$$

where p_i are as yet unknown functions. It can be concluded that the stability ratio, W, and the critical coagulation concentration, CCC, differ from the predictions of the classical DLVO theory (see Fig. 4.8). That the CCC increases with the scaled friction coefficient, b_i, of the ions in the double layer has been observed experimentally (see Fig. 4.9).

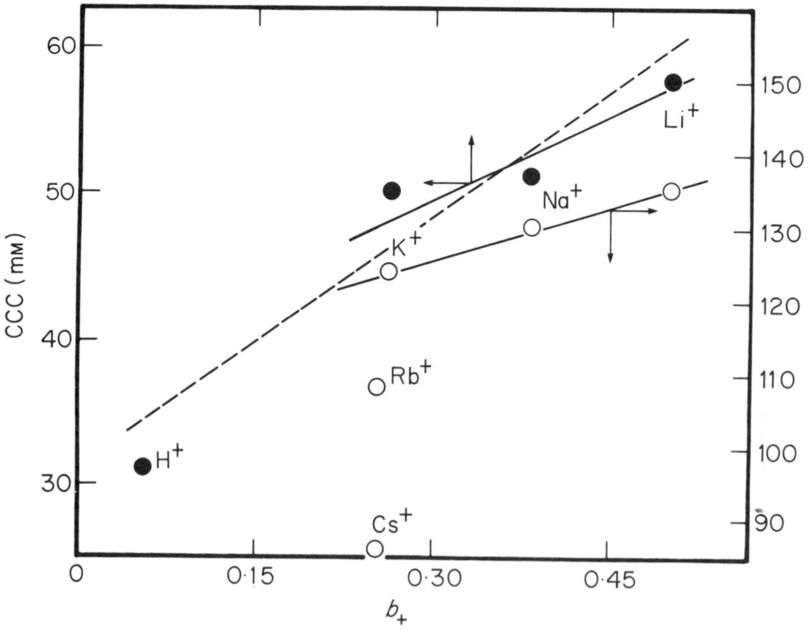

Fig. 4.9. The critical coagulation concentration (CCC) for (•) As_2S_3 - and (○) and AgI-sol (left and right axes, respectively), as a function of the scaled friction coefficient of the positive ions in the double layer. For K^+, Na^+ and Li^+ the data fall on straight lines (solid lines). For As_2S_3, all data, including H^+, fall approximately on a straight line (dashed). The ions Rb^+ and Cs^+ appear to be anomalous (after van de Ven, 1988a).

4.5 Diffusion of secondary doublets

For electrostatically stabilized dispersions we can distinguish between two different types of coagulation: (i) coagulation in a primary minimum, and (ii) coagulation in a secondary minimum. A doublet of two spheres captured in a primary energy minimum is called a primary doublet and a doublet of two spheres captured in a secondary minimum is called a secondary doublet (van de Ven and Mason, 1976b). In general, all doublets captured in an energy minimum occurring at a finite gap width and having no torques acting on the spheres, can be considered secondary doublets. Because the individual spheres in a secondary doublet are free to rotate, the behaviour of a secondary doublet differs from that of a primary doublet. A primary doublet can be considered to consist of two spheres in contact and behaves as a rigid axisymmetric body (cf. Section 2.1.2.2).

In essence, secondary doublets behave like the pairs of torque-free spheres, as discussed in Section 4.1.3. They are characterized by two translational-diffusion constants $D_{\|}$ and $D_{\perp}$ (given by eqs (4.31) and (4.32)), and two rotational-diffusion constants D_r and D_s (given by eqs (4.39) and (4.45)). The only difference is that the values given for a pair of torque-free spheres are instantaneous values, valid for a given separation between the spheres. In secondary doublets the distance between the spheres can vary because of the Brownian motion of the individual spheres and an average diffusion constant must be defined by averaging over all distances.

Let us consider a secondary doublet of two spheres captured in a secondary minimum occurring at a distance h_{eq}. Allowing for variations in the separation complicates the definition of a secondary doublet. How far can the spheres move away from h_{eq} before the doublet ceases to be a secondary doublet? It does not help to consider the steady-state probability-distribution function of an ensemble of particles originally at h_{eq}. As $t \rightarrow \infty$ the particles are either captured in a primary minimum or diffuse away towards infinity. It is possible to introduce arbitrary definitions, such as defining secondary doublets as two spheres whose potential energy of interaction differs by less than, say, $5kT$ from the energy at h_{eq}. This is the same as defining a secondary doublet as two spheres within an interval (h_{min}, h_{max}) centred around h_{eq} $(h_{min} < h_{eq} < h_{max})$. As an extreme case consider two spheres in a square potential well with infinite barriers at h_{min} and h_{max}, to prevent the particles from leaving the secondary minimum. When the probability of finding a particle at h in (h_{min}, h_{max}) is independent of h, the average rotary diffusion coefficient, $\overline{D}_r$, is given by:

$$\overline{D}_r = \frac{3}{r_{max}^3 - r_{min}^3} \int_{r_{min}}^{r_{max}} D_r(r)\, r^2 \mathrm{d}r \qquad (4.87)$$

To see how much $\overline{D_r}$ deviates from $D_r(r_{eq})$, let us look at a representative case: e.g. $a_1 = 5\,\mu$m, $a_2 = 1\,\mu$m and $h_{eq} = 25$ nm, so that $r_{eq} = 1\cdot205$ and $q = 0\cdot2$. Let the minimum allowed deviation of h_{eq} be 10 nm, so $r_{min} = 1\cdot203$ and $r_{max} = 1\cdot207$. In this case, integrating eq (4.87) yields $D_r = 2\cdot43 D_A$ as compared with $D_r(r_{eq}) = 2\cdot46 D_A$, from which it follows that the difference between D_r and $D_r(r_{eq})$ is about 1%. For more realistic potential energies it is more likely to find a particle near h_{eq} rather than near h_{min} or h_{max} and, therefore, for real secondary doublets the deviation will be less than 1%. In most cases we can expect that

$$\overline{D_r} \simeq D_r(h_{eq}) \tag{4.88}$$

with similar expressions for the other constants ($D_{\parallel}$, $D_{\perp}$, D_s). The coefficient D_r is of the most interest, firstly because the difference in D_r for secondary and primary doublets is large, and secondly because it is the easiest diffusion constant to measure experimentally (cf. Sections 2.4.3 and 2.4.4).

From Figs. 4.4 and 4.5 it can be seen that the maximum in D_r occurs at distances which are comparable to h_{eq}. From eq (4.88) it follows that, in principle, the equilibrium distance between two spheres can be obtained from measurements of the rotary-diffusion constant of secondary doublets.

Besides primary and secondary doublets, intermediate cases exist where two spheres forming a doublet do not touch but are not torque free. An example of this situation is a doublet consisting of two spheres linked by a flexible connection, e.g. a polymer chain. These are called flexible dumbbells (Takamura *et al.*, 1981c) and their rotary-diffusion constants, D_r, are intermediate between the D_r values of primary and secondary doublets.

4.6 Concentration dependence of diffusion constants of soft spheres

We have seen that the diffusion of interacting spheres can be described by three different types of diffusion constant. Here we will deal with the most important one, the mutual-diffusion constant, D_m. Recall that for hard spheres, according to Batchelor (1976),

$$D_m = D_0(1 + 1\cdot45c) \tag{4.89}$$

When interparticle forces are acting between the particles, the distribution of particles around a reference sphere is changed and this results in a different virial coefficient. We have seen that the mutual-diffusion constant can be found from an expansion of the osmotic compressibility, $\partial\pi/\partial c$, with the osmotic pressure given as

$$\pi = kT\,(n + B_2 n^2 + \ldots) \tag{4.90}$$

and an expansion of the sedimentation velocity

$$u = u_0\,(1 + Sc + \ldots) \tag{4.91}$$

For hard spheres, B_2 is given by eq (4.50) and $S = -6{\cdot}55$, resulting in

$$D_{\mathrm{m}} = D_0(1 + 8c)\,(1 - 6{\cdot}55c) \tag{4.92}$$

For soft spheres the values of B_2 and S must be adjusted:

$$\begin{aligned} B_2 &= B_2^0 + B_{\mathrm{int}} \\ S &= S^0 + S_{\mathrm{int}} \end{aligned} \tag{4.93}$$

where the superscript "0" refers to the hard-sphere values and the last terms are corrections due to particle interactions. B_{int} can be found from the MacMillan-Mayer (1945) virial expansion for the osmotic pressure (Hill, 1960). For equal spheres of radius a:

$$B_{\mathrm{int}} = \frac{3}{a^3} \int_{2a}^{\infty} f(V_{\mathrm{int}})\, r^2 \mathrm{d}r \tag{4.94}$$

where

$$f(V_{\mathrm{int}}) = \exp\left(\frac{-V_{\mathrm{int}}}{kT}\right) - 1 \tag{4.95}$$

The values of S_{int} can be found from the general expression for S given by Batchelor (1972). Van den Broeck *et al.* (1981) found:

$$S_{\mathrm{int}} = \frac{1}{a^3} \int_{2a}^{\infty} f(V_{\mathrm{int}})\,[g(r) - 3]\, r^2 \mathrm{d}r \tag{4.96}$$

where $g(r)$ is a known function of the distance between the sphere centres, derived from the coefficients of resistance between the two spheres.

Combining eqs (4.94) and (4.96) results in

$$D_{\mathrm{m}} = D_0(1 + k_{\mathrm{m}} c) \tag{4.97}$$

with

$$k_{\mathrm{m}} = 1\cdot 45 + \frac{1}{a^3}\int_{2a}^{\infty} f(V_{\mathrm{int}})\,[g(r) - 6]\,r^6 \mathrm{d}r \tag{4.98}$$

In eq (4.97), D_0 is the diffusion coefficient of a sphere given by eq (2.174). It can be shown that $[g(r) - 6]$ is negative for all r. Thus it follows from eq (4.98) that when V_{int} is positive (corresponding to repulsive forces between the spheres) k_{m} increases, while if V_{int} is negative (attractive forces between the spheres) k_{m} decreases. For charged spheres, the values of the constant 1·45 and the function $g(r)$ need modification to account for electroviscous effects.

Equation (4.98) shows that, in principle, information about particle-interaction energies can be obtained by measuring the concentration dependence of the diffusion coefficients.

4.7 Brownian motion of two non-spherical particles

We have seen that the grand diffusion matrix of a two-sphere system consists of an 8×8 matrix containing 10 independent components. In general for two particles of arbitrary shape, the diffusion matrix will be a 12×12 matrix. It can be proven (Happel and Brenner, 1983) that a unique point exists, called the centre of reaction, with the property that the diffusion matrix may be made symmetrical by specifying the torques relative to an origin that coincides with it. Hence we need, in general, 81 components (78 from the symmetrical matrix and three for the location of the centre of reaction) to completely describe the diffusion of a two-particle system. Furthermore, all these components are functions of the distance between the particles and their relative orientation. It follows that it is almost impossible to describe exactly the diffusion of a two-particle system. The more symmetry the particle possesses, the fewer components are needed to describe the diffusion, but even for spheroids very few of the components are known. Limited data has recently begun to appear for two spheroids at large separations (Kim, 1986).

From an experimental point of view, it is possible to define effective diffusion constants, obtained by assuming that the particles are spherical. As for spheres, various diffusion constants can be defined, such as for a single particle in the presence of a second, for a particle pair, etc.

As for spheres, we can expect the mutual translational-diffusion constant to be a function of the volume fraction:

$$\bar{D} = \bar{D}_0 \,(1 + k\phi) \tag{4.99}$$

where, for ellipsoids

$$\bar{D} = \frac{1}{3} \,(D_1 + D_2 + D_3)$$

where D_i is the component in the i direction. Similarly, for the rotational-diffusion constant

$$D_r = D_r^0 \,(1 + b\phi) \tag{4.100}$$

where for spheroids D_r^0 is the rotary-diffusion constant for an isolated spheroid, while for ellipsoids not too far from axisymmetry D_r^0 is given by eq (2.257). As yet no theories for predicting b or k are available.

Chapter 5

Two Colloidal Particles Subjected to an External Field

5.1 Two particles in a linear shear flow

5.1.1 Two hard spheres in a linear shear flow

To find the flow field around two spheres in a linear shear flow we must solve the creeping-flow equation with the no-slip boundary condition applied at the surface on both spheres, together with the condition that the flow approaches the undisturbed flow field at large distances from the spheres. The hydrodynamic forces and torques exerted on the particles can be calculated from the flow field and the trajectories of the spheres can be determined from the condition that the net force and the net torque must be zero. The solution of this problem has been given by Kao *et al.* (1977). For a linear two-dimensional flow field the relative trajectories of two hard spheres can be written as

$$\frac{\mathrm{d}r}{\mathrm{d}t}=G(1+\gamma)\,A(r,q)\,\sin^2\theta\sin 2\phi \tag{5.1a}$$

$$\frac{\mathrm{d}\theta}{\mathrm{d}t}=\tfrac{1}{4}G(1+\gamma)\,B(r,q)\,\sin 2\theta\sin 2\phi \tag{5.1b}$$

$$\frac{\mathrm{d}\phi}{\mathrm{d}t}=\tfrac{1}{2}G[(1-\gamma)+(1+\gamma)\,B(r,q)\,\cos 2\phi] \tag{5.1c}$$

where (r, θ, ϕ) are the polar coordinates (see Fig. 5.1) of one sphere relative to the other; r is rendered dimensionless by the mean radius $\bar{a}=\frac{1}{2}(a_1+a_2)$, a_1 and a_2 being the radii of spheres 1 and 2, respectively; G is the rate of shear; γ the flow parameter characterizing the type of flow, and q is the radius ratio

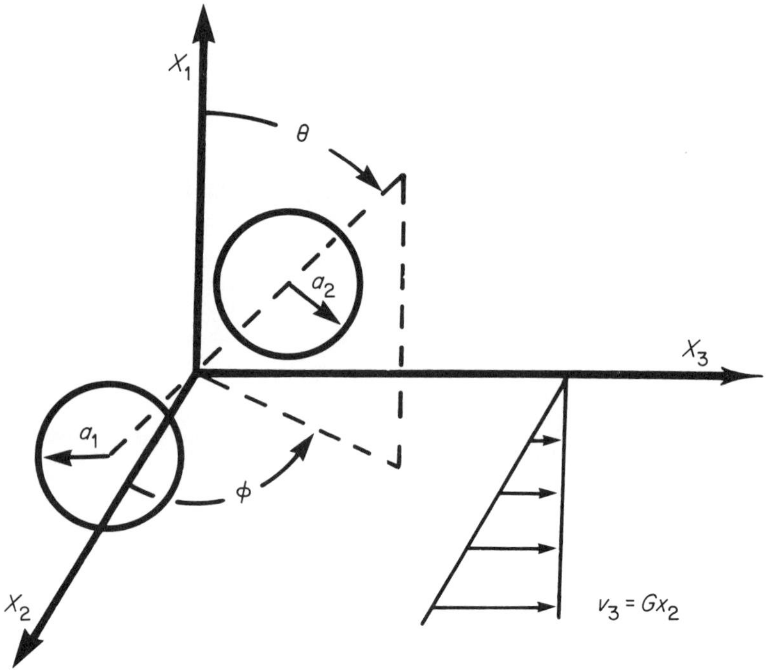

Fig. 5.1. Schematic representation of two spheres in simple shear.

($q = a_2/a_1$). $A(r, q)$ and $B(r, q)$ are known functions of r and q. Note the similarities with the streamline equations around a sphere in two-dimensional flow, which are obtained by setting $q = 0$ in the above equations. For equal-sized spheres ($q = 1$) which are close together, A and B are given by

$$A(r, 1) = 4\cdot077h + 3\cdot67h^2 \ln \frac{h}{23\cdot62} + \ldots \tag{5.2a}$$

and

$$B(r, 1) = 0\cdot5941 - \frac{0\cdot9589}{\ln h} - \frac{1\cdot125}{\ln^2 h} \tag{5.2b}$$

where $h = r - 2$. For two spheres far apart

$$A(r, 1) = \frac{r}{2}\left(1 - \frac{5}{r^3} + \frac{8}{r^5} - \frac{25}{r^6} + \ldots\right) \tag{5.3a}$$

and

$$B(r,1) = 1 - \frac{16}{3r^5} + \frac{10}{3r^8} - \ldots \tag{5.3b}$$

Intermediate values of A and B can be found in a paper by Arp and Mason (1977a) for equal-sized spheres and are given by Adler (1981a) for unequal-sized spheres.

As $r \to 2$, $\mathrm{d}r/\mathrm{d}t \to 0$ and the remaining equations reduce to Jeffery's equations (eq (3.86)) with $B = 0{\cdot}5941$ corresponding to a value of r_e of $1{\cdot}9817$ (for equal spheres). For unequal-sized pairs of touching spheres, values for r_e can be found in Nir and Acrivos (1973).

The set of eqs (5.1) can be integrated in the usual way (cf.eqs (3.191) and (3.192)) with the result

$$x_1 = r\,C f(r) \tag{5.4a}$$

$$x_2 = \pm r f(r)\left[E + \frac{\gamma f(r)^{-2}}{1-\gamma} + \frac{1-\gamma}{1+\gamma}\,g(r)\right]^{1/2} \tag{5.4b}$$

where

$$f(r,q) = \exp\left[-\int_r^\infty \frac{B(r,q)}{2A(r,q)}\,\mathrm{d}r\right] \tag{5.5a}$$

and

$$g(r,q) = \int_r^\infty \frac{B(r,q)-1}{2A(r,q)\,f^2(r,q)}\,\mathrm{d}r \tag{5.5b}$$

and C and E are the trajectory constants. The functions f and g are both positive and decrease monotonically with increasing r. As $r \to \infty$ (for equal-sized spheres)

$$f(r) = \frac{1}{r} + \frac{5}{3r^4} + \ldots. \tag{5.5c}$$

$$g(r) = \frac{16}{9r^3} + \frac{25}{27r^6} + \ldots. \tag{5.5d}$$

For small separations

$$f(r) = \frac{0 \cdot 479}{h^{0 \cdot 073}} \ln^{0 \cdot 118} h \left(1 - \frac{0 \cdot 138}{\ln h}\right) + \ldots . \tag{5.5e}$$

and

$$g(r) \to 0 \cdot 7197 \quad \text{as } r \to 2 \tag{5.5f}$$

The trajectories of two spheres in a linear flow field are qualitatively similar to the streamlines around a sphere in such a flow (cf. Section 3.1.9.2). When $\gamma = 0$ (simple shear), the trajectories become similar to the streamlines around a sphere in simple shear, as discussed in Section 3.1.1. Trajectories for simple shear are shown schematically in Fig. 5.2. Open trajectories are characterized by $E > 0$, closed ones by $E < 0$, and the surface of limiting trajectories corresponds to $E = 0$. Two spheres approaching from infinity are (in the absence of Brownian motion, external forces or multibody interactions) unable to penetrate the region of closed trajectories (shaded area of

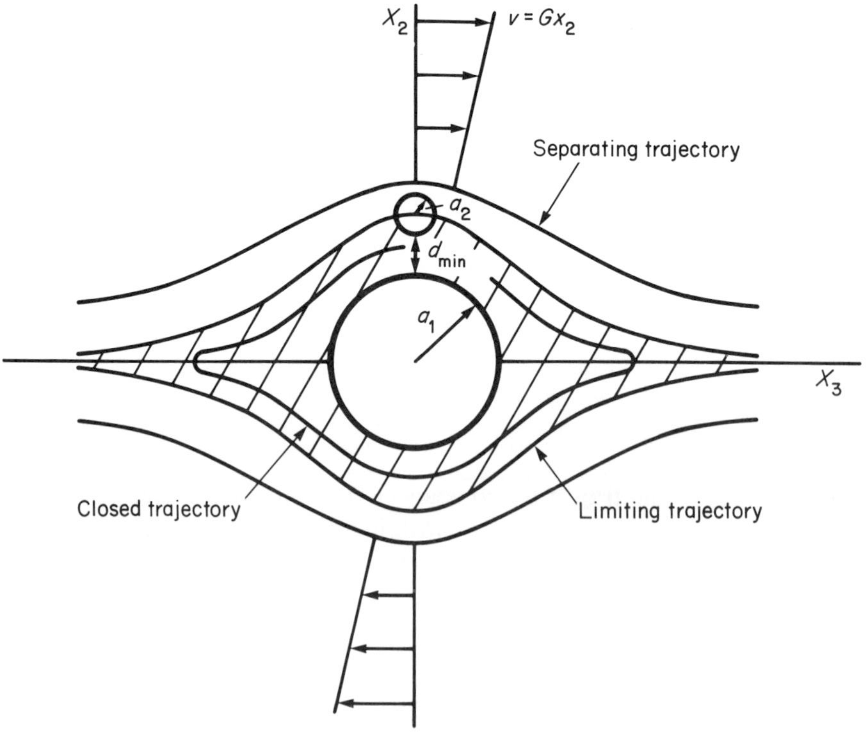

Fig. 5.2. Equatorial trajectories of two spheres in simple shear (schematic). The solid lines are the relative trajectories of a sphere of radius a_2 with respct to a reference sphere of radius a_1. Two kinds of trajectory exist: separating (or open) and closed ones, separated by a limiting trajectory. The shaded area is the region of closed trajectories.

Fig. 5.2) and, therefore, are unable to approach each other closer than the distance d_{min}, the minimum distance of approach of two spheres on the limiting trajectory $E=C=0$. The value of d_{min} is the main difference between the trajectories of equal-sized and unequal-sized spheres. Values of $d^*=d_{min}/a_1$ have been given by Adler (1981a) and are found to vary from $4\cdot2\times10^{-5}$ for $q=1$ (Arp and Mason, 1977a) to $0\cdot16$ for $q=0$ (Cox *et al.*, 1968). For equal-sized spheres of colloidal dimensions, d_{min} can be very small; e.g. when $a_1=1\,\mu$m, $d_{min}=0\cdot4$ Å. It is clear that some of the assumptions implicit in the theory start to break down at such small separations (continuum hypothesis, perfectly smooth spheres, etc.). For rough spheres the surface roughness prevents spheres from approaching one another closer than the maximum extension of the surface asperities (resulting in an apparent repulsion), but at large separations they behave as predicted by eq (5.1) (Arp and Mason, 1977b).

In the absence of external forces, the predicted hydrodynamic trajectories of two spheres in simple shear have been confirmed experimentally for both equal-sized (Arp and Mason, 1977b) and unequal-sized spheres (Adler, 1981b).

For a general linear two-dimensional flow the trajectories are more complicated. Relative trajectories for $C=0$ are summarized in Table 5.1.

5.1.2 Trajectories of two soft spheres in a linear shear flow

5.1.2.1 Two spheres subject to van der Waals forces

Van der Waals forces between two spheres usually act only along the line of sphere centres. For spheres made of the same material, the forces are attractive and increase the velocity with which the spheres approach and decrease the velocity at which the spheres separate. Equation (5.1a) can be regarded as a force balance along the line of sphere centres, in which two hydrodynamic forces are acting on the spheres: one due to the applied shear field

$$F_{hydr}=(1+\gamma)\,G\mu\bar{a}^2\,g(r,q)\,\sin^2\theta\sin2\phi \tag{5.6}$$

and one due to the hydrodynamic drag between the spheres

$$F_d=-\frac{3\pi\mu\bar{a}u}{C(r,q)} \tag{5.7}$$

where $g(r,q)$ is a function of the separation and the radius ratio, u is the relative particle velocity (dr/dt), and $C(r,q)$ is a correction factor for

TABLE 5.1

Summary of the trajectories of a pair of hard spheres in a linear shear flow

	$E < -\beta g(2)$[a]	$-\beta g(2) < E < E_0$	$E_0 < E < 0$[b]	$E > 0$
$\gamma = 0$	No trajectories are defined	(i) No open trajectories exist (ii) As for $0 < \gamma < 0{\cdot}2546$ except that closed trajectory region extends to infinity	($E_0 = 0$)	As for $0 < \gamma < 0{\cdot}2546$
$0 < \gamma < 0{\cdot}2546$	Open trajectory does not cut $\phi = \alpha$[c]; r_{min} occurs at $\phi = \pi/2$	(i) Open trajectory does not cut $\phi = \alpha$; r_{min} occurs at $\phi = \pi/2$ (ii) Closed trajectory cuts $\phi = \alpha$; r_{min} occurs at $\phi = 0$; r_{max} occurs at $\phi = \pi/2$; the closed trajectory region is infinite	Open trajectory cuts $\phi = \alpha$; r_{min} occurs at $\phi = 0$	Open trajectory does not cut $\phi = \alpha$; r_{min} occurs at $\phi = 0$
$0{\cdot}2546 < \gamma < 1$	Open trajectory does not cut $\phi = \alpha$; r_{min} occurs at $\phi = \pi/2$	Open trajectory cuts $\phi = \alpha$; r_{min} occurs at $\phi = 0$; E_0 does not exist	As for $-\beta g(2) < E < E_0$	As for $0 < \gamma < 0{\cdot}2546$
$\gamma = 1$	As for $0{\cdot}2546 < \gamma < 1$	($-\beta g(2) = 0$)		

[a] $\beta = (1-\gamma)/(1+\gamma)$.

[b] $E_0 = \left(\frac{\gamma}{1+\gamma}\right)\frac{1}{f^2(r_0)} - \beta g(r_0)$, with r_0 determined by $B(r_0) = \frac{2\gamma}{1+\gamma}$; E_0 does not exist when $0{\cdot}2546 < \gamma \leqslant 1$.

[c] α is defined by eq (3.182) (after Kao *et al.*, 1977).

Stokes law, taking into account the presence of the second sphere ($C(r,q)=d_{\|}(r,q)$, defined by eq (4.26b)). Combining eqs (5.6) and (5.7) yields eq (5.1a) with $A(r,q)=g(r,q)\,C(r,q)/3\pi$.

When van der Waals forces, F_{vdW}, are acting between the spheres, they must be included in the force balance:

$$F_{\mathrm{hydr}}+F_{\mathrm{d}}+F_{\mathrm{vdW}}=0 \tag{5.8}$$

which results in

$$\frac{\mathrm{d}r}{\mathrm{d}t}=G(1+\gamma)\,A(r,q)\,\sin^2\theta\sin 2\phi+\frac{C(r,q)\,F_{\mathrm{vdW}}}{3\pi\mu\bar{a}^2} \tag{5.9}$$

For spheres of the same material F_{vdW} is negative (attractive), while for spheres made of different material F_{vdW} can be positive or negative,

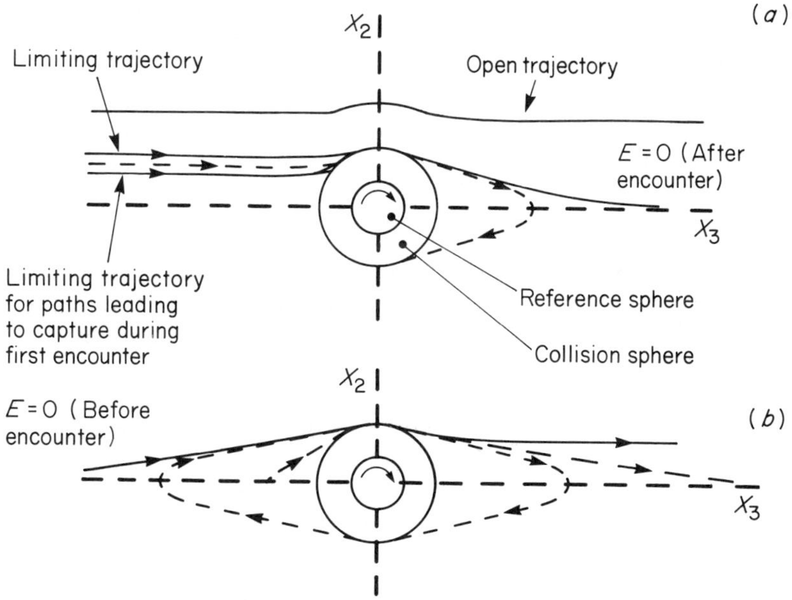

Fig. 5.3. Schematic trajectories of two spheres in simple shear subjected to a van der Waals force, F_{vdW}. The dashed trajectories are examples of trajectories that pass the reference sphere more than once before being captured *(a)* or separate to infinity *(b)*. When particles can be captured during the first encounter, the maximum number of encounters prior to capture is 2. If particles cannot be captured during the first encounter, the number of encounters can be increased to infinity.
(a) $F_{\mathrm{vdW}} < 0$.
(b) $F_{\mathrm{vdW}} > 0$.

depending on the Hamaker constants of the spheres and of the medium (cf. Section 1.4.2). The presence of the van der Waals term in eq (5.9) drastically changes the nature of the trajectories. When present, closed trajectories no longer exist and open trajectories become asymmetric about the X_2 axis. When $F_{vdW} < 0$ the possibility exists that two spheres come into contact during an encounter and form a permanent doublet. Schematic trajectories for spheres in simple shear are shown in Fig. 5.3 for both attractive and repulsive van der Waals forces. When F_{vdW} is attractive, a surface of limiting trajectories exists separating open trajectories from trajectories leading to capture. After the encounter, all trajectories on this limiting surface asymptotically approach a purely hydrodynamic trajectory with the trajectory constant $E = 0$. It is possible for a particle to orbit a reference particle several times before being captured and an example of this is shown schematically in Fig. 5.3a. An encounter in which two polystyrene spheres pass each other, return and are captured during the second encounter has been observed and filmed by the author using the travelling-microtube technique. On the whole, particles are usually captured during their first encounter.

In the case of repulsive van der Waals forces, the trajectories are in the reverse direction (Fig. 5.3b). Obviously no capture can occur and all encounters are transient. When F_{vdW} is weak the possibility exists that the particles orbit each other several times before separation takes place.

5.1.2.2 Two electrically charged spheres

In Section 1.4.3 we discussed the forces acting between two electrically charged spheres surrounded by electric double layers. At first sight it might appear that the trajectory equations in shear flow can be obtained from eq (5.8) by replacing F_{vdW} by the electrostatic force, $F_{e\ell}$, leading to the trajectory eqs (5.9) and (5.1b, c). This procedure was followed by van de Ven and Mason (1976a) and independently by Zeichner and Schowalter (1977). However, as shown by van de Ven (1988b), this procedure is incorrect. In Sections 2.2 and 3.1.2 we saw that the presence of a double layer changes the velocity field around a sphere and, for the same reason, we expect that the velocity field around two spheres will be different from that given by eq (5.1). For the case of a small neutral tracer particle encountering a charged sphere we have seen that the double layer introduces coupling forces and torques which, for symmetrical electrolytes and small surface potentials, are of a magnitude of $\bar{b}\Phi_s^2$ ($\bar{b}$ being the average scaled ion size and Φ_s the reduced ζ potential). It is reasonable to expect similar coupling forces and torques in the interaction between two charged spheres. The coupling forces (or electroviscous forces) will modify eq (5.9), while the electroviscous torques affect eq (5.1b, c). Including the coupling in the force and torque balance leads to

$$\frac{dr}{dt} = G(1+\gamma)\,A(r,q)\,\sin^2\sin 2\phi + \frac{C(r,q)\,F_{e\ell}}{3\,\pi\mu\bar{a}^2} + \bar{b}\Phi_s^2 f_1 \tag{5.10a}$$

$$\frac{d\theta}{dt} = \tfrac{1}{4}G(1+\gamma)\,B(r,q)\,\sin 2\theta \sin 2\phi + \bar{b}\Phi_s^2 f_2 \tag{5.10b}$$

$$\frac{d\phi}{dt} = \tfrac{1}{2}G[(1-\gamma)+(1+\gamma)\,B(r,q)\,\cos 2\phi] + \bar{b}\Phi_s^2 f_3 \tag{5.10c}$$

where the f_i are unknown functions of the relative orientation (r,θ,ϕ), of the flow field (G,γ), of the double-layer thickness $(\kappa\bar{a})$ and the radius ratio, q, and potential ratio, β. Analogously to the streamlines around spheres, from geometrical arguments we can expect that the orientation dependence of the coupling forces and torques is such that (including van der Waals forces)

$$\frac{dr}{dt} = G(1+\gamma)\,\tilde{A}(r,q)\,\sin^2\theta \sin 2\phi + \frac{\tilde{C}(r,q)\,(F_{e\ell}+F_{\mathrm{vdW}})}{3\,\pi\mu\bar{a}^2} \tag{5.11a}$$

$$\frac{d\theta}{dt} = \tfrac{1}{4}G(1+\gamma)\,\tilde{B}(r,q)\,\sin 2\theta \sin 2\phi \tag{5.11b}$$

$$\frac{d\phi}{dt} = \tfrac{1}{2}G[(1-\gamma)+(1+\gamma)\,\tilde{B}(r,q)\,\cos 2\phi] \tag{5.11c}$$

where $\tilde{A}=A+\bar{b}\Phi_s^2 f_4$, $\tilde{B}=B+\bar{b}\Phi_s^2 f_5$ and $\tilde{C}=C+b\Phi_s^2 f_6$, with f_4 to f_6 being unknown functions of $\kappa\bar{a}$ and r (and q and β).

For low potentials, the force $F_{e\ell}$ is given approximately by eq (1.131) or eq (1.132), from which it can be seen that $F_{e\ell} \propto \Phi_s^2$, i.e. the electrostatic force is of the same order as the coupling terms $\bar{b}\Phi_s^2 f_i$. Therefore, without a knowledge of the electroviscous forces it is difficult to make predictions about the trajectories of two charged spheres. From eqs (5.11) and (1.131), it can be seen that the trajectories depend on a large number of dimensionless parameters: b_1, b_2, $\tau=\kappa\bar{a}$, q, Φ_s, C_R and β. Here C_R is the ratio of electrostatic to hydrodynamic forces

$$C_R = \frac{2\epsilon\psi_0^2}{3\mu G\bar{a}^2} \tag{5.12}$$

When both electrostatic and van der Waals forces are present we have the additional dimensionless parameters (cf. eq (1.76)):

$$C_A = \frac{A}{36\pi\mu G\bar{a}^3} \tag{5.13}$$

and

$$\lambda^* = \lambda/2\pi\bar{a} \tag{5.14}$$

where A is the Hamaker constant and λ the retardation wavelength. Thus, in general (for simple electrolyte solutions), there are nine dimensionless parameters to be dealt with. For identical spheres, $q=\beta=1$ and only seven parameters remain. Assuming that the electroviscous terms do not qualitatively change the trajectory equations (analogous to the streamlines around a charged sphere), we can expect several types of relative trajectory, as shown schematically in Fig. 5.4 for the case of simple shear ($\gamma=0$). Assuming F_{coll} is of the form shown in the inset of Fig. 5.4, it can be shown (van de Ven and Mason, 1976a) that two limit cycles can exist (depending on V_{int}) around the collision sphere, one stable (corresponding to an orbit around the secondary minimum distance h_{eq}) and one unstable (corresponding to an orbit around the energy barrier). Any sphere whose centre is close to the stable-limit circle will approach that orbit, resulting in a system of two orbiting spheres, referred to as a secondary doublet. A sphere whose centre is close to the unstable limit cycle moves away from it, either towards the stable-limit cycle (if outside the unstable-limit cycle) or towards the reference sphere (if inside). Spheres approaching a reference sphere can be captured during their

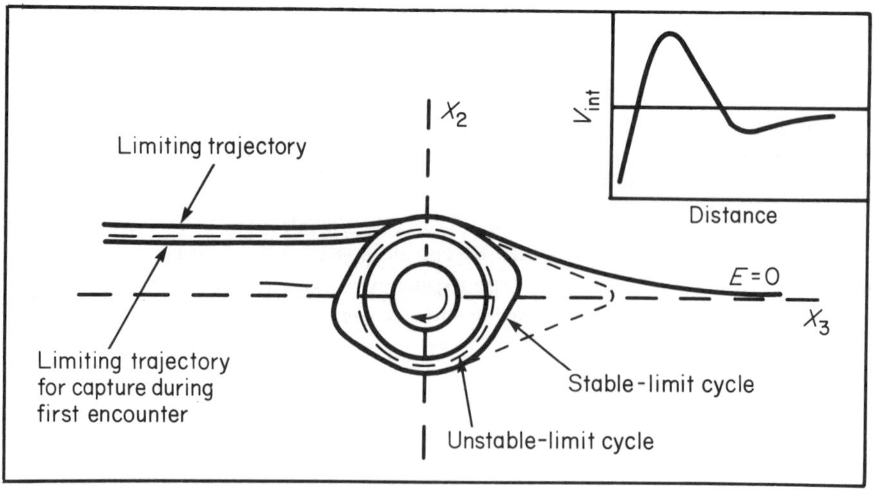

Fig. 5.4. Schematic representation of the possible relative trajectories of two spheres with an interaction energy as shown in the inset. The dashed trajectory is an example of a trajectory leading to capture in the secondary minimum during the second encounter.

first encounter, during subsequent encounters, or separate from each other. Again, a surface of limiting trajectories exists separating open trajectories from trajectories leading to capture (in the secondary minimum).

At large separations the forces F_{el}, F_{vdW} and the electroviscous forces and torques approach zero and thus the trajectories of two approaching particles can be described by eq (5.1) or (5.4). It follows that a collision between two colloidal particles can be characterized by the trajectory constants C_a and E_a of approach and by the constants C_r and E_r of recession. In the absence of colloidal forces $C_a = C_r$ and $E_a = E_r$ and the trajectories are symmetric. In the case of a net attraction between the spheres during an encounter $E_a > E_r$ and $C_a > C_r$ (except when $C = 0$), and for a net repulsion $E_a < E_r$ and $C_a \leqslant C_r$. In principle, from a measurement of the trajectory constants of approach and of recession, it is possible to determine the magnitude of the colloidal forces operating during an encounter (Takamura *et al.*, 1981a). Examples of encounters between latex spheres observed using the travelling-microtube technique are shown in Fig. 5.5.

Results like these were compared with the trajectory eqs (5.11) for $\gamma = 0$ (simple shear), neglecting the electroviscous terms $\bar{b}\Phi_s^2 f_i$. Such comparisons are made in Fig. 5.6 which shows the projected dimensionless distance between the sphere centres as a function of Gt, the zero time taken as the moment of closest approach. The different curves in Fig. 5.6a are calculated trajectories for various Hamaker constants, A, while the curves in Fig. 5.6b were obtained by calculating the trajectories for various surface potentials, ψ_0. The values which best fit the data are in agreement with the accepted Hamaker constants and the measured ζ potentials of the particles. The agreement between experiment and the theory (neglecting the coupling terms) indicates that, under the experimental conditions in Fig. 5.6, the electroviscous forces and torques were negligible. In these experiments the radius of the spheres was several micrometers and the salt concentration about 10 mM and, therefore, the experiments were performed under the condition $\kappa a \gg 1$. It thus appears that when $\kappa a \gg 1$ the coupling terms are negligible. In the other limit, $\kappa a \ll 1$, we can also expect the coupling terms to be negligible, since in the limit $\kappa a \to 0$ the concentration of ions in solution is nearly zero and the hydrodynamic flow field is unaffected by the presence of electrolyte. The coupling terms in eq (5.11) are probably only of importance when the double-layer thickness is comparable to the radius of the particle, and at distances κh of the order of unity.

5.1.2.3. Two spheres coated with polymer

Very few theoretical and experimental results are available for interactions between two spheres coated with a polymer subjected to a shear flow. A

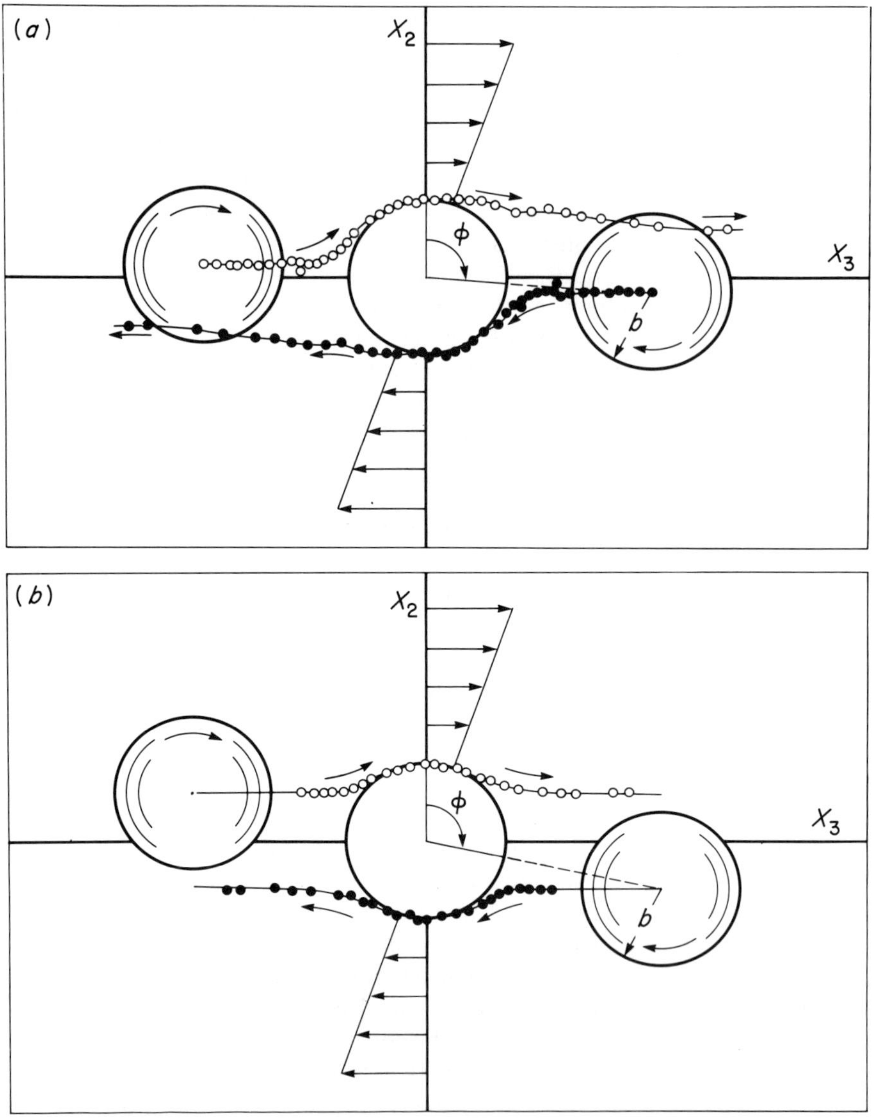

Fig. 5.5. Experimentally observed trajectories of equal-sized latex spheres in shear (Poiseuille flow) in 50% aqueous glycerol, observed using the travelling-microtube technique. (○), (●), Experimental observations of the (projected) positions of the spheres.
(a) Repulsion: 4·0-μm diameter spheres in l-mM KCl.
(b) Weak attraction: 2·6-μm diameter spheres in 10-mM KCl (after Takamura *et al.*, 1981a).

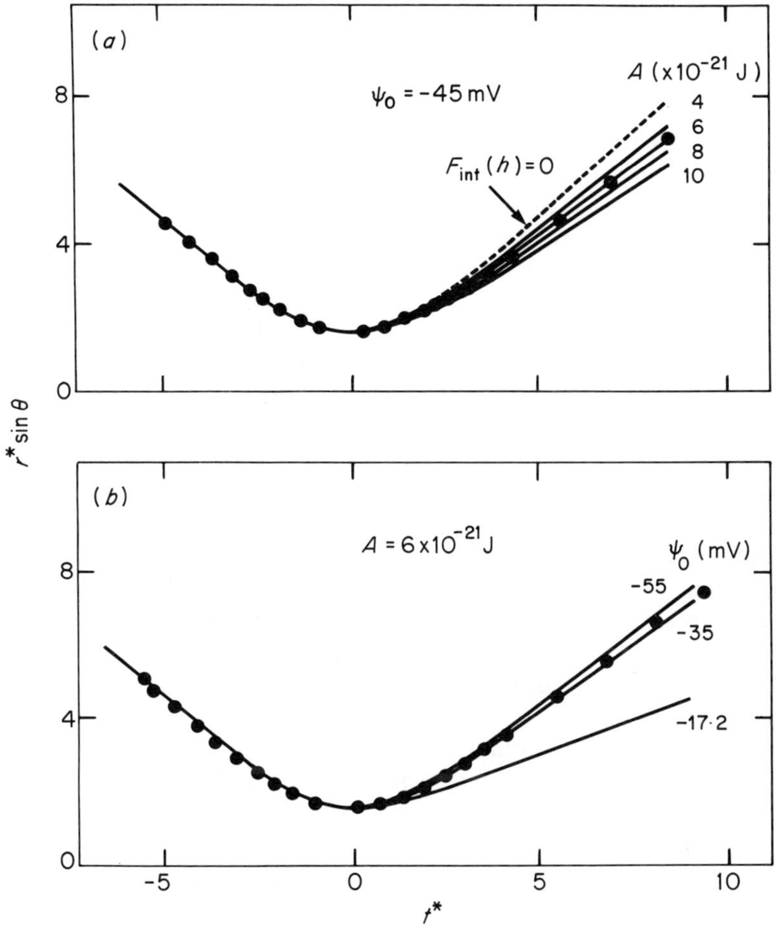

Fig. 5.6. Comparison of an experimentally observed trajectory (shown in Fig. 5.5b) with theory for $\lambda = 100$ nm. The dimensionless projected length $r^*\sin\theta (r^* = r/a)$ is plotted against $t^* = Gt$.
(a) Effect of varying the Hamaker constant A.
(b) Effect of varying the surface potential ψ_0.
The best fit ($\psi_0 = -45$ mV) corresponds to the measured ζ potential of the spheres in 50% glycerol. The value $\psi_0 = -17 \cdot 2$ mV refers to the potential in pure water (after Takamura *et al.*, 1981a).

natural approach would be to incorporate a force due to the presence of polymers in the trajectory equations discussed above. Such an approach neglects the effects of the flow field around the spheres on polymer configuration.

When the adsorbed polymer is a polyelectrolyte, several conclusions about two-sphere interactions can be drawn from the experiments performed by Takamura *et al.* (1981b), who studied collisions between negatively charged latex particles coated with cationic polyelectrolyte. When the amount of adsorbed polyelectrolyte is small (such that only a small fraction of the surface is covered), the main effect is a change in the effective ζ potential of the particles. The polymer is firmly attached to the surface and no evidence exists that loops or tails protrude from the surface because, in this case, the collisions between spheres are adequately described by eqs (5.11) (neglecting electroviscous terms when $\kappa a \gg 1$). When the amount of adsorbed polyelectrolyte is large (so that the surface of the spheres is completely covered), steric repulsion of the spheres can be observed at short distances. The collisions between such spheres are again described adequately by eq (5.11), provided that in the calculation the surface of slip (i.e. the surface at which the potential equals the measured ζ potential) is allowed to be about 20 nm away from the surface of the solid sphere. This distance corresponds closely to the thickness of the adsorbed polymer layer. For intermediate coverage by the cationic polyelectrolyte, collisions between two spheres are found to be irreproducible. It is believed that partial coverage causes an inhomogeneity in the surface-charge distribution, resembling a mosaic of negative and positive charges. The trajectories of the two spheres then depend strongly on the relative orientation of the spheres when close together. In the case of partial coverage, the formation of doublets was observed on several occasions.

Little information is available for spheres coated with neutral polymers subjected to shear. Recent experiments (De Witt and van de Ven, 1989) on collisions between poly(styrene) latex spheres coated with poly(vinyl) alcohol show a repulsion at large separations (a few hundred nanometers), which seem to indicate that the presence of long dangling ends may have a pronounced effect on the interaction.

5.1.3. Rotation of doublets in simple shear

5.1.3.1. Doublets in a simple electrolyte

We have seen that, according to the DLVO theory, two types of doublet exist: two particles captured either in a primary or a secondary minimum, referred to as primary and secondary doublets, respectively. Whether a primary or secondary doublet forms depends on the magnitude of the hydrodynamic and colloidal forces. Examples of trajectories leading to capture are shown in Fig. 5.7 which also shows the transition from secondary to primary

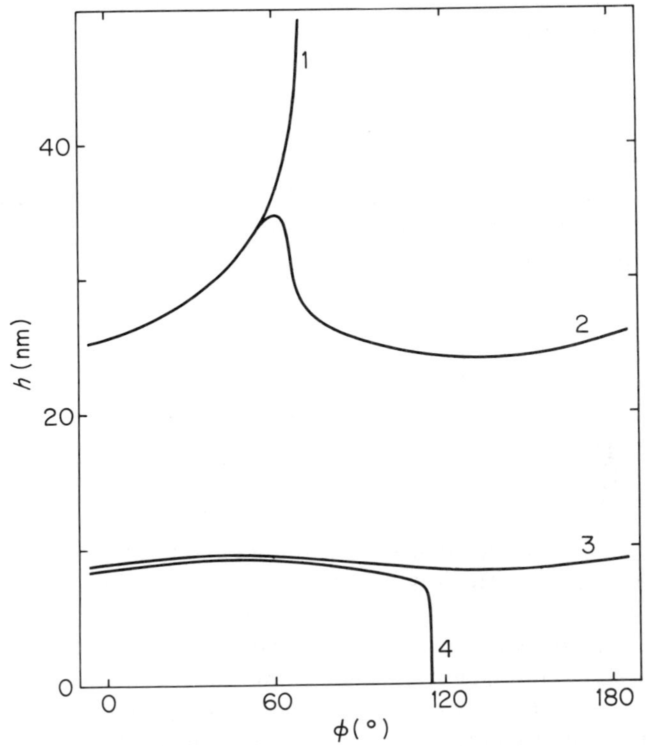

Fig. 5.7. Trajectory calculations showing the existence of primary and secondary doublets: (1) separating spheres ($\psi_0 = -41{\cdot}7$ mV); (2) and (3) secondary doublets ($\psi_0 = -41{\cdot}6$ and $-10{\cdot}4$ mV, respectively); (4) primary doublet ($\psi_0 = -10{\cdot}3$ mV). Calculations were done for the following conditions: $a_1 = a_2 = 1{\cdot}3$ μm; $\theta = 90°$; $G = 10$ s^{-1}; $\mu = 1$ mPa s; [KCl] = 10 mM; $\epsilon_r = 78$, $\lambda = 100$ nm; and $A = 10^{-20}$ J (after Takamura *et al.*, 1979).

doublets. Under the experimental conditons pertaining to Fig. 5.7, secondary doublets can exist consisting of spheres having surface potentials ranging from (±) 10·4 to 41·4 mV. As is shown in Fig. 5.7, the distance between the spheres in a secondary doublet is far too small to be observed directly using a microscope. However, it is possible to determine the distance between the spheres indirectly. After a doublet is formed it will rotate in simple shear indefinitely about the vorticity axis X_1 (see Fig. 5.1) with a well-defined period of rotation and it can be shown that this period depends strongly on the distance between the spheres. By numerically integrating the trajectory equations it can be shown that for equal- or near-equal-sized spheres under most conditions (van de Ven and Mason, 1976b)

$$T = \frac{2\pi}{G}(\bar{r}_e + \bar{r}_e^{-1}) \tag{5.15}$$

with an error of less than 1%. Here $\bar{r}_e$ is the equivalent axis ratio of two torque-free spheres at a distance h_{eq} corresponding to the secondary minimum distance. The value of $\bar{r}_e$ can be calculated from $B(r,q)$ for equal spheres given by eq (5.2b), from which it follows that h^*_{eq} $(=h_{eq}/a)$ is

$$h^*_{eq} = \exp\left[-\frac{1\cdot462 + (1\cdot255\,\bar{r}_e - 0\cdot351)^{1/2}}{\bar{r}_e - 1\cdot982}\right] \tag{5.16}$$

Using eqs (5.15) and (5.16), h_{eq} can be obtained from measurements of T and G.

For spheres of widely different sizes, eq (5.15) does not always hold. When the radius ratio, q, is approximately 1, the trajectory of the orbiting sphere falls partly outside the region of the closed trajectories of purely hydrodynamic interactions, and the orbit is nearly elliptical with maximum deviations of h_{eq} typically of the order of 10 nm. For small or intermediate values of q, h_{eq} may well be completely within the region of closed orbits.

As an example let us take the case of two poly(styrene) spheres of radii 5 μm and 1 μm with ζ potentials of -20 mV, subjected to a shear flow of gradient $G = 10\,s^{-1}$ in water ($\mu = 1$ m Pa s, $\epsilon_r = 78$, $A \simeq 0\cdot3 \times 10^{-20}$ J, and $\lambda = 200$ nm). From the DLVO theory it can be calculated that in a 1-1 electrolyte with a concentration of 10 mM the secondary minimum distance is about 28 nm, well within the region of closed trajectories because $d_{min} \simeq 120$ nm (cf. Fig. 5.2). By integrating eqs (5.11) numerically (neglecting the terms $\bar{b}\Phi_s{}^2 f_i$), it is found that during a rotation of the doublet the distance between the spheres remains almost constant, deviating from the equilibrium distance by not more than 0·1 nm. This is not surprising because in the absence of interparticle forces for closed trajectories with a gap width of about 25 nm, the gap width remains almost constant, deviating by not more than 1 nm from the average distance. If follows that, under these conditions, the period of rotation is also given by eq (5.15).

When we look at the same system at the lower electrolyte concentration of 1 mM, we find that the secondary minimum distance is increased to 98 nm. In this case the trajectories are very different, e.g. when $\theta = 90°$ the distance h varies from 86 to 1430 nm. Such a trajectory closely resembles a trajectory in the absence of interparticle forces. The role of F_{int} is mainly to select one

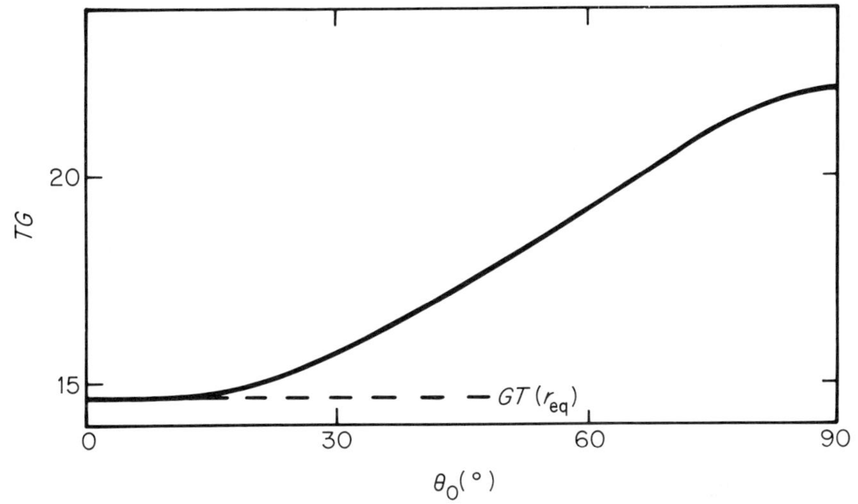

Fig. 5.8. Period of rotation, T, of a secondary doublet in a simple shear flow as a function of the initial θ orientation. Calculations were done for $a_1=5$ μm, $a_2=1$ μm, $\zeta_1=\zeta_2=-20$ mV, $\mu=1$ mPa s, $\epsilon_r=78$, $A=0{\cdot}3\times10^{-20}$ J, $\lambda=200$ nm, [KCl] = 1 mM, and $G=10$ s^{-1}. For low θ_0 angles, the spheres are almost aligned along the vorticity axis and the distance between the spheres varies little during an orbit. The period corresponding to the equilibrium separation distance is denoted by $T(r_{eq})$.

particular trajectory out of the infinite number of closed orbits and to perturb it slightly. The period of rotation of the doublet now depends strongly on its orientation with respect to the flow, as can be seen in Fig. 5.8, and is no longer given by eq (5.15). When $\theta=0$ (spheres aligned along the vorticity axis), TG is given by eq (5.15). For larger values of TG, variations in the distance between the sphere become important and the period increases.

It follows from this discussion that the submicroscopic distance between two spheres which make up a secondary doublet can be obtained by measuring the period of rotation of the doublet in simple shear.

Variations in the period of rotation of secondary doublets consisting of equal-sized spheres are shown in Fig. 5.9 as a function of the distance between the spheres. For comparison, the period of rigid dumbbells is also shown. It can be seen that the period of secondary doublets is considerably longer than that of rigid dumbbells.

Early experiments (van de Ven and Mason, 1976b) showed the existence of primary and secondary doublets, but agreement with eqs (5.15) and (5.16) was poor. The reason for this discrepancy is that primary doublets do not necessarily consist of two spheres in contact. Experiments by Takamura *et*

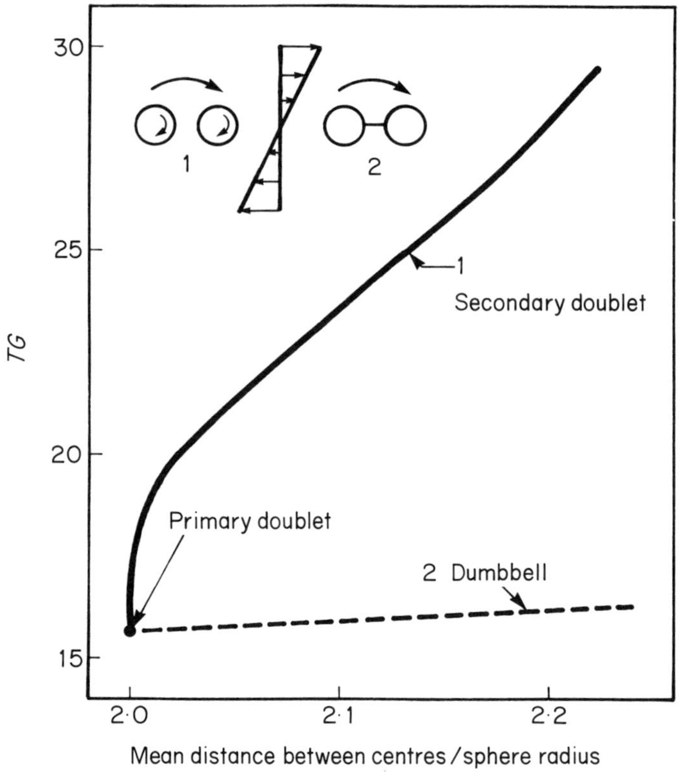

Fig. 5.9. The calculated period of rotation, T, of secondary doublets as a function of their equilibrium separation distance (———). Also included are the corresponding values for rigid dumbbells (----). Primary doublets correspond to the solid dot at $TG = 15 \cdot 62$ (after van de Ven and Mason, 1976b).

al. (1979) showed that in newly formed primary doublets the distance between the spheres is not zero and, as a result, the spheres are able to rotate individually. Strong evidence exists for the existence of both secondary and non-touching primary doublets, as is shown in Fig. 5.10. For the conditions of Fig. 5.10a the period of rotation of secondary doublets is calculated as $TG = 19 \cdot 3$, in excellent agreement with the data in the figure. At higher electrolyte concentrations, the secondary minima start to disappear slowly, as shown in Fig. 5.10b for which the theoretical value of TG of secondary doublets is $18 \cdot 4$. The doublets observed in Fig. 5.10 were suspended in an aqueous electrolyte solution and, therefore, were subject to the action of rotary Brownian motion. To obtain the average period of rotation about 20 doublets were observed, each executing 70 to 150 rotations.

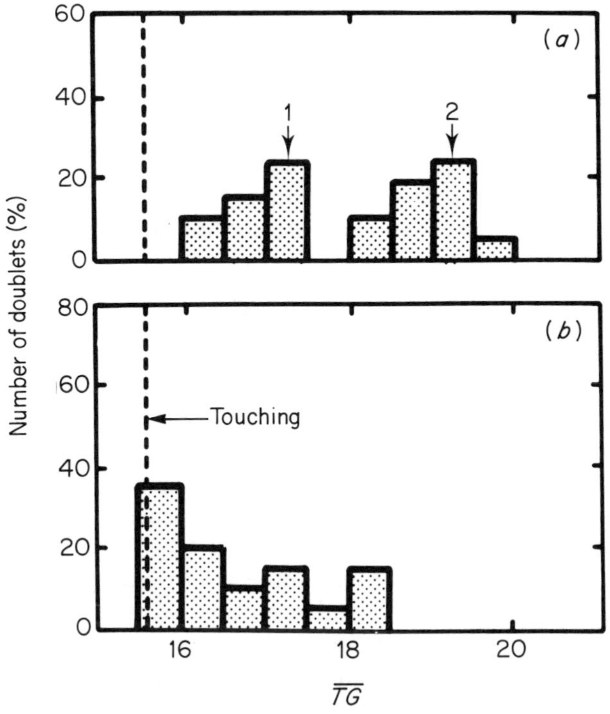

Fig. 5.10. Experimentally observed histograms of the number distribution in $\overline{TG}$ for doublets of poly(styrene) spheres in shear. The dashed line corresponds to the value of touching doublets ($TG = 15\cdot62$) (after Takamura *et al.*, 1979).
(a) In 10-mM KCl; the arrows indicate the peaks corresponding to primary and secondary doublets.
(b) In 30-mM KCl; the second peak has almost disappeared.

From the magnitude of the *TG* value for primary doublets (Fig. 5.10a) it can be concluded that the individual spheres are neither rigidly coupled nor freely rotating, but instead experience some intermediate and hindered rotation. In order to estimate the distance between spheres in a primary doublet, more details of this hindered rotation are needed than are yet available. The period of rotation of a primary doublet in which the spheres are not free to rotate is intermediate between those of rigid dumbbells and secondary doublets (cf. Fig. 5.9). Some models of doublets exhibiting hindered rotation can be found in the papers by Takamura *et al.* (1981c) and Adler *et al.* (1981).

Another point of interest is that evidence exists for the ageing of primary doublets. With time the period of rotation decreases, indicating an increased difficulty of the spheres to rotate individually. For aged suspensions (i.e. stored for several months or longer), observations show (van de Ven and Mason, 1976b; Takamura *et al.*, 1979) that the majority of doublets behave very similarly to two spheres in contact with $TG = 15 \cdot 62$, in contrast to newly formed doublets for which TG is larger. A possible explanation of this ageing process is the existence of an inhomogeneity in the surface properties of the spheres. Shear or rotary Brownian motion causes the spheres to rotate relative to each other. If the surfaces are not homogeneous, then during this rotation the parts of the spheres facing each other can have a larger attraction than for homogenous spheres, resulting in capture in a deeper potential minimum, which will lead to an additional hindrance to rotation. Experiments on the redispersion of coagulated melamine formaldehyde latex spheres by changing the electrolyte concentration or by sonic vibrations show (van de Ven, 1982) that aggregates formed during shear are more difficult to disperse than those formed in the absence of shear. The explanation could be that in shear flow more relative rotary motion of the captured spheres occurs than in the absence of shear, a process which is believed to accelerate ageing.

Doublets of colloidal-size particles subjected to shear are also subjected to translational and rotational Brownian motion. Most doublets behave as a single particle with two translational-diffusion constants $D_{\parallel}$ and $D_{\perp}$ (for motion along and perpendicular to the axis of revolution), and two rotational-diffusion constants D_s and D_r (for motion about the axis of revolution and about an axis perpendicular to it). For most secondary doublets the variation in the distance between the orbiting spheres is small and the diffusion constants are those discussed in Section 4.5 with the constants corresponding to the secondary-minimum distance, h_{eq}: e.g.

$$D_r \simeq D_r(h_{eq}) \tag{5.17}$$

with similar expressions for the other constants.

The only exception occurs when the distance between the spheres varies appreciably during an orbit. This will be the case for spheres of different sizes ($q < 0 \cdot 5$) at low electrolyte concentrations. In such cases an average diffusion constant can be assigned to a doublet, e.g. in simple shear when $\mathrm{Pe} \gg 1$,

$$\overline{D_r} = \frac{4}{TG} \int_0^{\pi} \frac{D_r(r)}{1 + B(r)\cos 2\phi} \, d\phi \tag{5.18}$$

which shows that $\overline{D_r}$ depends strongly on the initial orientation θ_0. The rotary diffusion of a pair of spheres with $q = 0 \cdot 2$ for conditions identical to

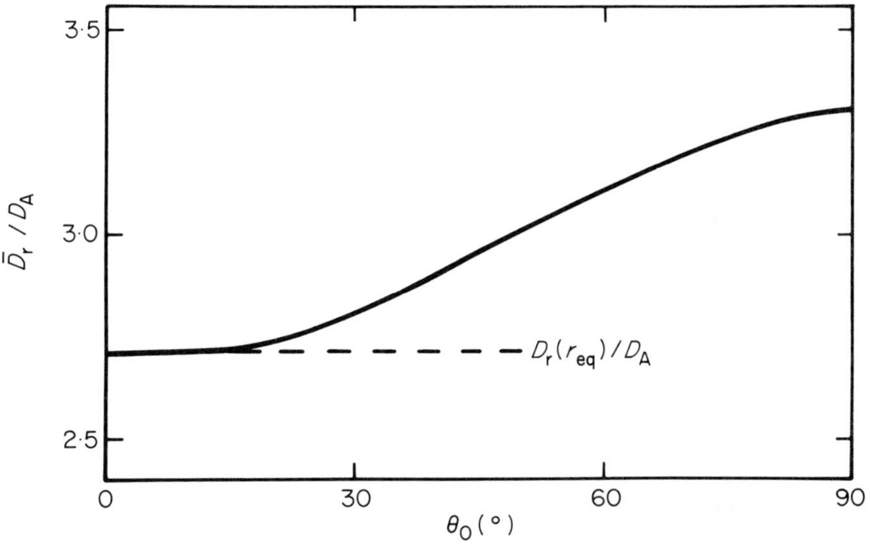

Fig. 5.11. Variation in the rotary-diffusion constant of a secondary doublet (normalized using the rotary-diffusion constant of the larger sphere) as a function of initial orientation. Conditions are identical to those described in Fig. 5.8.

those in Fig. 5.8 is shown in Fig. 5.11. It can be seen that the variations in $\overline{D_r}$ with θ_0 are very similar to those in *TG*.

5.1.3.2. Polymer-bridged doublets

When colloidal spheres are coated with a polymer, the formation of doublets by the mechanism of polymer bridging can occur. Data are available for latex spheres coated with a cationic polyelectrolyte (Catfloc) (Takamura *et al.*, 1979). As discussed in Section 5.1.2.3, at low coating densities only a few polymer segments protrude from the surface. In such a case primary and secondary doublets are observed, similar to doublets in a simple electrolyte (Fig. 5.12a). Increasing the coating density results in the formation of polymer-bridged doublets (Fig. 5.12b), most of which have a period of rotation close to that of two spheres in contact. However, some doublets were observed with dimensionless periods of up to 17·5, and it is of interest to note that during experiments such as those whose results are summarized in Fig. 5.12, doublets were occasionally observed in which the spheres were visibly separated by distances of up to a few micrometers, similar to doublets observed earlier (van de Ven, 1975). These doublets are believed to consist of two spheres connected by a long macromolecule which

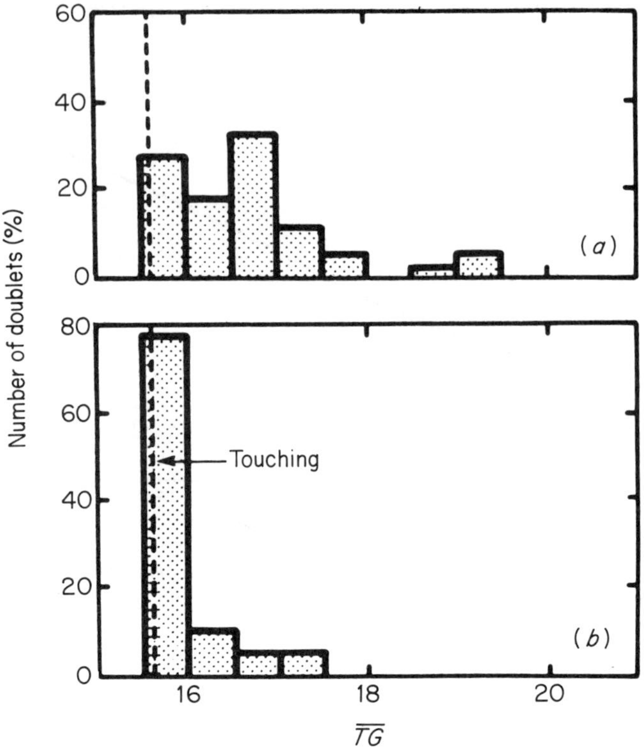

Fig. 5.12. Histogram of the distribution in the period of rotation of doublets in shear, in the presence of Catfloc, a cationic polyelectrolyte. In both cases the solution contains 10-mM KCl (after Takamura *et al.*, 1979).
(a) In 10-mM KCl, with 0·03 mg l^{-1} Catfloc.
(b) In 10-mM KCl, with 1·0 mg l^{-1} Catfloc.

is periodically stretched and compressed during the rotation of the doublet. By slightly extending the trajectory equations (eq (5.1)) to include the effects of external torques and taking the dimensions and the rigidity of the connecting filaments into account, Adler *et al.* (1981) succeeded in calculating the period of rotation of such flexible dumbbells. For rotations in the equatorial plane it was shown that the period of rotation is well defined and independent of the initial positions of the spheres and the filament. The results of such calculations, together with experimental data on model doublets of 3-mm diameter spheres connected by a flexible filament in Couette flow (Takamura *et al.*, 1981c) are shown in Fig. 5.13. The discrepancies between theory and experiment are most likely due to the fact that the filament and its connection to the spheres are not ideal; nevertheless, the

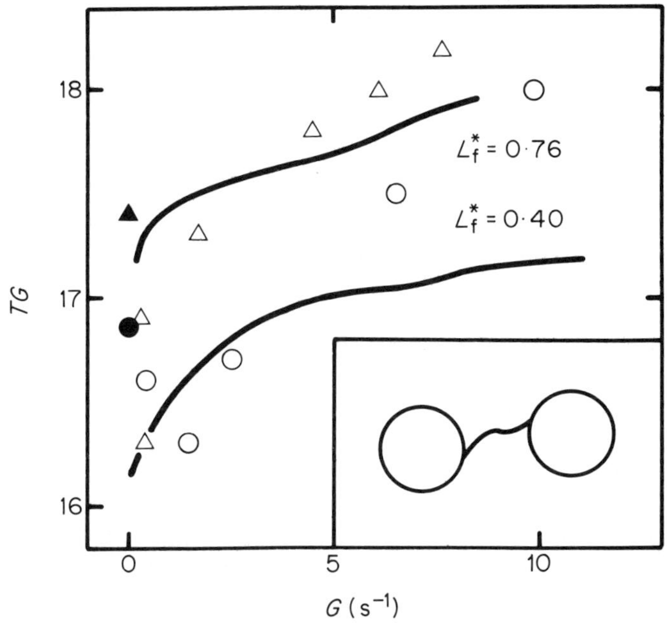

Fig. 5.13. Dimensionless period of rotation of flexible dumbbells as a function of the rate of shear. The flexible dumbbells consist of two 3-mm diameter spheres connected by a flexible filament as shown in the inset. Open symbols are experimental observations on model flexible dumbbells in Couette flow for two different filament lengths L_f ($L_f^*=L_f/a$); solid lines are theoretical predictions. The values for rigid dumbbells (i.e. two spheres rigidly connected by a solid connection of length L_f) are denoted by the solid symbols. At large G the experimental values are somewhat higher than predicted by theory (after Takamura *et al.*, 1981c).

agreement is reasonable, the largest discrepancy being only 5%. The two cases shown in the Fig. 5.13 correspond to two different filament lengths, L_f. It can be seen that for $L_f^*=L_f/a$ ranging from 0·4 to 0·8 and shear rates in the range 5–10 s^{-1}, the period of rotation, TG, is in the range 17–18. In view of this observation, it is possible that the distribution shown in Fig. 5.12b is a reflection of the fact that the doublets are connected by polymers of different (maximum) lenghts, L_f. This interpretation suggests that L_f varied from 0 to about 1·7 μm. As stated above, doublets with such a large separation distance are actually observed. The range of values of L_f may also be due to multiple strands and bonding sites (instead of one strand and two sites as in the inset of Fig. 5.12), preventing the complete stretching of the polymer during rotation. Such interactions of polymers with colloidal particles warrant further investigation.

5.1.4 Stability of colloidal dispersions subjected to shear

Because doublet formation is the first step in the coagulation process, it is important to know the conditions under which doublets are formed and at what rate. For large colloidal particles subjected to shear (in the absence of Brownian motion), integration of the trajectory equations tells us whether or not a doublet is formed during an encounter. Neglecting the coupling terms in the trajectory equations (probably allowed when $\kappa a \ll 1$ or $\kappa a \gg 1$) results in trajectory equations which, for electrically charged spheres, depend on six dimensionless constants C_A, C_R, λ^*, τ, q and β.

From numerical solutions of the trajectory equations the following general conclusions can be drawn, which are summarized in Fig. 5.14. We can distinguish three different regimes: (i) coagulation in a secondary minimum; (ii) coagulation in a primary minimum, and (iii) absence of coagulation. These three regimes are indicated schematically in Fig. 5.14. On the vertical axis is plotted some measure of the electrostatic repulsion between the particles. The repulsion can be increased by increasing the surface potential or by decreasing the ionic strength of the solution. On the horizontal axis the flow strength is plotted which can be changed by varying the rate of shear. It can be shown that the boundaries between the several

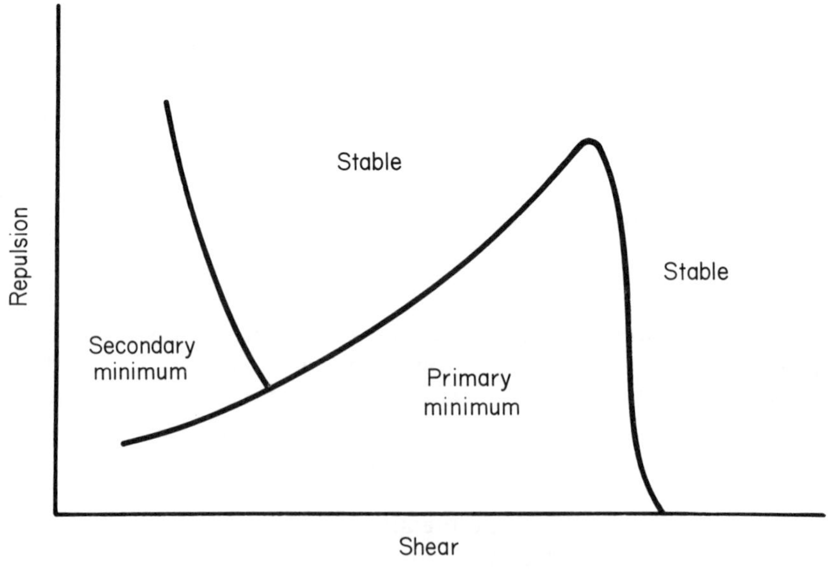

Fig. 5.14. Stability diagram for two spheres in shear (schematic). When there is little repulsion between the spheres, two spheres form doublets at all rates of shear (except very high ones). When there is appreciable repulsion, coagulation occurs at low and high shear rates, with a region of stability at intermediate shear rates. At very high shear rates, primary doublets can be broken apart and suspensions again become stable.

domains are not sharp. In the stable regime, the possibility of capture still exists, but the efficiency of capture may be one or more orders of magnitude lower than in the unstable regions. At the boundaries there are rapid decreases in capture efficiency to very low values.

Flowing dispersions are usually unstable at low and high shear rates, but stable in between (van de Ven and Mason, 1976a). At very high shear rates the possibility exists that even primary doublets can be broken apart. This will happen at some critical shear rate determined by the depth of the primary minimum. The deeper the minimum, the higher the shear rate at which a primary doublet breaks up. At relatively low shear rates and large electrostatic repulsions, spheres will be captured in a secondary minimum. By increasing the shear, a critical rate is reached at which most of the doublets break up (depending on their orientation). Increasing the shear rate further leads eventually to the formation of primary doublets, except for very shallow primary minima. When the electrostatic repulsion is weak (low

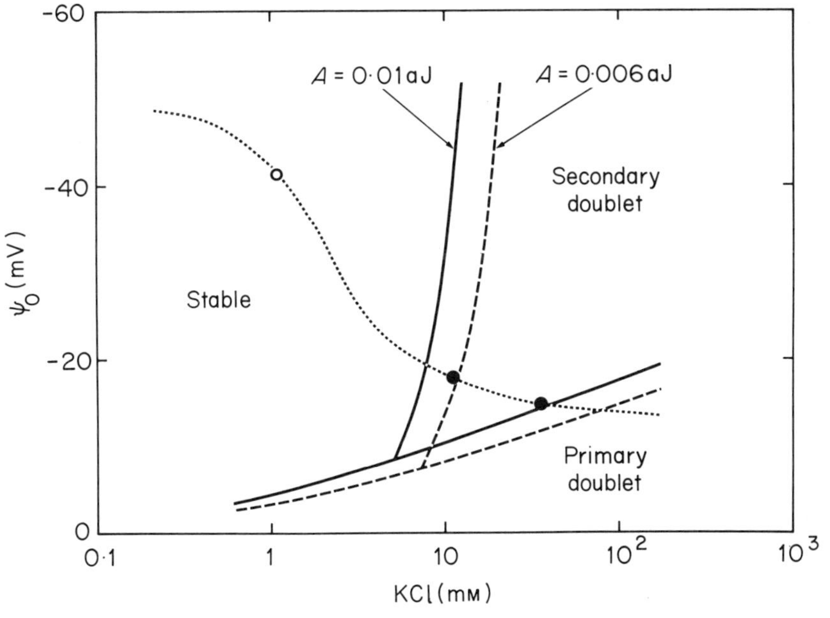

Fig. 5.15. Stability diagram of poly(styrene) latex spheres in simple shear, showing the stable regime and the regimes where primary and secondary doublets can be formed. Also shown is the observed dependence of the ζ potential on salt concentration (dotted line), which indicates that secondary doublets can only exist in a limited salt concentration region. ● Conditions under which secondary doublets were observed; ○, primary doublets, probably formed by ageing. The occurrence of secondary doublets is consistent with a Hamaker constant in the range 0·006–0·010 aJ (after Takamura *et al.*, 1979).

ζ potentials, high ionic strength), secondary doublets change into primary doublets at a critical shear rate. For small particles, secondary minima are often not deep enough for secondary doublets to stay together (because of Brownian motion) and for such particles the secondary minimum regime in Fig. 5.14 is absent. In such a case a dispersion can be stable at low shear rates and become unstable at a critical shear rate beyond which shear-induced coagulation occurs. By observing the transitions between the various regimes in Fig. 5.14, the magnitudes of the colloidal forces acting between the particles can be obtained. An example of these transitions is shown in Fig. 5.15.

5.1.5 Orthokinetic coagulation

5.1.5.1 Simple electrolyte solutions

Here we are interested in the rate at which doublets form in sheared dispersions. In the absence of Brownian motion the frequency of doublet formation can be calculated from the governing trajectory equations (eq 5.11)). The number of spheres captured by a reference sphere can be calculated from the capture cross-section, shown in Fig. 5.16 for simple

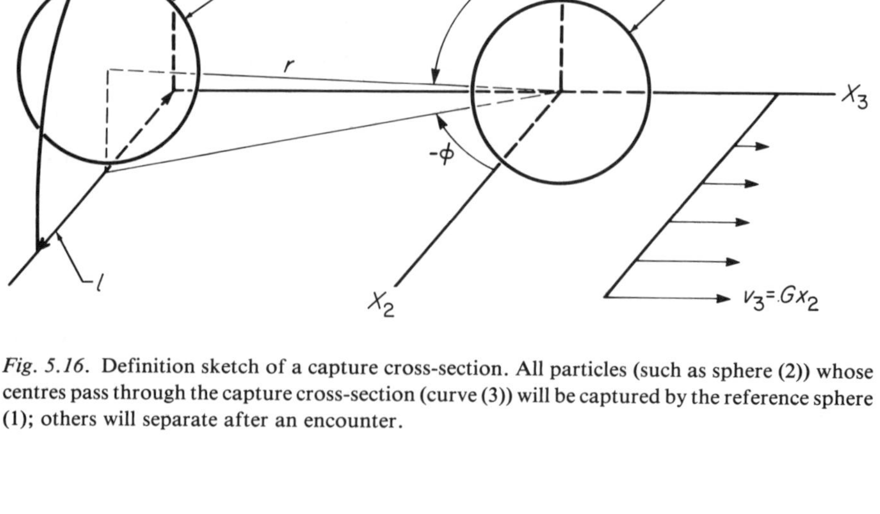

Fig. 5.16. Definition sketch of a capture cross-section. All particles (such as sphere (2)) whose centres pass through the capture cross-section (curve (3)) will be captured by the reference sphere (1); others will separate after an encounter.

shear. The number of particles whose centres pass through the capture cross-section per unit time is

$$J = 4 \int_0^{\ell} un\, z(x_2)\, dx_2 \tag{5.19}$$

where ℓ is the extension of the cross-section in the X_2 direction, $z(x_2)$ is the height of the cross-section at position x_2, u is the particle velocity given by $u = Gx_2$, and n is the number concentration of particles. In general, the number of particles of radius a_2 captured by a reference sphere of radius a_1 is

$$J = 4n_2 G\, a_1^3 \int_0^{\ell} x_2\, z(x_2)\, dx_2 \tag{5.20}$$

where x_2, z and ℓ are made dimensionless by dividing by a_1, and n_2 is the number of particles of radius a_2. The boundary $z(x_2)$ of the capture cross-

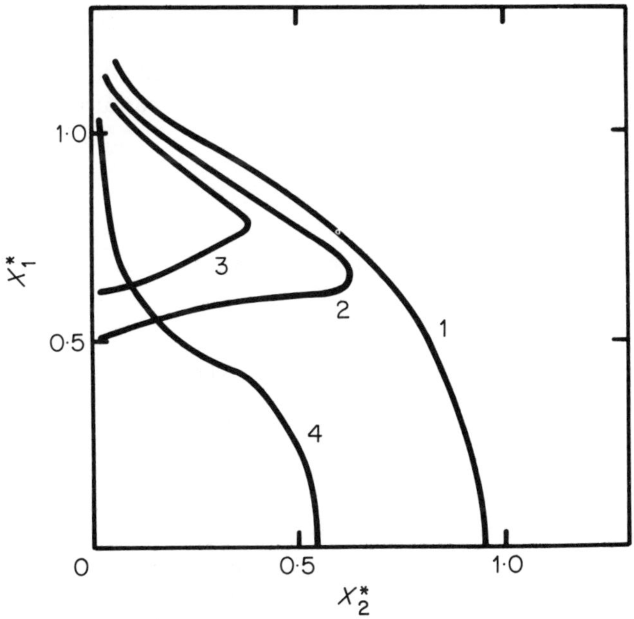

Fig. 5.17. Examples of calculated capture cross-sections. The following values of the various dimensionless constants were assumed: $q = 0{\cdot}5$, $\beta = 1$, $\tau = 225$, $\lambda^* = 2{\cdot}4 \times 10^{-3}$, $C_R/C_A = 0{\cdot}14$. (1) $C_A = 0{\cdot}95 \times 10^{-2}$; (2) $C_A = 4{\cdot}7 \times 10^{-3}$; (3) $C_A = 2{\cdot}37 \times 10^{-4}$; (4) $C_A = 2{\cdot}37 \times 10^{-4}$ (after Adler, 1981b).

section can be found by integrating the trajectory equations, either forward in time by trial and error (van de Ven and Mason, 1977a), or backward in time by starting the integration on the limiting trajectory $E=0$ (Adler, 1981b). Examples of capture cross-sections (neglecting electroviscous effects) are given in Fig. 5.17. It can be seen that $z(x_2) \to \infty$ as $x_2 \to 0$. However, the value of $z(x_2)$ near $x_2=0$ contributes little to J (cf. eq (5.20)). Another important feature is the existence of capture cross-sections (curves 2 and 3 in Fig. 5.17) for which capture in the equatorial plane (X_2X_3 plane) is impossible. These capture cross-sections are the reason why, in the stable region of Fig. 5.14, the capture frequency is not exactly zero.

Traditionally, coagulation rates are compared to the rates predicted by Smoluchowski (1917), who assumed that all particles move along linear trajectories until they collide with a sphere, in which event they are captured. This is equivalent to neglecting all hydrodynamic and colloidal interactions, in which case the capture cross-section is simply a circle of radius a_1+a_2. With these approximations, ℓ and z in eq (5.20) are given by $\ell=1+q$ and $z=[(1+q)^2-x_2^2]^{1/2}$. Equation (5.20) then reduces to

$$J_s=\frac{4}{3}(1+q)^3 n_2 G a_1^3 \tag{5.21}$$

The actual doublet-formation rate can be compared with the rate J_s calculated by Smoluchowski and define an orthokinetic capture efficiency α_0 can be defined by

$$J=\alpha_0 J_s \tag{5.22}$$

which, using eqs (5.20) and (5.21), can be expressed as

$$\alpha_0=\frac{3}{(1+q)^3}\int_0^{\ell} x_2\, z(x_2)\, dx_2 \tag{5.23}$$

Capture efficiencies have been calculated by van de Ven and Mason (1977a) for equal-sized spheres and by Adler (1981b) for unequal-sized spheres. Some typical results are shown in Fig. 5.18. Several interesting observations can be made from this figure. For equal- or nearly-equal-sized particles a plot of log α_0 versus log C_A is (in the absence of electrostatic repulsion) approximately linear. For equal-sized spheres it follows, from Fig. 5.18, that when $10^{-5} < C_A < 10^{-1}$:

$$\alpha_0 \propto C_A^{0.18} \tag{5.24}$$

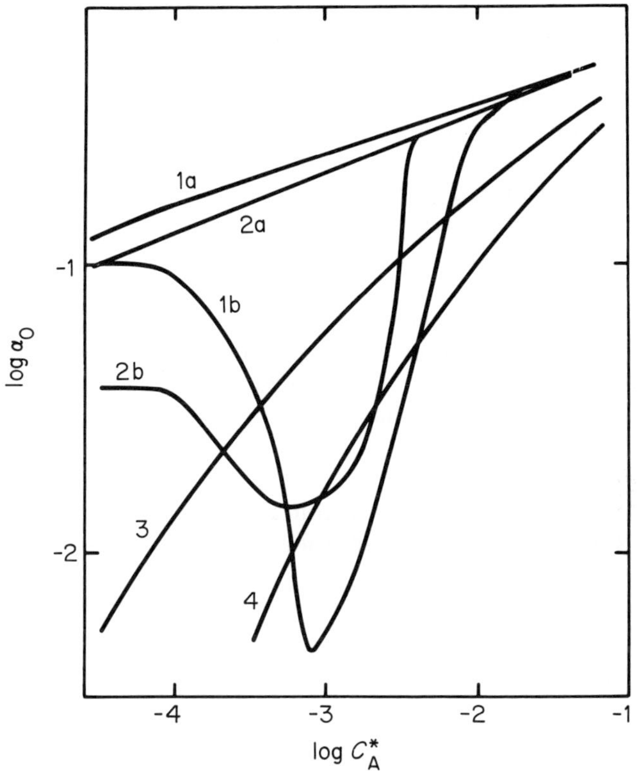

Fig. 5.18. Orthokinetic capture efficiency as a function of the parameter $C_A^* = A/36\pi\mu G a_1^3$. The following values are used in the calculation: $\beta = 1$, $\kappa a_1 = 300$, $\lambda_L/2\pi a_1 = 3 \cdot 2 \times 10^{-3}$; (1a) $q = 1$, $C_R = 0$; (1b) $q = 1$, $C_R/C_A = 1 \cdot 14$; (2a) $q = 0 \cdot 5$, $C_R = 0$; (2b) $q = 0 \cdot 5$, $C_R/C_A = 0 \cdot 321$; (3) $q = 0 \cdot 2$, $C_R = 0$; (4) $q = 0 \cdot 1$, $C_R = 0$ (adapted from Adler, 1981b).

When $q < 0 \cdot 5$ no such linear relationship exists. A second important observation is that for equal or nearly-equal-sized particles, α_0 is much larger than for particles widely different in size. This difference is due to the fact that d^*_{min} (see Fig. 5.2) depends strongly on q. Unequal-sized particles do not readily approach to within distances at which the colloidal forces are appreciable. As a result, in shear flow the coagulation of equal-sized particles is, in general, favoured over coagulation of unequal-sized particles.

Curves 1b and 2b in Fig. 5.18 show the effect of electrostatic repulsion on the capture efficiency. It can be seen that as C_A decreases from large values (low G), the repulsive forces have little effect until they cause a substantial sudden decrease in α_0. At low values of C_A (high G), the capture efficiency increases and slowly approaches the value for $C_R = 0$ when $C_A \to 0$.

Experimental evidence of the variation of α_0 with G (for $C_R=0$) is provided by Curtis and Hocking (1970) (cf. van de Ven and Mason, 1977a) and Zeichner and Schowalter (1979). An advantage of measuring α_0 from the frequency of orthokinetic doublet formation is that Hamaker constants can be derived without a knowledge of the repulsive forces, provided measurements are made at shear rates for which the effects of repulsive forces are negligible (see Fig. 5.18).

The above discussion on orthokinetic coagulation was restricted to simple shear flow. It is straightforward to extend the analysis to arbitrary values of the flow parameter γ (cf. eq (5.11)). Calculations for extensional flow ($\gamma=1$) have been made by Zeichner and Schowalter (1977), which show that in such a flow α_0 is larger than for simple shear. Typically (in the absence of electrostatic repulsion), α_0 is 30% higher at high shear rates ($C_A \simeq 10^{-4}-10^{-5}$) and more than twice the value for simple shear at low shear rates ($C_A \simeq 1$).

It is of interest to investigate the effects of salt concentration on orthokinetic coagulation. As can be seen from Fig. 5.14, at low salt concentration (high electrostatic repulsion) a dispersion is usually stable over a wide range

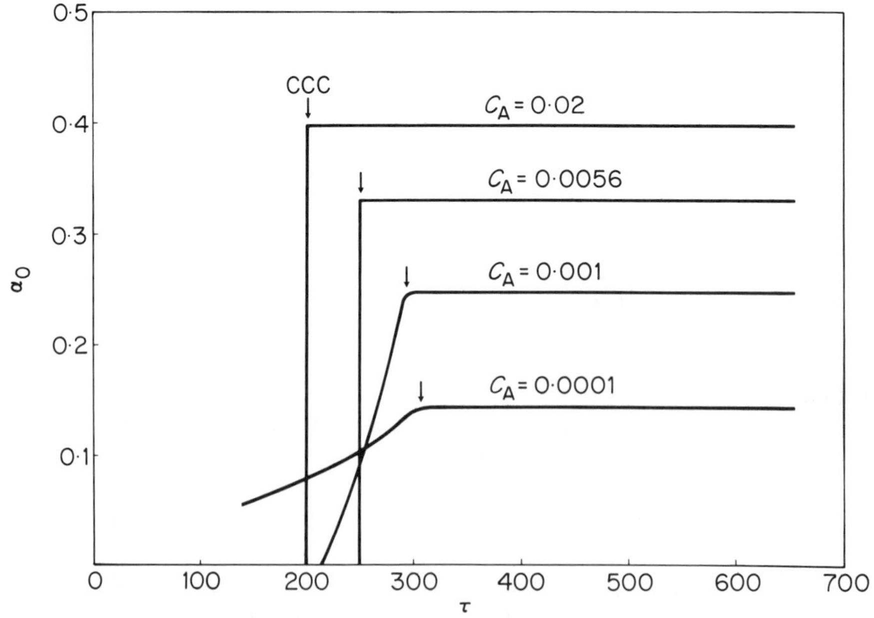

Fig. 5.19. Orthokinetic capture efficiency, α_0, for equal-sized spheres as a function of τ (proportional to the square root of the electrolyte concentration) for various values of C_A. $C_R/C_A=393$ and $\lambda^*=7\cdot96\times10^{-3}$. The arrows indicate the location of the CCC. Notice that the CCC depends on the shear rate, G.

of shear rates. Increasing the salt concentration (decreasing the electrostatic repulsion) at a given rate of shear results in coagulation, either in a secondary or in a primary minimum. As in the case of perikinetic coagulation (cf. Section 4.4), a critical coagulation concentration (CCC) exists beyond which fast coagulation occurs. Examples are shown in Fig. 5.19 which gives the orthokinetic capture efficiency, α_0, as a function of κa (proportional to the square root of the electrolyte concentration). It is clear that for orthokinetic coagulation the CCC depends on the rate of shear. Although the observation of CCC values in perikinetic coagulation is common, only recently have CCC values been observed in sheared suspensions (Stein *et al.*, 1986). Analogously to perikinetic coagulation, we can expect ion size to have an effect on the coagulation rate (due to the terms $\bar{b}\Phi_s^2 f_i$ in eq (5.11)): the larger the dissolved ions, the larger the CCC. (These effects are not included in the calculations of Fig. 5.19.)

It can readily be seen from the trajectory equations (eqs (5.11)) that when the shear rate (and thus the Péclet number $\mathrm{Pe} = Ga^2/D$) becomes large, the capture efficiency, α_0, approaches zero because the interaction term in eqs (5.11) becomes negligible. In the absence of this interaction term, the trajectory of approach is the mirror image of the trajectory of recession and no capture occurs. At the same time, as $G \to \infty$, the number of collisions per reference particle, J_{SM}, approaches infinity. The number of particles captured, J, is the product of the capture efficiency and J_{SM}. It has been shown (O'Brien, 1975) that for equal-sized spheres, as $G \to \infty$, the capture frequency approaches the value

$$J_\infty = 3{\cdot}7 \times 10^5\, C_A\, J_{SM} \tag{5.25}$$

Since $C_A \propto G^{-1}$ and $J_{SM} \propto G$, J_∞ approaches a constant value. Equation (5.25) is valid when $C_A < \sim 10^{-8}$.

The value of J_∞ depends strongly on the radius ratio, q. From the nature of the streamlines around a sphere one can expect that $J_\infty = 0$ when $q = 0$. At high shear rates one can use the interception model to estimate coagulation rates. In this model the particles follow the streamlines around a reference sphere until the gap between a particle and the reference sphere reaches a certain value corresponding, for example, to the secondary-minimum distance. At this distance the particle is captured (intercepted) by the reference sphere. On the basis of this model one expects that $J_\infty = 0$ for small values of q because a small particle that encounters a large one never reaches the interception distance. It follows that J_∞ is a strong function of q, as can be seen from the trends shown in Fig. 5.18. It must be noted that the conditions required to reach J_∞ are very extreme. Very large particles and very large shear rates are required to obtain $C_A < \sim 10^{-8}$.

5.1.5.2 Coagulation in the presence of polymer

When colloidal particles are coated with polymer, the interactions between them are strongly affected by the polymer at the surface of the particles, especially when the particles are close together. In principle, an additional force, F_{pol}, can be included in the trajectory equations eq (5.11)) and the trajectory equations integrated to obtain the orthokinetic capture efficiency, α_0, given by eq (5.23). As yet no such calculations have been performed, probably because of the uncertainty in the magnitude and form of F_{pol}. Several approximate approaches have been used to take into account the effects of polymers on orthokinetic coagulation.

At low coverage and in the absence of loops and tails, the main effect of the polymer is to make the surface of the colloidal particles heterogeneous. Patches coated with polymer will have different electrical charges than uncoated patches. The probability that two spheres will be captured during an encounter will depend strongly on the relative orientations of the spheres when they are close together. If there are two types of patches (with and without polymer), denoted by A and B, there are three possibilities to be considered: a patch A can face a patch A during closest approach, a patch A can face a patch B or a patch B can face a patch B. For each possibility there exists a capture efficiency α (α_{AA}, α_{AB}, α_{BB}) and therefore

$$\alpha_0 = \frac{\eta^2\alpha_{AA} + \eta(1-\eta)\,\alpha_{AB} + (1-\eta)^2\,\alpha_{BB}}{\eta^2 - \eta + 1} \tag{5.26}$$

where η is the fraction of the surface coated with polymer. In writing down eq (5.26) it is assumed that the probability of one patch facing another is, on average, independent of the history of particle approach and that partial overlap of patches is unimportant. It can be seen from eq (5.26) that α_0 is maximum for given values of α_{AA}, α_{AB} and α_{BB} when the surface coverage is given by

$$\eta_{opt} = \alpha_1 \pm (\alpha_1^2 + \alpha_2)^{1/2} \tag{5.27}$$

where

$$\alpha_1 = \frac{\alpha_{AA} - \alpha_{AB}}{\alpha_{AA} - \alpha_{BB}} \quad \text{and} \quad \alpha_2 = \frac{\alpha_{AB} - \alpha_{BB}}{\alpha_{AA} - \alpha_{BB}}$$

The minus sign applies when $\alpha_1 > 0$ and $\alpha_2 < 0$, otherwise the plus sign applies. Equation (5.27) does not apply when $(\alpha_{AB}, \alpha_{AA}) \ll \alpha_{BB}$, in which case eq (5.27) corresponds to a minimum in α_0. If it is assumed that two surfaces covered with polymer repel each other due to steric repulsion, then $\alpha_{AA} = 0$ and eq (5.27) reduces to:

$$\eta_{\text{opt}} = \frac{\alpha_{\text{AB}}}{\alpha_{\text{BB}}} \pm \frac{(\alpha_{\text{AB}}^2 - \alpha_{\text{AB}}\alpha_{\text{BB}} + \alpha_{\text{BB}}^2)^{1/2}}{\alpha_{\text{BB}}} \tag{5.28}$$

In most cases of interest $a_{\text{AB}} \gg \alpha_{\text{BB}}$ and eq (5.28) reduces further to

$$\eta_{\text{opt}} = \frac{1}{2}\left(1 - \frac{\alpha_{\text{BB}}}{\alpha_{\text{AB}}}\right) \tag{5.29}$$

which shows that if polymer bridging is the dominant coagulation mechanism, then the fastest coagulation rate is obtained for coverages of slightly less than 50%, in agreement with most experimental observations. When polymer bridging is the only factor, then $\eta_{\text{opt}} = 1/2$ and $\alpha_0 = \alpha_{\text{AB}}/3$.

When the polymer configurations on the surface of a colloidal particle are such that many loops and tails extend into the surrounding solution, then the coagulation rate is increased because of an increase in the radius of the collision sphere. If we assume that two spheres cannot approach closer than the distance ℓ, where ℓ is the maximum extension of the tails into the solution, then the radius of the collision sphere will be $2a + \ell$. Hence the number of captures per unit time will increase to

$$J = J_0 \left[1 + \frac{3}{2}\frac{\ell}{a} + \frac{3}{4}\left(\frac{\ell}{a}\right)^2 + \frac{1}{8}\left(\frac{\ell}{a}\right)^3\right] \tag{5.30}$$

When ℓ is comparable to the particle radius the increase can be more than 200%.

For two spheres of very different sizes, the effects of loops and tails may be very important. In the absence of the polymer the capture efficiency for the spheres is very low because the hydrodynamic forces prevent the spheres from approaching one another closely (cf. Fig. 5.2). Long tails extending from the surface of the spheres can capture particles at much larger distances, thus significantly increasing the capture efficiency (Van de Ven, 1981).

5.1.6 Combined perikinetic and orthokinetic coagulation

5.1.6.1 Governing equations

Calculating the coagulation rates from trajectory equations is only allowed when Brownian motion can be neglected. For a complete description of the doublet-formation process it is required that the convective-diffusion equation be solved

$$\frac{\partial n}{\partial t} = \nabla \cdot (\boldsymbol{D} \cdot \nabla n) - \nabla \cdot (\boldsymbol{u}_{\text{hydr}} n) - \nabla \cdot (\boldsymbol{u}_{\text{coll}} n) \tag{5.31}$$

where n is the number concentration of particles around a reference particle located at the origin and t is the time. The components of the relative diffusion tensor $\boldsymbol{D}$ are given by eq (4.27); $\boldsymbol{u}_{\text{hydr}}$ is the relative velocity between the spheres, given by eq (5.1); and $\boldsymbol{u}_{\text{coll}}$ the colloidal velocity given by

$$\boldsymbol{u}_{\text{coll}} = \frac{\boldsymbol{D} \cdot \boldsymbol{F}_{\text{coll}}}{kT} \tag{5.32}$$

The relative velocity between two spheres, $\boldsymbol{u} = \boldsymbol{u}_{\text{hydr}} + \boldsymbol{u}_{\text{coll}}$, is given by eq (5.11). For charged spheres, $\boldsymbol{u}_{\text{hydr}}$ must contain the contributions of the electroviscous forces and torques. As yet no general solution of eq (5.31) is available. In most problems one is interested in steady-state conditions for which $\partial n/\partial t = 0$.

When the only colloidal force between the spheres is a van der Waals attraction, a useful approximation is to neglect both the hydrodynamic and the (attractive) van der Waals interactions. This can be done because of compensating effects: van der Waals forces accelerate the rate of coagulation, while the hydrodynamic resistance during approach retards coagulation and these effects roughly cancel each other. In essence, this is the same assumption as that made by Smoluchowski in deriving eqs (4.75) and (5.21) and by Levich (1962) in calculating the rate of deposition on rotating disks. For this reason we refer to this approximation as the Smoluchowski–Levich approximation. In simple shear and for steady states, this approximation leads to

$$D\nabla^2 n = Gx_2 \frac{\partial n}{\partial x_3} \tag{5.33a}$$

or, expressed in terms of the translational Péclet number $\text{Pe} = Ga^2/D$ and in distances rendered dimensionless by the sphere radius,

$$\nabla^2 n = \text{Pe}\, x_2 \frac{\partial n}{\partial x_3} \tag{5.33b}$$

Equations (5.33) must satisfy the boundary conditions $n(r=2)=0$ and $n(r=\infty)=n_\infty$. Unfortunately no complete solution of the Smoluchowski–Levich approximation with these boundary conditions is available. When $\text{Pe}=0$ (diffusion-controlled coagulation), the right-hand side of eq (5.33) vanishes and eq (5.33) reduces to eq (4.73), resulting in a coagulation rate given by eq (4.75). When $\text{Pe}=\infty$ (orthokinetic coagulation), the left-hand side of eq (5.33) vanishes, resulting in a coagulation rate given by eq (5.21).

It is often assumed that for finite values of Pe, the coagulation rate is simply the sum of the peri- and orthokinetic coagulation rates, i.e. it is assumed that (when $q=1$)

$$J=16\pi\alpha_p D n_0 a+\frac{32}{3}\alpha_0 G n_0 a^3 \tag{5.34a}$$

or

$$J=16\pi\alpha_p D n_0 a(1+\alpha \mathrm{Pe}) \tag{5.34b}$$

with $\alpha=2\alpha_0/3\pi\alpha_p$, where α_p and α_0 are the peri- and orthokinetic capture efficiencies, respectively. But, from known asymptotic solutions of eqs (5.31) and (5.33) valid at low and high Péclet numbers, it can be concluded that the assumption of additivity is invalid.

5.1.6.2 Effect of shear on perikinetic coagulation

When $\mathrm{Pe} \ll 1$, it is possible to find an expansion solution of the steady-state convective-diffusion equation:

$$\nabla\cdot(\boldsymbol{D}\cdot\nabla n)=\nabla(\boldsymbol{u}_{\mathrm{hydr}}+{}_{\mathrm{coll}})\,n \tag{5.35}$$

Whether or not a steady state will be reached depends on the nature of the colloidal forces between the particles. It can be shown that, when the forces are repulsive at short distances, then a necessary condition for reaching a steady state is $F_{\mathrm{coll}}(\infty)=0$. If the colloidal force becomes infinitely repulsive as the gap width, h, approaches zero, a steady state will be reached only if $hF_{\mathrm{coll}}\rightarrow-\infty$ as $h\rightarrow 0$. Since van der Waals forces usually dominate at short separations ($F_{\mathrm{coll}}\propto h^{-2}$), this condition is satisfied for most unstable colloidal systems.

If these conditions are satisfied and a steady-state solution exists, then the nature of the solution itself depends on the form of F_{coll}. In attempting a solution of the form

$$n=n_0+\mathrm{Pe}\,n_1+\ldots. \tag{5.36}$$

it is found that it is impossible to satisfy the boundary condition $n(\infty)=1$ (normalized concentration), except for forces that are repulsive or finite at contact (corresponding to $\alpha_p=0$) and which decay sufficiently quickly to zero at large separation (e.g. an exponential decay). When $\alpha_p=0$ an expansion up to order Pe is allowed when $r^2F_{\mathrm{coll}}\rightarrow 0$ as $r\rightarrow\infty$. At large separations van der Waals forces usually dominate and the retarded dispersion energy decays as r^{-7} and hence the force as r^{-8}. From this it can be concluded that for stable dispersions ($\alpha_p = 0$) an expansion of the type described by eq (5.36) is allowed. In such a case no particles are captured by a reference particle and $J = 0$. For unstable dispersions ($\alpha_p \neq 0$) it can be shown (van de Ven and Mason, 1977b) that a singular perturbation expansion must be made. At short separations (inner expansion) the expansion is of the type

$$n = n_0 + \mathrm{Pe}^{1/2}\, n_1 + \ldots . \tag{5.37}$$

while at large separations (outer expansion) it is

$$n(\hat{r}, \theta, \phi) = n_0 + \mathrm{Pe}^{1/2}\, n_1 + \ldots . \tag{5.38}$$

where the variable $\hat{r}$ is defined as $\hat{r} = \mathrm{Pe}^{1/2} r$. Solving for the inner and outer expansion and determining the integration constants by matching both expansions yields:

$$n = 1 - \frac{\alpha_p}{r} - 0{\cdot}2568\ \mathrm{Pe}^{1/2}\, \alpha_p \left(1 - \frac{\alpha_p}{r}\right) \tag{5.39}$$

provided $rV_{\mathrm{coll}} \to 0$ as $r \to \infty$.

From eq (5.33) and Fick's first law applied at the surface of the collision sphere, it is found that

$$J = J_0\,(1 + 0{\cdot}5136\ \alpha_p\, \mathrm{Pe}^{1/2} + \ldots) \tag{5.40}$$

where J_0 is the flux due to diffusion alone, given by eq (4.72). This result shows that the assumption of additivity, expressed by eq (5.34b), is invalid. It can be seen that eq (5.40) also applies when $\alpha_p = 0$ because, in this case, $J = J_0 = 0$. Thus a stable dispersion remains stable when subjected to low rates of shear. For unstable suspensions, the coagulation rate increases with increasing shear rates, the increase being more pronounced for highly unstable dispersions.

5.1.6.3 Effect of Brownian motion on orthokinetic coagulation

When coagulation is dominated by shear forces, an attempt can be made to solve the convective-diffusion equation in terms of an expansion in Pe^{-1}. This was done by Feke and Schowalter (1985) who showed that the governing equation could be solved as a regular expansion in Pe^{-1}:

$$n = n_0 + \mathrm{Pe}^{-1}\, n_1 + \ldots . \tag{5.41}$$

They found analytical solutions for n_1 which are valid for short and large separations and performed numerical calculations to obtain the solution in the intermediate regime. From eq (5.41) it follows that

$$J = J_0 + \mathrm{Pe}^{-1} J_1 + \ldots . \tag{5.42}$$

where J_0 is the orthokinetic coagulation rate given by eqs (5.22) and (5.23).

From eq (5.42) we can again conclude that the assumption of additivity of peri- and orthokinetic coagulation rates (eq (5.34b)) is invalid. Values for $j_1 = J_1/nGa^3$ (for equal-sized spheres) are shown in Fig. 5.20 in the case where electrostatic forces between the spheres are absent ($C_R = 0$). It can be seen that at high shear rates $J_1 < 0$, while for low shear rates $J_1 > 0$. This shows that not only are the effects of shear and Brownian motion not additive, but including the effects of diffusion in shear-induced coagulation can sometimes decrease the coagulation rate. The same is true for particles subjected to extensional flow ($\gamma = 1$).

It is of interest to look at the magnitude of the increase or decrease in the

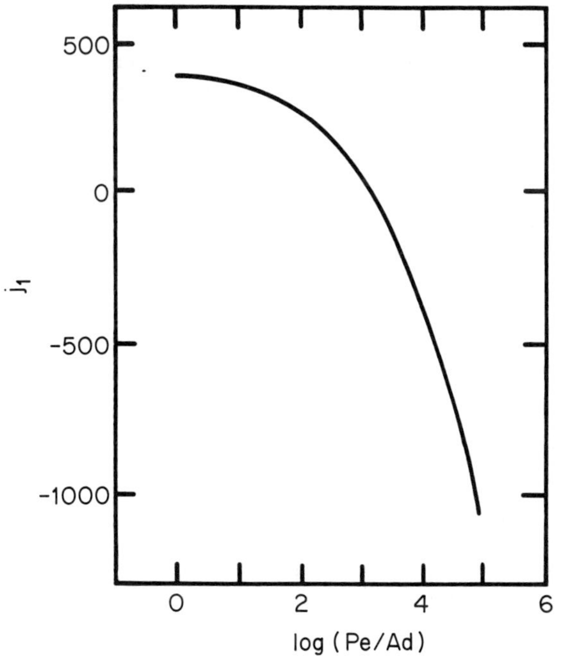

Fig. 5.20. Variation of j_1 with Pe/Ad, showing the effect of Brownian motion on orthokinetic coagulation ($C_R = 0$) (after Feke and Schowalter, 1983).

coagulation rate due to diffusion. From the definitions of C_A and Pe it follows that

$$\mathrm{Pe} = \frac{\mathrm{Ad}}{C_A} \tag{5.43a}$$

where Ad is the adhesion number

$$\mathrm{Ad} = A/6kT \tag{5.43b}$$

Typically, Ad is in the range 10^{-1}–10. At high shear rates at a value of $j_1 = -1000$, Pe is typically in the range 10^4–10^6 and hence $\mathrm{Pe}^{-1}j_1$ is in the range of -10^{-3} to 10^{-1}. Since j_0 is of order of unity, the decrease is in the range 0·1–10%. In the regime of low shear rates, when $j_1 \simeq 300$ and Pe is in the range 1–10^2, the theory holds only when the adhesion number, Ad, is large, because it is only valid when $\mathrm{Pe} \gg 1$. It predicts an increase in the coagulation rate of about 30% when $\mathrm{Pe} \simeq 100$.

It must be noted that J_0 in eq (5.42) is itself a function of Pe because of its dependence on C_A, and Ad is a constant for a given system (cf. eqs (5.42) and (5.43)). However, in calculating J_0 it is assumed that Brownian motion does not affect the coagulation rate. In deriving eq (5.42) it is tacitly assumed that conditions under which the effects of colloidal forces are important do occur, while those of Brownian motion can be neglected. Because the adhesion number, Ad, is usually of order unity, it follows that the effects of Brownian motion are often of the same importance as van der Waals forces. Strictly speaking, when making an expansion in Pe, the zero-order solution must be taken to be J_∞, given by eq (5.25).

5.1.6.4 General trends for arbitrary Péclet numbers

From the nature of the solutions at low and high Péclet numbers we can establish general trends in the coagulation rate at all Péclet numbers. These trends are shown schematically in Fig. 5.21 for the case $C_R = 0$ (no electrostatic forces). It can be seen that for equal-sized spheres the number of particles captured per reference particle increases monotonically from its value at $\mathrm{Pe} = 0$ (given by eq (4.72)) to its value at $\mathrm{Pe} = \infty$ (given by eq (5.25)). For unequal-sized spheres of widely different size, the number captured first increases, then goes through a maximum and finally decays to zero as $\mathrm{Pe} \rightarrow \infty$.

When $C_R \neq 0$ and an electrostatic repulsion occurs between the particles, J_0 is lowered because of a decrease in the capture efficiency. When $\mathrm{Pe} \ll 1$, J_0 is lowered because of a decrease in α_p. In the range $10 \lesssim \mathrm{Pe} \lesssim 10^2$ the effects

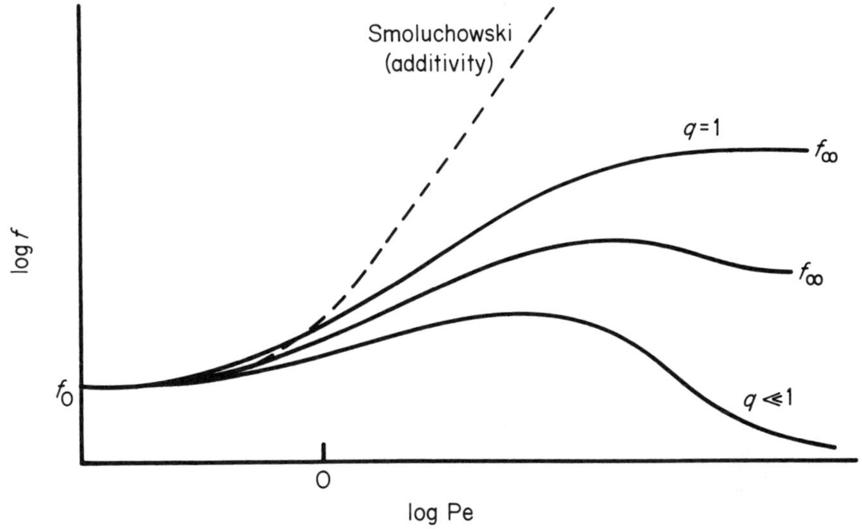

Fig. 5.21. Trends in doublet-formation frequency, f, as a function of the Péclet number, Pe. At Pe $\ll 1$, $f = f_0$ (the rate governed by Brownian motion). As Pe $\rightarrow \infty$, f approaches a constant depending on the radius ratio, q. The three examples shown correspond to equal-sized spheres, widely different-sized spheres ($q \ll 1$) and an intermediate case. The additivity approximation (----) is also shown.

of repulsive forces are negligible, while in the range $10^2 \lesssim \text{Pe} \lesssim 10^4$ they can result in near-zero orthokinetic efficiencies, α_0, corresponding to very low values of J. For Pe $> \sim 10^4$ the effects of electrostatic repulsion again become negligible.

For comparison, the additivity approximation is also shown in Fig. 5.21 (dashed line). It can be seen that the additivity approximation underestimates J at low Pe and overestimates it at high Pe.

5.1.7 Two spheroids in simple shear

Very few results are available for the hydrodynamic interactions between two particles other than spheres. Recently Kim (1986) solved the problem of two spheroids subjected to simple shear in the case where the particles are far apart. He found that the hydrodynamic interaction depends strongly on the initial relative orientation of the spheroids and that the interaction between prolate spheroids is qualitatively different from the interaction between oblate ones. His method is general and applies to spheroids of arbitrary size in any configuration, but he considered only the case of two equal-sized spheroids, whose centres were located on the vorticity axis, with polar angles (ϕ_0, θ_0) and $(\phi_0, \pi - \theta_0)$. The orbits of two such spheroids differ from the angular motion predicted by Jeffery for single spheroids. Oblate spheroids ($r_e < 1$) must push fluid out of the space between them as they rotate and their orbits become larger than the Jeffery orbits, while at the same time the period of rotation increases. For prolate spheroids in the same position and orientation, only small deviations from Jeffery's orbits occur. Spheroids which initially have the same orientation drift away from their inital position on the vorticity axis, or their centres move in a closed limit cycle, depending on orientation.

For arbitrary initial relative positions and orientations we may expect that the period of rotation of each spheroid will change during a two-particle encounter. In most cases the period will probably increase because of increased hydrodynamic resistance, but it is conceivable that for certain interactions the period decreases. The amount ΔT by which a period changes during an encounter can, in principle, be calculated from the solution of the flow field around two spheroids. In general, ΔT will be a function of the initial relative position of the spheroids, described by the polar coordinates (r_0, θ_0, ϕ_0) and by the initial orientation of each spheroid with respect to the flow (θ_1, ϕ_1) and (θ_2, ϕ_2). In a suspension of particles the average period of rotation (neglecting three- and more-body interactions) is given by

$$\bar{T} = kT_0 + (1-k)\,(T_0 + \overline{\Delta T}) \tag{5.44}$$

where T_0 is the period of a single isolated spheroid, given by eq (3.91), and $\overline{\Delta T}$ is the mean value of ΔT, obtained by averaging over all initial orientations and positions; k is the fraction of spheroids that undergo no collisions during time T, and $(1-k)$ the fraction undergoing one collision. The value of k depends on the shape and the volume fraction of the particles.

Because the tip of a rotating prolate spheroid describes an orbit on a sphere with a radius equal to the major semi-axis, we may define a collision sphere of radius $2a$ around a reference spheroid. In a manner similar to that of an encounter between spheres, we can define a collision as an encounter in which the centre of a spheroid would have penetrated the collision sphere of the reference spheroid if it had continued to move along a rectilinear trajectory. Hence, for simple shear, the number of collisions per reference spheroid is given by

$$J_s = \frac{32}{3}\, nGa^3 \tag{5.45a}$$

or, expressed in terms of the volume fraction c

$$J_s = \frac{8G}{\pi} r_e^2 c \qquad r_e > 1 \tag{5.45b}$$

where r_e is the axis ratio a/b. For oblate spheroids, the collision sphere is of radius $2b$ and we have

$$J_s = \frac{8G\,c}{\pi\,r_e} \qquad r_e < 1 \tag{5.45c}$$

The number of collisions which a reference spheroid undergoes during a rotational period T is $j = J_s T$ or, stated otherwise, on average a spheroid experiences one collision every $1/j$ rotations and hence

$$k = (j^{-1} - 1)/j^{-1} = 1 - j = 1 - J_s T \tag{5.46}$$

and thus

$$\bar{T} = T_0(1 + Ac) \tag{5.47}$$

where

$$A = (8/\pi) r_e^2 \overline{\Delta T G} \qquad r_e > 1$$

or

$$A = (8/r_e \pi) \overline{\Delta T G} \qquad r_e < 1$$

The increase in $\bar{T}$ with increasing volume fraction was observed by Jayasuriya and van de Ven (1987) from measurements of the frequency of rheo-optical transients (cf. Section 3.1.12).

Besides changing the frequency of the (damped) oscillations in optical and rheological properties, particle interactions also affect the damping rate in these oscillations. In the absence of particle interactions the characteristic relaxation time of the attenuation is determined by rotary Brownian motion and the polydispersity in the particle shapes. During a collision between two particles the angular velocities of the particles change, resulting in so-called phase mixing and hence shorter relaxation times. For a suspension of identical spheroids, and in the absence of Brownian motion, the relaxation time of the transient oscillations due to two-body interactions can be estimated as (Okagawa *et al.*, 1973):

$$\tau \simeq \frac{\pi \ln r_e}{4G\,r_e\,c} \tag{5.48}$$

Experimental data on transient oscillations in the viscosity of a suspension of rods (Ivanov *et al.*, 1982) give a value for τ of about twice the value predicted using eq (5.48), indicating that it predicts the correct order of magnitude of τ reasonably well.

5.1.8 Interactions between drops or between drops and solid spheres in simple shear

The interactions between two drops or a hard sphere and a drop are more complicated than those between two solid spheres. The only dimensionless parameter entering into the problem of two hard spheres is the radius ratio, q (provided Brownian motion is absent and the Reynolds number is Re $\ll$ 1). In interactions between drops there are additional dimensionless parameters: the capillary number Ca, defined as

$$Ca = Ga_1\mu_0/\sigma_1 \tag{5.49}$$

two viscosity ratios

$$p_1 = \mu_1/\mu_0; \; p_2 = \mu_2/\mu_0 \tag{5.50}$$

and one interfacial tension ratio

$$\tilde{\gamma} = \sigma_1/\sigma_2 \tag{5.51}$$

Here μ_0, μ_1 and μ_2 are the viscosity of the suspending medium and of drop 1 and drop 2, and σ_1 and σ_2 are the interfacial tensions of the 01 and 02 interfaces, respectively.

When $Ca \ll 1$, which as we have seen (cf. Section 3.1.9.4) is usually the case, the problem simplifies considerably. In this case the drops maintain their spherical shape in a shear flow and it is possible, as for hard spheres, to solve the creeping-flow and continuity equations for the flow field around and inside the drops. The trajectory equations describing the relative motions of the drops are the same as those given by eqs (5.1), with the exception that the functions A and B depend, not only on r and q, but also on p_1 and p_2 (but not on Ca and $\tilde{\gamma}$). No general results for the functions A and B are available, except for equal-sized drops ($q = 1, p_1 = p_2$) at large separation, r. When both p_1 and p_2 become very large the problem reduces to that of the interaction between hard spheres. When only one of p_1 or p_2 approaches infinity, the equations describe the interaction between a fluid drop and a solid sphere. We may expect that all these types of interaction are qualitatively similar to the interactions between hard spheres: two drops are either on an open or a closed trajectory, depending on whether the trajectory constant, E, is larger or smaller than zero (cf. Fig. 5.2).

It must be noted that even when the capillary number, Ca, is sufficiently small for deformations of a single drop to be absent, it is still possible for deformations to occur during a collision with another drop. When two particles in a simple shear flow are close together, the actual rate of shear in the gap can be 10 to 100 times larger than the velocity gradient of the undisturbed flow and this implies that the hydrodynamic forces acting on drops are also much larger. Furthermore, additional forces, such as van der Waals attractions, can cause drop deformation during close contact. When two drops deform during an encounter, their trajectories will no longer be symmetric about the X_1X_2 plane. Often the deformation results in a repulsion between the particles, corresponding to an increase in the trajectory constants during a collision. This increase is even more pronounced when the drops are already deformed prior to the collision. Asymmetric trajectories were observed by Bartok and Mason (1959) for two fluid drops and by Smith and van de Ven (1985b) for the collision between a fluid drop and solid sphere. An example of a collision between a poly(styrene) sphere and a drop of Pale-4-Oil in silicone oil is shown in Fig. 5.22.

When colloidal forces are acting between the drops, orthokinetic coalescence can occur. In an emulsion the coalescence rate is given by eqs (5.20) to (5.22). However, it is impossible to calculate the coalescence efficiency, α_0, without taking drop deformation during an encounter into account. As yet no theories for calculating α_0 for coalescing drops are available.

An encounter between a drop and a solid particle can lead to the formation of a solid–liquid pair. Depending on the surface free energies of the drop and the particle, the encounter can lead to partial or complete engulfment of the solid particle by the fluid drop (Smith and van de Ven, 1985b), a phenomenon closely related to phagocytosis.

Examples of partially engulfed solid–liquid pairs rotating in simple shear are shown in Fig. 5.23. It can be shown that (when $Ca \ll 1$) the period of rotation depends mainly on the radius ratio, q, of the solid–liquid pair ($q = a_d/a_s$, a_d and a_s being the radii of the drop and the sphere, respectively). When $q = 0$, the period is that of a solid sphere ($T = 4\pi/G$), while if $q = \infty$ the period is that of the period of rotation of a surface element of a fluid drop which, from eqs (3.173) to (3.175), can be shown to be (Torza *et al.*, 1982)

$$T = \frac{4\pi}{G} \frac{p+1}{(p^2+2)^{1/2}} \tag{5.52}$$

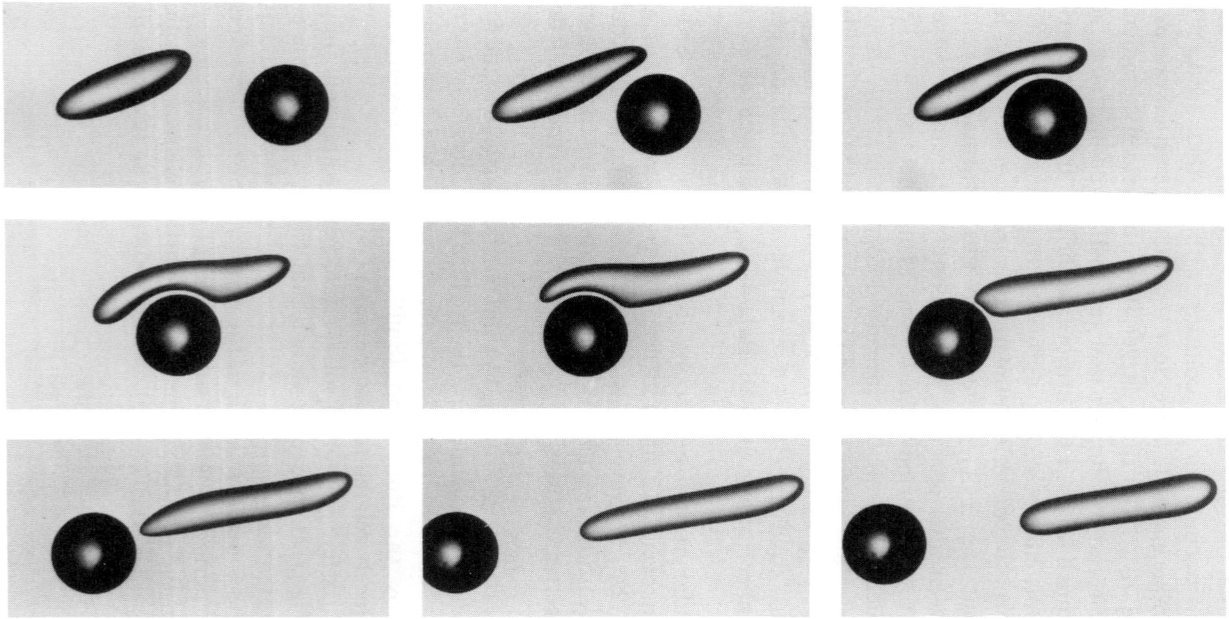

Fig. 5.22. A sequence of photographs of a collision in simple shear between a poly(styrene) sphere and a liquid drop (Pale-4-Oil) in silicone oil. $Ca = 0{\cdot}124$, $q = 0{\cdot}89$, $p = 0{\cdot}49$ (after Smith and van de Ven, 1985b).

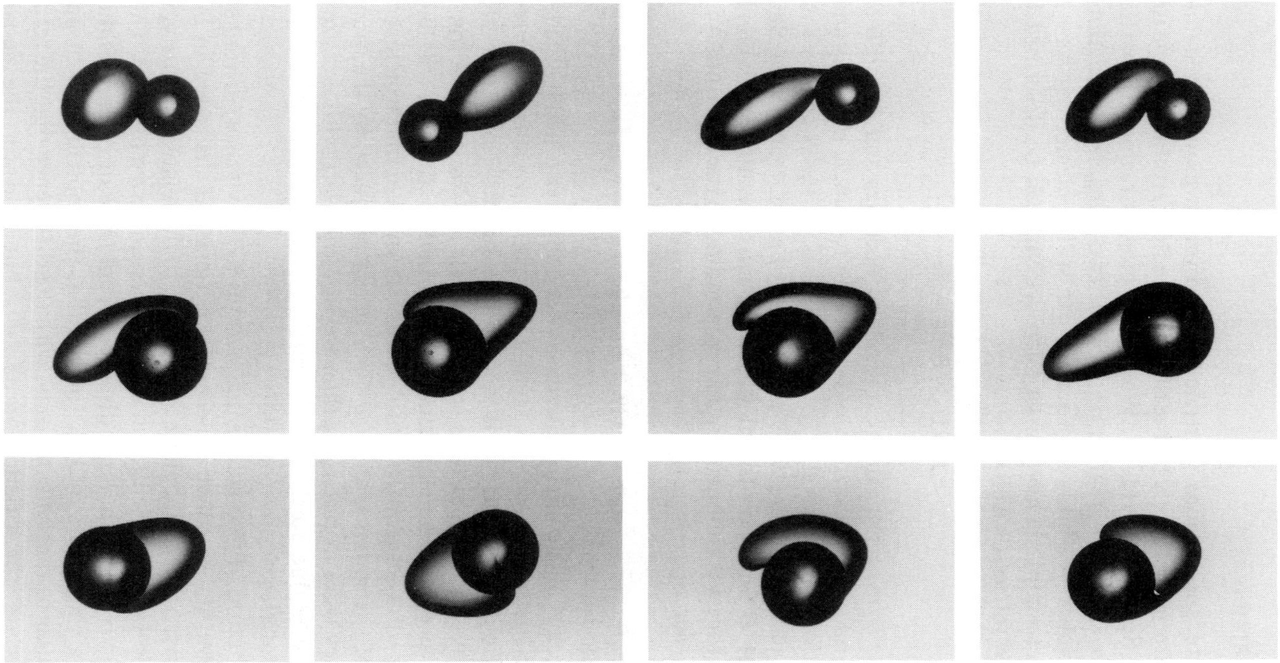

Fig. 5.23. Behaviour of solid–liquid pairs suspended in silicone oil and subjected to a simple shear. First sequence: poly(styrene)/water, Ca = 0·216. second sequence: poly(styrene)/Pale-4-Oil; Ca = 0·258. Third sequence: poly(styrene)/Ucon oil; Ca = 0·240 (after Smith and van de Ven, 1985b).

5.2 Viscosity of suspensions with two-body interactions

5.2.1 Hard spheres

In very dilute suspensions particle interactions are negligible and the viscosity is determined solely by the disturbance flow around a single isolated particle (cf. Section 3.1.11.1). When the concentration of particles is increased, we reach a regime in which two-particle interactions are important, while three-body interactions can still be neglected. As we have seen, the macroscopic behaviour of such systems can be described by the stress tensor $\boldsymbol{P}$ (cf. eq (3.259)):

$$\boldsymbol{P} = \boldsymbol{\Pi}_0 + \boldsymbol{A}c + \boldsymbol{B}c^2 \tag{5.53}$$

Here $\boldsymbol{\Pi}_0$ is the stress tensor in the absence of particles and $\boldsymbol{A}$ and $\boldsymbol{B}$ are the first and second virial stress tensors accounting for single particles and two-body interactions, respectively. Formal expressions for $\boldsymbol{B}$ can be found in papers by Cox and Brenner (1971) and Batchelor and Green (1972). Since the calculation of $\boldsymbol{B}$ is a non-trivial problem, most results available in the literature are for its component B_{32}, corresponding to the contribution to the viscosity of order c^2:

$$\mu = \mu_0(1 + [\mu]c + kc^2) \tag{5.54}$$

We have seen that the intrinsic viscosity, $[\mu]$, is determined by the leading term of the disturbance flow far from a particle, when it is immersed in a linear flow field. Similarly, k is determined by the flow field at large distances from a pair of particles in the same flow. Consider the total flow field around two hard spheres embedded in a flow field that is linear at infinity. If we subtract from it the undisturbed flow, i.e. that which occurs in the absence of the spheres, and the disturbance produced by the individual spheres, if they were in isolation, then we would be left with a disturbance flow, v, which is due solely to two-particle interactions. As the distance between the spheres approaches infinity, $v \to 0$.

The contribution of two-sphere interactions to the viscosity is very similar to the contribution of single spheroidal particles, as discussed in Section 3.1.11.5. Such particles contribute a term, $[\mu]$, to the relative viscosity, where $[\mu]$ is a function of the orientation of the spheroid (cf.eq (3.283)). The flow field around two spheres at large distances from the pair resembles the flow field around a single spheroid. It follows that the contribution of pairs of spheres is similar to that of spheroids, with the difference that the result not only depends on the orientation (θ, ϕ), but also on the separation distance between the spheres. The contribution to the viscosity must also be multiplied by the number of particles having the relative position (r, θ, ϕ), which is proportional to the volume fraction, c, and so it follows that the contribution of two-body interactions is of the order of c^2. To obtain k from eq (5.54) we must average over all the relative positions of the two spheres, hence

$$k = \int_2^{\infty} \int_0^{\pi} \int_0^{2\pi} H(r,\theta,\phi)\, n(r,\theta,\phi)\, r^2 \sin\theta \, \mathrm{d}r \, \mathrm{d}\theta \, \mathrm{d}\phi \tag{5.55}$$

where $H(r, \theta, \phi)$ is a function which can be derived from the known velocity field v and n is the normalized concentration of particles around a reference particle. The quantity n is often referred to as the pair distribution function because it is the probability of finding a particle at position $\boldsymbol{r}$, given that there is a particle at some position $\boldsymbol{r}_0$. Without loss of generality we can choose the location of the reference particle to be at the origin. When Brownian motion is absent, the function $H(r, \theta, \phi)$ is entirely determined by the hydrodynamic velocity field around two spheres. Allowing for Brownian motion introduces two additional terms to H, one due to a Brownian stress and another arising from the random (or thermodynamic) force $F_r = -kT\nabla(\ln\, n)$ (Russel, 1978). The distribution of particles around a reference particle can be found by solving the convective-diffusion equation:

$$\frac{\partial n}{\partial t} = \nabla \cdot (\boldsymbol{D} \cdot \nabla n) - \nabla \cdot (\boldsymbol{u}_{\mathrm{hydr}}\, n) \tag{5.56}$$

with the boundary conditions $n(2) = 0$ and $n(\infty) = 1$. The velocity $\boldsymbol{u}_{\mathrm{hydr}}$ is given by eq (5.1).

Without solving eq (5.56) for n and evaluating the integral in eq (5.55), it can be seen that, similar to the intrinsic viscosity $[\mu]$, k goes through a transient regime before reaching a steady-state value. In contrast to $[\mu]$, the steady-state value of k depends on $\boldsymbol{u}_{\mathrm{hydr}}$ (cf. eq (5.56)) and thus on the type of flow (i.e. on γ). Furthermore, the value of k depends on the translational Péclet number, $\mathrm{Pe} = Ga^2/D$, which determines the pair distribution function, $n(r,\theta,\phi)$.

For extensional flow ($\gamma = 1$) and $\mathrm{Pe} = \infty$ (no Brownian motion), the steady-state solution for n can be found and the integral in eq (5.55) can be evaluated. For equal-sized spheres this results in $k = 7{\cdot}6$ (Batchelor and Green, 1972).

For simple shear flow, no unique steady-state solution of eq (5.56) exists. This is due to the occurrence of closed orbits around a reference particle,

TABLE 5.2
Second virial coefficient, k, for the viscosity of suspensions of hard spheres

Type of flow	Péclet number	k
$-1 \leqslant \gamma \leqslant 1$	0	6·2
$\gamma = 0$	∞	~5·2
$\gamma = 1$	∞	7·6

which can be shown to enclose an infinite volume. When $\mathrm{Pe} = \infty$, the number of particles in closed orbits is solely determined by the initial distribution of particles. As a result, $n(r, \theta, \phi)$ also depends on the inital distribution, even as $t \to \infty$. By making some plausible assumptions about the number of particles in closed orbits, Batchelor and Green found (when $\gamma = 0$, $q = 1$ and $\mathrm{Pe} = \infty$) that $k \simeq 5 \cdot 2$.

When Brownian motion is dominant ($\mathrm{Pe} \to 0$), eq (5.56) has a steady-state solution which is independent of the type of flow. For equal-sized spheres, Batchelor (1977) found that ($\mathrm{Pe} = 0$, $-1 \leqslant \gamma \leqslant 1$) $k = 6 \cdot 2$. The values of k for various flows and Péclet numbers are summarized in Table 5.2. Notice the similarities with suspensions of prolate spheroids. Such suspensions are shear thinning in simple shear and shear thickening in extensional flow (for $r_e < \sim 15$). The same is true for suspensions of hard spheres, but here the shear thinning and shear thickening are caused by two-body interactions. This shows clearly that pairs of spheres in a suspension behave in a way which is qualitatively similar to prolate spheroids.

No results are available for the values of k for suspensions of non-spherical particles.

5.2.2 Soft spheres: secondary electroviscous effect

When colloidal forces act between the particles, the trajectory equations contain additional particle-interaction terms (cf. eqs (5.11)), which result in additional contributions to the function $H(r, \theta, \phi)$ in eq (5.55). Furthermore, interparticle forces contribute a term $\nabla \cdot (\boldsymbol{u}_{\mathrm{coll}} n)$ to the convective-diffusion equation (eq (5.56)). Solutions of the convective-diffusion equation for soft spheres were discussed in Section 5.1.6. For the viscosity of a suspension of soft spheres, results are available only for systems subjected to weak flows in which the effects of Brownian motion dominate ($\mathrm{Pe} \ll 1$). In such cases a solution of the pair distribution function can be obtained as a series expansion in Pe (cf. Section 5.1.6.2):

$$n = n_0 + \mathrm{Pe}\, n_1 + \ldots \tag{5.57}$$

Russel (1978) found an approximate solution for n_1 by simplifying the convective-diffusion equation to

$$D\nabla^2 n = Gx_2 \frac{\partial n}{\partial x_3} + \nabla\cdot(\boldsymbol{u}_{\text{coll}}\, n) \tag{5.58}$$

with

$$\boldsymbol{F}_{\text{el}} = 3\pi\mu a \boldsymbol{u}_{\text{coll}}$$

where $\boldsymbol{F}_{\text{el}}$ is the electrostatic force between the spheres. This approximation is, in essence, the Smoluchowski–Levich approximation discussed earlier (cf. Section 5.1.6.1). In determining the function H, the effects of the terms $\bar{b}\Phi_s^2 f_i$ in eqs (5.11) were approximately taken into account by considering the flow field around a single charged sphere only.

With these approximations it was found that

$$k = \frac{5}{2} + 2\bar{b}\Phi_s^2 Q(\tau) + \frac{3}{40\tau^5}\left[\ln\left(\frac{\alpha}{\ln(\alpha/\ln\alpha)}\right)\right]^4 \ln\left(\frac{\alpha}{\ln\alpha}\right) \tag{5.59}$$

with $Q(\tau)$ given by eq (3.35f); the parameter α describes the importance of the double-layer interactions (relative to Brownian motion):

$$\alpha = 4\pi\epsilon\psi_0^2 a^3 \tau e^{2\tau}/kT \tag{5.60}$$

We may expect that including the effects of double layers on friction coefficients will result in

$$n = n^0\,[1 + \bar{b}\Phi_s^2 h(\tau, r, \theta, \phi)] \tag{5.61}$$

with n^0 being the solution of eq (5.58) and h an, as yet unknown, function. Taking corrections of this nature into account again leads to eq (5.59), but with a modified function Q, which must now be considered an unknown function of τ.

From eq (5.59) it follows that as $\psi_0 \to 0$ (i.e. the charge on the spheres approaches zero) k approaches 2·5. We saw earlier (cf. Table 5.2) that when Pe$=0$ then $k=6\cdot2$. The reason for this discrepancy is that in deriving eq (5.59) the hydrodynamic flow field was taken into account only approximately. Hence it seems better (when Pe $\ll 1$) to express k as

$$k = 6\cdot2 + \frac{3}{40\tau^5}\left[\ln\left(\frac{\alpha}{\ln(\alpha/\ln\alpha)}\right)\right]^4 \ln\left(\frac{\alpha}{\ln\alpha}\right) + \bar{b}\Phi_s^2\, g(\tau) \tag{5.62}$$

with $g(\tau)$ as yet unknown.

The contribution of the second term of the right-hand side of eq (5.62) is usually dominant. The theoretical predictions are in fair agreement with the experimental data (Russel, 1980).

The increase in k due to double-layer interactions is usually referred to as the secondary electroviscous effect. From eq (5.62) we can see that the secondary electroviscous effect consists of two contributions, one arising from the electrostatic repulsion between two spheres (electrostatic forces) and the other due to the modification of the flow field around two spheres when they are electrically charged (electroviscous forces). No results are available for particles other than spherical ones.

5.3 Two spheres in a gravity field

5.3.1 Relative sedimentation of two hard spheres

Consider two spheres of radii a_1 and a_2 and densities ρ_1 and ρ_2, respectively, freely suspended in a Newtonian liquid of density ρ and viscosity μ (cf.

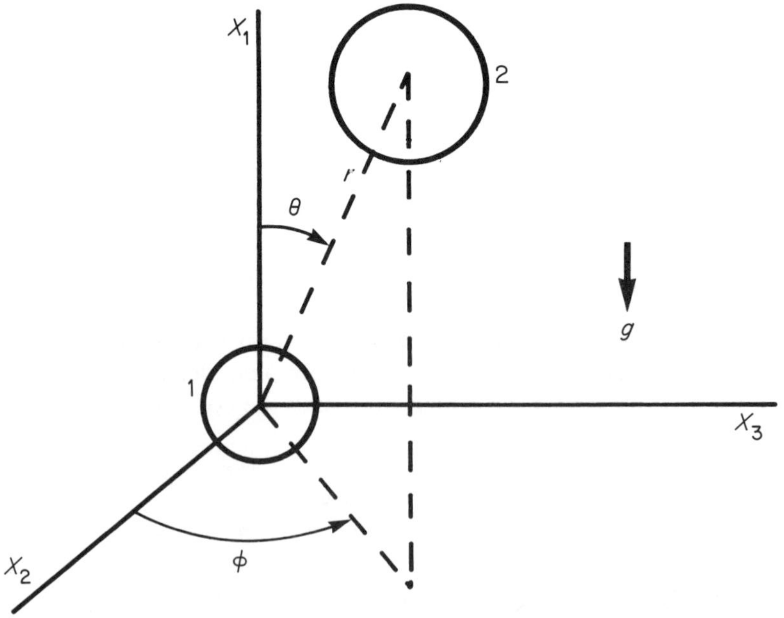

Fig. 5.24. Schematic diagram of two spheres in a gravity field. The relative orientation of sphere 2 with respect to sphere 1 is described by the polar coordinates (r, θ, ϕ). Sphere 2 moves in a plane $\phi = \text{constant}$ and, without loss of generality, we can choose $\phi = 0$ and consider motion in the X_1X_2 plane only.

Fig. 5.24). Since gravitational forces act only in the negative X_1 direction, the velocity component $v_\phi = 0$ and the relative motion of sphere 2 with respect to sphere 1 occurs in a plane for which ϕ is constant. Without loss of generality, we can choose $\phi = 0$ and consider motion in the X_1X_2 plane only. Suppose a gravitational force $\boldsymbol{F}_1 = 6\pi\mu a_1 \boldsymbol{u}_1^0$ acts on the centre of sphere 1 and a force $\boldsymbol{F}_2 = 6\pi\mu a_2 \boldsymbol{u}_2^0$ on sphere 2 ($\boldsymbol{u}_1^0$ and $\boldsymbol{u}_2^0$ being the sedimentation velocities of the spheres in isolation), then the velocities of the spheres, denoted by $\boldsymbol{u}_1$ and $\boldsymbol{u}_2$, are determined by the generalized Stokes equation:

$$\begin{pmatrix} \boldsymbol{F}_1 \\ \boldsymbol{F}_2 \end{pmatrix} = \begin{pmatrix} \boldsymbol{f}_{11} & \boldsymbol{f}_{12} \\ \boldsymbol{f}_{21} & \boldsymbol{f}_{22} \end{pmatrix} \cdot \begin{pmatrix} \boldsymbol{u}_1 \\ \boldsymbol{u}_2 \end{pmatrix} \tag{5.63}$$

The 16 components of the resistance matrix $\boldsymbol{f}$ can be found from eq (4.2) by equating all torques to zero. In doing so, all the angular velocities can be expressed in terms of translational velocities (using $\boldsymbol{c}\cdot\boldsymbol{u} + \boldsymbol{f}_r\cdot\omega = 0$) and eq (4.2) reduces to eq (5.63). Inverting eq (5.63) results in

$$\begin{pmatrix} \boldsymbol{u}_1 \\ \boldsymbol{u}_2 \end{pmatrix} = \boldsymbol{f}^{-1}\cdot \begin{pmatrix} \boldsymbol{F}_1 \\ \boldsymbol{F}_2 \end{pmatrix} = \begin{pmatrix} \boldsymbol{M}_{11} & \boldsymbol{M}_{12} \\ \boldsymbol{M}_{21} & \boldsymbol{M}_{22} \end{pmatrix} \cdot \begin{pmatrix} \boldsymbol{F}_1 \\ \boldsymbol{F}_2 \end{pmatrix} \tag{5.64}$$

where $\boldsymbol{M}_{ij}$ are the mobility tensors. By following the above outlined procedure for determining $\boldsymbol{f}$ it can be shown that

$$\boldsymbol{M}_{ij} = \frac{1}{3\pi\mu(a_1 + a_2)} \begin{pmatrix} A_{ij} & 0 \\ 0 & B_{ij} \end{pmatrix} \tag{5.65}$$

where the two-sphere mobility functions A_{ij} and B_{ij} are functions of the dimensionless separation distance r (the distance divided by $\frac{1}{2}(a_1 + a_2)$) and the radius ratio $q = a_2/a_1$. Values for the functions A_{ij} and B_{ij} can be found in Batchelor (1982). When $i = j$, A_{ij} and $B_{ij} \to 1$ as $r \to \infty$, and when $i \neq j$ they approach zero. As $r \to 2$, the A_{ij} become proportional to $(r-2)$, while the B_{ij} approach a constant value. From eq (5.64) it follows that the relative velocity between the spheres is

$$\boldsymbol{u}_{12} = \boldsymbol{u}_2 - \boldsymbol{u}_1 = 6\pi\mu a_1 \boldsymbol{u}_1^0\cdot(\boldsymbol{M}_{21} - \boldsymbol{M}_{11}) + 6\pi\mu a_2 \boldsymbol{u}_2^0\cdot(\boldsymbol{M}_{22} - \boldsymbol{M}_{12}) \tag{5.66}$$

which can be written as

$$\boldsymbol{u}_{12} = \boldsymbol{u}_{12}^0 \cdot \begin{pmatrix} P(r,q) & 0 \\ 0 & Q(r,q) \end{pmatrix} \tag{5.67}$$

with

$$P(r)=\frac{q^2\gamma A_{22}-A_{11}}{q^2\gamma-1}+\frac{2(1-q^3\gamma)A_{12}}{(1+q)(q^2\gamma-1)} \tag{5.68a}$$

and

$$Q(r)=\frac{q^2\gamma B_{22}-B_{11}}{q^2\gamma-1}+\frac{2(1-q^3\gamma)B_{12}}{(1+q)(q^2\gamma-1)} \tag{5.68b}$$

where

$$\gamma=(\rho_2-\rho)/(\rho_1-\rho) \tag{5.69}$$

In polar coordinates, eq (5.67) can be written as

$$\begin{aligned}\frac{\mathrm{d}r}{\mathrm{d}t}&=-u_{12}^0P(r)\cos\theta\\ \frac{\mathrm{d}\theta}{\mathrm{d}t}&=u_{12}^0\frac{Q(r)}{r}\sin\theta\\ \frac{\mathrm{d}\phi}{\mathrm{d}t}&=0\end{aligned} \tag{5.70}$$

Although the trajectory equations (eqs (5.70)) appear to be simple, they show a fascinating variety of possible relative trajectories. This is due to the fact that, under certain conditions (certain values of γ and q), the functions $P(r)$ and $Q(r)$ can change sign. From graphs of the functions $P(r)$ and $Q(r)$ given by Wacholder and Sather (1974), it can be seen that for small values of γ the functions $P(r)$ and $Q(r)$ are positive for all r, but for higher γ values $P(r)$ is negative at small r and positive at large r, and there is a distance, say r_1, at which $P(r_1)=0$. At even larger γ values, $Q(r)$ also becomes negative at small r, and there is a distance, say r_2, at which $Q(r_2)=0$. From the definitions of the functions P and Q it can be proved that in all cases $r_2 < r_1$. It follows that we can distinguish four different types of relative trajectory:

(A) P and Q positive for all r;
(B) P negative for some r, Q positive for all r;
(C) Both P and Q negative for some r;
(D) P and Q negative for all r.

Figure 5.25 shows under which conditions the various types of trajectory occur. Examples of trajectories are shown in Fig. 5.26. Trajectories of type (A) consist of a heavier sphere approaching the lighter reference sphere from above. All these trajectories are open and extend from $-\infty$ to $+\infty$. The

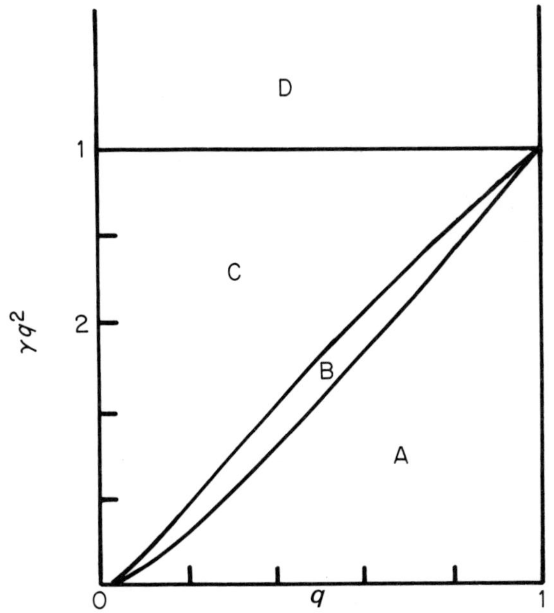

Fig. 5.25. Conditions for the occurrence of various types of relative trajectory between two sedimenting spheres (after Wacholder and Sather, 1974).

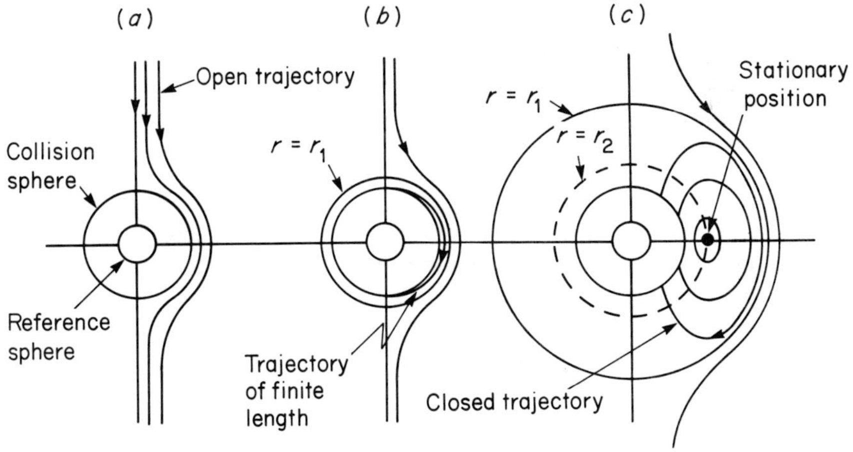

Fig. 5.26. Examples of the possible relative trajectories for two sedimenting spheres.
(a) Type (A): open trajectories from above; type (D) is identical with arrows reversed.
(b) Type (B): open trajectories and trajectories of finite length.
(c) Type (C): open and closed trajectories (after Wacholder and Sather, 1974).

minimum distance of approach is $r=2$ (contact) for a trajectory on $x_2=0$. Trajectory type (B) is characterized by the property that $P(r_1)=0$, i.e. at distance r_1, $dr/dt=0$ and particles never cross the circle $r=r_1$. It follows that the minimum distance of approach for an open trajectory is r_1. At $r < r_1$ trajectories of finite length occur, which start from and end on the collision sphere ($r=2$). Trajectories of type (C) are characterized by two distances r_1 and r_2 ($r_2 < r_1$) where $P(r_1)=0$ and $Q(r_2)=0$. As in case (B), open trajectories remain outside the circle of radius $r=r_1$. For $r < r_1$ there is a distance r_2 at which $d\theta/dt=0$, i.e. for which the only velocity component is in the (outward) radial direction. At $r < r_2$, $d\theta/dt$ is decreasing, while at $r > r_2$, $d\theta/dt$ is increasing. As a result, closed orbits occur within the region $r < r_1$. When $r=r_2$ and $\theta=90°$, both dr/dt and $d\theta/dt$ are zero and the particle remains fixed in position. Case (D), for which P and Q are negative for all r, is similar to case (A), but with the direction reversed, i.e. a lighter sphere approaches from below and moves up passing the heavier reference sphere.

The trajectory equations (eqs (5.70)) can be integrated with the result that

$$\sin\theta = C\,e^{-f(r)} \tag{5.71}$$

where C is a trajectory constant and

$$f(r) = \int_{r^*}^{r} \frac{Q(r)}{rP(r)}\,dr$$

where $r^* = r_1$ for open trajectories of type (B) or (C) and $r^* = 2$ for either a finite-length or a closed trajectory of type (B) or (C) or for trajectories of type (A) or (D). Because $x_2 = r\sin\theta$ when $\phi = 0$, we can also write

$$x_2 = C\,r\,e^{-f(r)} \tag{5.72}$$

For open trajectories, a value of zero for the trajectory constant C corresponds to sphere 2 approaching sphere 1 along the X_1 axis, either coming into contact with it after an infinite amount of time has elapsed ((A) or (D)), or stopping at $r=r_1$ ((B) or (C)). An infinite value of this trajectory constant corresponds to sphere 2 moving down in a straight line infinitely far away from sphere 1 (as $r \to \infty$, $f(r) \propto \ln r$ and $x_2 \to$ constant).

5.3.2 Sedimentation of two soft spheres: gravity-induced coagulation

By analogy with eq (5.8) for the interaction between two spheres subjected to shear and van der Waals dispersion forces, in the case of interaction between two spheres subjected to gravitational and van der Waals forces, we can write

$$F_{gr} + F_{d} + F_{vdW} = 0 \tag{5.73}$$

Rendering distances dimensional by dividing them by $\bar{a} = (a_1 + a_2)/2$, this equation can be written in component form as

$$\begin{aligned} \frac{dr}{dt} &= -u_{12}^{0} P(r) \cos\theta + \frac{C(r,q) F_{coll}}{3\pi\mu\bar{a}^2} \\ \frac{d\theta}{dt} &= u_{12}^{0} \frac{Q(r) \sin\theta}{r} \end{aligned} \tag{5.74}$$

For electrostatically charged spheres we must take into account the body force $\rho\nabla\psi$ per unit volume of fluid (ρ being the charge density and ψ the potential) in the creeping-flow equation which describes the flow field around the spheres. This results in electroviscous corrections to the functions $P(r)$, $Q(r)$ and $C(r)$, which for low potentials are of the form

$$\tilde{P}(r) = P(r)\,[1 + \bar{b}\Phi_s^2 f(\kappa\bar{a}, q, r)] \tag{5.75}$$

with f an as yet unknown function. For spheres with unequal potentials, f also depends on the potential ratio β. Thus for charged spheres, eq (5.74) must be modified to

$$\begin{aligned} \frac{dr}{dt} &= -u_{12}^{0} \tilde{P}(r)\cos\theta + \frac{\tilde{C}(r,q) F_{coll}}{3\pi\mu\bar{a}^2} \\ \frac{d\theta}{dt} &= u_{12}^{0} \frac{\tilde{Q}(r)\sin\theta}{r} \end{aligned} \tag{5.76}$$

Including the colloidal forces in the trajectory equations results in asymmetric trajectories. When the colloidal force is attractive at short separations, two spheres can come into contact and form an aggregate. In this case, the limiting trajectories (separating open trajectories from those leading to capture) can be calculated by integrating eq (5.76) backwards, starting at $r = 2$ (or at $r = r_1$ for type (B) and (C) trajectories), $\theta = -\pi$, and calculating the distance x_2 as $r \to \infty$. Denoting this distance by ℓ, the number of particles of radius a_2 colliding with a sphere of radius a_1 is simply

$$J = \pi\ell^2 u_{12}^{0} n_2 \tag{5.77}$$

n_2 being the number concentration of spheres of radius a_2. If we assume that the hydrodynamic forces are absent, in which case a sphere would fall in a straight line until it hits another sphere, then the number of collisions per unit time would be equal to

$$J_r = \pi(a_1 + a_2)^2 u_{12}^0 n_2 \tag{5.78}$$

J_r is a convenient reference-collision frequency. By comparing eqs (5.77) and (5.78) we can define a gravity-induced capture efficiency, α_G, by

$$\alpha_G \equiv J/J_r = \ell^2/(a_1 + a_2)^2 \tag{5.79}$$

The set of eqs (5.76) was solved numerically by Melik and Fogler (1984a) for the case of two spheres with the same potential, ψ_0, and (implicitly) assuming that $\tilde{P}=P$, $\tilde{Q}=Q$ and $\tilde{C}=C$, by integrating eq (5.76) forward in time and calculating the radius of the capture cross-section ℓ by trial and error.

Defining a dimensionless van der Waals force by

$$\mathrm{Ad}\, F_{\mathrm{vdW}} = -\frac{1}{kT}\frac{\partial V_{\mathrm{vdW}}}{\partial r} \tag{5.80}$$

with V_{vdW} given by eq (1.76) and a dimensionless electrostatic repulsive force by

$$\mathrm{Dl}\, F_{e\ell} = -\frac{1}{kT}\frac{\partial V_{e\ell}}{\partial r} \tag{5.81}$$

with $V_{e\ell}$ given by eq (1.133), eq (5.76) can be written as

$$\frac{dr}{d\theta} = -\frac{r[\tilde{P}(r)\cos\theta + \tilde{C}(r,q)\,\mathrm{Pe}^{-1}(\mathrm{Ad}F_{\mathrm{vdW}} + \mathrm{Dl}F_{e\ell})]}{\tilde{Q}(r)\sin\theta} \tag{5.82}$$

Here the Péclet number Pe is defined as

$$\mathrm{Pe} = \frac{3\pi\mu\bar{a}^2 u_{12}^0}{kT} \tag{5.83}$$

the adhesion number, Ad, as

$$\mathrm{Ad} = \frac{A}{6kT} \tag{5.84}$$

and the double-layer number, Dl, as

$$\mathrm{Dl} = \frac{4\pi\epsilon\epsilon_0\psi_0^2\bar{a}}{kT} \tag{5.85}$$

Besides these three dimensionless parameters, the colloidal forces also depend on the reciprocal double-layer thickness $\tau = \kappa\bar{a}$, the retardation parameter $\lambda^* = \lambda_L/2\pi\bar{a}$, and the radius ratio q. More generally, the solution also depends on the scaled friction coefficients, b_i, of the ions in solution, on the reduced zeta-potential, Φ_s, and the potential ratio, β. (These parameters are related to those defined by Melik and Fogler by: $\mathrm{Ad} = [(1+q)^2/12q]N_A$, $\mathrm{Dl} = [(1+q)^3/2q^2]N_A N_R$ and $\mathrm{Ad/Pe} = [(1+q)^2/12q]/N_G$.) Because of the way Pe, Ad and Dl enter into eq (5.82), only two dimensionless parameters need specification, which we can choose to be Ad/Pe (the ratio of dispersion to gravity forces) and Dl/Ad (the ratio of electrostatic to dispersion forces).

Because the colloidal forces act only when the spheres are close together, the spheres will be on a purely hydrodynamic trajectory when they are sufficiently far apart, both prior to and after an encounter. Thus for the trajectory of approach we can write (cf. eq (5.72))

$$x_2 = C_a\, r\, e^{-f(r)} \tag{5.86a}$$

and for the trajectory of recession

$$x_2 = C_r\, r\, e^{-f(r)} \tag{5.86b}$$

where C_a and C_r are the trajectory constants of approach and recession, respectively. The difference between C_a and C_r is a measure of the colloidal force between the spheres; when $C_a > C_r$ a net attraction acts between them, while when $C_a < C_r$ a net repulsion exists.

From the numerical solutions of eq (5.82) it can be concluded that the coagulation behaviour of two spheres sedimenting under the action of gravity is qualitatively very similar to the orthokinetic coagulation in simple shear, as discussed in Section 5.1.5. In Section 5.1.5 it was shown that, in general, suspensions are unstable at low and high shear rates, with a region of relative stability in between. The same is true for gravity-induced coagulation. For small and large values of the relative sedimentation velocity, u_{12}^0, suspensions are unstable, while there is a regime of stability at intermediate u_{12}^0 values. This can be seen from Fig. 5.27 which shows the coagulation efficiency, α_G, as a function of Pe/Ad for various values of τ. At low Péclet numbers (low u_{12}^0) coagulation in a secondary minimum occurs, while at high Péclet numbers (high u_{12}^0) coagulation in a primary minimum

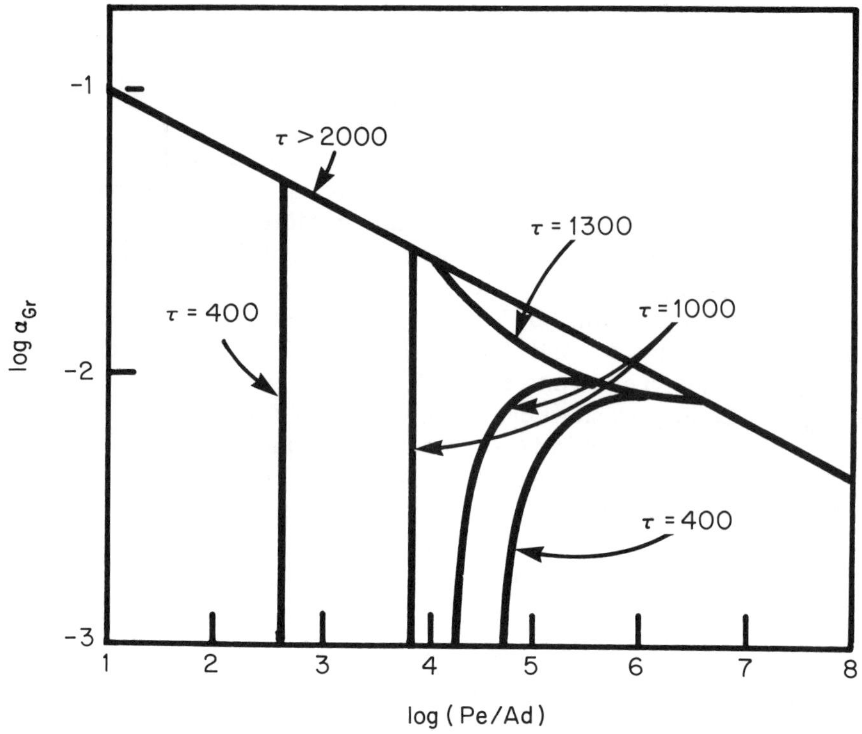

Fig. 5.27. Effect of the relative sedimentation velocity, u_{12}^0 (proportional to Pe), on the gravity-induced capture efficiency for various electrolyte concentrations. Dl/Ad = 3600, $\lambda_L/a = 0{\cdot}1$, $\gamma = 1$, $q = 2$ (after Melik and Fogler, 1984a).

takes place. At intermediate values of Pe/Ad the efficiency reaches very low values. When $\mathrm{Dl} < \mathrm{Dl}_{\mathrm{crit}}$ or $\tau > \tau_{\mathrm{crit}}$ (corresponding to the critical coagulation concentration), fast coagulation occurs, i.e. the electrostatic repulsive forces have no effect on the kinetics of coagulation. Under such conditions

$$\alpha_G \propto (\mathrm{Ad/Pe})^{0{\cdot}20} \tag{5.87}$$

As in the case of peri- and orthokinetic coagulation, we can expect the critical coagulation concentration to depend on the size of the ions in the double layer surrounding the particles, an effect not included in the numerical calculations discussed in the above paragraph. The above results are for systems in which Brownian motion can be neglected. When Brownian motion is important, the coagulation rate can be determined from the solution of the steady-state convective-diffusion equation

$$\nabla \cdot (\boldsymbol{D} \cdot \nabla n) = \nabla \cdot (\boldsymbol{u}_{\text{sed}}\, n) \tag{5.88}$$

where $\boldsymbol{D}$ is the relative diffusion tensor, given by eq (4.26) and $\boldsymbol{u}_{\text{sed}}$ is the relative sedimentation velocity of the spheres in which the components $\mathrm{d}r/\mathrm{d}t$ and $r\mathrm{d}\theta/\mathrm{d}t$ are given by eq (5.76) and in which the velocity due to colloidal forces is included.

No general solution of eq (5.88) exists but, similar to combined peri- and orthokinetic coagulation in shear (Section 5.1.6), in principle it is possible to find expansion solutions for high or low values of the Péclet number, Pe, defined by eq (5.88). So far, only an expansion in small Péclet numbers is available (Brownian motion dominating over sedimentation). In sedimentation the expansion also turns out to be a singular perturbation expansion, which consists of an inner expansion valid at small separations and an outer expansion valid at large separations, similar to the small Péclet number expansion for coagulation in Shear. By matching both expansions it was found that (Melik and Fogler, 1984b)

$$\mathrm{J} = \mathrm{J}_0\,(1 + \alpha_{\mathrm{p}}\mathrm{Pe}) \tag{5.89}$$

where J_0 is the perikinetic doublet-formation frequency (given by eq (4.77)) and α_{p} the perikinetic capture efficiency (given by eq (4.80)).

5.3.3 Concentration dependence of mean sedimentation velocities

In Sections 4.2 and 4.6, where the mutual diffusion constant for a two-sphere system was discussed, we saw that, according to Batchelor (1972), the mean velocity of a sedimenting dispersion of hard Brownian spheres is

$$\bar{u} = u_0(1 + Sc) \tag{5.90}$$

where, for equal-sized spheres, $S = -6{\cdot}55$. We have also seen that the inclusion of interparticle forces affects the value of S:

$$S = -6{\cdot}55 + \int_2^{\infty} g(r)\,(\mathrm{e}^{-V_{\text{int}}/kT} - 1)\, r^2\mathrm{d}r \tag{5.91a}$$

with $g(r) = A_{11} + 2B_{11} - 3 + A_{12} + 2B_{12}$, A_{ij} and B_{ij} being the mobility functions defined by eq (5.65). Equation (5.91a) can be simplified by noting that $g(r)$

varies only by a few percent in the range $2 < r < 2{\cdot}2$; for larger values of r the interaction energy, V_{int}, is essentially zero. We then have

$$S \simeq -6{\cdot}55 + 1{\cdot}32 \int_2^\infty (\mathrm{e}^{-V_{\text{int}}/kT} - 1)\, r^2 \mathrm{d}r \tag{5.91b}$$

It must be noted that eq (5.91) is not applicable to dispersions of electrically charged spheres, where the coupling between hydrodynamic flow and double-layer deformation must be taken into account. This electroviscous coupling affects all the coefficients of the resistance or mobility tensor. For small potentials, eq (5.91) must be modified to

$$S = -6{\cdot}55 - \bar{b}\Phi_s^2 f_1(\kappa a) + \int_2^s [g(r) + \bar{b}\Phi_s^2 f_2(\kappa a, r)]\, (\mathrm{e}^{-V_{\text{int}}/kT} - 1)\, r^2 \mathrm{d}r$$

or

$$S = S^0 - \bar{b}\Phi_s^2 f(V_{\text{int}}) \tag{5.92}$$

with S^0 given by eq (5.91) and

$$f = f_1 - \int_2^\infty f_2(\mathrm{e}^{-V_{\text{int}}/kT} - 1)\, r^2 \mathrm{d}r$$

For spheres which repel each other at all distances, $V_{\text{int}} > 0$ and the interaction term is always negative. It follows that the mean velocity of repelling spheres is less than that of hard spheres. The reason is that there are fewer close pairs which fall faster than two spheres far apart. The above results can be extended to the more general case of spheres of unequal radii. For the mean velocity of spheres of species i:

$$\bar{u}_i = u_i^0 \left(1 + \sum_{j=1}^{m} S_{ij} c_j\right) \tag{5.93}$$

where m is the number of species present. General expressions for S_{ij} are given by Batchelor and Wen (1982). For large Péclet numbers (defined by eq (5.83)) and in the absence of interparticle forces:

$$S_{ij} = \int_2^\infty \left[\left(\frac{1+q}{2q} \right)^3 (A_{11} + 2B_{11} - 3)\, n_{ij} + \frac{\gamma}{4}\,(1+q)^2 \left\{ (A_{12} + 2B_{12})\, n_{ij} - \frac{3}{r} \right\} \right]$$

$$\times\, r^2 \mathrm{d}r - \gamma (q^2 + 3q + 1) \tag{5.94}$$

Here n_{ij} is the pair distribution function which denotes the probability of finding a particle of species j at a given relative position with respect to a particle of species i and which is governed by the convective-diffusion equation. Under steady-state conditions, in the absence of Brownian motion and any interparticle forces, n_{ij} is determined by

$$\nabla \cdot (n_{ij} \boldsymbol{u}_{ij}) = 0 \tag{5.95}$$

with the relative particle velocity given by eqs (5.67) or (5.70). The solution of this equation is

$$\ln n_{ij} = \int_r^\infty \left[\frac{2(P-Q)}{rP} + \frac{1}{P} \frac{\mathrm{d}P}{\mathrm{d}r} \right] \mathrm{d}r \tag{5.96}$$

with P and Q given by eq (5.68). S_{ij} can be obtained by numerical intergration of eqs (5.94) and (5.96). Some selected values of S_{ij} are given in Fig. 5.28, which shows S_{ij} versus the radius ratio q for various values of the relative density parameter γ for trajectories of type (A) or (D). It can be seen that S_{ij} can take on all kinds of values, both positive and negative, depending on the values of q and γ. Some asymptotic values of S_{ij} are of interest: when $q \ll 1$

$$S_{ij} \simeq -2{\cdot}5 - \gamma \tag{5.97}$$

When $\gamma = 0$, it turns out that S_{ij} is close to $-2{\cdot}5$ for most values of q. When $q \gg 1$ we have the relation

$$S_{ij} \to -\gamma (q^2 + 3q + 1) \tag{5.98}$$

For $q = 1$ we have

$$S_{ij} = -2{\cdot}52 - 0{\cdot}13\, \gamma \tag{5.99}$$

which would suggest a value of S_{ij} of $-2{\cdot}65$ at $\gamma = 1$. However eq (5.99) is not valid when $\gamma = 1$. The reason for the singular behaviour at $\gamma = q = 1$ is that

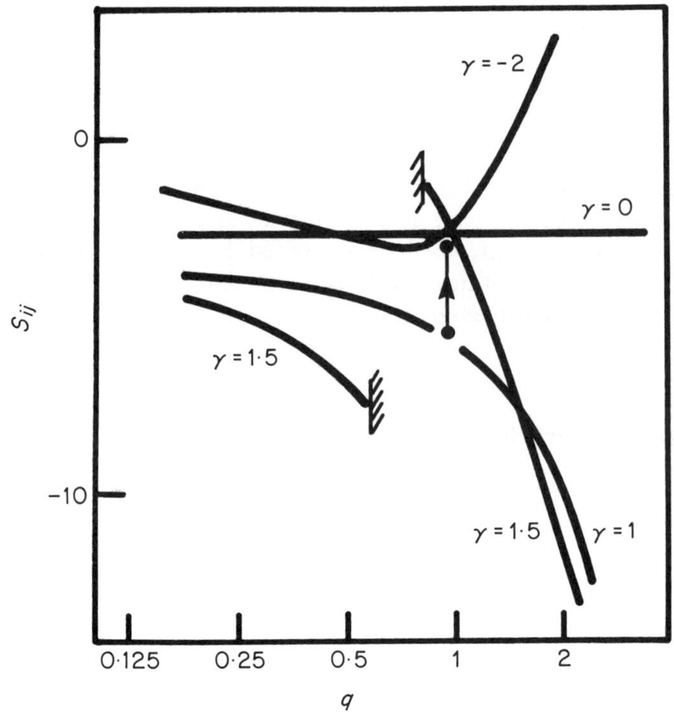

Fig. 5.28. The sedimentation coefficient, S_{ij}, as a function of the radius ratio, q, for various values of γ (Pe $\gg 1$, $F_{\text{coll}}=0$). For $\gamma=1\cdot5$ for some interval of q values no values of S_{ij} are calculated because of the occurrence of trajectories of finite length. The point $\gamma=q=1$ is singular and the steady-state value lies well above the curve $\gamma=1$ (after Batchelor and Wen, 1982).

there is no relative motion between two identical spheres falling under gravity.

In the limit of q (or γ) $\to 1$, it takes an infinite time to reach the steady-state pair distribution function (eq (5.96)); at short times the distribution is close to the one corresponding to q (or γ)$=1$, but at longer times the distribution starts to deviate from the one governed by the initial position of the spheres.

It is also possible to determine S_{ij} for small Péclet numbers, when it can be decomposed into two contributions, one due to gravity and one due to Brownian motion:

$$S_{ij}=S'_{ij}+\gamma\, S''_{ij} \tag{5.100}$$

with $S'_{ij} = S'^{\mathrm{G}}_{ij} + S'^{\mathrm{B}}_{ij}$ and $S''_{ij} = S''^{\mathrm{G}}_{ij} + S''^{\mathrm{B}}_{ij}$, where the superscripts G and B refer to gravity and Brownian motion, respectively. To evaluate the values of the various contributions to S_{ij} we again need to know the pair distribution function n_{ij}, determined from

$$\nabla \cdot (\boldsymbol{D}_{ij} \cdot \nabla n_{ij}) = \nabla \cdot (\boldsymbol{u}_{ij} n_{ij}) \tag{5.101}$$

$\boldsymbol{D}_{ij}$ being the relative diffusion tensor. For Pe $\ll 1$, the solution of this equation is

$$n_{ij}(r, \theta) = 1 + \mathrm{Pe}\, Z(r) \cos\theta \tag{5.102}$$

with $Z(r)$ determined by the differential equation

$$\frac{\mathrm{d}}{\mathrm{d}r}\left(r^2 \mathrm{d}_{\parallel} \frac{\mathrm{d}Z}{\mathrm{d}r}\right) - 2d_{\perp} Z = r^2 \left[\frac{2(P-Q)}{r} + \frac{\mathrm{d}P}{\mathrm{d}r}\right] \tag{5.103}$$

($d_{\parallel}$ and $d_{\perp}$ being the non-dimensionalized components of $\boldsymbol{D}_{ij}$) subject to the boundary conditions $Z \to 0$ as $r \to \infty$ and $d_{\parallel}\mathrm{d}Z/\mathrm{d}r = 0$ at $r = 2$.

In this case the various contributions to S_{ij} can be expressed as

$$S'^{\mathrm{G}}_{ij} = \left(\frac{1+q}{2q}\right)^3 \int_2^{\infty} (A_{11} + 2B_{11} - 3)\, r^2 \mathrm{d}r \tag{5.104a}$$

$$S''^{\mathrm{G}}_{ij} = \left(\frac{1+q}{2}\right)^2 \int_2^{\infty} \left(A_{12} + 2B_{12} - \frac{3}{r}\right) r^2 \mathrm{d}r - (q^2 + 3q + 1) \tag{5.104b}$$

and

$$S^{\mathrm{B}}_{ij} = \left(\frac{1+q}{2q}\right)^2 \int_2^{\infty} \left[\frac{A_{11} - B_{11}}{r} + \frac{1}{2}\frac{\mathrm{d}A_{11}}{\mathrm{d}r} - \frac{2(A_{12} - B_{12})}{(1+q)r} - \frac{1}{1+q}\frac{\mathrm{d}A_{12}}{\mathrm{d}r}\right] \times (\gamma^2 q - 1)\, Z r^2 \mathrm{d}r \tag{5.104c}$$

When $\gamma = q = 1$, numerical integration yields $S'^{\mathrm{G}}_{ij} = -1{\cdot}83$; $S''^{\mathrm{G}}_{ij} = -4{\cdot}72$; and $S^{\mathrm{B}}_{ij} = 0$ (which follows directly from eq (5.104c)), and we recover the result $S_{ij} = -6{\cdot}55$ for identical spheres. Values for other values of γ and q can be found in Batchelor and Wen (1982).

Finally, we will discuss the effects of interparticle forces on S_{ij}. In this case $\boldsymbol{u}_{ij}$ in eq (5.101) is given by eq (5.76) and the solution for n_{ij} is

$$n_{ij} = e^{-V_{\text{int}}/kT}\,[1 + \text{Pe}\, Z(r)\cos\theta] \tag{5.105}$$

with $Z(r)$ determined by

$$\begin{aligned}\frac{d}{dr}\left(r^2 d_{\|}\frac{dZ}{dr}\right) - \frac{r^2 d_{\|}}{kT}\frac{\partial V_{\text{int}}}{\partial r}\frac{dZ}{dr} - 2d_{\perp} Z \\ = r^2\left[\frac{2(P-Q)}{r} + \frac{dP}{dr}\right] - \frac{r^2 P}{kT}\frac{\partial V_{\text{int}}}{\partial r}\end{aligned} \tag{5.106}$$

subject to the same boundary conditions as eq (5.103). Notice that now $d_{\|}$ and $d_{\perp}$ are functions of b_i, Φ_s and τ.

The sedimentation coefficient now contains an additional contribution, S^{I}_{ij}, due to interparticle forces. For equal spheres, and neglecting the electroviscous contributions to A_{ij} and B_{ij}, the result for S_{ij} reduces to eq (5.91a). For other values of q and γ, results can again be found in Batchelor and Wen (1982). In all cases the trend is the same as for equal-sized spheres, repulsion resulting in lower S_{ij} values and attraction in larger ones.

Experimental values for sedimentation coefficients for monodisperse spheres are often found to lie in the range -5 to -6, i.e. greater than the theoretical value $-6{\cdot}55$. One possible explanation is that van der Waals attractive forces are responsible for the difference. Another possibility is that the discrepancy is due to experimental difficulties. Most often the velocity of the top front of a sedimenting dispersion is measured, as this is most readily observed. It is assumed that the velocity of the front equals the mean velocity in the bulk. However, the sedimentation behaviour of spheres near the front is quite complicated and differs from that in the bulk (Janse van Rensburg, 1988). Furthermore, observations are often performed at volume fractions where more than two-body interactions cannot be fully neglected and the results are then extrapolated to low concentrations, which introduces complications in interpreting sedimentation data. Experimental data on bimodal dispersions show interesting instabilities (Batchelor and Janse van Rensburg, 1986).

The preceding analysis shows that it is possible, in principle, to obtain information about interparticle forces by measuring the sedimentation coefficients of sedimenting particles. However, the theory has so far only been developed for hard spheres that bear no electrical charge. No exact results are available for charged spheres or particles of other shapes. To make additional progress, the theory must be developed further and reliable experimental data must be obtained. In principle, one could measure the velocities of individual test particles (using, for example, the travelling microtube) in a dispersion in which most particles are optically invisible (due to refractive-index matching), or in which certain particles are labelled by some means (radioactivity, phosphorescence, etc.).

5.4 Two spheres in an electric field

5.4.1 General remarks

When two spheres, suspended in a liquid, are subjected to an external electric field, they experience electric forces which cause them to move. For charged spheres we can distinguish three different types of electric force: (i) one due to electrostatic repulsion or attraction between the spheres (Coulombic-type forces); (ii) one associated with the electrophoretic mobility of the particle, which arises as a result of the response of a charged particle to a uniform electric field; and (iii) electric forces which stem from induced dipoles in the particles, which occur in charged and uncharged particles alike. These induced forces can be thought of as being consequences of the fact that a single dielectric particle experiences a force in a non-uniform electric field which causes it to move, a phenomenon usually referred to as dielectric electrophoresis. A single particle in a uniform field locally distorts the field and a second particle placed in its neighbourhood experiences a non-uniform electric field. As a result there is relative motion between two dielectric spheres placed in an electric field. The two spheres can either move towards or away from each other, depending on the relative orientation.

In order to determine the relative velocity u_{12} of the spheres, the forces $\boldsymbol{F}$ acting parallel and perpendicular to the line of sphere centres must be considered:

$$\boldsymbol{F} = \boldsymbol{f} \cdot \boldsymbol{u}_{12} \tag{5.107}$$

where the relative resistance tensor, $\boldsymbol{f}$, is related to the relative diffusion tensor, $\boldsymbol{D}$, by $\boldsymbol{D} = \boldsymbol{f}^{-1}\, kT$. As mentioned above, there are three types of force acting on two particles in an electric field:

$$\boldsymbol{F} = \boldsymbol{F}_{\mathrm{de}} + \boldsymbol{F}_{\mathrm{em}} + \boldsymbol{F}_{\mathrm{coll}} \tag{5.108}$$

where $\boldsymbol{F}_{\mathrm{de}}$ is the force due to dielectric electrophoresis, $\boldsymbol{F}_{\mathrm{em}}$ the force related to the electrophoretic mobility of the spheres, and $\boldsymbol{F}_{\mathrm{coll}}$ the colloidal force (which includes the electrostatic Coulombic force) between the particles. Inversion of eq (5.107) gives the relative velocity between the two spheres:

$$\boldsymbol{u}_{12} = \frac{\boldsymbol{D} \cdot \boldsymbol{F}}{kT} \tag{5.109}$$

Next we consider the consequences of eq (5.109) starting with hard spheres and subsequently introducing additional contributions for soft spheres.

5.4.2 Two hard spheres in an electric field

The problem of finding the electrical forces between two spheres in an electric field was solved by Davis (1964). For equal-sized spheres of radius a the force acting on one sphere along the direction of sphere centres is:

$$F_1 = 4\pi\epsilon_0\epsilon_2 E^2 a^2 \,[f_1(r,q)\cos^2\theta_2 + f_2(r,q)\sin^2\theta_2] \qquad (5.110\text{a})$$

and the force perpendicular to the line of centres:

$$F_2 = 4\pi\epsilon_0\epsilon_2 E^2 a^2 f_3(r,q)\sin 2\theta_2 \qquad (5.110\text{b})$$

Here f_1, f_2 and f_3 are functions of the dimensionless distance between the sphere centres and the dielectric constant ratio $q=\epsilon_1/\epsilon_2$, ϵ_1 being the dielectric constant of the particle and ϵ_2 that of the medium. (In this section q stands for the dielectric constant ratio, not the radius ratio, to be consistent with the usage in Sections 3.32 and 3.34.) The angle θ_2 is the angle between the line of sphere centres and the direction of the field (cf. Fig. 3.73) of magnitude E, and ϵ_0 is the permittivity of free space.

Inserting eq (5.110) into eq (5.109) and writing the relative velocity in terms of the polar components $\mathrm{d}r/\mathrm{d}t$ and $r\mathrm{d}\theta_2/\mathrm{d}t$ gives

$$\frac{\mathrm{d}r}{\mathrm{d}t} = -\frac{1}{\tau_{\mathrm{el}}}\,[L_1(r,q)\cos^2\theta_2 - L_2(r,q)\sin^2\theta_2] \qquad (5.111\text{a})$$

$$\frac{\mathrm{d}\theta_2}{\mathrm{d}t} = -\frac{1}{\tau_{\mathrm{el}}}\,\frac{L_3(r,q)\sin 2\theta_2}{r} \qquad (5.111\text{b})$$

where L_1, L_2 and L_3 are functions of the separation distance and the dielectric constant ratio (and for unequal-sized spheres of the radius ratio as well). The charactristic time, τ_{el}, is defined as

$$\tau_{\mathrm{el}} = \frac{\mu}{8\pi\epsilon_0\epsilon_2 E^2} \qquad (5.112)$$

The values of L_1, L_2 and L_3 for conducting spheres ($q=\infty$) have been tabulated by Arp and Mason (1977c), together with the asymptotic expressions for large and small separation distances. When $h=r-2 \ll 1$,

$$L_1(r,\infty) = \frac{\pi^3}{54\ln^2(h/23{\cdot}689)} + \ldots \qquad (5.113\text{a})$$

$$L_2(r,\infty)=0\cdot0113h+0\cdot0100h^2\ln(h/23\cdot6)+\ldots \tag{5.113b}$$

$$L_3(r,\infty)=0\cdot0160+0\cdot359/\ln(h/12\cdot689)+\ldots \tag{5.113c}$$

When $r \gg 1$,

$$L_1(r,\infty)=\frac{1}{\pi}\left[\frac{1}{r^4}-\frac{3}{2r^5}\right]+\ldots \tag{5.114a}$$

$$L_2(r,\infty)=\frac{1}{2\pi}\left[\frac{1}{r^4}-\frac{3}{2r^5}\right]+\ldots \tag{5.114b}$$

$$L_3(r,\infty)=\frac{1}{2\pi}\left[\frac{1}{r^4}-\frac{3}{4r^5}\right]+\ldots \tag{5.114c}$$

It is of interest to note that as $r\to2$, both $L_1(r)$ and $L_2(r)\to0$, i.e. $dr/dt\to0$ and $L_3(r)$ approaches a constant. In this limit eq (5.111b) becomes identical to eq (3.419), which describes the alignment of a non-spherical particle in an electric field. Thus the alignment time for a touching doublet is:

$$\tau_{al}=\tau_{el}/2L_3(2,q) \tag{5.115}$$

or $P(r_e,q)=-8\pi L_3(2,q)$. By comparing the value of $L_3(2,\infty)$ with $P(r_e,q)$, given by eq (3.410), it follows that $r_e=2\cdot16$. Thus the equivalent axis ratio for a touching doublet of equal-sized spheres for alignment in an electric field ($r_e^E=2\cdot16$) is different from the actual geometrical axis ratio ($r_p=2$) and from the equivalent axis ratio for the rotation of a doublet in shear flow ($r_e=1\cdot9817$). Differences between r_e^E, r_e and r_p were also found by Okagawa and Mason (1974) for the rotation of cylindrical rods.

Dividing eq (5.111a) by eq (5.111b) yields the trajectory equation for the relative motion of hard spheres in an electric field.

$$\frac{dr}{d\theta_2}=\frac{r[L_1\cos^2\theta_2-L_2\sin^2\theta_2]}{L_3\sin2\theta_2} \tag{5.116}$$

It does not seem possible to integrate eq (5.116) analytically in a closed form for the entire range $2\leqslant r<\infty$, but analytical solutions for large and small separation distances can be found. It is interesting to note that as $t\to-\infty$, all trajectory equations converge to the same point, $r=2$ and $\theta_2=90°$. Thus all trajectories have a common source. This can be seen by integrating eq (5.111b) backwards in time. After sufficiently long times, the separation distance will be small enough for the approximate equation (eq (5.113c)) for L_3 to hold. Hence, when $h\ll1$,

$$\frac{\mathrm{d}\theta_2}{\mathrm{d}t} = -\frac{L_3(2,q)}{2\tau_{\mathrm{el}}}\sin 2\theta_2 \tag{5.117}$$

where $L_3(2,q)$ is a constant, equal to 0·0160 when $q=\infty$. Integrating eq (5.117) yields

$$\tan\theta_2 = K\,\mathrm{e}^{-L_3 t/\tau_{\mathrm{el}}} \tag{5.118}$$

where K is an integration or trajectory constant. It follows that as $t \to -\infty$, $\theta_2 \to 90°$, independent of the value of K. Inserting eq (5.118) into eq (5.111a) yields

$$\frac{\mathrm{d}r}{\mathrm{d}t} = -\frac{1}{\tau_{\mathrm{el}}}\left[\frac{L_1}{1+K^2\,\mathrm{e}^{-2L_3 t/\tau_{\mathrm{el}}}} - \frac{L_2 K^2\,\mathrm{e}^{-2L_3 t/\tau_{\mathrm{el}}}}{1+K^2\,\mathrm{e}^{-2L_3 t/\tau_{\mathrm{el}}}}\right] \tag{5.119a}$$

As $t \to -\infty$ and $h \ll 1$, this reduces to

$$\frac{\mathrm{d}h}{\mathrm{d}t} = \frac{L_2(2,q)}{\tau_{\mathrm{el}}} = \frac{K_2 h}{\tau_{\mathrm{el}}} \tag{5.119b}$$

where K_2 is a constant which equals 0·0113 when $q=\infty$. Integrating the above equation yields

$$h = K_1\,\mathrm{e}^{K_2 t/\tau_{\mathrm{el}}} \tag{5.120}$$

which indicates that $h \to 0$ as $t \to -\infty$, independent of the trajectory constant K_1. Combining eqs (5.118) and (5.120) yields for the trajectory equation in the limit $t \to -\infty$:

$$\tan\theta_2 = C\,h^{-K_1/L_3} \tag{5.121}$$

C being the trajectory constant. Equation (5.121) shows clearly that $\theta_2 \to 90°$ as $h \to 0$, independent of C. It follows that all trajectories start off at the common point $(2, \pi/2)$, and their initial trajectories are given by eq (5.121), which is a limiting solution of eq (5.116).

Also, as $t \to \infty$ the spheres come together. In this case it can be seen from eq (5.118) that $\theta_2 \to 0$, i.e. the two spheres end up as a doublet aligned along the external electric field. However, the trajectories do not converge to the same point. For long time periods, eq (5.119a) reduces to

$$\frac{\mathrm{d}h}{\mathrm{d}t} = -\frac{L_1(h,q)}{\tau_{\mathrm{el}}} \tag{5.122}$$

which has the solution, when $h \ll 1$,

$$h = -C_1 t/\tau_{el} + C_2 \tag{5.123}$$

where, for $q = \infty$, $C_1 = \pi^3/108$ and C_2 is an integration constant. It follows from eq (5.123) that the two spheres come into contact in finite time. The time at which contact is reached is different for different trajectories, characterized by different constants C_2.

At large separations ($r > \sim 3$), the asymptotic expressions (eqs (5.114)) for the functions L_i can be used to integrate eq (5.116). For conducting spheres the solution is

$$\sin^2\theta_2 \cos\theta_2 = C^* r(r^2 - 3/2) \tag{5.124}$$

C^* being the trajectory constant. Because the spheres move in a plane of constant ϕ_2, we can, without loss of generality, describe particle motion in the X_2X_3 plane of a Cartesian coordinate system, the X_2 axis coinciding with the direction of the electric field. When $C^* = 0$, the spheres move along the

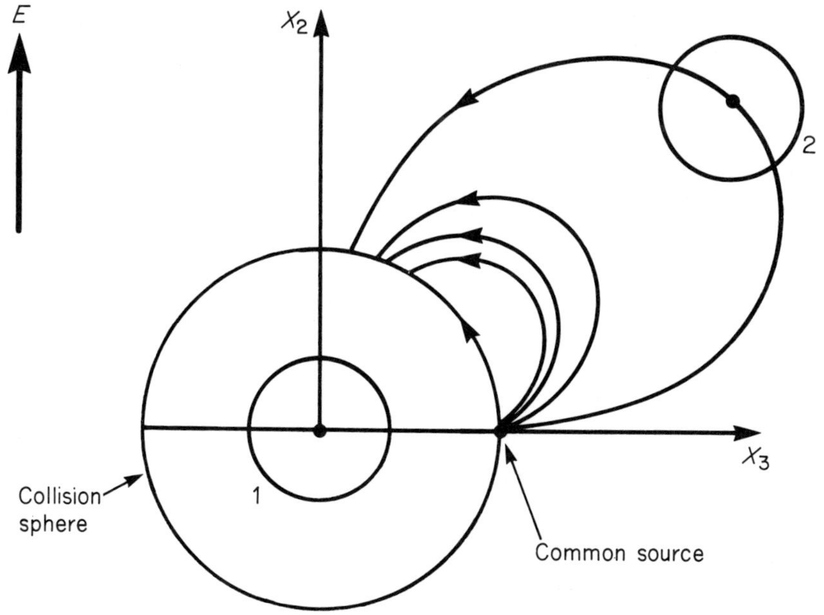

Fig. 5.29. Examples of the relative trajectories of two hard spheres subjected to an electric field. The collision sphere around the reference sphere towards which all trajectories converge is shown (spheres are not drawn to scale) (after Arp and Mason, 1977c).

X_2 or X_3 axis. For a trajectory characterized by $C^* \neq 0$, the spheres, starting from $r = 2$, $\theta = \pi/2$, first separate (as can be concluded from the solution for $t \to -\infty$), reach a maximum separation distance, r_{max}, at an angle θ_{max}, and subsequently approach one another, making contact at an angle θ_c. Afterwards the touching doublet aligns in the electric field varying its orientation from θ_c to 0°. The maximum separation, r_{max}, occurs when $\sin^2\theta_2 \cos\theta_2$ reaches its maximum value, i.e. when $\theta_2 = \theta_{max} = \sin^{-1}(\sqrt{2/3}) = 54.74°$. The value of r_{max} decreases with increasing C^*. The general features of the relative trajectories of two spheres in an electric field can be seen by the various examples shown in Fig. 5.29. The figure illustrates the common source of all trajectories as well as the absence of a common sink.

5.4.3 Relative trajectories of two soft spheres

A single suspended charged sphere, subjected to a uniform electric field of magnitude E, will move with an electrophoretic velocity u_i^0, say, which is proportional to E; $u_i^0 = mE$. The mobility m is given by eq (3.396) or a modification of it (cf. Fig. 3.71). Two charged spheres acted on by an electric field move with velocities u_1^0 and u_2^0, say, when they are in isolation. When they approach one another, their velocities are influenced by each other's presence. Their relative motion is very similar to that of two spheres in a gravity field, in which case the velocities u_1^0 and u_2^0 are their sedimentation velocities. Neglecting the body force $\rho\nabla\psi$ in the creeping-flow equation as well as colloidal forces and dielectric electrophoresis, the trajectory equations for two charged spheres in an electric field have the same form as for two sedimenting spheres (cf. Eq (5.70))

$$\frac{dr}{dt} = -u_{12}^0 P(r) \cos\theta_2$$
$$\frac{d\theta_2}{dt} = u_{12}^0 \frac{Q(r)}{r} \sin\theta_2 \qquad (5.125)$$

Here P and Q are basically the same as the functions defined in eq (5.68), the only difference being that the combination $q^2\gamma$ must be replaced by the mobility ratio m_2/m_1. (For sedimentation $u_2^0/u_1^0 = q^2\gamma$, while for electrophoresis $u_2^0/u_1^0 = m_2/m_1$.) As before, u_{12}^0 denotes $u_2^0 - u_1^0$. Introducing the body force $\rho\nabla\psi$ results in the modified functions $\tilde{P}(r)$ and $\tilde{Q}(r)$, which now also depend on b_i, Φ_s, κa and the electric field E. It follows from eq (5.125) that, when $u_{12}^0 = 0$, i.e. the isolated spheres move with the same velocity, there can be no relative motion between the spheres, no matter what their position or orientation is. This was proven by Reed and Morrison (1976) for two identical spheres either aligned along the electric field or perpendicular to it. However, eq (5.125) shows that this is true for all pairs of spheres with the same electrophoretic mobility. Two spheres with different ζ potentials and different radii can, in principle, have the same mobility, because they have different κa values (cf. Fig. 3.71). However, for unequal-sized spheres, the relative velocity, u_{12}^0, is usually non-zero (except for special circumstances); u_{12}^0 is also non-zero for equal-sized spheres of different potentials.

The relative trajectories of two spheres in electrophoretic motion are the same as those for two sedimenting spheres. As in the case of two sedimenting spheres four types of relative trajectories can be distinguished for two spheres in electrophoretic motion, examples of which are shown in Fig. 5.26 labelled (A), (B), (C) and (D). The conditions under which a trajectory of a certain type occurs are indicated in Fig. 5.25, where the vertical coordinate $q^2\gamma$ must be replaced by the mobility ratio m_2/m_1. Note that in Fig. 5.25, q denotes the radius ratio and not the ratio of the dielectric constants.

The relative velocity (eqs (5.125)) applies for electric fields, E, of small magnitude, when the relative motion due to induced dipoles (given by eq (5.111)) can be neglected. As $E \to 0$, $\tau_{\rm el} \to \infty$ and the relative velocity given by eq (5.111) becomes small compared to that given by eq (5.125). The relative importance of these velocities can be described by the dimensionless number V defined as

$$V = \frac{\bar{a}}{u_{12}^0 \tau_{\rm el}} \tag{5.126}$$

The parameter V^{-1} is equivalent to the Mason number, Mn, which is defined below eq (3.454). When $V \ll 1$, eq (5.125) applies, while for $V \gg 1$, eq (5.111) must be used. In general, both motions are present and we can write:

$$\frac{dr}{dt} = -\frac{1}{\tau_{\rm el}}[\tilde{L}_1(r)\cos^2\theta_2 - \tilde{L}_2(r)\sin^2\theta_2] - u_{12}^0 \tilde{P}(r)\cos\theta_2 \tag{5.127a}$$

$$\frac{d\theta_2}{dt} = -\frac{\tilde{L}_3(r)\sin 2\theta_2}{\tau_{\rm el} r} + u_{12}^0 \frac{\tilde{Q}(r)}{r}\sin\theta_2 \tag{5.127b}$$

As mentioned above, $u_{12}^0 = 0$ for identical spheres. The tildes over the various functions in eq (5.127) indicate that, to be rigorous, these functions must be corrected for electroviscous effects, whose magnitude is, as yet, unknown. It is of interest to note that two equal-sized spheres carrying opposite electrical charges become a permanent dipole on contact and eq (5.127b) reduces to eq (3.412) for $r = 2$.

In writing eqs (5.127) the colloidal forces acting between the particles have been neglected. Since u_{12}^0 is approximately proportional to the ζ potential (cf. eqs (3.390) and (3.394)) and the electrostatic force to the square of the potential (cf. eq (1.124)), eq (5.127) applies at small potentials. In general, the colloidal forces must be included in the equations of motion

$$\frac{dr}{dt} = -\frac{1}{\tau_{el}}[\tilde{L}_1(r)\cos^2\theta_2 - \tilde{L}_2(r)\sin^2\theta_2] - u_{12}^0\tilde{P}(r)\cos\theta_2 + \frac{\tilde{C}(r)F_{coll}}{3\pi\mu\bar{a}^2} \tag{5.128a}$$

$$\frac{d\theta_2}{dt} = -\frac{\tilde{L}_3(r)\sin 2\theta_2}{\tau_{el}r} + u_{12}^0\frac{\tilde{Q}(r)}{r}\sin\theta_2 \tag{5.128b}$$

When $\tau_{el} \to \infty$, these equations become identical to those describing the relative trajectories between sedimenting spheres (eq (5.76)). The functions $\tilde{L}_i$, $\tilde{P}$, $\tilde{Q}$ and $\tilde{C}$ are all functions of the radius ratio as well as of the separation distance. Furthermore, the functions $\tilde{L}_i$ depend on the dielectric-constant ratio. It can be seen that the first term on the right-hand side of eq (5.128a) is approximately proportional to E^2 (neglecting electroviscous contributions), the second term to E, while the third is independent of E. Thus, by varying the magnitude of the electric field the relative importance of the various contributions can be modified. Also the amount of charge on the spheres affects the various terms in different ways. The first term on the right-hand side of eq (5.128a) is independent of surface charge or potential, the second is approximately proportional to the surface potential and the third to its square.

5.4.4 Electrocoagulation

Because there are numerous contributions to the relative velocity of two spheres in an electric field, their motions can be quite complicated. By dividing eq (5.128a) by eq (5.128b), we obtain the trajectory equation

$$\frac{1}{r}\frac{dr}{d\theta_2} = \frac{f(r,\theta_2)}{V\tilde{L}_3(r)\sin 2\theta_2 - \tilde{Q}(r)\sin\theta_2}$$

with

$$\begin{aligned} f(r,\theta_2) = {} & V[\tilde{L}_1(r)\cos^2\theta_2 - L_2\sin^2\theta_2] + \tilde{p}(r)\cos\theta_2 \\ & + \tilde{C}(r)\,\mathrm{Pe}^{-1}\,[\mathrm{Ad}F_{vdW} + \mathrm{Dl}F_{el}] \end{aligned} \tag{5.129}$$

which shows that the relative trajectories depend on a large number of dimensionless parameters. Here V is as defined by eq (5.126) and Pe, Ad and Dl by eqs (5.83) to (5.85). The colloidal forces depend further on the parameters $\tau(=\kappa\bar{a})$, λ^* and the potential ratio β. All functions appearing in eq (5.129) depend on the radius ratio and some on the dielectric-constant ratio as well. The inclusion of electroviscous effects requires the additional parameters b_i and Φ_s. The behaviour of two spheres depends on the magnitude of the various dimensionless constants.

When $V \gg 1 \gg$ (Ad/Pe, Dl/Pe), the trajectory equation reduces to eq (5.16) with the limiting solutions given by eqs (5.121), (5.123) and (5.124). Under such conditions, coagulation occurs because two spheres, no matter what their initial relative positions, will always come into contact in a finite time. From the nature of the trajectory equation it can be concluded that the doublet-formation frequency depends strongly on the initial number and distribution of particles in the suspension. The half time of coagulation, T (i.e. the time at which the initial number of single particles has halved), can be estimated from the time it takes for two spheres, initially separated by the mean distance of separation corresponding to a given volume fraction, to reach contact. This time can be found by integrating eq (5.111) numerically. Typically, T/τ_{el} can vary from 10^{-2} for short separations to 10^6 for separations some 20 radii away.

When (Ad/Pe, Dl/Pe) $\gg V$, we reach the regime of electrocoagulation induced by relative electrophoretic mobilities. This regime is identical to that of gravity-induced coagulation discussed in Section 5.3.2. We can define a capture efficiency, α_E, which for fast coagulation, when $\mathrm{Dl} < \mathrm{Dl}_{crit}$ or $\tau > \tau_{crit}$, is approximately proportional to $(\mathrm{Ad/Pe})^{0.20}$. Similar to the case of sedimentation, suspensions will be unstable at low and high values of Pe/Ad and stable at intermediate values.

So far we have discussed the motion of two spheres in an electric field in the absence of Brownian motion. To predict coagulation rates for Brownian spheres, we must solve the steady-state convective-diffusion equation

$$\nabla \cdot (\boldsymbol{D}_{12} \cdot \nabla n) = \nabla \cdot (\boldsymbol{u}_{12} n) \tag{5.130}$$

with $\boldsymbol{D}_{12}$ given by eq (4.26) and $\boldsymbol{u}_{12}$ by eq (5.128). No general solution of this equation is available. When Ad/Pe $\gg V$ and Pe $\ll 1$, we have for the flux towards a reference sphere, in analogy with sedimentation (cf. eq (5.89)):

$$\mathrm{J} = \mathrm{J}_0(1 + \alpha_p \mathrm{Pe}) \tag{5.131}$$

For more general conditions, eq (5.130) must be solved numerically.

5.4.5 Concentration dependence of electrophoretic mobilities

If the electrophoretic mobilities of particles are measured in a suspension at a volume fraction c where two-particle interactions are important, but three- and more-body interactions can be neglected, we expect that the mean electrophoretic mobility of a particle will be given by

$$\bar{u} = u_0(1 + Mc) \tag{5.132}$$

where M is the first virial coefficient, which we will call the mobility coefficient. When $V \gg 1$ (cf. eq (5.126)) and Pe $\gg$ (Ad, Dl) the relative trajectory equations are the same as those for two sedimenting spheres. Therefore, under these conditions the mobility coefficient, M, is expected to be equal to the sedimentation coefficient, S. Hence, for identical spheres at small Pe and in the absence of colloidal forces we expect that $M = -6{\cdot}55$. For equal-sized spheres of different potential the mobility coefficient will be equal to the value of $S(\gamma)$, with $\gamma = m_2/m_1$.

At large Pe the value of S is singular for identical spheres since in the limit $q = a_2/a_1 \to 1$ it takes an infinite time for the pair distribution function to reach a steady state. In such a case the value of M will depend on the form of the pair distribution function at the moment the mobility is being measured.

At first sight it might be thought that, when $u_{12}^0 = 0$ and Pe $= \infty$, the mobility coefficient M equals -1 because, when there is no relative motion between spheres, the only contribution to M is the backflow of liquid which replaces the volume of the particles moving forward. (This contribution is equal to the term k_1 in eq (4.43).) However, even when $u_{12}^0 = 0$ for a two-sphere system, there is still relative motion of the particles in a suspension. For a given distribution of single particles in the suspension there will be a certain number of close pairs, i.e. two particles relatively close together. Such a pair moves faster than the mean velocity of the individual spheres and, thus will eventually collide with a third sphere. These three-body interactions determine the pair distribution function and thus should never be completely neglected.

The electrophoretic mobility was determined by Zukoski and Saville (1987) as a function of the volume fraction, by measuring the velocity of red blood cells in a suspension of ghost cells. All cells were made spherical and hardened and behaved as solid spheres. The ghost cells were identical to the red blood cells, except that the cell content had been removed. In this way the ghost cells were made invisible (whence their name) and the motion of the red cells readily observed. Within experimental error they found, for all the conditions they studied, that $M = -1$, implying that the only contribution to the mobility coefficient arises from the backflow of fluid past the moving particles. It seems that all other contributions cancel each other more or less exactly. The interpretation of this result is complicated by the fact that the ghost cells were polydisperse, with diameters varying from about 3 to 7 μm, and the experiments were done in a flat parallel-plate chamber in which the mobility is measured at the stationary layer (cf. Section 3.4.7.1). However, under such conditions the relative velocity between the spheres is given by eq (5.125) only at the stationary plane, everywhere else there is a contribution due to shear (cf. eq (3.508)). The sedimentation coefficients were calculated on the assumption that the suspension was uniform all the way to $r \to \infty$ and it is known that a large contribution comes from this far-field region (cf., for example, eq (5.104)). This assumption is not valid for electrophoresis experiments, where the pair distribution function depends on the position in the channel. The cancellation of the various contributions to the mobility coefficient, M, appears to be fortuitous.

Chapter 6

Particle–Wall Interactions in Colloidal Systems

6.1 General considerations

The interaction between a sphere subjected to flow with a neighbouring wall is a limiting case of the interaction between two spheres, discussed in the previous chapter. If the radius of one of the two interacting spheres is allowed to approach infinity, then the two-sphere problem becomes the sphere–wall problem. As such, all the theory discussed for two spheres applies equally well to a sphere near a flat solid surface. For a hard sphere near a hard wall, the hydrodynamic flow field can be calculated from the creeping-flow and continuity equations, subject to the no-slip boundary condition on the sphere and wall surfaces and the condition that far from the sphere the fluid flow approaches the undisturbed flow field. In general, the translational motion of the sphere differs from that of the fluid, the difference being described by the hydrodynamic correction functions, F_i, discussed below. In the case of soft spheres near soft surfaces, e.g. when the surfaces are electrically charged, these functions require modification, but, as in the case of two spheres, their form is, as yet, unknown. These electroviscous contributions will be ignored, except in the section dealing with critical deposition concentrations (Section 6.3.3).

From a practical point of view, the most important sphere–wall interactions are those leading to the deposition of a sphere onto the surface and its removal after having been deposited. As such, the majority of this chapter deals with the deposition of particles onto various surfaces in a variety of flows. The detachment of particles is treated to a lesser extent, mainly because it is difficult to make predictions about detachment rates because the depth of the energy minimum in which a particle is captured is usually unknown. On the other hand, measuring detachment rates provides information about the strength of the bond between the particle and the surface.

The aim of this chapter is to review the various theoretical approaches which have been developed to predict deposition rates, as well as to discuss and compare the relevant experimental data and techniques reported in the literature. We will focus our attention mainly on problems associated with interactions between one or more spherical particles and a solid surface, usually referred to as a collector.

The transport of a particle towards a collector of radius of curvature R is shown in a schematic way in Fig. 6.1. As we will show below, the flow near

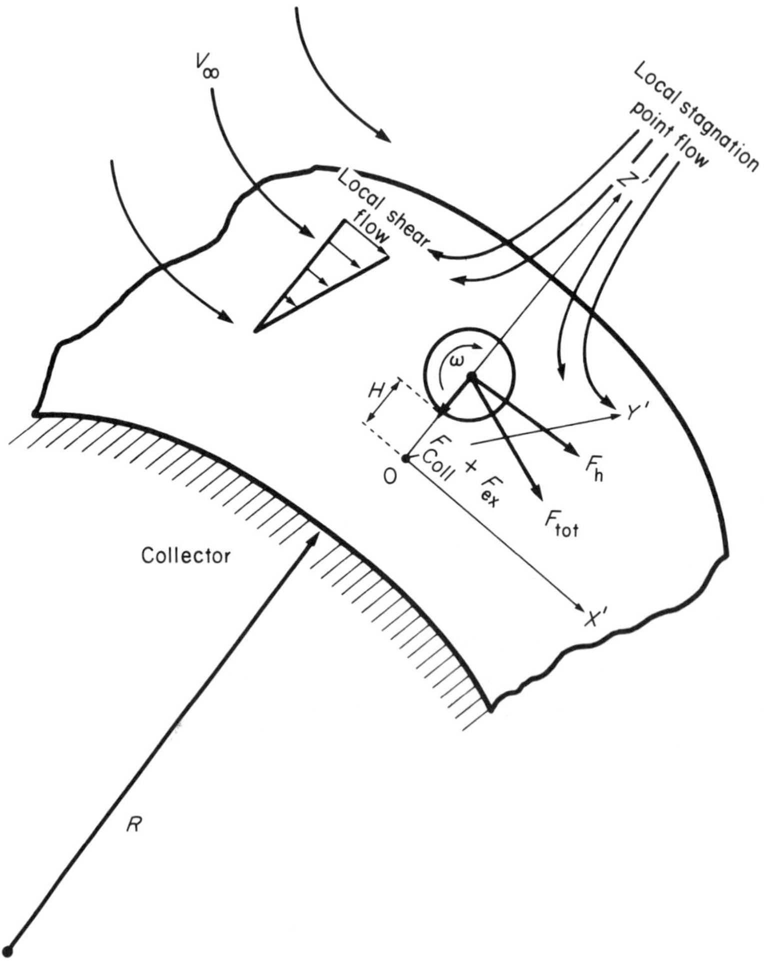

Fig. 6.1. Schematic representation of a particle in the neighbourhood of a collector. In most situations a particle is subjected to a flow which can be decomposed locally into a stagnation-point flow and a simple-shear flow.

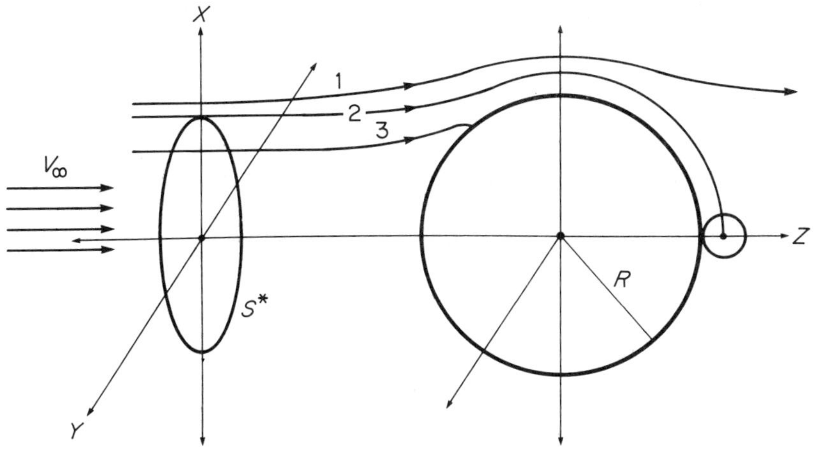

Fig. 6.2. Scheme to show the limiting-trajectory concept. Three different trajectories are shown of particles moving in the vicinity of a spherical collector of radius R: (1) an open (separating) trajectory; (3) a trajectory leading to capture; and (2) the limiting trajectory, separating both types of trajectory. The capture cross-section is also shown schematically. Particles whose centre passes through this cross-section are captured by the collector.

the collector can be decomposed into a local stagnation-point flow and a local simple-shear flow. In order to describe deposition, all forces, $\boldsymbol{F}$, and torques, $\boldsymbol{L}$, on a particle must be considered. In most cases particles also spin around an axis perpendicular to the collector, but for homogeneous spherical particles this rotation does not affect deposition.

The different theoretical approaches from which deposition rates can be calculated and which we will discuss in detail are shown schematically in Schemes 6.1 and 6.2. Two different concepts can be distinguished: Lagrangian methods in which a particle's motion is followed along its trajectory (Scheme 6.1), and Eulerian methods in which the particle concentration (or the probability density) in space is considered (Scheme 6.2).

6.2 Lagrangian methods for calculating deposition rates

6.2.1 Limiting-trajectory approach in creeping flow

Historically, the limiting-trajectory approach was the first attempt at a quantitative description of particle-deposition rates on collector surfaces, most notably on cylindrical or spherical collectors, motivated by their great

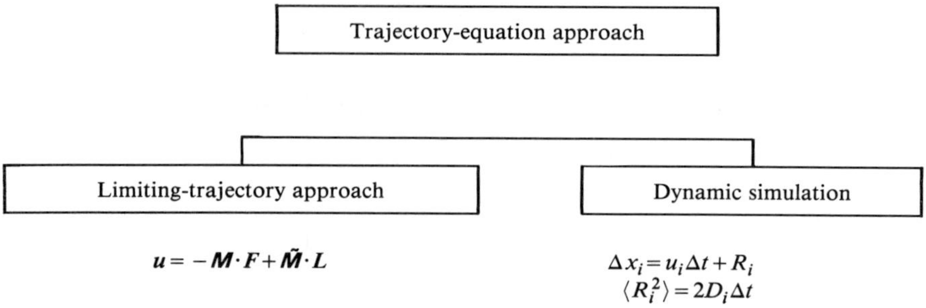

SCHEME 6.1
Lagrangian methods for calculating deposition rates (for meaning of symbols see Section 6.2.2).

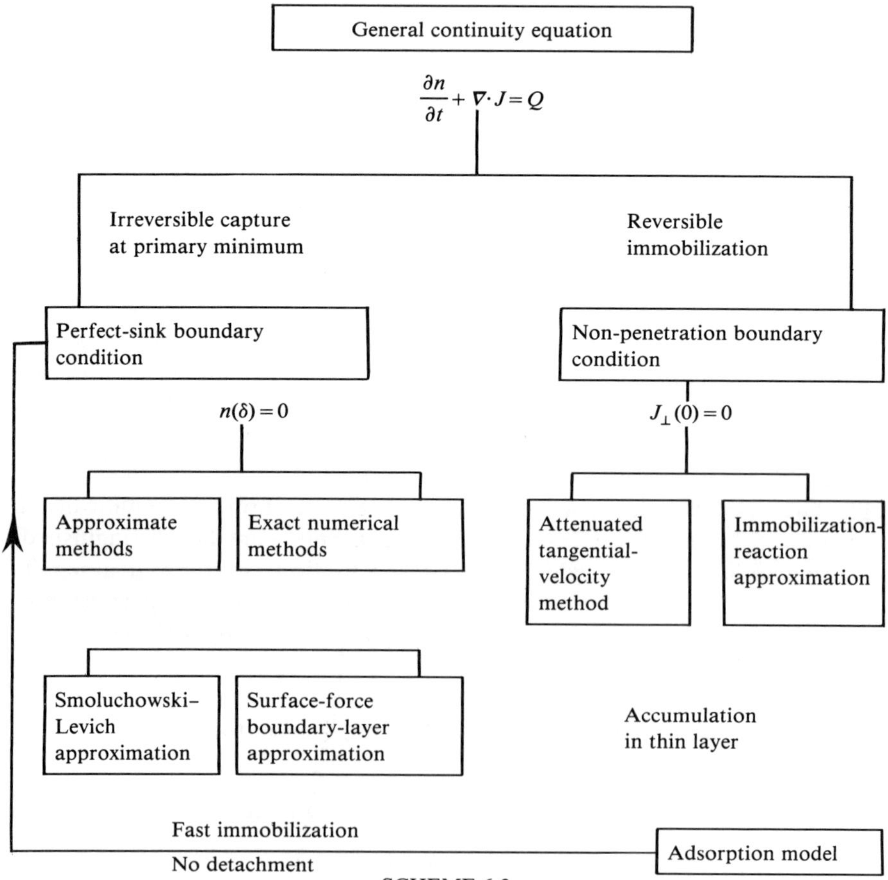

SCHEME 6.2
Eulerian methods for calculating deposition rates

practical significance in filtration and other industrial processes. The method is based on the Lagrangian description of particle motion, i.e. attention is focused on a single-particle trajectory which can be described by the second law of mechanics. In most cases the analysis neglects the random thermal force on the particles treating their motion as deterministic, and so applies only to relatively large particles and/or to strong flows. There is no conceptual difficulty in incorporating diffusion effects into the theory and when this is done the equation of motion becomes a Langevin equation. In principle, the Langevin approach yields the same results as those obtained from the convective-diffusion equation discussed below.

In the absence of diffusion, the solution of the force–torque balance equations are sufficient for predicting particle trajectories near a collector. The exact path depends on the initial conditions (particle position at time $t=t_0$). The particular trajectory that divides the approaching particle paths terminating in capture on the collector surface from those which pass by, is called the limiting trajectory (see Fig. 6.2).

In the absence of closed trajectories (e.g. a spherical collector in a uniform flow), the limiting trajectory can be determined by an integration backwards in time, starting from the point at the rear of the collector (cf. Fig. 6.2) where the separation between the particle and the collector surface is zero, back to where the particle is sufficiently far from the collector for its motion to coincide with that of the fluid element. In the case of a spherical or cylindrical collector in a simple shear flow, and in the absence of colloidal and external forces there exists a surface which separates the open and closed trajectories (cf. Section 5.1.1). In the presence of colloidal forces, all limiting trajectories approach this surface as the distance approaches infinity at the rear of the collector (after an encounter). Hence, limiting trajectories separating open trajectories from those leading to capture can be found by a backwards integration starting sufficiently far away on this limiting surface. In the case of more complicated flows, e.g. flows with closed streamlines not extending to infinity, limiting trajectories can be calculated only by trial and error, integrating forwards in time. Once the limiting trajectories are found, the rate of particle capture as well as the collector efficiency can be evaluated.

The motion of particles suspended in liquids in the absence of inertial effects (the Reynolds number based on particle dimension Re $\ll$ 1) can be found from the forces and torques exerted on the particles (cf. eqs (2.105) and (2.106)):

$$\begin{aligned} \boldsymbol{F} &= -(\boldsymbol{f}\cdot\boldsymbol{u}+\boldsymbol{c}\cdot\omega) \\ \boldsymbol{L} &= -(\boldsymbol{c}^{\dagger}\cdot\boldsymbol{u}+\boldsymbol{f}_{\mathrm{r}}\cdot\omega) \end{aligned} \tag{6.1}$$

where $\boldsymbol{f}$, $\boldsymbol{f}_{\mathrm{r}}$ and $\boldsymbol{c}$ are the translation, rotation and coupling tensors, † denoting the transpose operation, ω is the particle angular velocity, $\boldsymbol{F}$ is the net force acting on the particles and consists of external, colloidal and hydrodynamic contributions, and $\boldsymbol{L}$ is the torque vector. Equation (6.1) includes particle–wall interactions as well as the coupling between translational and rotational particle motions. As a result the hydrodynamic resistance coefficients (and also the force and torque exerted by the fluid stream on a particle in the vicinity of the boundary) are functions of the particle–wall separation and the direction of particle motion (parallel or perpendicular to the collector).

Eliminating $\boldsymbol{\omega}$ from eq (6.1) we obtain the following explicit solution for the vector $\boldsymbol{u}$:

$$\boldsymbol{u}=\frac{\mathrm{d}\boldsymbol{r}}{\mathrm{d}t}=\boldsymbol{M}\cdot\boldsymbol{F}-\tilde{\boldsymbol{M}}\cdot\boldsymbol{L} \tag{6.2}$$

where $\boldsymbol{M}=(\boldsymbol{f}-\boldsymbol{c}\cdot\boldsymbol{f}_{\mathrm{r}}^{-1}\cdot\boldsymbol{c}^{\dagger})^{-1}$ is the mobility tensor and $\tilde{\boldsymbol{M}}=\boldsymbol{M}\cdot\boldsymbol{c}\cdot\boldsymbol{f}_{\mathrm{r}}^{-1}$. Equation (6.2), when formulated in an appropriate coordinate system, leads to a set of non-linear first-order ordinary differential equations. The solution of such a system of equations with the initial condition $\boldsymbol{r}=\boldsymbol{r}_0$ at $t=t_0$, determines the limiting trajectories and hence particle deposition rates and collector efficiencies.

Results obtained by this method are reported by Spielman and Fitzpatrick (1973) and Spielman and Cukor (1973) for the spherical, rotating disk and cylindrical collectors where gravity as well as dispersion and electrical double-layer forces were taken into consideration. Zebel (1965), Hochrainer *et al.* (1969), and Payatakes *et al.* (1974) used the method for calculating filtration rates onto cylindrical fibres from a gaseous suspension, and Payatakes and co-workers (Payatakes and Chi-Tien, 1976; Payatakes, 1977; Payatakes and Gradon, 1980) extended it to the treatment of multilayer deposition of particles in the form of a dendrite structure during the filtration of aerosol particles.

The limiting trajectory approach was used by van de Ven and Mason (1977a) for predicting particle aggregation rates in sheared monodisperse suspensions (orthokinetic coagulation, cf. Section 5.1.5.1) and was extended to unequal-sized spheres by Adler (1981a). The same method applies to the deposition onto spherical or cylindrical collectors in simple shear. In the case of spherical collectors of radius R, when the particle radius $a \ll R$, the total number of particles captured in time t is (eq (5.21) and (5.22)):

$$N_{\mathrm{t}}=\frac{4}{3}\,\alpha n_{\infty}\,G\,R^{3}\,t \tag{6.3}$$

where G is the rate of shear and α the collector efficiency, defined as

$$\alpha = \frac{3}{R^3} \int_0^{\ell} y\, x_c(y)\, \mathrm{d}y \tag{6.4}$$

where $x_c(y)$ is the extension of the capture cross-section in the X direction at y and ℓ is the maximum extension in the Y direction.

6.2.2 Dynamic-simulation method

In principle it is possible to extend the trajectory equation approach to allow for particle diffusion. When a flowing particle is also subjected to Brownian motion, its trajectory is no longer deterministic and the limiting trajectory approach cannot be used because limiting trajectories no longer exist. Instead, for a particle being initially at a given position, we can assign the probability that the particle will be captured. Consider a plane XY sufficiently far from the collector such that in this plane the particle velocity $\boldsymbol{u}$ is equal to the undisturbed fluid-flow velocity v (in the absence of an external force). Let $p(x,y)$ be the probability that a particle passing through the plane XY at the point (x,y) will be captured, then in the case of uniform flow the total number of particles captured, N_t, is

$$N_t = n_\infty u_\infty t \iint_{-\infty}^{\infty} p(x,y)\, \mathrm{d}x\mathrm{d}y \tag{6.5}$$

where n_∞ and u_∞ are the particle number concentration and velocity far from the collector, respectively. In the absence of diffusion $p(x,y) = 1$ inside the capture cross-section, S^*, and zero outside, hence the double integral appearing in eq (6.5) is equal to S^* and eq (6.5) reduces to

$$N_t = n_\infty u_\infty S^* t \tag{6.6}$$

In the case of deposition on a cylinder or sphere in a simple shear flow, eq (6.3) is no longer applicable when the particles are subjected to Brownian motion. However, eq (6.3) can be used in unchanged form if $x_c(y)$ in eq (6.4) is defined as

$$x_c(y) = \int_0^{\infty} p(x,y)\, \mathrm{d}x \tag{6.7}$$

The problem of predicting deposition rates now reduces to calculating the probability of capture, given the initial position of the particle. This can be done by dynamic simulation methods, similar to those used to calculate stochastic particle trajectories in an unbounded simple shear flow (van de Ven, 1977) or near solid walls (Foister and van de Ven, 1980). For non-spherical particles, rotary Brownian motion must also be considered (van de Ven and Mason, 1976b; Zuzovski *et al.*, 1980a). The dynamic-simulation method consists of integrating the particle-motion equations, including in every integration step a random displacement (translational or angular) chosen (randomly) from a normal (Gaussian) distribution function, the width of which is determined by the local diffusion coefficient. The stochastic trajectory equations (Langevin equations) can be written as:

$$x_i = u_i \Delta t + R_i \tag{6.8}$$

where u_i are the components of $\boldsymbol{u}$ given by eq (6.2). The integration step Δt must be chosen small enough for additivity of flow and Brownian motion to apply; R_i is a random displacement chosen from a normal distribution of variance $2D_i\Delta t$. The velocity along a trajectory can usually be decomposed into a component perpendicular to the collector surface for which the random displacement must be chosen such that $\langle R_\perp^2 \rangle = 2F_1 D_0 \Delta t$, and into a component tangential to the surface, chosen such that $\langle R_\parallel^2 \rangle = 2F_4 D_0 \Delta t$. For flat collectors, F_1 and F_4 are universal correction functions equal to $d_\parallel(r, o)$ and $d_\perp(r, o)$ defined in eq (4.26b).

By integrating eq (6.8) numerically, stochastic trajectories can be generated. Starting from an initial position a generated trajectory will either lead to capture or it will not. By repeating this procedure a sufficient number of times for the same initial position (x,y), $p(x,y)$ can be obtained. By applying the same scheme to a sufficient number of initial positions, the integrals appearing in eqs (6.5) and (6.7) can be evaluated. So far dynamic-simulation calculations have been performed for spheres near solid walls in shear or Poiseuille flow in the absence of colloidal or external force fields (Foister and van de Ven, 1980), in which case no deposition occurs. Extension to include additional force fields is straightforward and this method could be useful for situations for which no solutions of the convective-diffusion equation are available, for example, the deposition of Brownian particles on a sphere in a simple shear flow. Obviously such "brute-force" calculations consume far more computer time than do numerical solutions of differential equations.

6.3 Eulerian methods

6.3.1 Convective-diffusion equation

In contrast to Lagrangian methods in which the trajectories of individual particles are followed, Eulerian methods are concerned with obtaining the distribution of particles in space and, for non-spherical particles, the orientation distribution as well. In their most general form these distributions are described by a set of coupled ordinary differential equations, known as Fokker–Planck equations. If the distribution of particles throughout the suspension is known, the flux of particles towards the surface can be readily calculated. In this chapter we investigate various solutions and methods of solving the governing transport equations. In most cases of interest, especially for colloidal particles dispersed in fluids (or gases of sufficiently high pressure), the Fokker–Planck equation reduces to the usual convective-diffusion equation. Most of this chapter will deal with this fundamental equation.

We consider a suspension of spherical non-interacting particles of radius a suspended in a fluid undergoing macroscopic motion characterized by the fluid velocity vector $\boldsymbol{v}$ relative to a space-fixed frame of reference. The convective-diffusion equation can be written as

$$\frac{\partial n}{\partial t} + \nabla \cdot \boldsymbol{J} = Q \tag{6.9a}$$

$$\boldsymbol{J} = -\boldsymbol{D} \cdot \nabla n + \boldsymbol{u} n \tag{6.9b}$$

where J is the particle flux vector, Q a source term, $\boldsymbol{D}$ the diffusion tensor, and $\boldsymbol{u}$ the velocity which can consist of several contributions

$$\boldsymbol{u} = \boldsymbol{u}_{\text{hydr}} + \boldsymbol{u}_{\text{coll}} + \mathrm{u}_{\text{ext}} \tag{6.10}$$

The hydrodynamic velocity is usually obtained by solving the Navier–Stokes equation, neglecting the presence of the dispersed phase, and hence the results apply only to dilute dispersions. Equation (6.9a) also applies under turbulent flow conditions; however in this case the determination of $\boldsymbol{u}$ is very difficult, but useful approximations can be made (see below).

In the vicinity of solid surfaces, the hydrodynamic forces and torques exerted by the fluid on the particles are altered. The presence of boundaries also gives rise to additional forces stemming from colloidal particle–wall interactions. These interactions are of very short range (usually a small fraction of the particle size) and are usually described by the classical DLVO theory, i.e. by a combination of dispersion and double-layer forces. At still smaller distances (comparable with molecular dimensions) some kind of Born-type repulsive forces appear which prevent the penetration of particles into collector surfaces. All these forces are believed to act perpendicularly to the interacting surfaces assumed to be of regular geometrical shape. Such a picture of colloidal forces may be relevant for perfectly smooth and homogeneous surfaces, but it breaks down in real systems in which geometrical irregularities (surface roughness), non-uniformity of surface charge, steric forces, hydration layers and also chemical-type forces (hydrogen bond formation, polymer-chain interactions, etc.) cause the colloidal force vector to possess components tangential to the interacting surfaces. This is the reason why particles form permanent contacts with collector surfaces (even in the presence of strong hydrodynamic shearing forces) and are immobilized at given surface sites (as experimental observations show). Therefore, in real systems, the distribution of the interaction potential of colloidal forces is, in general, very complex and depends on coordinates both normal and tangential to the surface (and also on particle orientation relative to the collector). If we were able to specify these forces explicitly and formulate the appropriate boundary conditions, eqs (6.9) could, in principle, be solved and hence the particle concentration distribution and the flux at the collector surface would be known as a function of the position vector, $\boldsymbol{r}$, and time, t.

Because of insufficient knowledge about conditions near the surface, we are forced to make some simplifying assumptions regarding colloidal-particle–collector interactions. Until recently, an approach used exclusively in calculations of particle deposition rates was the "perfect-sink" model, based on a highly idealized picture of tangential interactions. In this model it is assumed that all particles arriving at a position sufficiently close to the collector surface (the primary minimum distance) are irreversibly consumed by an unspecified but extremely fast immobilization reaction. In other words, it is assumed that particles are captured in an infinite energy sink with respect to both perpendicular and tangential motion and subsequently disappear from the system. This approach has proved to be extremely useful in performing numerical calculations and results were obtained for a large variety of collectors and flow patterns.

However, a major disadvantage of this model is that it does not take into account the formation of a separate phase of immobilized particles at the interface, not even in an implicit way. Therefore, the kinetics of particle removal from surfaces cannot be investigated. Moreover, the possibility of a finite-rate immobilization reaction (and in the limiting case no reaction at all) is outside the scope of this model. To overcome these limitations, we can replace the perfect-sink boundary condition by a non-penetration boundary condition and consider the motion or immobilization of particles near the collector surface explicitly (Adamczyk and van de Ven, 1984). We will discuss the consequences of both boundary conditions in detail and show that the perfect-sink boundary condition can be considered to be a limiting case of the non-penetration boundary condition.

6.3.2. Collectors and flow fields

Before discussing solutions of the convective-diffusion equation, it is useful to look at some of the collectors and flow fields for which solutions of this equation are available. The various collectors studied are shown schematically in Fig. 6.3.

These collectors can be divided into four different groups: (i) collectors

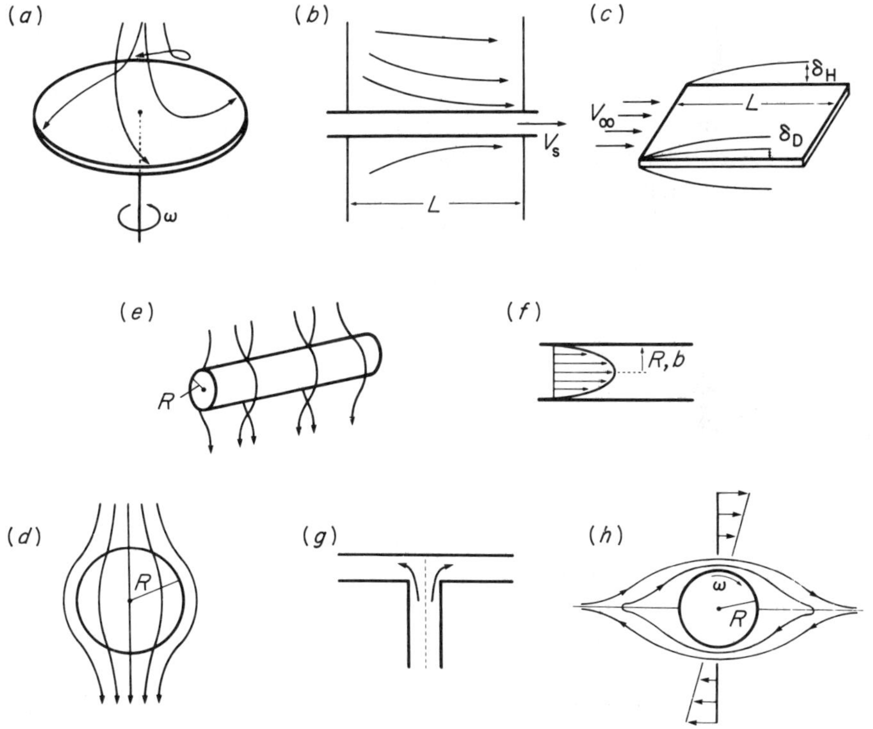

Fig. 6.3. Schematic representation of various collectors for which solutions of the convective-diffusion equation are available: (a) rotating disk; (b) continuous moving wire or sheet; (c) plate in uniform flow; (d) sphere in uniform flow; (e) cylindrical collector in uniform flow; (f) parallel-plate or cylindrical channel; (g) stagnation-point flow collector; (h) sphere or cylinder in simple shear flow.

TABLE 6.1

Undisturbed flow fields near collectors ($z \ll R$)

Collector	Velocity components[i]	G_{st}	G_{sh}	Pe
Rotating disk[a]	$v_r = 0{\cdot}5102\ \omega^{3/2}\nu^{-1/2} r\, z$	$0{\cdot}5102\ \omega^{3/2}\nu^{-1/2}$	$0{\cdot}5102\ \omega^{3/2}\nu^{-1/2} r_p$	$\dfrac{1{\cdot}02\ \omega^{3/2}\ a^3}{\nu^{1/2}\ D_0}$
	$v_z = -0{\cdot}5102\ \omega^{3/2}\nu^{-1/2} z^2$ $v_\theta = \omega\, r(1-0.616\ \omega^{1/2}\nu^{-1/2} z)$		$0{\cdot}616\ \omega^{3/2}\nu^{-1/2} r_p$	
Stagnation-point flow collector[b]	$v_r = \alpha_s r\, z$ $v_z = \alpha_s z^2$	α_s	$\alpha_s r_p$	$(2\alpha_s a^3)/D_0$
Plate in uniform flow[c]	$v_x = 0{\cdot}332\ v_\infty^{3/2}\nu^{-1/2} z\ x^{-1/2}$	$-\dfrac{0{\cdot}332\ v_8^{3/2}\nu^{-1/2}}{4x_p^{3/2}}$	$\dfrac{0{\cdot}332\ v_\infty^{3/2}\nu^{-1/2}}{x_p^{1/2}}$	$\dfrac{0{\cdot}64\ v_\infty a^3}{\nu^{1/2} L^{3/2} D_0}$
	$v_z = \dfrac{0.332}{4}\ v_\infty^{3/2}\nu^{-1/2} z^2 x^{-3/2}$			
Cylinder in uniform flow[d]	$v_r = -2v_\infty A_f \left(\dfrac{z}{R}\right)^2 \cos\theta$	$\dfrac{2v_\infty A_f \cos\theta_p}{R^2}$	$\dfrac{4v_\infty A_f \sin\theta_p}{R}$	$\dfrac{4A_f v_\infty a^3}{R^2 D_0}$
	$v_\theta = 4v_\infty A_f \dfrac{z}{R} \sin\theta$			
Sphere in uniform flow[e]	$v_r = -\dfrac{3}{2} v_\infty A_f \left(\dfrac{z}{R}\right)^2 \cos\theta$	$\dfrac{3v_\infty A_f \cos\theta_p}{2R^2}$	$\dfrac{3v_\infty A_f \sin\theta_p}{2R}$	$\dfrac{3A_f v_\infty a^3}{R^2 D_0}$
	$v_\theta = \dfrac{3}{2} v_\infty A_f \dfrac{z}{R} \sin\theta$			
Parallel-plate channel[f]	$v_x = \dfrac{3v_m z}{b}$		$\dfrac{3v_m}{b}$	$\dfrac{3v_m a^3}{2b^2 D_0}$

Cylindrical channel[f]	$v_z = \frac{3v_m z}{R}$		$\frac{4v_m}{R}$	$\frac{3v_m a^3}{R^2 D_0}$
Continuous moving plate[g]	$v_x = v_S - 0{\cdot}444\, v_S^{-2/3} \nu^{-1/2} z\, x^{-1/2}$ $v_z = \frac{0{\cdot}444}{4} v_S^{3/2} \nu^{-1/2} z^2 x^{-3/2}$	$\frac{0{\cdot}444\, v_S^{3/2} \nu^{-1/2}}{4x_p^{3/2}}$	$\frac{0{\cdot}444\, v_S^{3/2} \nu^{-1/2}}{x_p^{1/2}}$	$\frac{0{\cdot}222\, v_S^{3/2} a^3}{\nu^{1/2} L^{3/2} D_0}$
Continuous moving wire[g]	$v_r = v_S \left(1 - \frac{1}{\beta(\bar{x})}\right) \bar{z}$ $v_r = -\frac{1}{2} v_S \frac{R}{L} \frac{\beta^1(\bar{x})}{\beta^2(\bar{x})} \bar{z}$	$\frac{v_S R\, \beta'(\bar{x})}{L\, \beta^2(\bar{x})}$	$\frac{v_S}{\beta(\bar{x})}$	$\frac{0{\cdot}222\, v_S^{3/2} a^3}{\nu^{1/2} L^{3/2} D_0}$
Cylinder in simple shear[h]	$v_r = 2\frac{G}{R} z^2 \sin 2\phi$ $v_\phi = \frac{1}{2} G \left(1 + \frac{4z}{R} \cos 2\phi\right)$	$\frac{-2G \sin 2\phi_p}{R}$	$-2G \cos 2\phi_p$	$\frac{G\, R^2}{D_0}$

[a] Levich (1962).
[b] Dabros and van de Ven (1983c).
[c] Adamczyk and van de Ven (1982).
[d] Natanson (1957), Adamczyk and van de Ven (1981a).
[e] Levich (1962), Prieve and Ruckenstein (1974).
[f] Adamczyk and van de Ven (1981b).
[g] Adamczyk *et al.* (1982a, b).
[h] Cox *et al.* (1968).
[i] Velocity compounds are expressed in terms of a space-fixed coordinate system.
A^f *and* β, flow parameters; $\alpha_s = 0{\cdot}164\, \beta^{3/2} \ldots^{-1/2}$; v_∞, uniform flow velocity; v_m, mean fluid velocity; b, half width of channel; L, length of plate; $\beta(\bar{x})$, function defined by Adamczyk and van de Ven (1982); v_s, velocity of continuous moving plate/wire; G, rate of shear; subscript p refers to coordinate of point on collector surface.

moving with constant angular or translational velocity in or through a contained dispersion (rotating disk, continuous moving sheet or wire, Fig. 6.3a,b); (ii) collectors fixed in a given flow (or sedimenting in a quiescent dispersion) such as a plate, sphere or cylinder in a uniform flow (Fig. 6.3c,d,e); (iii) collector surfaces constituting the inside of channels or other flow cells such as parallel-plate or cylindrical channels (Fig. 6.3f) or the stagnation-point flow collector (Fig. 6.3g); and (iv) collectors freely moving in flow fields such as a cylinder or sphere in a simple shear flow (Fig. 6.3h). By inhibiting the motion of collectors in simple shear, collectors intermediate between (ii) and (iv) might also be considered.

The corresponding undisturbed flow fields in the vicinity of these collectors are summarized in Table 6.1. These flow fields can be decomposed in local Cartesian coordinates (x,y,z) into a stagnation flow, v_{st}, and a simple shear flow, v_{sh}, with flow strengths of G_{st} and G_{sh}, respectively. Values of G_{st} and G_{sh} are included in Table 6.1. For the rotating disk (Spielman and Fitzpatrick, 1973)

$$v_{st} = G_{st}\,(xz\,\boldsymbol{i}_x + \boldsymbol{i}_y - z^2\,\boldsymbol{i}_z) \tag{6.11}$$

where (x,y,z) is the local-coordinate system for the rotating disk, the origin being located at the point $(r_p, \theta_p, 0)$.

Values of $G_{sh,y}$ and $G_{sh,x}$ are given in Table 6.1. Besides the translational motion, a particle near the rotating disk also spins, but this spin does not affect the particle transport. For all other collectors in Table 6.1

$$v_{st} = G_{st}(2zx\,\boldsymbol{i}_x - z^2\boldsymbol{i}_z) \tag{6.12a}$$

and

$$v_{sh} = G_{sh} z\,\boldsymbol{i}_x \tag{6.12b}$$

Also included in Table 6.1 are the definitions of the Péclet number, Pe, denoting the ratio of flow to diffusion effects.

From the velocity field near the collector the particle velocity can be found from

$$\begin{aligned} u_{\parallel} &= F_3 v_{\parallel} \\ u_{\perp} &= F_1 F_2 v_{\perp} \end{aligned} \tag{6.13}$$

where $v_{\parallel}$ and $v_{\perp}$ are the parallel and perpendicular components of the velocity field, respectively, shown in Table 6.1. F_1, F_2 and F_3 are universal correction factors, shown graphically in Fig. 6.4, together with the correction factors for diffusional motion, i.e. $D_{\perp} = F_1 D_0$ and $D_{\parallel} = F_4 D_0$

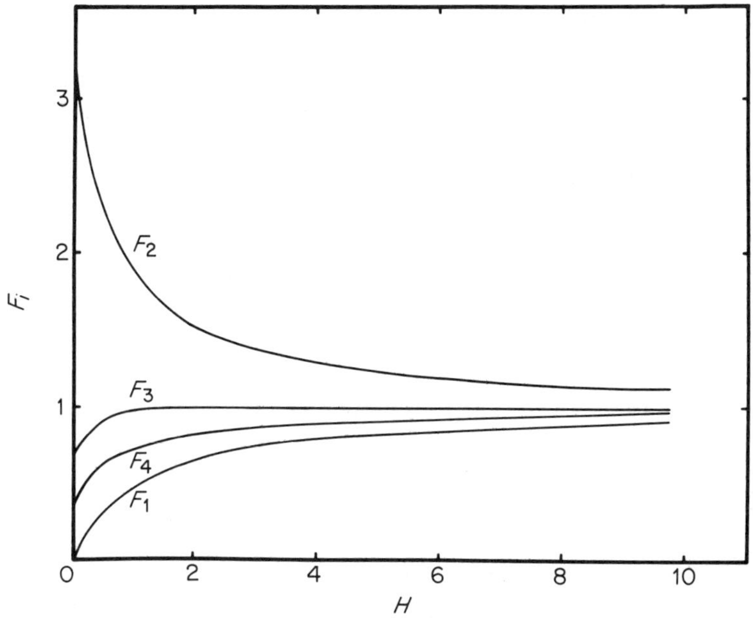

Fig. 6.4. Universal hydrodynamic correction factors for diffusion and particle motion near a flat solid wall.

(Brenner, 1964; Goldman *et al.*, 1967; Bart, 1968; Goren and O'Neill, 1971).

In general, the deposition rate on a given collector of area S_c can be found by solving the convective-diffusion equation (eq (6.9)) for n and calculating the flux $J_\perp$ towards the surface. The total flux is then given by

$$j = \iint_{S_c} \boldsymbol{J}_\perp \cdot \mathrm{d}\boldsymbol{S} \qquad (6.14)$$

and the number of particles deposited in time t is given by

$$N_t = \mathrm{j}t \qquad (6.15)$$

which is valid when j does not depend on time.

6.3.3 Perfect-sink model

6.3.3.1 General remarks

In this approximation it is assumed that the source term Q in eq (6.9a) is zero and that the boundary condition at the collector surface can be stated as:

$$n = 0 \text{ at } h = \delta \tag{6.16}$$

where h is the separation between the particle and collector surface (gap width) and δ is the primary minimum distance (sometimes assumed to be zero). As mentioned previously, this condition is equivalent mathematically to the assumption that all particles arriving at the distance $h = \delta$ disappear from the system.

Far from the collector the suspension is assumed to have a uniform concentration of particles, hence:

$$n \to n_\infty \text{ as } h \to \infty \text{ for } t \geqslant 0 \tag{6.17}$$

The initial condition at $t = t_0$ is specified as:

$$n = n_0(\boldsymbol{r}) \tag{6.18}$$

where $n_0(\boldsymbol{r})$ is the initial concentration distribution throughout the suspension, in most cases taken to be uniform, i.e. $n_0 = n_\infty$.

Even with these simple boundary conditions, the mass-balance equation (eq (6.9)) cannot be solved analytically when the hydrodynamic corrections and colloidal forces are taken into account. Such solutions are feasible when numerical methods are used which have recently become a standard and inexpensive way of dealing with differential equations of the type under consideration. Before these numerical techniques were widely accessible, several approximate analytical results were derived which we will discuss here in some detail.

In general, the steady-state convective-diffusion equation ($\partial n/\partial t = 0$) consists of two terms, a diffusion term, $-\boldsymbol{D} \cdot \nabla n$, and a convective term, $\boldsymbol{u}n$, which can be split further into two parts, one which accounts for hydrodynamic forces and torques and another due to colloidal and external force fields. The relative magnitude of hydrodynamic and diffusion effects is denoted by the Péclet number, Pe (see Table 6.1). The relative magnitude of colloidal and external force fields can be expressed by a variety of dimensionless parameters, the most important ones being:

(i) The adhesion number $\mathrm{Ad} = A_{123}/6kT$, where A_{123} is the Hamaker constant for the dispersion interaction of particle 1 with a wall 2 through a medium 3.
(ii) The double-layer number $\mathrm{Dl} = 4\pi\epsilon\epsilon_0\zeta_1\zeta_2 a/kT$, where ϵ is the dielectric constant of the medium, ϵ_0 the permittivity of the vaccum, and ζ_1 and ζ_{22} the zeta potentials of the particle and the wall, respectively.
(iii) The dimensionless double-layer thickness $\tau = \kappa a$, κ being the reciprocal Debye length.
(iv) The double-layer asymmetry parameter $\mathrm{Da} = (\zeta_1 - \zeta_2)^2/2\zeta_1\zeta_2$.
(v) The retardation parameter $\bar{\lambda} = \lambda/a$, λ being the London retardation wavelength.
(vi) The gravity number $\mathrm{Gr} = 4\pi(\rho_1 - \rho_3)ga^4/3kT$, ρ_1 and ρ_3 being the densities of the particle and the medium, respectively, and g the gravity acceleration constant.
(vii) The electric-field number $\mathrm{El} = E_0 q\, a/kT$, E_0 being the strength of the electric field and q the charge on the particle.

Of the above dimensionless numbers, only Ad depends exclusively on the material properties of the collectors, medium and particles. For a given system it can only be changed by changing the temperature, but then only by a limited amount. Usually Ad is in the range 0·1–10. The remaining dimensionless parameters can be controlled by varying the particle size, ζ potential, ionic strength, electric field strength, etc. If we let Ex stand for either of the external fields Gr or El, then depending on the relative magnitudes of Pe, Dl, and Ex, we can differentiate several regimes in which useful approximations of the convective-diffusion equation can be made.

6.3.3.2 Smoluchowski–Levich approximation

Here we consider the case where colloidal forces, with the exception of the dispersion forces, can be neglected, i.e. $\mathrm{Dl} \ll \mathrm{Pe}$. The reason for treating dispersion forces separately is that they can almost never be neglected. In the absence of attractive forces at short separations, the deposition rate is always zero because the components of the translational-diffusion tensor become zero as the distance between particle and collector approaches zero. Therefore, attraction at short separations is essential for deposition to occur and, in most cases, the dominant attractive force at short separations is the van der Waals dispersion force.

When $\mathrm{Dl} \ll \mathrm{Pe}$, the convective-diffusion equation cannot be solved analytically, but approximate solutions are available. A useful approximation is the Smoluchowski–Levich approximation. In this approach it is assumed that the increase in hydrodynamic drag that a particle experiences while approaching a solid wall is balanced by attractive dispersion forces. It has been found from exact solutions of the convective-diffusion equation for the rotating disk (Bart, 1968), that neglecting both factors (hydrodynamic drag and dispersion forces) yields results very close to the exact solution. The Smoluchowski–Levich approximation consists of letting Ad = 0, treating all components of the diffusion tensor as scalars (equal to the bulk diffusion coefficient), and assuming that the particle moves with the undisturbed fluid velocity (in the absence of external fields), i.e. all hydrodynamic correction factors, F_i (defined above), are assumed to equal 1. This approximation was first made by Smoluchowski in calculating the coagulation rate of spherical particles, both in the presence and absence of a simple shear flow (cf. Sections 4.4 and 5.1.5). Later it was adopted by Levich (1962) in calculating ion transfer to the rotating disk and spherical collector.

By making these assumptions, eq (6.9) is reduced, for a steady state, to the form:

$$D_0 \nabla^2 n = v \cdot \nabla n \tag{6.19}$$

where the diffusion coefficient of particles, D_0, is assumed to be constant. The boundary conditions are the same as those expressed by eqs (6.15) to (6.18). Equation (6.19) has been solved analytically for the rotating disk (Levich, 1962; Dabros and Adamczyk, 1979), spherical and cylindrical collectors (Natanson, 1957; Levich, 1962), parallel-plate channel (Adamczyk and van de Ven, 1981b), a plate in uniform flow (Levich, 1962; Adamczyk and van de Ven, 1982), and continuous moving surfaces (Adamczyk *et al.*, 1982a,b). To obtain analytical results with this approximation, it is usually also assumed that interception plays no role in the deposition process, i.e. $a \ll \delta_D$, where δ_D is the thickness of the diffusion boundary layer. The results can be expressed most conveniently in terms of the dimensionless mass transfer Sherwood number, Sh, defined as:

$$\mathrm{Sh} = Ja/D_0 n_\infty \tag{6.20}$$

where $J = J_\perp$ is the normal component of particle flux at the collector surface. Since J depends on the position on the collector surface, Sh is a local quantity. A global measure of deposition rate can be obtained by averaging Sh over the entire collector surface, i.e.

$$\overline{\mathrm{Sh}} = \frac{1}{S_c} \iint_{S_c} \mathrm{Sh}\, \mathrm{d}S \tag{6.21}$$

By introducing these definitions, the results obtained for the collectors listed earlier can be expressed as:

$$\begin{aligned} \mathrm{Sh} &= f(q_{\parallel})\,\mathrm{Pe}^{1/3} \\ \overline{\mathrm{Sh}} &= C\,\mathrm{Pe}^{1/3} \end{aligned} \tag{6.22}$$

where $f(q_{\parallel})$ is a dimensionless function of the coordinate $q_{\parallel}$ parallel to the collector surface and C is a positive dimensionless constant. Explicit values of $f(q_{\parallel})$ and C are given in Table 6.2 for various collectors. Note that for moving continuous surfaces the mass-transfer rate (Sh number) also depends on the Schmidt number $\mathrm{Sc} = \nu/D_0$ where ν is the fluid kinematic viscosity. It is interesting to observe that for the rotating disk and the stagnation point flow collectors, Sh does not depend on $q_{\parallel}$, which means that the disk surface is uniformly accessible to particles. The same conclusion applies for the region near the forward stagnation point of spherical and cylindrical

TABLE 6.2
Deposition rates from the Smoluchowski–Levich approximation

Collector	Sh	$\overline{\mathrm{Sh}}$
Rotating disk or stagnation-point flow collector	$0{\cdot}616\ \mathrm{Pe}^{1/3}$	$0{\cdot}616\ \mathrm{Pe}^{1/3}$
Cylinder in uniform flow[a]	$0{\cdot}538\ f_1(\theta)\ \mathrm{Pe}^{1/3}$	$0{\cdot}460\ \mathrm{Pe}^{1/3}$
Sphere in uniform flow[b]	$0{\cdot}682\ f_2(\theta)\ \mathrm{Pe}^{1/3}$	$0{\cdot}435\ \mathrm{Pe}^{1/3}$
Channel[c]	$\dfrac{0{\cdot}678\ \mathrm{Pe}^{1/3}}{\tilde{x}^{1/3}}$	$\dfrac{1{\cdot}017}{\tilde{L}^{1/3}}\ \mathrm{Pe}^{1/3}$
Plate in uniform flow[d]	$\dfrac{0{\cdot}616\ \mathrm{Pe}^{1/3}}{\tilde{x}^{1/2}}$	$1{\cdot}232\ \mathrm{Pe}^{1/3}$
Continuous moving wire/plate	$\dfrac{0{\cdot}932\ \mathrm{Pe}^{1/3}\mathrm{Sc}^{1/6}}{\tilde{x}^{1/2}}$	$1{\cdot}964\ \mathrm{Pe}^{1/3}\mathrm{Sc}^{1/6}$

[a] $-f_1(\theta) = \sin^{1/2}\theta \Big/ \left[\int_0^\theta \sin^{1/2}\theta\, d\theta\right]^{1/3}.$

[b] $-f_2(\theta) = \sin\theta \Big/ \left(\theta - \frac{1}{2}\sin 2\theta\right)^{1/3}.$

[c] $-\tilde{x} = x/b$ or x/R; $\tilde{L} = L/b$ or L/R.

[d] $-\tilde{x} = x/L.$

collectors where $f(q_{\parallel})$ varies very little. The limiting solutions listed in Table 6.2 can be used for $\mathrm{Pe} < \sim 10^{-2}$ when there is no external force or energy barrier present (Dabros and Adamczyk, 1979).

The efficiency of the first six collectors listed in Table 6.2 is very similar, because in all cases the order of magnitude of Sh is similar. The deposition on a continuous moving plate or wire (through a cell of length L containing the dispersed particles) is, however, larger by a factor of $\mathrm{Sc}^{1/6}$. As Sc is typically of the order of 10^6 for colloidal particles in aqueous media, it can be seen from Table 6.2 that the deposition on continuously moving surfaces is about 10 times more efficient than on other surfaces.

It is possible to consider in a quantitative manner the finite particle-size effects (interception effects) and the presence of an external force for uniformly accessible surfaces. The results obtained by solving eq (6.19) analytically including a term for the external force, can be expressed as:

$$\mathrm{Sh} = \overline{\mathrm{Sh}} = I^{-1} \exp\left(-\frac{\mathrm{Pe}}{6}(\bar{\delta}+1)^3 + \mathrm{Ex}\,\bar{\delta}\right)$$

$$I = \int_{\bar{\delta}}^{\infty} \exp\left(-\frac{\mathrm{Pe}}{6}(H+1)^3 + \mathrm{Ex}\,H\right) \mathrm{d}H \qquad (6.23)$$

where $H = h/a$ is the dimensionless gap width, $\bar{\delta}$ is the dimensionless minimum distance, and Ex is the dimensionless parameter characterizing the magnitude of the uniform external force, i.e. Ex = Gr or El.

It has been shown (Dabros and Adamczyk, 1979) that eq (6.23) holds even for fairly large values of Pe and Ex up to 10 (which corresponds to particle sizes up to a few micrometers in water). Again the reason is a compensation of the effects of a decrease in particle mobility near surfaces by attractive dispersion forces which act in the opposite direction. Both these contributions were neglected in deriving eq (6.23).

The Smoluchowski–Levich approximation can also be used in describing deposition in turbulent flow (Dabros and van de Ven, 1983a). In this case the flow in the viscous sublayer near the surface of the collector can be considered to fluctuate between inrushes of fluid towards the wall (downsweeps) and bursts of fluid away from the wall (upsweeps). These sweeps can be approximated as local stagnation-point flows, the velocity field of which is given in Table 6.1. Thus the flow near the wall can be regarded as a stagnation-point flow of fluctuating intensity α_s, where $\alpha_s > 0$ denotes downsweep and $\alpha_s < 0$ denotes upsweep. In principle, the flux towards the collector can be calculated from the convective-diffusion equation for a given value of α_s. The total flux toward the collector can be found by averaging over all intensities, i.e.

$$J = \int_{-\infty}^{\infty} J(\alpha_s)\, p(\alpha_s)\, d\alpha_s \tag{6.24}$$

where $p(\alpha_s)$ is the probability distribution of α_s values. Because deposition only occurs during downsweep ($\alpha_s > 0$), eq (6.24) can be written as

$$J = \int_{0}^{\infty} J(\alpha_s)\, p(\alpha_s)\, d\alpha_s \simeq \frac{1}{2}\, J(\bar{\alpha}_s) \tag{6.25}$$

where $\bar{\alpha}_s$ is the root mean square (r.m.s.) average value of α_s. From theoretical and experimental data it can be estimated that $\bar{\alpha}_s \simeq 0{\cdot}003\ v^{*3}/\nu^2$, v^* being the friction velocity of the fluid. Using eq (6.19) and substituting the velocity field given by Table 6.1 as the stagnation-point flow for v, it is possible to obtain analytical expressions for $J(\alpha_s)$ (Dabros and van de Ven, 1983a).

6.3.3.3. Limiting solutions for zero Péclet numbers

We now consider the case where Pe $\ll$ (Dl, Ex), so that the effects of flow can be neglected and the problem reduced to deposition due to colloidal and external forces alone. We will first treat the case when the colloidal and external forces can be derived from a potential V_s, which is a function of r only. For a spherical collector of radius R, the convective-diffusion equation reduces to, when Pe = 0,

$$\frac{d}{dr}\left[r^2 \tilde{F}_1 \left(\frac{dn}{dr} + n\, \frac{dV_s}{dr} \right) \right] = 0 \tag{6.26}$$

where n, r and V_s are non-dimensionalized with respect to n_∞, R and kT, respectively, and $\tilde{F}_1$ is a hydrodynamic correction function (approaching F_1 as $R \to \infty$). The solution of eq (6.26), satisfying the perfect-sink boundary condition at $r = 1 + (a+\delta)/R = r_0$, is (cf. eqs (4.68) and (4.67))

$$n = \alpha e^{-V_s} \int_{r_0}^{r} \frac{e^{V_s}}{\tilde{F}_1 r^2}\, dr \tag{6.27}$$

where

$$\alpha^{-1} = \int_{r_0}^{\infty} \frac{e^{V_s}}{\tilde{F}_1 r^2} \, dr \qquad (6.28)$$

From eq (6.27) it follows that

$$\text{Sh} = \frac{1}{2} \alpha \frac{a}{r} \qquad (6.29)$$

In the case of the Smoluchowski–Levich approximation, when $a = R$ and $\delta = 0$, $\alpha = 2$.

For infinite collectors like a plane or an infinitely long cylinder, the steady-state deposition rate is zero because it takes an infinite time to develop the steady-state distribution. For finite collectors it is to be expected that results similar to eq (6.29) apply, but Sh will depend on the precise geometry of the collector.

Another case of interest is when the deposition is due to an external force field only, acting normally to the collector, i.e. (Dl, Pe) $\ll$ Ex. In this case the solution of the convective-diffusion equation for the rotating disk, a plate in uniform flow, a continuously moving plate, and a parallel-plate channel is

$$\overline{\text{Sh}} = \text{Ex} \qquad (6.30)$$

and for a cylindrical collector, a cylindrical channel and a continuously moving wire

$$\overline{\text{Sh}} = \text{Ex}/\pi \qquad (6.31)$$

6.3.3.4 Surface-force boundary-layer approximation

The surface-force boundary-layer approximation (SFBLA) applies when an energy barrier exists due to a superposition of attractive dispersion forces and repulsive double-layer forces. This method is based on the assumption that in a thin layer adjacent to the collector surface the only mechanism of particle transport is diffusion in a potential force field, V_{int} (derived from colloidal interactions acting only perpendicularly to the collector everywhere). In this layer the fluid convection effects are neglected and the transport equation is given by eq (6.26), which for flat surfaces can be written as

$$\frac{\mathrm{d}J}{\mathrm{d}H} = -\frac{\mathrm{d}}{\mathrm{d}H}\left[F_1\left(\frac{\mathrm{d}n}{\mathrm{d}H} + n\,\frac{\mathrm{d}V_{\mathrm{int}}}{\mathrm{d}H}\right)\right] = 0 \tag{6.32}$$

which can be immediately integrated once with the result:

$$J = \text{constant} \tag{6.33}$$

Thus the normal component of the particle flux has a constant value throughout the region where the energy barrier is present (surface-force layer). Integrating eq (6.33) once more (and assuming the perfect-sink condition at $H = \bar{\delta}$) yields

$$J = -n_e k' \tag{6.34}$$

with

$$k'^{-1} = \int_B \frac{\mathrm{e}^{V_{\mathrm{int}}}}{F_1(H)}\,\mathrm{d}H \tag{6.35}$$

where n_e is the particle concentration at the outer edge of the surface-force layer whose value must be determined. In eq (6.35) the integral is evaluated over the width B of the energy barrier.

Outside the interaction layer eq (6.19) applies because the colloidal forces as well as the hydrodynamic corrections are neglected. In order to match these two solutions, the boundary condition for the outer region is specified in the form:

$$J_{\mathrm{inner}} = J_{\mathrm{outer}} \tag{6.36}$$

or (6.36)

$$-k' n_e = \left(\frac{\partial n}{\partial H}\right)_e$$

Far from the collector the boundary condition expressed by eq (6.17) applies. The solution of eq (6.19) with the boundary condition of eq (6.36) assumes a very simple form for the rotating disk (Ruckenstein and Prieve, 1973):

$$\mathrm{Sh} = \overline{\mathrm{Sh}} = \mathrm{Sh}_0 k'/(k'+1) \tag{6.37}$$

where Sh_0 denotes the value of Sh given by eq (6.22).

For other collectors the expressions for Sh are more complicated because Sh depends on the tangential coordinate, and will not be reported here. More results are given by Kim and Rajagopalan (1982).

The SFBLA has been extended for the rotating disk to account for the presence of the secondary energy minimum, external forces and radial loss of particles (Prieve and Lin, 1980).

6.3.3.5 Additivity of deposition rates

When Dl, Ex and Pe are of comparable magnitude, no exact or approximate analytical solutions of the convective-diffusion equation are available. The only analytical approach often attempted assumes the additivity of the various deposition rates caused by the flow and the external or colloidal force fields. The additivity rule can be expressed as

$$\mathrm{Sh} = \mathrm{Sh}_0 + C\,\mathrm{Pe}^{1/3} \tag{6.38}$$

where Sh_0 is the value of the Sherwood number in the limit $\mathrm{Pe} \to 0$, i.e. as given by eqs (6.29), (6.30) or (6.31); C is a numerical constant given in Table 6.2. No theoretical foundation exists for eq (6.38), but it nevertheless provides useful estimates in several instances. However, in some cases, eq (6.38) can be incorrect by more than one order of magnitude.

It should also be mentioned that the $\mathrm{Pe}^{1/3}$ dependence only applies to collectors whose motion is unaffected by the flow field. It does not apply to collectors that are free to move in the flow field, such as freely rotating cylinders and spheres in a simple shear flow. For freely rotating spherical collectors, an exact expansion solution of the convective-diffusion equation leads, when $\mathrm{Pe} \to 0$, to (cf. eq (5.40))

$$\mathrm{Sh} = \mathrm{Sh}_0(1 + C\,\mathrm{Pe}^{1/2} + \ldots .) \tag{6.39}$$

where $C = 0{\cdot}2568\alpha$, where α is given by eq (6.28).

It is also of interest to note that the result for such collectors when $\mathrm{Pe} \to \infty$ (and with negligible particle dimensions), is (cf. eq (5.25))

$$\mathrm{Sh} = \text{constant} \tag{6.40}$$

From eqs (6.39) and (6.40) it can be concluded that the additivity rule does not apply to spherical and cylindrical collectors in simple shear because these equations are not compatible with eq (6.38). The same conclusion applies to the coagulation rate of the colloidal particles themselves (cf. Section 5.1.6)

The range of applicability of eq (6.38) can be seen from some examples, presented below, of exact numerical solutions of the convective-diffusion equation.

6.3.3.6 Numerical solutions of the convective-diffusion equation

Because of the large number of dimensionless numbers involved in describing particle transport to collectors, discussing their effects on deposition rates for all of them is prohibitive. Instead, we will look at some important and representative cases.

First we look at numerical solutions for deposition on cylindrical collectors in a uniform flow. Figure 6.5 shows the average Sherwood number as a function of the Péclet number for weak, medium and strong attraction between the particles and collector surfaces. The solid line (lower curve) corresponds to Ad$=0\cdot04$, $\bar{\lambda}=0\cdot3$, $A=a/R=0\cdot01$, Dl$=0$; the middle curve

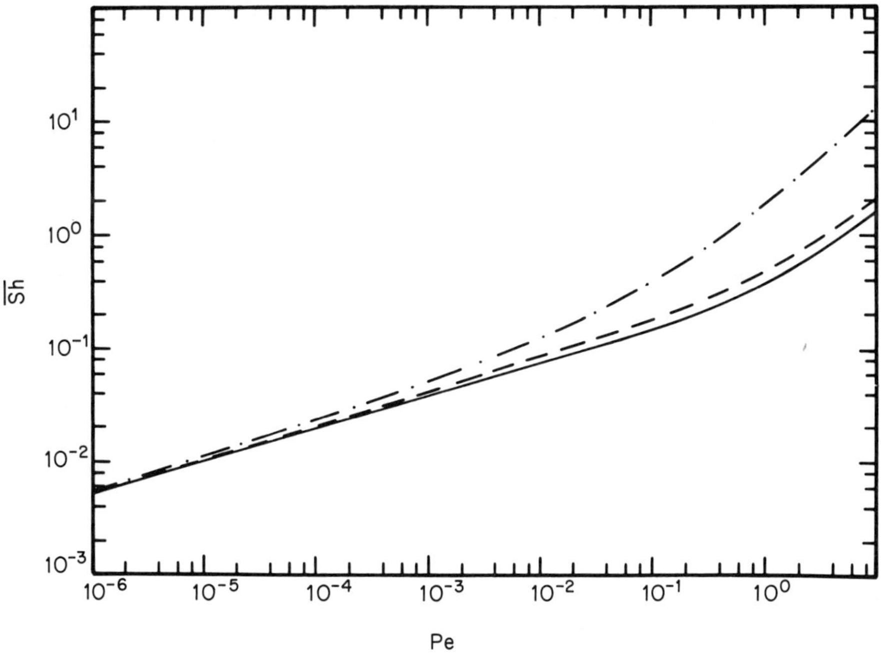

Fig. 6.5. The averaged mass-transfer number, $\overline{\text{Sh}}$ (dimensionless deposition rate per unit length of collector), for a cylindrical collector in uniform flow as a function of Pe for Ex$=0$ (no external force), $A=10^{-2}$, $\bar{\lambda}=0\cdot3$. (———) Ad$=0\cdot04$, Dl$=0$; (----) Ad$=0\cdot4$, Dl$=0$;(-·-·-·) Ad$=0\cdot4$, Dl$=-4\times10^3$,$\tau=5$ (after Adamczyk and van de Ven, 1981b).

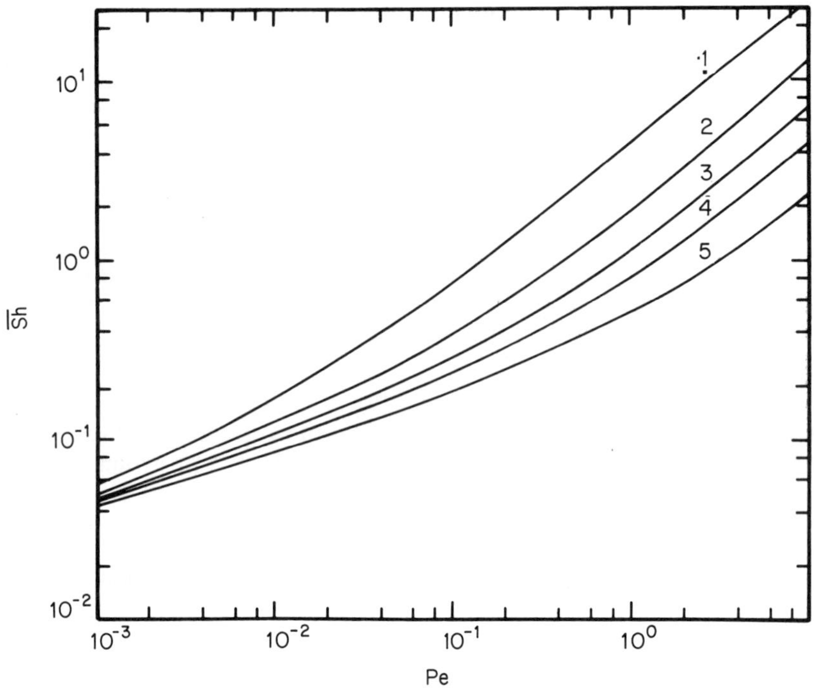

Fig. 6.6. Dependence of $\overline{\mathrm{Sh}}$ on Pe for a cylindrical collector in uniform flow for various values of τ. $\mathrm{Gr}=0$, $\mathrm{El}=0$, $A=1\times10^{-2}$, $\mathrm{Ad}=0\cdot4$, $\bar{\lambda}=0\cdot3$, $\mathrm{Dl}=-10^3$. (1) $\tau=2$, (2) $\tau=5$, (3) $\tau=10$, (4) $\tau=20$, (5) $\tau=100$ (after Adamczyk and van de Ven, 1981b).

corresponds to $\mathrm{Ad}=0\cdot4$ and the upper curve to $\mathrm{Ad}=0\cdot4$, $\mathrm{Dl}=-4\times10^3$ and $\tau=5$ (electric double-layer attraction, due to opposite charges). Figure 6.6 shows the variation of Sh with Pe for a variety of double-layer thicknesses characterized by the parameter τ. It can be seen that, at moderately large Pe values, strong attractive forces can increase the deposition rate by more than one order of magnitude. This conclusion is valid for all types of collector. It can be seen that at small Péclet numbers the curves approach a straight line of slope 1/3, in agreement with the analytical results of Table 6.2.

Figure 6.7 shows the effects of gravity on the deposition rate and the deposition on a cylinder is shown as a representative example. It can be seen that at low Péclet numbers and relatively large gravity numbers, the deposition rate becomes constant, i.e. as given by eq (6.31). Also of interest, is that it can be seen in Fig. 6.7 that the additivity rule does not apply. For curve 1 and $\mathrm{Pe}>10^{-2}$, Sh decreases with increasing Pe, in contradiction to eq (6.38). For deposition on a continuously moving wire the

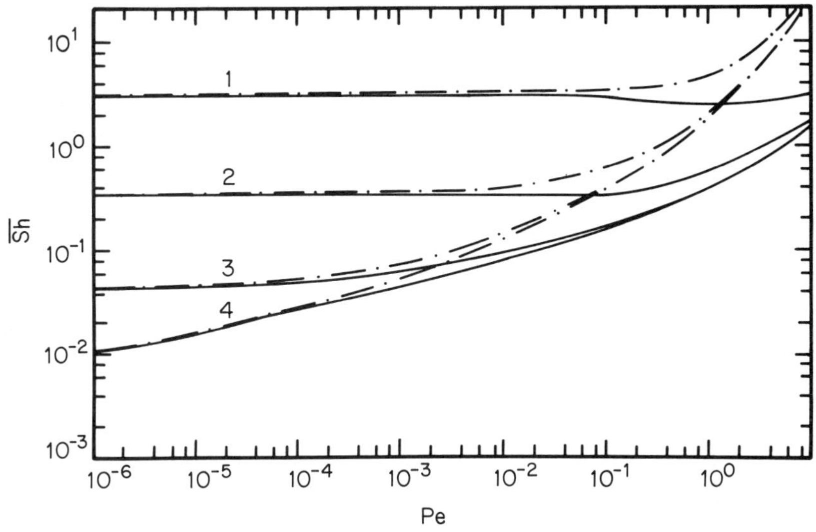

Fig. 6.7. Plots of $\overline{\text{Sh}}$ versus Pe for cylinders in uniform flow for various values of Gr (gravity directed parallel to the flow) and for $A = 10^{-2}$, $\bar{\lambda} = 0{\cdot}3$. (1) Gr = 10, (2) Gr = 1, (3) Gr = 0·1, (4) Gr = 0·01. (———) Ad = 0·04, Dl = 0; (----) Ad = 0·4, Dl = 0; (-·-·-·) Ad = 0·4, $\text{Dl} = -4 \times 10^3$, $\tau = 5$ (after Adamczyk and van de Ven, 1981b).

decrease in Sh with Pe is even more pronounced; under certain conditions (large Gr) Sh can decrease by one order of magnitude with increasing Pe.

Another important feature is the large effect a small change in the double-layer parameter, Dl, can have on the deposition rate. This is shown for the stagnation point flow collector in Fig. 6.8, where the Sherwood number, Sh, is given as a function of Dl, for three different values of the Reynolds number, Re (or Péclet number Pe). It can be seen that at a critical number, Dl_c, the deposition rate is suddenly reduced by several orders of magnitude. Dl_c depends only weakly on Pe. The reason for this sudden decrease is that as Dl increases the energy barrier increases and the height of the barrier becomes the rate-determining factor. This sudden drop in deposition rates is characteristic of all collectors and also of the coagulation rate of colloidal particles.

Alternatively, the salt concentration, c, in the suspending medium can be varied. For electrostatically stabilized colloidal particles and a surface having a charge of the same sign, no deposition will occur at low salt concentrations. As soon as a critical salt concentration, referred to as the critical deposition concentration (CDC), is reached, deposition occurs (described by eqs (6.22) or (6.23)). For salt concentrations below the CDC a deposition efficiency, α_d, can be defined:

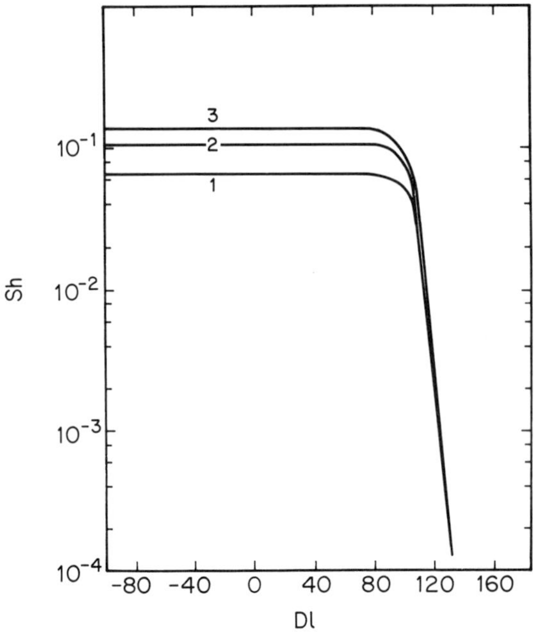

Fig. 6.8. The Sherwood number, Sh, versus the double-layer parameter, Dl, for various Reynolds numbers, Re, for the stagnation-point flow collector. Curves 1–3 correspond to Reynolds numbers of 10, 20 and 30, respectively. $\text{Ad} = 0{\cdot}8$, $\bar{\lambda} = 0{\cdot}4$, $\text{Gr} = 0{\cdot}002$ and $\tau = 58{\cdot}2$ (after Dabros and van de Ven, 1983c).

$$\text{Sh} = \alpha_d \, \text{Sh}_{\text{fast}} \tag{6.41}$$

The inverse of α_d can be called the stability ratio $W_d = 1/\alpha_d$, in analogy with peri- or orthokinetic coagulation. Experimental plots of log W_d versus log c are approximately linear below the CDC (cf. Fig. 6.9), similar to coagulation (cf. Sections 4.4 and 5.1.4). Again, slopes of log W_d versus log c plots, as well as the location of the CDC, are found to deviate from standard theory. In all cases the observed CDC is larger than the theoretical one and the slopes are less steep than predicted and depend on the size of the ions in solutions (Boluk and van de Ven, 1989). It is found that the location of the CDC and the variation in the slope corresponds well to the ion size, as one would expect, by incorporating finite ion-size effects in the theory. Including these electroviscous contributions makes all the hydrodynamic correction terms F_1, F_2, F_3 and F_4 functions of b_i, Φ_s and κa and, therefore, the deposition rate must also depend on these parameters.

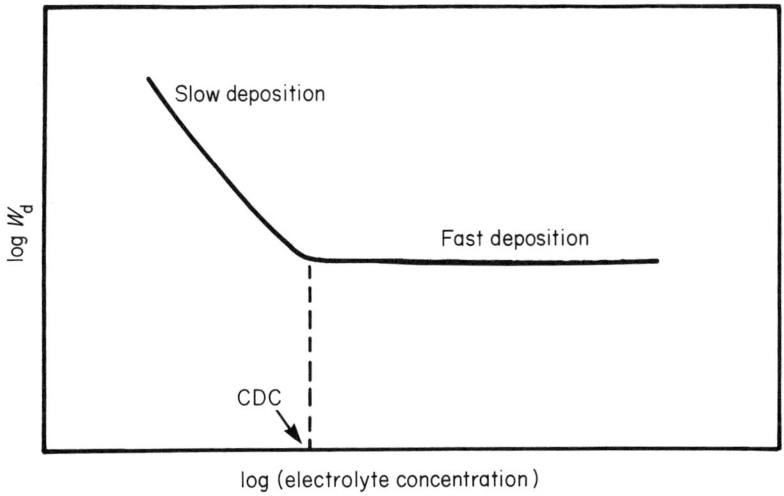

Fig. 6.9. Stability ratio for particle deposition as a function of salt concentration (schematic). At $c >$ CDC fast deposition occurs.

6.3.4 Non-penetration boundary condition

6.3.4.1 General remarks

The perfect-sink boundary condition, as we have seen, assumes that particles disappear when they arrive at the distance $H = \bar{\delta}$. To keep track of particles which arrive at the primary minimum, we can no longer assume that they disappear. Instead, we specify as a boundary condition that particles cannot penetrate solid walls, i.e. we assume that $J(0) = 0$, in which case the motion of particles or their immobilization near the collector can be taken into account explicitly. Two limiting approaches can be distinguished and these are described below.

6.3.4.2 Attenuated tangential velocity model

Real surfaces can be regarded as consisting of a collection of surface sites of variable size, characterized by local differences in surface properties (i.e. charge, topology, composition, etc.) Hence the energy of interaction of a particle with such a surface will exhibit many local energy minima in which particles can be captured. According to the DLVO theory, for small particles ($a < 0{\cdot}1$ μm), such local energy minima tend to be shallow, because both van der Waals and electrostatic forces are proportional to a and the possibility exists that particles deposited on a given site will desorb and subsequently be captured at another site. Hence particle motion tangential to the surface can be regarded as a series of fast adsorption–desorption transitions (jumps) between neighbouring surface sites. Particle motion in these circumstances can be described by a model in which the average tangential velocity of the particles is attenuated (damped) by some factor. The decrease in tangential velocity could be calculated if the distribution of energy minima over the collector surface were known. The model can be stated as

$$\bar{u}_{\parallel} = \sigma(z) u_{\parallel} \tag{6.42}$$

where $u_{\parallel}$ is the tangential particle velocity which would occur in the absence of surface irregularities and $\sigma(z)$ is a function of the surface characteristics of the collector and the particle and of the separation distance. So far no use has been made of this model because surfaces are usually too poorly characterized to allow the determination of σ.

6.3.4.3. Immobilization reaction approximation

A more general approach has been developed (Adamczyk and van de Ven, 1984) in which the tangential surface-specific interactions are accounted for in a more realistic way by making use of the reaction term Q in the mass-balance equation, thereby covering all specific interactions that are not explicitly known. Also, the presence of the immobilized phase of particles is explicitly included. Accordingly, the transport equation takes the form:

$$\frac{\partial n}{\partial t} + \nabla \cdot \boldsymbol{J} = -Q(n, n_s, \boldsymbol{r}, t) \tag{6.43a}$$

$$\frac{\partial n_s}{\partial t} = Q(n, n_s, \boldsymbol{r}, t) \tag{6.43b}$$

where n_s is the number concentration of the immobilized phase and

$$Q = Q_s(n, n_s, \boldsymbol{r}, t) - Q_r(n, n_s, \boldsymbol{r}, t) \tag{6.43c}$$

where $Q_s(n, n_s, \boldsymbol{r}, t)$ is the immobilization-rate term and Q_r the particle removal-rate term (characterizing the rate of particle transition from the immobile phase to the suspension). Although eq (6.43) also represents an idealization of a more complex reality, it is nevertheless much more general than the perfect-sink model, which can, in principle, be derived from eq (6.43) when the immobilization rate is assumed to be large and in the absence of particle detachment from the surface, i.e. $Q_r = 0$. This point is discussed extensively below.

In the immobilization reaction approach, we distinguish two separate phases; one composed of particles having translational degrees of freedom (mobile phase) and another formed by particles immobilized at the collector surface sites at local energy minima with respect to both perpendicular and tangential motion. However, the immobilized particles may retain some degree of rotational motion. If the depth of the local minimum is denoted by $\Delta V_\ell(x,y)$, then the time t_i which a given particle spends at a surface site (the immobilization time) is of the order of $(\Delta x^2/D_\parallel)\exp(-\Delta V_\ell/kT)$ seconds (where Δx is a characteristic dimension of the surface site). When ΔV_ℓ is sufficiently large (in absolute value), then the value of t_i considerably exceeds that of $\Delta x/u_\parallel$ ($u_\parallel$ being the tangential component of particle velocity at the minimum), and the particles can be treated as being permanently localized. This model is expected to be well suited for describing the adsorption of larger particles (0·1–1 μm) when the local energy minima are deep. (The colloidal force increases linearly with particle dimension and for large particles the diffusion coefficient is small.)

The temporal and spatial change in particle concentration (adsorption or desorption) of the mobile and immobile phases can be determined if we formulate explicitly the reaction terms in eq (6.43c) and solve eqs (6.43a and b) with the appropriate boundary and initial conditions. Even for relatively simple collector geometries (such as the rotating disk) this leads to a complicated partial differential equation which can only be solved by elaborate numerical techniques (Adamczyk *et al.*, 1984).

By assuming that the immobilization and accumulation of particles occurs within a very thin layer near the collector, referred to as the accumulation layer, eq (6.43) can be expressed in terms of surface concentrations of adsorbed particles. We will now discuss this adsorption model in detail.

6.3.5 Adsorption model

6.3.5.1 General remarks

A considerable simplification of eq (6.43) can be attained by noting that accumulation occurs around energy minima, whose extensions are extremely small in comparison with particle dimensions. Thus, the tangential component of the particle velocity varies very little over the region of the minimum and can be treated as being constant. Moreover, due to the symmetry of all the collectors discussed in this chapter, two spatial coordinates denoted by x and y are sufficient for formulating the transport equation, x being the tangential component and y the normal one. For the remainder of this section (6.3.5.1), we render distances dimensionless by dividing them by the particle radius a, concentrations by n_0 and surface concentrations by $n_0 a$. The dynamic and immobilized surface concentrations are defined as

$$N = \int_0^{\ell} n \mathrm{d}H \; ; \; N_s = \int_0^{\ell} n_s \mathrm{d}H \tag{6.44}$$

where $H = y - 1$ and ℓ is the thickness of the accumulation layer. N is the number of particles per unit area within a distance ℓ from the surface which are free to move along the surface (dynamically adsorbed particles) and N_s is the number within the same distance that are immobilized. Expressing time and the term Q in non-dimensional form, by dividing by a^2/D_0 and $D_0 n_0/a$, respectively, then integrating eq (6.43) over the accumulation region yields:

$$\begin{aligned} \frac{\partial N}{\partial t} &= \Delta \mathrm{Sh}(N, N_s, x, t) - \frac{\bar{f}}{f(x)} \frac{\partial}{\partial x}\left[f(x) \left\{ -A^2 \frac{\partial N}{\partial x} \right.\right. \\ &\qquad \left.\left. + (\mathrm{Pe}\; g(x) + A\; p(x)\; \mathrm{Ex}) N \right\} \right] - Q(N, N_s, x, t) \\ \frac{\partial N_s}{\partial t} &= Q(N, N_s, x, t) \end{aligned} \tag{6.45}$$

where $\Delta \mathrm{Sh} = \mathrm{Sh}_2 - \mathrm{Sh}_1$, Sh_1 and Sh_2 are the mass-transfer (Sherwood) numbers at $y = \delta_1$ and $y = \delta_2$ and $(\delta_2 - \delta_1)$ is the width of the energy minimum; $\bar{f}$ is a dimensionless constant, $A = a/R$, R being a characteristic dimension of the collector; and $f(x)$, $g(x)$ and $p(x)$ are dimensionless functions of the x coordinate alone, explicit expressions of which are given by Adamczyk and van de Ven (1984).

Equation (6.45) is very general and can be used to predict the kinetics of particle accumulation (or desorption) both in the primary minimum region, or when a primary minimum together with a secondary minimum occurs. Its applicability is limited by the restriction that the aspect ratio is $A \ll 1$ (in order to use the known expressions for resistance coefficients of a particle approaching a planar solid surface). It can be shown easily that for colloidal-sized particles the inequality $A^2 \ll (\mathrm{Pe} + A\ \mathrm{Ex})$ is generally true, so that the term describing the longitudinal dispersion of N can be neglected in eq (6.45) (for the rotating disk this requirement is always met). In consequence, this equation simplifies to a first-order partial-differential equation, the solution of which can be obtained relatively easily in an analytical form for various collectors.

For particle deposition in a primary minimum the following phenomenological relationship may be assumed for the reaction term Q:

$$Q(N, N_s, x, t) = k_s N - k_r N_s \tag{6.46}$$

The values of k_s and k_r cannot at present be calculated *a priori* and are treated as semi-empirical constants. Equation (6.46) is supposed to hold well for low surface coverages when the reaction term can always be linearized in terms of N and N_s. For larger surface concentrations of particles (being of the order of a few percent of a monolayer and higher), geometrical blocking effects and ageing phenomena may lead to a deviation from linearity. These aspects will be discussed later.

6.3.5.2 Kinetic equation for rotating disk

It is possible to solve eq (6.45) analytically (by means of Laplace transforms) with the appropriate initial conditions. The general expressions for N_1, N_2 and N_s are given by Adamczyk and van de Ven (1983), where N_1 is the surface concentration of particles in the primary minimum and N_2 the concentration in the secondary minimum. If no secondary minimum is present, the result for $N = N_1$, for the rotating disk collector, is:

$$N = N_\infty [1 - \exp(-t/T_1)] + N_0 \exp(-t/T_1) \tag{6.47}$$

where $T_1 = 1/(\bar{f}_1 \mathrm{Pe} + k_s)$, $\bar{f}_1$ being a constant of order unity, $N_\infty = \mathrm{Sh}_0/(\bar{f}_1 \mathrm{Pe} + k_s)$ is the steady-state surface concentration of dynamically adsorbed particles, and Sh_0 is the Sherwood number obtained from the perfect-sink boundary condition. N_0 is the surface concentration of particles at zero time.

The immobilized number of particles is given by

$$N_s = k_s N_\infty \{t + T_1[\exp(-t/T_1) - 1] + k_s N_0 T_1 [1 - \exp(-t/T_1)]\} \tag{6.48}$$

The dependence of N on $t_1 = t/T_1$ is plotted in Fig. 6.10 for different values of the initial dimensionless surface concentration N_0, while in Fig. 6.11 values of N_s are shown. It can be seen from Fig. 6.10 that for $t_1 \ll 1$ the change in N with time is linear (which can be easily deduced from eq (6.47)). On the other hand, for $t_1 \gg 1$ a stationary value of N is attained (N_∞), which is independent of the initial surface coverage. The surface concentration of immobilized particles, however, increases linearly with time (see Fig. 6.11) according to the equation:

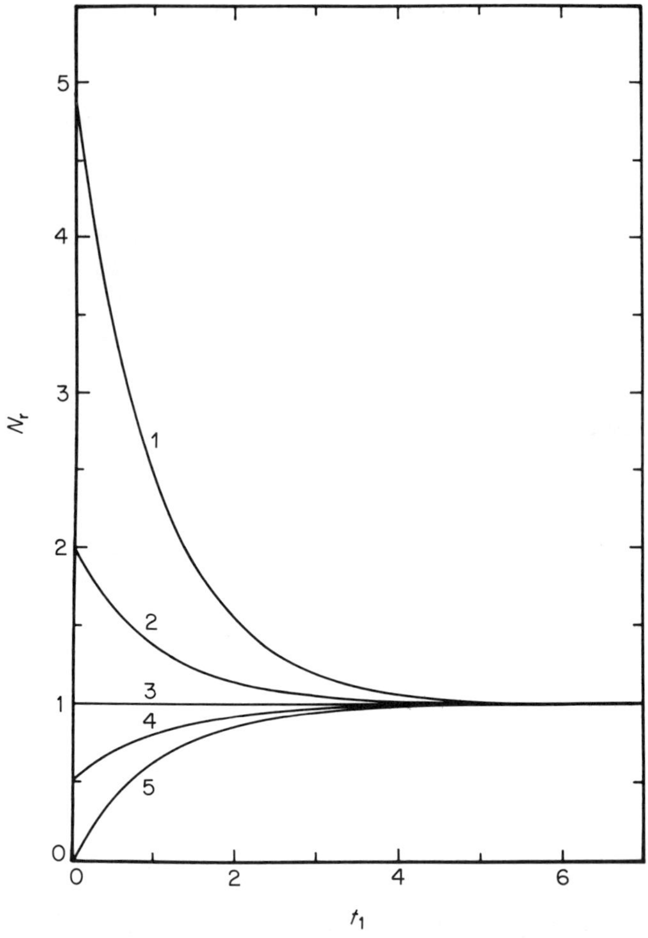

Fig. 6.10. Time dependence of the relative surface concentration, $N_r = N/N_\infty$, of non-immobilized particles in the primary energy minimum for various initial surface concentrations, N_0 (normalized by N_∞). Curves 1 to 5 correspond to values of N_0 of 4, 2, 1, 0·5 and 0, respectively (after Adamczyk and van de Ven, 1984).

$$N_s = k_s N_\infty t - k_s N_\infty T_1 + k_s N_0 T_1 \qquad (6.49)$$

When $k_s \gg \bar{f}_1\text{Pe}$ (very fast immobilization reaction), $N_\infty \to \text{Sh}_0/k_s$ and $N_s - N_\infty = \text{Sh}t$ (for $t \gg T_1$). This result is identical to the perfect-sink boundary-condition approximation. For the opposite extreme, i.e. when $k_s = 0$, the steady-state value of N becomes $\text{Sh}/\bar{f}_1\text{Pe}$ and $N_s = 0$.

It is interesting to note that, in this model, the number of particles dyna-

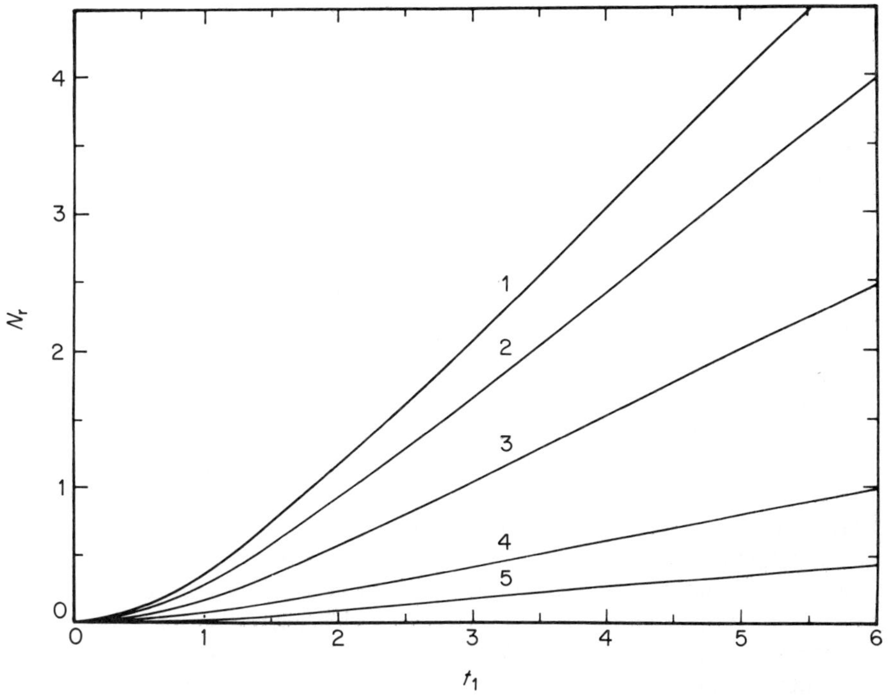

Fig. 6.11. Time dependence of the relative surface concentration, $N_r = N_s/N_\infty$, of particles immobilized in the primary energy minimum for various immobilization rates $\beta = k_s/\bar{f}_1 \mathrm{Pe}$. Curves 1 to 5 correspond to values of β of ∞, 4, 1, 0·25 and 0·1, respectively (after Adamczyk and van de Ven, 1984).

mically adsorbed at the primary minimum region (as well as the value of N_s) does not depend on the depth or extension of the minimum.

The value of N_∞ (for $k_s = 0$) (the largest possible value) can be expressed in terms of a monolayer coverage, N_{rm}, as:

$$N_{rm} = n_\infty a\, N_\infty / N_m = n_\infty a^3 \mathrm{Sh}/\bar{f}_1 \mathrm{Pe} \tag{6.50}$$

Considering that for Pe < 10^{-2} (and no external force), Sh = 0·616 $\mathrm{Pe}^{1/3}$ (see Table 6.2), eq (6.49) becomes:

$$\mathrm{N}_{rm} = \frac{2 \cdot 46}{\bar{f}_1}\, n_\infty a^3 \mathrm{Pe}^{-2/3} = C_d n_0 a^{1/3}/\omega \tag{6.51a}$$

where

$$C_{\mathrm{d}}=\frac{2\cdot 46}{\bar{f}_1}\left(\frac{1\cdot 02\ 6\pi\mu}{\nu^{1/2}\ kT}\right)^{-2/3} \tag{6.51b}$$

Thus, for a fixed value of the particle number concentration, n_∞, N_{rm} is proportional to $a^{1/3}$ and inversely proportional to the disk angular velocity.

When a strong external force acts towards the disk, Sh = Ex (cf. eq (6.30)), and eq (6.50) takes the form:

$$N_{\mathrm{rm}}=n_\infty a^3\mathrm{Ex}/\bar{f}_1\mathrm{Pe} \tag{6.52}$$

When Ex = Gr, eq (6.52) indicates that N_{rm} is proportional to $a^3/\omega^{3/2}$.

These formulae show that the number of particles adsorbed dynamically in the primary minimum region can be considerably decreased by an increase in the disk angular velocity. It is also interesting to note that the kinetics of particle desorption from the primary minimum region (i.e. when the initial surface concentration $N_0 \gg N_\infty$ and $k_{\mathrm{s}}=0$) is described, for all Sherwood numbers, by an exponential decay with a characteristic transition time $t_{\mathrm{d}}=(a^2/D_0)\bar{f}_1\mathrm{Pe}$, which is independent of the depth of the primary minimum. This shows that the particle desorption rate is solely controlled by the tangential component of the fluid velocity and is generally much larger than particle adsorption rates in the primary minimum region, which are governed by the value of Sh.

A typical value of N_∞, obtained from eq (6.50) by assuming $a=0\cdot 1\ \mu\mathrm{m}$, $n=10^{15}\ \mathrm{m}^{-3}$, and $\mathrm{Pe}=10^{-5}$ (corresponding to $\omega\simeq 12\ \mathrm{rad\ s}^{-1}$ when $\nu=10^{-6}\ \mathrm{m}^2\ \mathrm{s}^{-1}$) is 0·8% of a monolayer coverage and is rather insensitive to the particle radius. This is a very small value and cannot account for experimentally observed coverages which can attain 10% or more of a monolayer. Thus, for perfectly smooth and homogenous surfaces, in the absence of immobilization reactions, the fraction of particles adsorbed dynamically under normally encountered experimental conditions is small and cannot be detected easily. Detection is possible, however, in direct observations of particle deposition onto the collector surface in an impinging jet apparatus, discussed below. Obviously, when Pe → 0 (which, for a constant particle size, can be obtained by a decrease in the disk angular velocity), the value of N may become a considerable fraction of a monolayer, but such particles are not localized; they form a kind of cloud around the primary minimum region and can hardly be detected in indirect experiments in which the disk is washed and withdrawn from the suspension prior to particle counting.

In order to account for experimentally observed coverages and the linear dependence of N versus time, t, we can conclude that particles are immobilized at the collector surface and that the dimensionless rate of this process is larger than $\bar{f}_1\mathrm{Pe}$. It is worthwhile noting that in all previous theoretical approaches aimed at predicting particle transfer rates (in particular the perfect-sink approximation), this conclusion was taken as a basic assumption (often not explicitly articulated) and, as a result, no estimates of the reaction rate were made.

Similar, although much more complicated, kinetic equations have also been formulated for other collectors, such as the spherical and cylindrical collector and the parallel-plate channel. There are no fundamental differences from the rotating disk. For all these collectors the number of particles accumulated dynamically in the primary minimum region is of the order of $\mathrm{Sh}(\bar{u}_{\parallel}+k_s)$ (where $\bar{u}_{\parallel}$ is the dimensionless tangential component of velocity at the minimum), and the steady-state value of N is attained after a transition time of the order of $1/(\bar{u}_{\parallel}+k_s)$. If the immobilization rate constant, k_s, is much larger than $\bar{u}_{\parallel}$, then the number of particles immobilized is a linear function of time and the perfect-sink approximation applies.

6.3.5.3 Concentration distribution in accumulation layers

The introduction of the surface concentration, N, as a primary variable instead of n proved to be very advantageous in formulating and solving the kinetic equation that describes particle accumulation in energy minima. However, it is also interesting to know explicitly the particle-concentration distributions within these regions, especially in relation to the equilibrium Boltzmann distribution, which is universally determined by the energy distribution. Because the spatial extension of the energy minimum region is very small in comparison with the particle dimensions, the diffusion relaxation time, $t_d=\delta^2/D$ (where δ is the distance between the surface and the location of the primary minimum), is, for colloidal particles, a very small fraction of a second. Consequently, a local quasi-equilibrium must be established rapidly throughout this region and the concentration distribution must closely resemble the Boltzmann distribution with a normalization constant depending on the particle concentration at the region's boundary. It is also obvious, however, that the concentration distribution must deviate from the Boltzmann distribution in order to produce a finite value of the particle flux. In view of this, it is assumed that the particle-concentration distribution in the accumulation layer (i.e. of particles whose distance to the surface is less than ℓ) can be expressed as:

$$n=F(x,t)\exp(-V_{\mathrm{int}})+n^*(x,H,t)=n_{\mathrm{B}}+n^* \tag{6.53}$$

where the tern n_{B} denotes the quasi-Boltzmann contribution and n^* is the perturbation term which is assumed to be very small (in comparison with n_{B})

around the minimum. (The interaction energy, V_{int}, is expressed in non-dimensional form.)

It can be shown that when only a primary (and no secondary) minimum is present, the function F is equal to $N(x,t)\, I_{pm}^{-1}$, where N is the surface concentration of particles in the mobile phase which can be calculated from the previously mentioned kinetic equations, and $I_{pm} = \int \exp(-V_{int})\mathrm{d}H$, the integral being over the width of the minimum. The perturbation term n^* can be expressed as:

$$n^* = \mathrm{Sh}\exp(-V_{int}) \int_{H_1}^{H} \frac{\Gamma_1(H)}{F_1(H)} \exp(V_{int})\,\mathrm{d}h \tag{6.54}$$

with

$$\Gamma_1(H) = I_{pm}^{-1} \int_{H_1}^{H} \exp - (V_{int})\,\mathrm{d}H$$

where H_1 is the lower edge of the minimum. The ratio of the quasi-Boltzmann term to the perturbation term is approximately given by

$$\frac{n^*}{n_B} = \frac{\mathrm{Sh}}{F_m} \left(\frac{\mathrm{d}^2 V_{int}}{\mathrm{d}H^2}\right)_m^{-1} N(x,t)) \tag{6.55}$$

where $F_m = F_1(H_m)$, which indicates that n^*/n_B is indeed a very small quantity, usually of the order of 10^{-4} to 10^{-6} for $N \simeq 1$ because the curvature of the energy distribution at the minimum, i.e. $(\mathrm{d}^2V_{int}/\mathrm{d}H^2)$, is extremely large. However, at larger separations from the minimum (further away from the surface), n^* may well exceed the quasi-Boltzmann contribution (this point is discussed later).

6.3.5.4 Comparison with numerical solutions

The range of validity of the kinetic equations and the expressions for the particle-concentration distribution can be estimated by comparing them with numerical solutions of the exact transport equation (Adamczyk *et al.*, 1984). For the rotating disk, eq (6.43) can be written as

$$\frac{\partial n}{\partial t} = \frac{\partial J}{\partial H} - \mathrm{Pe}F_3(H)\,(H+1)\,n - Q(n, n_s t) \tag{6.56}$$

with

$$J = F_1(H)\left[\frac{\partial n}{\partial H} + \frac{\mathrm{d}V_{\mathrm{int}}}{\mathrm{d}H}\,n + \frac{1}{2}\,\mathrm{Pe}F_2(H)\,(H+1)^2\,n + \mathrm{Ex}\,n\right]$$

where $F_3(H)$, $F_2(H)$ are the dimensionless correction functions. Numerical solutions of eq (6.56), with appropriate boundary and initial conditions, were obtained when either a primary minimum or a primary and a secondary minima are present. For the sake of brevity we only report here the results for an isolated energy minimum and when $k_s = k_r = 0$.

Figure 6.12 shows some characteristic results for $N_r = N/N_\infty$ as a function of the dimensionless time $t_1 = \bar{f}_1 \mathrm{Pe}\, t$. In the same figure the reduced mass-transfer number, i.e. $\mathrm{Sh}_r = \mathrm{Sh}(t)/\mathrm{Sh}_\infty$ (where Sh_∞ is the steady-state value of Sh), is also plotted for $\mathrm{Pe} = 10^{-3}$ and 10^{-1}. It can be seen from Fig. 6.12 that the steady-state conditions are attained after a transition time that is a small fraction of $T_1 = 1/\bar{f}_1\mathrm{Pe}$, the transition time becoming smaller as Pe decreases. It can also be seen that the agreement between the analytical and numerical

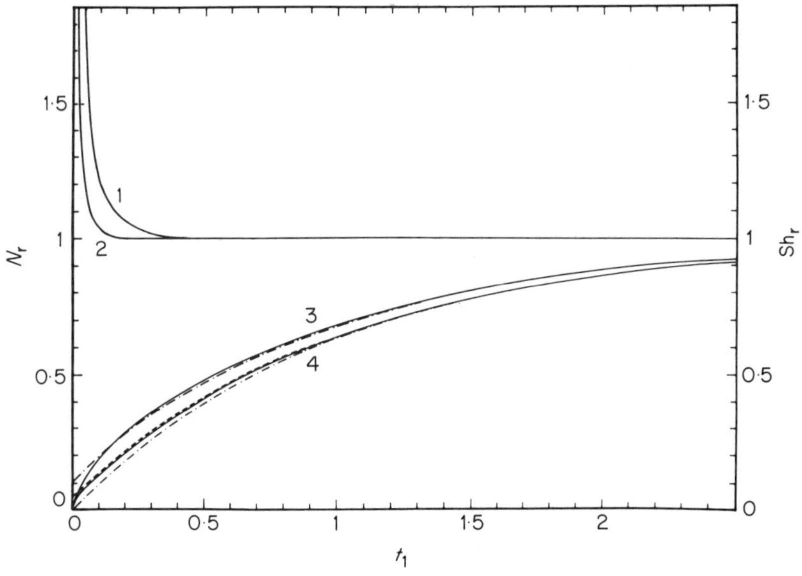

Fig. 6.12. A comparison of numerical and analytical results for an isolated energy minimum and $k_s = k_r = \mathrm{Ex} = 0$. Curves 1 and 2 show the dependence of the reduced particle flux at the primary minimum region $\mathrm{Sh}_r = \mathrm{Sh}(t)/\mathrm{Sh}_\infty$ for $\mathrm{Pe} = 0 \cdot 1$ and 10^{-3}, respectively, and curves 3 and 4 represent the dependence of N for the same values of Pe. (----) Values of N_r calculated from the equation $N_r = 1 - \exp[-(t_1 - t_0)] + N_0 \exp[-(t_1 - t_0)]$ where $t_1 = \bar{f}_1 \mathrm{Pe}t$, $t_0 = 0 \cdot 1$ and $N_0 = N(t_0)/N_\infty$. (-·-·-·)Values obtained from this equation when $N_0 = 0$ and $t_0 = 0$ (after Adamczyk *et al.*, 1984).

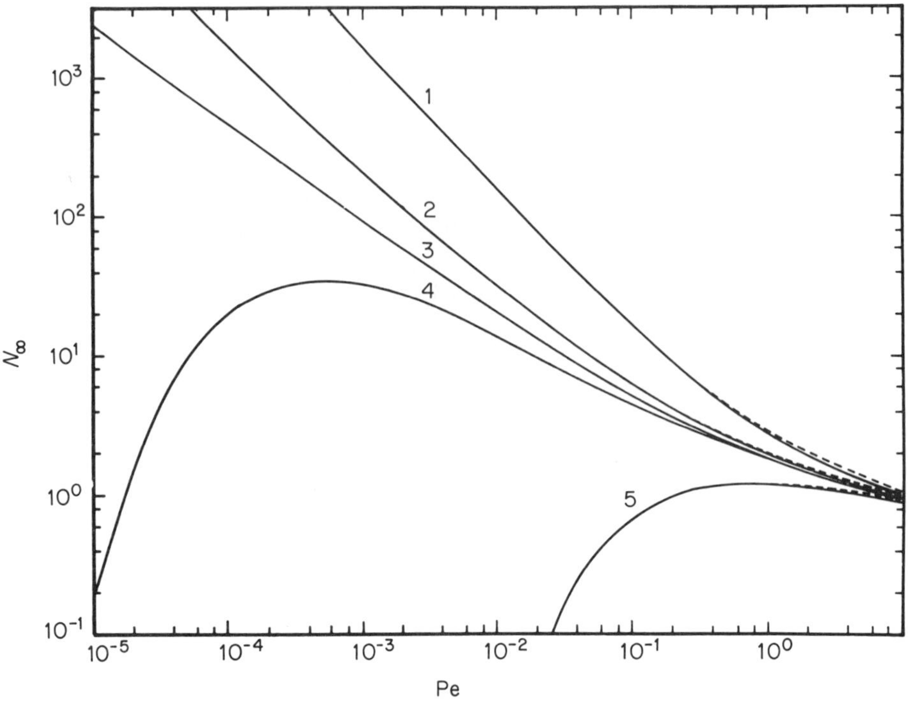

Fig. 6.13. The steady-state surface coverages, N_∞, at the primary minimum as a function of Pe obtained by numerical solution of the transport equation; $k_s = 0$, $k_r = 0$. Curves 1 to 5 correspond to values of Ex of 1, 0·1, 0, −0·1 and −1, respectively. (----)The analytical approximation, i.e. $N_\infty = \mathrm{Sh}_0/\bar{f}_1\mathrm{Pe}$ (after Adamczyk *et al.*, 1984).

solutions is good, especially when the number of particles accumulated during the transition time is treated as an initial condition in eq (6.47). Good agreement was also observed for values of Pe (or Ex) of the order of unity. It was also found that varying the shape and depth of the primary minimum within wide limits did not affect the kinetics of particle accumulation (in accordance with theoretical predictions).

For $t_1 \gg 1$, a steady-state value of N is established, which is shown in Fig. 6.13 as a function of Pe and at various values of the external-force parameter, Ex. Again, Fig. 6.13 shows that the numerical and analytical results agree within 5%, even for Pe values as large as 10.

In Fig. 6.14a the concentration profile obtained from numerical solutions is compared with the analytical expression (eq (6.53)). It can be seen that around the minimum, the deviation from the quasi-Boltzmann distribution is very small. However, when we proceed further from the minimum towards

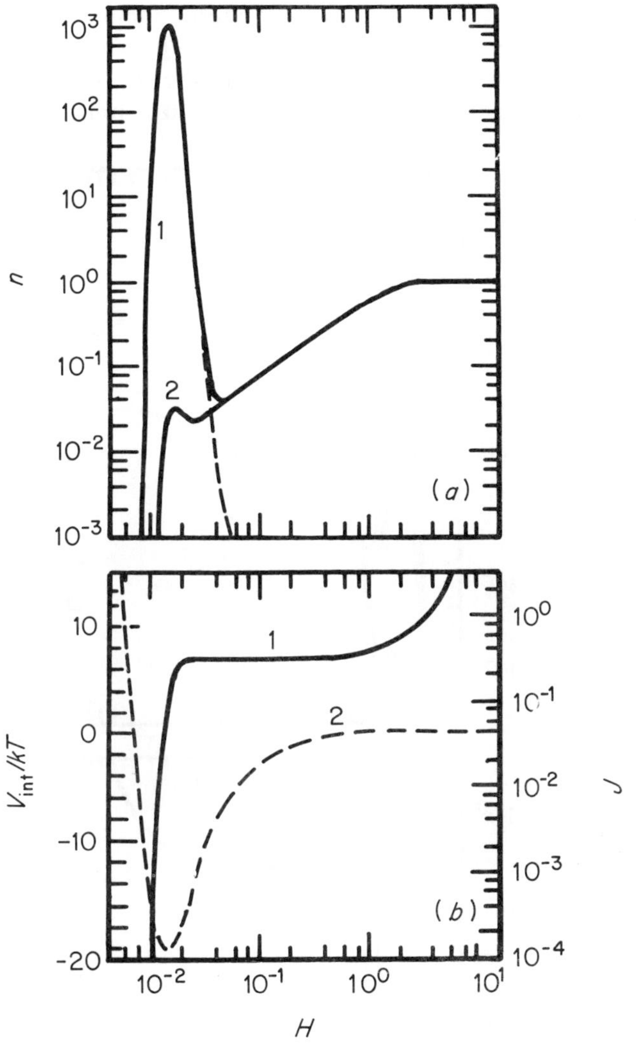

Fig. 6.14. *(a)* Steady-state concentration profile for (mobile) particles accumulated in the primary energy minimum ($k_s = 0$). H is the non-dimensionalized gap width h/a. Curve 1 is the complete numerical solution of eq (6.56). Curve 2 is the perturbation term (second term on right-hand side of eq (6.53)). (----)The quasi-Boltzmann distribution; n calculated from eq (6.53) is indistinguishable from Curve 1 up to H = 1.
(b) The dependence of the normal component of the particle flux J (non-dimensionalized by $D_0 n_\infty / a$) on the separation distance $H = h/a$ (curve 1). Also shown is the interaction energy, V_{int} (non-dimensionalized by kT) (curve 2), for which the flux J and the concentration profile n is calculated (after Adamczyk *et al.*, 1984).

the bulk of the suspension, the perturbation term dominates the concentration profile and eq (6.53) should be used. In Fig. 6.14b the normal flux component, J, and the energy distribution are shown. Note that the flux is constant over large distances ranging from $H \simeq 10^{-2}$ up to $H=2$. Since the

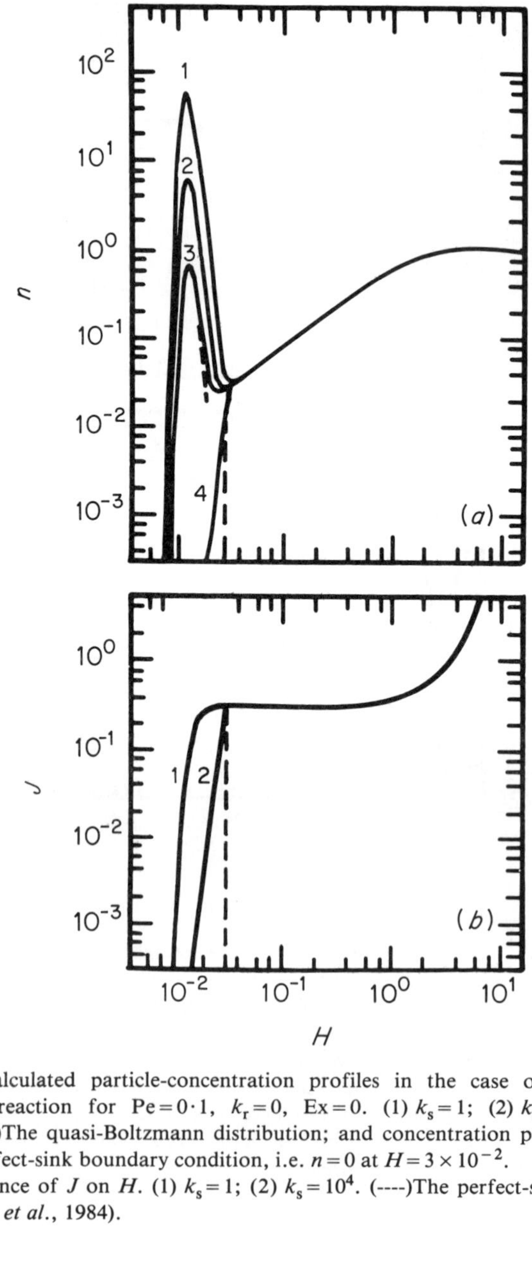

Fig. 6.15. (a) Calculated particle-concentration profiles in the case of an homogenous immobilization reaction for $\mathrm{Pe}=0{\cdot}1$, $k_r=0$, $\mathrm{Ex}=0$. (1) $k_s=1$; (2) $k_s=10$; (3) $k_s=10^2$; (4) $k_s=10^4$. (----)The quasi-Boltzmann distribution; and concentration profiles obtained by adopting the perfect-sink boundary condition, i.e. $n=0$ at $H=3\times10^{-2}$.
(b) The dependence of J on H. (1) $k_s=1$; (2) $k_s=10^4$. (----)The perfect-sink approximation (after Adamczyk *et al.*, 1984).

concentration peaks are well pronounced and the flux is effectively constant, the boundary of the accumulation layer can be chosen arbitrarily within wide limits and the choice will not affect the results.

Figure 6.15 shows the numerical solutions for situations where particle immobilization occurs. These results (obtained with the boundary condition $J=0$ at the collector surface) are compared with those obtained by adopting the perfect-sink model. It can be seen that the adsorption model produces very similar concentration and flux distributions to those obtained with the perfect-sink model when $k_s > 10^3$; this value is much larger than that required to attain similarity of particle deposition kinetics (i.e. for $k_s \gg \bar{f}_1\mathrm{Pe}$). Thus the perfect-sink boundary condition postulated at $H=\bar{\delta}$ is equivalent to the assumption that an irreversible immobilization of particles takes place within the region $0 < H < \bar{\delta}$ (not only at $H=\bar{\delta}$), the rate constant of which tends to infinity. This condition is much more restrictive than the assumptions in the adsorption model.

The numerical results indicate that the approximate analytical solutions are useful for predicting particle-accumulation kinetics even when Pe and Ex are of the order of unity. The adsorption model may fail for very large Péclet numbers (large particles and intense flows) when strong attractive forces such as the double-layer interactions are acting over distances comparable with the particle dimensions, for example when the ionic strength is low.

All results referred to so far apply when the surface concentration of the immobile (and mobile) phase remains low. When the surface coverage is higher than approximately 10% of a monolayer the interaction among immobilized particles and those flowing in their vicinity becomes increasingly important. No adequate quantitative theory exists at present to account for these effects. Some approximate results can be derived if the value of Sh in eq (6.45) is modified to account for the effective area available for particle deposition (due to geometrical blocking effects).

An approximate theoretical model, aimed at describing surface-blocking effects and the influence of long-lasting changes in the particle–collector adhesion force on the kinetics of particle deposition on collectors, will be discussed in the next section.

6.4 Factors affecting deposition kinetics

6.4.1 General remarks

The kinetics of deposition of colloidal particles depends, in a complicated way, on the conditions at the collector surface. As deposition proceeds, the fraction of surface coated by particles increases, thus changing the geometry of the collector surface and the hydrodynamic conditions near the surface.

As a result, blocking effects occur which, when the colloidal dispersion is stable, decrease the flux to the collector surface. The blocking effects may also depend on the nature of the collector surface. Another factor affecting deposition (already discussed in relation to eq (6.46)), is the detachment of particles from the collector surface due to their thermal motion and to external hydrodynamic forces, especially in the case of turbulent flow. The removal rate strongly depends on the depth and shape of the primary energy minimum which, in turn, often depends on time, due to the time dependence of dispersion, van de Waals and electrical double-layer interaction forces caused by particle or surface deformation, rearrangement of charges, etc. (Dabros and Czarnecki, 1980). Moreover, any kind of surface roughness or non-uniformity of the surface charge distribution can produce significant tangential forces which act on the particles locally; this has a large effect on the number of particles immobilized at the surface. Finally, when the dispersion is unstable, coagulation will lead to the formation of aggregates. Under such conditions the coating density will be a non-linear function of time due to a decrease in the particle number concentration and variations in the mass-transfer rates of the aggregates (Waligora and Dabros, 1978).

The results presented in this section can be derived directly from the general kinetics equation (eqs (6.43) and (6.46)) if it is assumed that the removal-rate constant, k_r, depends explicitly on time. Blocking can, in principle, be incorporated by assuming that k_s is a function of N_s. In this case the linear relation (eq (6.46)) no longer holds.

6.4.2 Particle escape and ageing of particle–collector bonds

For the moment let us assume that blocking effects are absent; they will be considered later. Let N^* be the number of particles deposited per unit area between the times τ and $\tau + \Delta\tau$, which still remain at the surface at some later time $t(\Delta\tau \ll t - \tau)$. Then the removal flux, J_r, at time t of the population N^* can be written as

$$J_r(t) = k_r(t - \tau)N^* \tag{6.57}$$

where k_r is the escape-rate coefficient. The time dependence of k_r is justified by variations in the double-layer interaction forces, and by changes in the strength of the adhesive bonds formed between the collector and the particle. The latter may be attributed to a variety of effects such as sintering, mixing by mutual diffusion, and surface creeping. It has been observed that adhesion forces can rise to several times their initial value after a contact time of the order of 1 hr (Krupp, 1967). This observation suggests the possibility that the particle detachment is a function of the contact time. The total number of particles per unit area can be calculated by integrating the contributions from the deposition in the time periods $(\tau, \tau + d\tau)$ multiplied by the probability $P(t-\tau)$ of the survival of these particles on the collector surface. This probability can readily be calculated, assuming that the escape-rate coefficient, k_r is a function of the duration of the particle–collector contact:

$$P(t-\tau) = \exp\left[-\int_0^{t-\tau} k_r(\xi)\, d\xi\right] \tag{6.58}$$

The surface-coating density (i.e. the number of particles per unit area) can then be expressed as

$$N(t) = \int_0^t J_0\, P(t-\tau)\, d\tau \tag{6.59}$$

where J_0 is the flux of the particles to the collector, which may be a function of time.

The net flux of particles to the collector surface will thus be

$$J(t) = J_0 - \int_0^t J_0 k_r(t-\tau) P(t-\tau)\, d\tau \tag{6.60}$$

The integral term in this equation describes the escape flux of particles deposited prior to time t.

When the escape coefficient, k_r is a constant, corresponding to the case that each particle on the surface has the same probability of escape (due to its thermal motion), the solution of the above integral is

$$J = J_0 e^{-t/\tau_{esc}} \tag{6.61a}$$

where $\tau_{esc} = 1/k_r$ and thus the number of particles on the surface per unit area is

$$N(t) = N_\infty \left(1 - e^{-t/\tau_{esc}}\right) \tag{6.61b}$$

with $N_\infty = J_0 \tau_{esc}$. From a solution of the convective-diffusion equation, taking the full energy of interaction, V_{int}, between a particle and the surface into account, it can be shown that (Dabros and van de Ven, 1982)

$$\tau_{esc} = \frac{a \int_{h_1}^{h_2} e^{-V_{int}/kT} \, dh}{\text{Sh } D_0} \tag{6.61c}$$

where $\Delta h = h_2 - h_1$ is the width of the energy minimum.

An alternative way of interpreting the escape time, τ_{esc}, is to say that the probability p of a particle escaping during an interval Δt is $p = \Delta t / \tau_{esc}$, i.e. $\Delta N = -pN$, which is equivalent to eq (6.57). It follows that τ_{esc} can be interpreted as the average time that a particle spends on the surface.

When, due to ageing, the escape-rate coefficient, k_r, changes exponentially from some initial value k_0 to a final value k_k with relaxation time $1/\delta$:

$$P(t-\tau) = \exp - \left[k_k(t-\tau) - \frac{(k_0 - k_k)}{\delta} \{\exp[-\delta(t-\tau)] - 1\} \right] \tag{6.62}$$

It can be seen from eqs (6.59) and (6.62) that, when $k_k = 0$, the coating density, N, reaches a steady-state value $J_0 \exp(-k_0/\delta)$ as $t \to \infty$.

It is also of interest to consider the removal of particles from the surface in the absence of freely suspended particles. If a unit surface on which N particles are deposited is exposed to a fluid flow field, the number of particles will decrease exponentially with time, provided that all particles adhere to the surface with the same strength:

$$N = N_0 e^{-t/\tau_{esc}} \tag{6.63a}$$

Again τ_{esc} can be considered as the average time that a particle spends on the surface. It is important to note that τ_{esc} during deposition (i.e. given by eq (6.61c)), is different from τ_{esc} during particle removal. The reason is that the probability of escape is different in both cases. This can be seen by considering a particle arriving at the energy maximum after leaving the primary minimum. During a deposition process, the position of the energy barrier corresponds to the position of low particle concentration. To move over the barrier, particles must diffuse against a concentration gradient and thus the detachment rate during deposition depends on the number of particles outside the barrier, which is governed by the number of particles reaching the barrier from the outside. This explains why τ_{esc}^{dep} depends on the Sherwood number, Sh (cf. eq (6.61c)). In contrast, during a removal process, a particle must diffuse down a concentration gradient when it arrives at the barrier

barrier, because there are few or no particles beyond the barrier. It follows that, in all cases, $\tau_{esc}^{dep} > \tau_{esc}^{removal}$ (for identical hydrodynamic conditions); τ_{esc} during removal can be estimated from (Dahneke, 1975)

$$\tau_{esc} = \frac{e^{\Delta V/kT}}{D_0} \int_{h_1}^{h_2} \int_{s}^{\infty} F_1(s)\, e^{-[V(h)-V_0]/kT}\, e^{-[V_1-V(s)]/kT}\, dh\, ds \qquad (6.63b)$$

provided that the height of the energy barrier, $\Delta V = V_1 - V_2$, is sufficiently large.

6.4.3 Blocking by deposited particles

When blocking effects are important it is possible to incorporate them in the following way: if J_0 is the inital flux of particles to the collector, a fraction of this flux becomes ineffective due to masking effects by deposited particles. So in the time period τ to $\tau + d\tau$ the number of new particles that can reach the collector surface is equal to $J_{eff} d\tau$, with J_{eff} given by

$$J_{eff} = J_0 - J_0 \gamma \int_0^{\tau} J(s) ds \qquad (6.64)$$

where the integral in this expression represents the actual value of the coating density. The coefficient γ is the area blocked per particle. In the simplest case, γ may be assumed to be constant. In general, however, γ may also depend on the hydrodynamic conditions, the coating density and other parameters. Multiplying the number of particles reaching the surface by the probability $P(t-\tau)$ that these particles will stay there, one obtains the number of particles per unit area that were deposited in the time $(\tau, \tau + d\tau)$ and that remain on the collector surface at time t. This value multiplied by the escape-rate constant, $k_r(t-\tau)$, yields the total value of the escape rate. Thus the coating density at any moment t may be expressed as:

$$N(t) = \int_0^t J_0 \left[1 - \gamma \int_0^{\tau} J(s)\, ds\right] P(t-\tau)\, d\tau \qquad (6.65)$$

The flux can be obtained by differentiating eq (6.65) and can be formally expressed as

$$J = J_0 - J_0\gamma \int_0^t J(s)\,\mathrm{d}s - \int_0^t J_0\left[1 - \gamma \int_0^\tau J(s)\,\mathrm{d}s\right] P(t-\tau)\,k_\mathrm{r}(t-\tau)\,\mathrm{d}\tau \tag{6.66}$$

When the escape-rate coefficient, k_r, and the masking coefficient, γ, are constants equal to k_0 and γ_0, respectively, it may be checked, by substitution, that solutions of eqs (6.65) and (6.66) are given by:

$$J(t) = J_0 \mathrm{e}^{-\beta t} \tag{6.67a}$$

$$N(t) = N_\infty(1 - \mathrm{e}^{-\beta t}) \tag{6.67b}$$

where

$$\beta = \gamma_0 J_0 + k_0 \tag{6.68a}$$

and

$$N_\infty = J_0/\beta \tag{6.68b}$$

6.4.4 Monte Carlo simulation of the coating process

It is often difficult to obtain analytical solutions of eqs (6.59) and (6.65). In such cases the coating density can be calculated by Monte Carlo simulation methods (Dabros and van de Ven, 1983b). To this end, the number of particles at the surface at time t can be found by dividing t into n time intervals Δt (i.e. $t = n\Delta t$). During each time interval Δt_i, $J_0\Delta t_i$ particles are deposited. In the absence of blocking, and without loss of generality, it can be assumed that $J_0 = 1$. For all time intervals prior to t_i ($= i\Delta t_i$), the probability of escape must be calculated for every deposited particle, e.g. depending on the conditions, according to eq (6.62). If this probability is larger than a randomly generated number in the range (0, 1), the deposited particle is assumed to escape from the collector. In this way, the number of particles that will disappear during a time interval Δt_i can be calculated. This process is repeated for all times $t_i = i\Delta t_i$, with i going from 1 to n.

In the case of modelling the deposition process with blocking effects, the probability of deposition is also a random function which depends on the total number of previously deposited particles. If the calculated fraction J_eff/J_0 is less than a randomly generated number in the range (0,1), then it is assumed that deposition does not occur.

Figure 6.16 shows results of Monte Carlo simulations as well as comparisons with the analytical model discussed above. Curves 1 and 2 show the results when only particle escape is considered. In curves 3 to 5, blocking effects are also taken into account. When the escape-rate coefficient, k_r, is

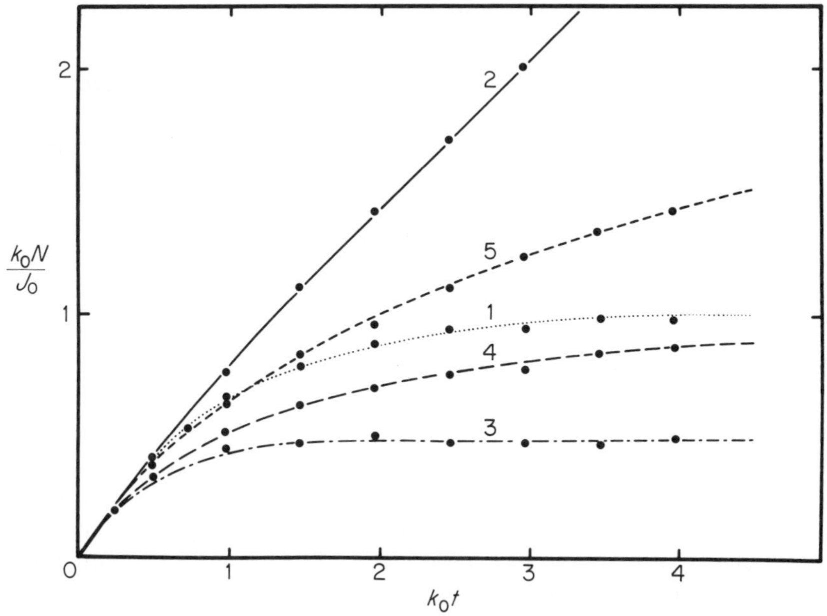

Fig. 6.16. Dimensionless coating density, k_0N/J_0, as a function of dimensionless time, k_0t. Points are obtained from Monte Carlo calculations. $k_0 = 0.005$ and $\gamma = 0.01$. Curves 1 and 2 are calculated from eqs (6.59) and (6.62) for $k_k = k_0 = 0{\cdot}005$ and $k_k = 0$, respectively (i.e. no blocking). In curves 3 to 5 both blocking ($\gamma J_0 = 0{\cdot}005$) and ageing are taken into account. Curve 3 is calculated from eq (6.67b). For curves 4 and 5 (for $k_k = 0{\cdot}0025$ and 0, respectively) no solutions of eqs (6.62) and (6.65) are available.

time independent, the coating density can be described by eq (6.67b) (curve 3). When k_r varies exponentially, no solution of eq (6.66) is available and only the Monte Carlo calculations are presented (curves 4 and 5). It can be seen from Fig. 6.16 that, when comparison with the analytical model is possible, agreement between both models is excellent.

Modelling by Monte Carlo simulations has the advantage that it can be used when the escape rate constant, k_r, and the blocking coefficient, γ, are complicated functions of the coating density and/or time, in which case analytical solutions of eqs (6.59) or (6.65) are difficult or impossible to obtain.

6.4.5 Influence of colloid stability

The influence of the colloid stability on the coating process has been discussed by Davis (1973) and, in more detail, by Dabros (1977) and Waligora and Dabros (1978). Let us assume that at time $t = 0$ the colloidal dispersion consists solely of spherical particles of radius a_1 at a concentration n_1^0. For $t > 0$, coagulation occurs and aggregates consisting of two, three, etc., primary particles appear. The concentrations of the aggregates, n_2, n_3, etc., are functions of time.

When the concentration is sufficiently small the fluxes are linear functions of the concentrations:

$$J_i(t) = B_i n_i(t) \tag{6.69}$$

where the functions B_i depend on the hydrodynamic conditions in the vicinity of the wall, on the geometry of the collector and the particles, and on the external and colloidal forces acting on the particles. It can also be a function of time.

When the half-time of coagulation, T_s, is much larger than the time τ_d for the development of the diffusion boundary layer, the deposition process may be regarded as quasi-stationary; τ_d is of the order of $\delta_d^2/2D_0$, where δ_d is the thickness of the diffusion boundary layer and D_0 is the diffusion coefficient of particles in the bulk. For perikinetic coagulation, the half-time of coagulation can be shown to be

$$T_s = \frac{W}{8\pi a D_\infty n_1^0} \tag{6.70}$$

where W is the stability ratio, which is the inverse of the perikinetic collision efficiency (i.e. $W = 2\alpha^{-1}$, given by eq (6.27)). The concentration of i-fold particles in the dispersion can be expressed by the equation

$$n_i = n_1^0 \frac{\tau^{i-1}}{(1+\tau)^{i+1}} \tag{6.71}$$

and the total number of particles and aggregates as

$$n = n_1^0 \frac{1}{1+\tau} \tag{6.72}$$

where $\tau = t/T_s$.

Let us consider the simplest case when the deposition rate for single particles and aggregates is equal, i.e. $f_i = B_i/B_1 = 1$. The mean value of the translational-diffusion coefficient for aggregates consisting of i spherical particles can be expressed as.

$$D_i = \frac{kT}{6\pi\,\mu\,a_i} \tag{6.73}$$

where a_i is the hydrodynamic radius of the aggregate, given by $a_i = a_1 i^p$ with $1/4 < p < 1/3$. Thus $D_i = D_1^0/i^p$. This equation suggests that the number of particles comprising an aggregate only weakly affects the value of the diffusion coefficient; e.g. for doublets the diffusion coefficient is about 20% less than for a single particle. The decrease in the diffusion coefficient for larger aggregates depositing on the surface is partially cancelled by the increase in the collision radius of the aggregates, which suggests that the assumption $f_i = 1$ may describe experimental data reasonably well. With this assumption and assuming additivity of the fluxes of various aggregates, the total flux equals

$$J(t) = J_1^0 \frac{1}{1+\tau} \tag{6.74}$$

and after integration

$$N(t) = J_1^0 T_s \ln(1+\tau) \tag{6.75a}$$

For $\tau < 0{\cdot}1$, this equation can be written (with better than 1% accuracy) as

$$\frac{N}{t} = J_1^0 \left(1 - \frac{t}{2T_s}\right) \tag{6.75b}$$

On the basis of this equation it is possible to estimate the values of J_1^0 and T_s from experimental data. Thus, for some systems, studies of deposition kinetics may also be used for the determination of the coagulation half-time, if blocking and other effects are negligible.

When the assumption that $f_i = 1$ is not valid, it is possible, in principle, to obtain values for f_i by fitting a polynomial of an appropriate degree through the experimental $N(t)$ dependence. Data on f_i may provide useful information for developing a simplified theory of aggregate behaviour in the vicinity of a wall.

Counting each aggregate as a single particle, eqs (6.69) and (6.71) give

$$J(t) = J_1^0 \frac{1}{(1+\tau)^2} \tag{6.76}$$

for the flux and

$$N(t) = J_1^0 T_s \frac{\tau}{1+\tau} \tag{6.77}$$

for the coating density.

6.4.6 Dynamic aspects of particle–wall interactions

As we have seen from the above discussion, much more detailed information is needed, in general, to predict deposition rates accurately. Factors which influence deposition rates and which are not usually included in the transport equations are: (i) surface roughness, (ii) surface chemical heterogeneity, (iii) non-uniformity of surface charge, (iv) ageing effects, (v) blocking effects, and (vi) dynamic effects occurring when relative motion occurs inside an electrical double layer. Factors (i) to (v) are discussed above, but the dynamics of electrical double layers has not yet been addressed. The latter can, however, often be an important factor in determining deposition and stability, which can be illustrated by a few crude calculations.

During the approach of a particle towards a collector, the respective surface charge densities can be altered by an exchange of ions with the suspending medium. The characteristic time over which such changes take place is

$$\tau_\sigma = \sigma / i_0 \tag{6.78}$$

where σ is the surface charge density and i_0 the exchange-current density. This current is known to vary within very wide limits, depending on the systems, ranging from 10^{-6} to 10^4 A m^{-2}. Assuming a reasonable value for the surface charge, $\sigma \simeq 0{\cdot}1$ C m^{-2}, it follows from eq (6.78) that 10^{-5} s $< \tau_\sigma < 10^5$ s. The collision time, τ_c, is approximately equal to

$$\tau_c \simeq 1/(2\kappa^2 D_0) \tag{6.79}$$

When the double-layer thickness is $\kappa^{-1} = 10$ nm the collision time is $\tau_c \simeq 10^{-3}$ s, and it can be concluded that for some systems $\tau_\sigma < \tau_c$, while for others the opposite is true. When $\tau_\sigma \ll \tau_c$ the surface can be treated as being at constant potential, and when $\tau_\sigma \gg \tau_c$ as being at constant charge. Most systems will lie between these two extremes, i.e. the collision will proceed under non-equilibrium conditions and will be accompanied by a flow of ions across the solid–liquid interface and/or inside the particle body. Associated with this current is an energy dissipation, with a corresponding electric work, W, given by

$$W \simeq \sigma \Delta V \tag{6.80}$$

where ΔV is the change in surface potential; for a collision of a 1-μm particle with a collector W is typically of the order of 10 mV (Usui, 1972, 1973). Assuming that $\sigma \simeq 0{\cdot}1$ C m^{-2}, yields $W \simeq 10^{-5}$ J m^{-2}. When $a = 0{\cdot}5$ μm and $\kappa^{-1} = 10$ nm, only about 1% of the particles will feel the electrical double-layer overlap (because most of the volume of the particle is too far from the gap to feel its effects). Thus for a single particle the energy dissipation will be $W \simeq 10^{-5}$(J m^{-2}) $\times$ 10^{-13}(m^2) $\simeq 10^{-18}$ J. The corresponding force is $F_{dyn} \simeq \kappa W \simeq 10^{-10}$ N.

For the particle size and double-layer thickness considered, the magnitude of other forces acting at, say, 10-nm separation, are: (i) the dispersion force $F_d \simeq 10^{-11}$ N; (ii) the double-layer interaction force $F_{d\ell} \simeq -10^{-11}$ N (assuming $\zeta_1 = \zeta_2 = 10$ mV); (iii) the hydrodynamic drag force $F_n \simeq 10^{-13}$ N (assuming a velocity of 10 μm s^{-1}; and (iv) a gravity force $F_{gr} \simeq 10^{-14}$ N (assuming $\Delta\rho = 1$ kg dm^{-3}). For a $0{\cdot}1$ μm particle: $F_{dyn} \simeq 10^{-11}$ N; $F_d \cong -F_{d\ell} \simeq 10^{-12}$ N; and $F_{gr} \simeq 10^{-17}$ N.

From these numbers it can be seen that F_{dyn} can be an important contribution to the net force acting on approaching particles. It must be noted that this force arises because of an energy dissipation due to an electrical current, and hence this force always retards particle motion. It acts as a repulsive force on approach and as an attractive force on recession, in a similar manner as the hydrodynamic resistance and electroviscous forces.

Besides the transport of charges across the solid–liquid interface, a redistribution of surface charges at the interface can also occur during a collision; this is observed, for example, with AgI-sols (Lyklema and van Leeuwen, 1982). Such a current will also retard the particle velocity. Little has been done so far to include these dynamic effects in the theory of colloid stability and they warrant more investigation in the future.

6.5 Experimental techniques

6.5.1 General remarks

To interpret experimental data or theoretical predictions of the deposition rate in a given system, the hydrodynamic conditions in the vicinity of the collector surface and the colloidal and external forces acting on the particles must be defined and characterized as accurately as possible. Different experimental methods may introduce different systematic errors that must also be taken into consideration. We will start our discussion with a short classification of the experimental methods based on these factors and, next, along the same lines, we will discuss some experimental data in the literature which deal with various aspects of particle-deposition kinetics.

6.5.2 Hydrodynamic conditions in the vicinity of the collector

In some experimental systems the hydrodynamic conditions can be fully controlled and the number of parameters that uniquely describe the flow is small. The rotating disk and the stagnation-point flow collector are examples of such systems. The flow field can also be found for any kind of collector of sufficiently simple geometry if it is placed in a well-defined flow. Examples of such collectors are spherical and cylindrical objects placed in a uniform or a simple shear flow. The flow field is also simple for various kinds of channel flow. The situation becomes more complex if an ensemble of collectors is used in the form of a packed bed, e.g. in filters; however, even then some simple correction factors can be introduced to characterize the flow around a single element of the system. Finally, hydrodynamic conditions can be kept under control but must be described by a large number of parameters in a system with random geometry, e.g. porous materials. In some experiments both the flow field and the geometry of the collector is complex. Such conditions exist, e.g. when deposition of fines (small fibre fragments) on fibres is studied in a container in which the suspension is stirred mechanically.

To understand the deposition process, a knowledge of the forces acting between a particle and a collector, as well as the force of interaction of colloid particles between themselves, is essential. Instability of the colloidal dispersion results in changes in the particle concentration and formation of aggregates, usually of a complex geometry. Moreover, surface aggregates can occur due to collisions of suspended particles with particles deposited on the collector surface. The influence of an energy barrier and of external forces on deposition was discussed in detail earlier. From a practical point of view we usually know the Hamaker constant and the ζ potentials of the surfaces which are used to calculate the energy of interaction on the basis of the DLVO theory. It should be borne in mind, however, that regardless of our knowledge of these parameters, the theory of interaction usually breaks down in the region vitally important for deposition processes, namely at very small particle–collector separations. As discussed above, this is due to a number of factors such as: the breakdown of continuity, surface roughness, surface heterogeneity, dynamical effects in the electric double layer, etc. For example we know that the dispersion interaction cannot be described solely by the Hamaker constant and that other parameters which are dependent on the surface roughness (Czarnecki and Dabros, 1980) and the heterogeneity, must be investigated to provide more detailed information on the magnitude of this force. Also, the influence of electric double-layer interactions on the deposition process cannot be predicted properly if only mean values of the ζ potential are taken into account. An improvement, in order to provide a better description of a real system, is to consider the standard deviation about the mean potential (Prieve and Lin, 1982).

6.5.3 Experimental methods

Various collectors were discussed in Section 6.3.2 (cf. Fig. 6.3). All these collectors can be used to study the deposition (or detachment) of colloidal particles. For many of these collectors it is difficult to observe the deposition process directly and particle counting must be done by examining the surface after a deposition experiment, usually by microscopy. Since microscopy has certain disadvantages, we will discuss here experimental techniques in which the number of particles can be observed directly as a function of time during the deposition process.

Examples of devices which allow direct observation are various channels, such as cylindrical or parallel-plate channels, or a Couette apparatus, made of transparent material (or having transparent windows), in which the deposition process can be observed microscopically. Alternatively, we can measure the amount of light scattered from deposited particles or, by using radioactive tracers, we can measure the amount of radiation, both measurements being proportional to the number of particles on the surface, after corrections for the flowing particles have been made. All these flow cells have a constant rate of shear along the collector surface (the wall shear rate), but the deposition rate is usually a function of position (cf. Table 6.2), resulting in non-uniform deposition. It is often advantageous to have data for a range of shear rates. This can be achieved, for example, in a converging or diverging flat-plate channel, for which the wall shear rate is a function of position, but the deposition will again be non-uniform.

An experimental technique in which the deposition rate is constant, but the wall shear rate is a function of position, is the impinging-jet technique. With this technique a stagnation-point flow is created in the impingement area, given by (cf. Table 6.1):

$$v_r = \alpha_s r z \quad ; \quad v_z = -\alpha_s z^2 \tag{6.81}$$

and thus the wall shear rate is

$$G_w = \alpha_s r \tag{6.82}$$

i.e. the wall shear rate is zero at the stagnation point ($r=0$) and increases linearly with radial distance. The strength of the stagnation-point flow, α_s, can be obtained experimentally by measuring the amount of fluid, Q, passing through the cell per unit time. The quantity α_s is related to the Reynolds number, $Re = \bar{v}R\rho/\mu$, where $\bar{v}$ is the mean flow velocity in the jet, R the jet radius, and ρ and μ the density and viscosity of the fluid. The relation between α_s and Re can be found by comparing numerically generated flow fields in the cell with eq (6.81). It is found that for $r/R < \sim 0 \cdot 15$, eq (6.81) is a very good description of the flow field and this is the best regime in which to make observations. For $Re > 100$, it is found that $\alpha_s \propto Re^{1/2}$, with the proportionality constant depending on the geometry of the flow cell. From the cell dimensions and the measurement of Q $(= \pi\bar{v}R^2)$, Re can be calculated

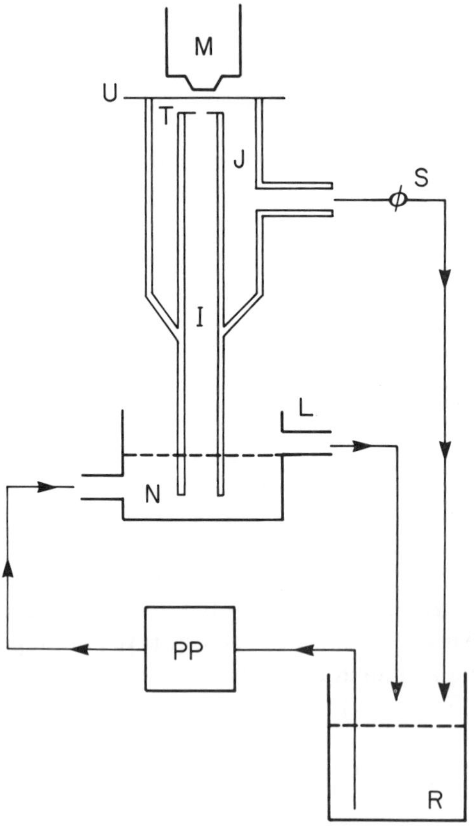

Fig. 6.17. Impinging-jet apparatus. The suspension flows from container (N) to (R) through the deposition cell. After leaving inner tube (I) through the hole in plate (T) it impinges upon transparent plate (U), which is observed by the microscope (M). The suspension flows into the outer tube (J) and passes through valve (S), which controls the velocity of the suspension. The peristaltic pump (PP) returns the suspension from (R) to (N). Any excess liquid overflows and enters (R) through (L) (after Dabros and van de Ven, 1987).

and α_s determined. An example of an impinging-jet apparatus is shown in Fig. 6.17.

All the techniques discussed here make it possible to study either particle deposition or particle removal. Other techniques exist with which particle motion or particle distributions near a surface can be measured. One such method is to observe the velocity of a Brownian particle, captured in a secondary minimum, along the surface in a parallel-plate channel (Prieve and Alexander, 1986). It is then possible to determine the distance between the particle and the wall from eq (6.13) ($u_{\|}$ is known from measurement and F_3 and $v_{\|}$ are known functions of distance). Brownian motion causes variations in this distance and from the observed variations in velocity it is possible to determine the potential field in which the particle is moving.

As a last method, we mention here the measurement of particle distributions near walls using evanescent waves (Lan *et al.*, 1986). This can be achieved by projecting a laser beam onto the bounding surface of a dispersion (see Fig. 6.18). For incident angles, θ_i, larger than the critical angle for total reflection, $\theta_c = \sin^{-1}(n_d/n_g)$, n_d being the refractive index of

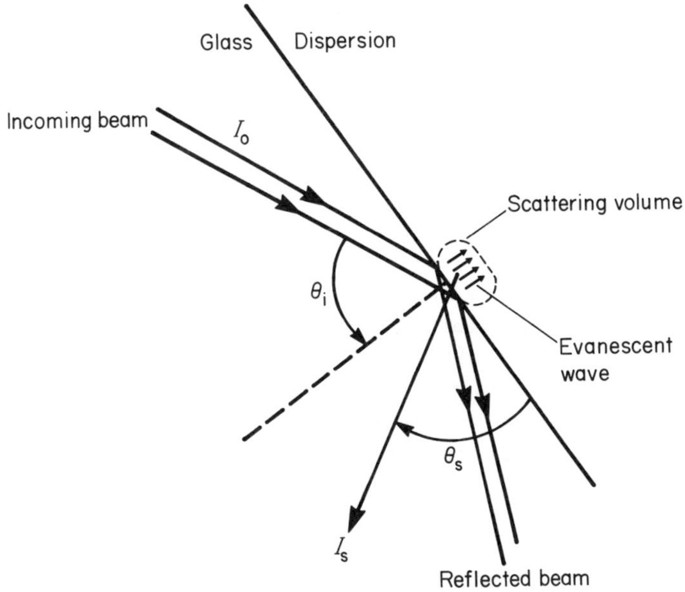

Fig. 6.18. Schematic representation of the evenascent-wave technique. A laser beam impinging on a glass/dispersion interface is reflected. At the impingement point, an evanescent wave penetrates into the dispersion. Light scattered from the evanescent wave (I_s) by dispersed particles can be detected with a photomultiplier.

the dispersion and n_g of the glass (or any other transparent material), a wave penetrates the dispersion normal to the surface. The intensity of this evanescent wave, I_{ev}, decays quickly to zero as the distance from the wall increases:

$$I_{ev} = I_0 e^{-2z/\xi} \tag{6.83a}$$

where the penetration depth, ξ, is given by:

$$\xi = \frac{\lambda}{2\pi n_g} \sqrt{\sin^2\theta_i - \sin^2\theta_c} \tag{6.83b}$$

(λ being the wavelength of the incident beam) and can be varied by varying θ_i. Usually $\xi < \sim 1$ μm. Light from this evanescent wave is scattered by small particles at or near the wall and measurement of the intensity (static light scattering) or of fluctuations in the intensity (dynamic light scattering, PCS) can provide information about the number and distribution of particles near the wall.

The evanescent-wave technique can be combined with the impinging-jet technique to determine the number of particles deposited on the surface which, at low coverages, is simply proportional to the intensity of the light scattered from the evanescent wave (Albery *et al.*, 1985).

6.6 Discussion of the literature data on deposition

6.6.1 General remarks

The first experimental data on deposition on a collector with a well-defined geometry under well-defined hydrodynamic conditions were reported by Marshall and Kitchener (1966), and Hull and Kitchener (1969). In their studies of the deposition process these authors used the rotating-disk technique. Because these papers can be regarded as classical, we will discuss them in some detail. Other relevant papers will be discussed insofar as they contribute to a better understanding of the deposition process in allowing certain generalizations to be made, or when they contain characteristic features of the subject. However, due to the large number of publications in this field, our listing of experimental papers is not exhaustive.

6.6.2 Deposition in the absence of an energy barrier

Marshall and Kitchener (1966) studied the deposition of graphitized carbon black particles with an average diameter of 0·45 μm and a concentration of about 6×10^{14} m^{-3} on various collectors such as glass, poly(styrene), poly(vinyl formaldehyde), poly(vinyl pyridine) and cellulose. The angular velocity of the disk was usually 200 rpm and all experiments were made under laminar-flow conditions. An optical microscope at $480\times$ magnification with dark-field illumination was used to determine the coating density. It was found that a regular deposition of individual particles was observed only in dispersions that were very slowly aggregating. Highly stabilized dispersions gave negligible deposition while rapidly aggregating dispersions gave irregular, weakly adhering deposits containing a wide range of aggregate sizes and showing patterns determined by the flow field near the disk.

Considering their data on deposition rates in the absence of energy barriers or, strictly speaking, when the ζ potentials of the particles and the collector surface were of opposite sign, the value of the flux to the collector surface was found to be one order of magnitude smaller than expected theoretically. This discrepancy (which was beyond experimental uncertainty) was explained quantitatively by the instablity of the colloidal dispersion. The influence of coagulation on deposition was discussed in Section 6.4.5. From that discussion it can be shown that attributing the shortfall to coagulation is indeed justified. The half-time, T_s, of the fast coagulation of the colloidal dispersion used in the experiments under consideration, calculated from eq (6.70) (with $W = 1$), is 4·5 min. The flux of particles and aggregates to the collector surface can be described approximately by eq (6.74). It follows that the mean value of the flux found from measurements in a time interval (t_1, t_2) is

$$j = \frac{j_0 T_2}{t_2 - t_1} \ln \frac{1 + t_2/T_s}{1 + t_1/T_s} \tag{6.84}$$

Taking $T_s = 4 \cdot 5$ min, and estimating $t_1 = 5$ min and $t_2 = 30$ min for the experimental conditions, it follows that coagulation is able to reduce the mean value of the flux by a factor of 5 or 6. Some corrections probably also need to be made for blocking effects. It must be noted that in these experiments only single particles or small unresolvable aggregates were counted. Taking all these effects into account, it can be concluded that the value of the experimentally observed flux was close to what can be expected theoretically.

A much better quantitative agreement with results of the Smoluchowski–Levich formula (see Table 6.2) was reached in subsequent experiments (Hull and Kitchener, 1969) with the deposition of 0·308-μm diameter latex particles of poly(vinyl) pyridine copolymer. The ζ potentials of the latex and collector surfaces were -70 mV and $+72$ mV, respectively. The medium was distilled water and the angular velocity of the disk was 2 to 6 rps. The question arises as to whether it is justified to use the Levich formula for systems in which so many additional effects influence deposition. The thickness of the diffusion boundary layer, δ_D (as defined by Levich, 1962), in these experiments was about 3–5 μm. As is well known, hydrodynamic particle–collector interactions become important at distances of the order of 10-particle radii (1–3 μm for the system under consideration), significantly decreasing the diffusion coefficient and changing the velocities of the particles in the vicinity of the surface. Assuming that the electrolyte concentration was in the range 10^{-6} to 10^{-4} mol dm^{-3}, the Debye length, κ^{-1}, ranged from 30 to 300 nm.

Solving the transport equation numerically, taking into account hydrodynamic, dispersion, electric double-layer and gravitational interactions, the parameter $\tau = \kappa a$ is found to have a relatively strong influence on the flux. The calculations show that for low electrolyte concentrations (of the order of 10^{-5} mol dm^{-3}) the flux should be about 15–30% higher than predicted by the Levich formula, depending on the angular velocity of the disk. Over a wide range these results are insensitive to the value of the Hamaker constant and other parameters, including the ζ potentials of the particles and collectors which can be changed by 10% without significantly affecting the flux. The difference in the fluxes between the results of the Smoluchowski–Levich approximation and exact numerical solutions is a function of the Péclet number, and may be expected to decrease as the Péclet number decreases (see e.g. Fig. 6.5). It can be shown that, when Pe $\leqslant 10^{-5}$, these deviations become negligible. Also for Pe$\simeq 10^{-4}$–10^{-3}, the deviations should be smaller at higher ionic concentrations when the ratio of $\kappa\delta_D \gg 1$. It may be concluded that the observed good experimental agreement with the Levich formula (within 1–2%) must have been a result of the fortuitous mutual compensation of several factors, some of them not explicitly included in the present mass-transport theory. In a broader sense, the results obtained experimentally are consistent with results obtained from the complete convective-diffusion equation within an error range of about 10–20%.

A similar conclusion can be reached by analysing the experimental data obtained by Bowen and Epstein (1979) who studied the deposition of uniform spherical silica particles in a parallel-plate channel. A radioactive-tracer technique was used to count the deposition of particles. In a series of experiments, the deposition rates of particles of 0·402, 0·645 and 0·6 μm on negatively and positively charged plastic substrates was observed under laminar-flow conditions. Under experimental conditions similar to those discussed above (counter-ion concentration $\sim 7 \times 10^{-6}$ mol dm^{-3}, ζ potential of the particles -75 mV and of the collector $+15$ mV), the initial deposition rate was found to be $J = 7{\cdot}87 \times 10^{6}$ particles m^{-2} s^{-1}. A numerical calculation of the flux, with corrections for hydrodynamic, van der Waals and double-layer interactions under similar conditions yields $J=8\cdot14\times10^6$ particles $m^{-2}\,s^{-1}$ (Adamczyk and van de Ven, 1981a). Although the theoretical prediction is in fair agreement with experiment as compared with the results of other studies, the observed difference seems to be typical. Because of the neglect of interception and long-range double-layer interactions (for the system under consideration $\kappa^{-1}=0\cdot12$ μm and $a=0\cdot325$ μm), the analytical approximation gives values of the flux which are smaller than those observed, while the solution of the complete transport equation slightly overestimates the flux. Nevertheless, the observed differences are within experimental error.

In 21 experiments of this type, in which the channel wall was positively charged and the particles negatively charged, theory and experiments yielded very close results, especially at intermediate counter-ion concentrations. For 14 sets of data in which the double-layer thickness, κ^{-1}, ranged from 40 to 100 nm, the average difference between the theoretical and experimental values of the flux was about 10%. However, it was observed that as κ^{-1} decreases from 50 to 5 nm, the discrepancy between the theoretical and experimental flux values increased to above 10%. The authors attributed these differences to the influence of hydrodynamic forces and colloidal interactions. However, numerical solutions of the transport equation, which take these effects into account, show that no better agreement can be achieved. Because the experimental method is based on an indirect measurement of the deposition rate, it is hard to provide an unambiguous explanation of the discrepancy. One possible explanation is that smooth particles on smooth plastic surfaces can be easily removed from the surface by hydrodynamic forces, unless strong tangential forces act on the particles. In similar independent experiments it was shown that the escape of deposited particles was negligible. Because the colloidal dispersion used was stable, the only explanation of the observed discrepancy seems to be that it stems from the heterogeneity of the surface. Surface heterogeneity is likely to have a larger effect at higher salt concentrations when the double layer is thin (and comparable to the size of the surface asperities) and will result in a reduction in the mean value of the flux. This hypothesis can also explain why the dependence of the coating density versus time was found to be non-linear.

Other relevant experiments have been reported by Dabros and van de Ven (1983c, 1987) in which deposition in the stagnation-point flow collector was studied. Figure 6.19 shows the dependence of the Sherwood number, Sh, on the Reynolds number, Re, obtained with this technique. The deposition of $0\cdot5$ μm latex particles on cover glass slips in a laminar stagnation-point flow was studied. The solid line was obtained from a numerical solution of the transport equation, including hydrodynamic, gravitational and dispersion forces. As we can see, the agreement is within the previously

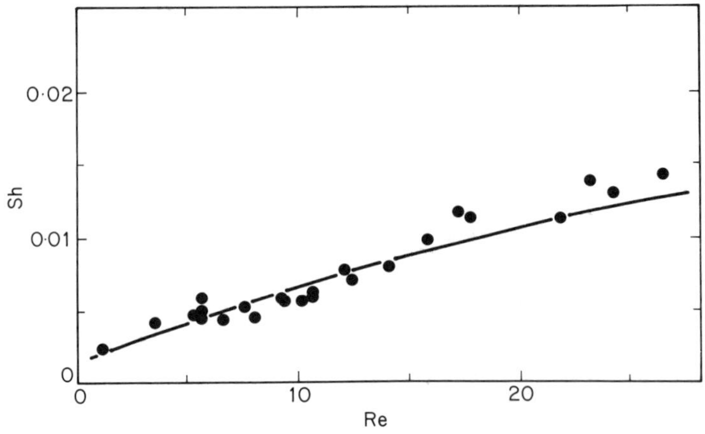

Fig. 6.19. Experimental value of Sh as a function of Re for deposition of 0·5-μm latex particles in 10^{-3}M NaCl on a stagnation-point flow collector. The solid line is the numerical solution of the convective-diffusion equation for Dl = 0, Ad = 9·8, $\lambda = 0 \cdot 4$ and Gr = − 0·002 (after Dabros and van de Ven, 1983c).

estimated range of errors. It must be noted that, according to DLVO theory, in these experiments an energy barrier should exist. However the barrier seems to be absent, probably because of surface roughness of the latex spheres which are presumed to be "hairy" (cf. Varennes and van de Ven, 1987).

In these experiments a characteristic behaviour of the deposited latex particles was observed. A significant percentage of the particles that were stuck to the collector surface executed some type of oscillatory motion when subjected to turbulent flow. Typically, a particle moves irregularly over a certain site, its lateral motion (parallel to the surface) varying from a fraction of a micrometer to several micrometers. Whatever factors are responsible for this surface motion of the particles, they might have a large influence on the particle deposition, especially when flexible bonds could be formed by the process of polymer bridging.

All the results discussed so far seem to indicate that under well-defined hydrodynamic conditions in the absence of energy barriers, the deposition rate can usually be predicted quite well from theoretical considerations. However, differences between theory and experiment are observed, presumably due to the influences of surface heterogeneity, roughness and lack of sufficient information about the real energy of the interactions, especially at small distances. Even for latex particles, regarded by many as a model colloidal system with well-defined surface properties, "anomalous" behaviour (in the sense of the classical colloid-interaction theory) is noted (Varennes and van de Ven, 1987).

The experimental data on the deposition rates for systems with more complex geometries and hydrodynamic conditions, such as packed beds and columns (Fitzpatrick and Spielman, 1973; Kuo and Matijevic, 1980), show that when the particles and collector surfaces are oppositely charged the flux can be predicted fairly well theoretically, but usually the scattering of experimental points around theoretical curves is more pronounced. This is understandable because, besides the problems discussed previously, in these cases we are dealing with less well-defined and non-uniform hydrodynamic conditions.

6.6.3 Deposition controlled by an energy barrier

Theoretical calculations on the effects of the double-layer parameter, Dl, on the deposition rate show that if Dl is larger than some critical value Dl_c, the flux is strongly reduced, by several orders of magnitude (see Fig. 6.8). Of course this is simply related to the height of the energy maximum which, if higher than about 5 kT units, controls the deposition. A theoretical prediction of the deposition is only useful when the theory of colloidal interactions provides a good approximation of the actual energy. Due to the sensitivity of the deposition rate to the interaction energy, experiments in which there is an energy barrier can provide a direct verification of the theory.

As before, we will start with the analysis of experimental data from the work of Hull and Kitchener (1969). These authors studied the deposition of 0·31 μm poly(styrene) particles on poly(vinyl formaldehyde) in the presence of sodium dodecyl sulfate (a surfactant) at various concentrations of sodium chloride. The ζ potentials of the colloidal particles and collectors varied with surfactant and salt concentration and were in the range of -78 to -38 mV for the particles and -51 to -32 mV for the collector (rotating disk). An evident dependence of the stability ratio, W ($=1/\alpha_d$, cf. eq (6.41)) on salt concentrations was observed. However, the theoretical values of the stability ratio were many orders of magnitude higher than the experimental values, even for the highest plausible value of the Hamaker constant for the interactions between the particles and collector. It was shown that the particles were not retained in a secondary energy minimum and so this minimum could not have been the origin of the discrepancy. A number of other effects that could influence the deposition rate, such as the influence of (i) surface roughness, (ii) the approximate nature of the formulae for Hamaker and double-layer interactions used in the calculations, and (iii) colloid stability, were excluded. Therefore the only possibility is to ascribe the observed discrepancy to surface heterogeneities.

Two complicating factors were noted in the experiments: the coating density was a non-linear function of time, and the initial deposition rate did not seem to be proportional to the particle concentration. A strong non-linear dependence of the coating density on time can conceivably result in incorrect estimates of the initial slope, giving rise to apparent deposition rates which are not proportional to particle concentration. This non-linearity in the stable colloidal systems used in the experiments was presumably due to the implied heterogeneity of the collector and particle surfaces, and possibly to particle detachment as well. In using the rotating-disk apparatus, it is difficult to observe the behaviour of the particles in the vicinity of the collector, and the particle escape will be particularly difficult to observe. The escape rate is correlated with the initial flux because both these quantities are dependent on the shape of the curve of the interaction energy for small particle–collector separations.

The microscopic observations of a large uneveness in the distribution of deposited particles on some plastic films supports the heterogeneity hypothesis, which can be responsible for early saturation of the surface by blocking or masking of its active sites. Although the use of an anionic surfactant eliminated the heterogeneity there was no means of ensuring that it produced a uniformity in the potential on a sub-microscopic scale. Weakly charged areas do not need to be large in size to provide a "hole" in a general potential-energy barrier. Therefore, even small spots with lower than average potentials might ensure deposition.

The data presented seem to be characteristic of most experiments using particles and collectors whose surfaces were similarly charged. Relatively good agreement with theory was obtained when the potentials of the surfaces were small (Clint *et al.*, 1973) and the deposition was controlled only to a small extent by the energy of interaction. However, it appears that even then the flux is extremely sensitive to surface potentials; varying the potential by 0·1 mV can result in significant changes in the particle flux. On the other hand, it is known that even in model polymer latexes the standard deviation of the electrophoretic mobility is typically 10–20% of the mean value. Prieve and Lin (1982) showed that if the standard deviation in the Stern potential is 10% of the mean, it can reduce the stability ratio, W, by several orders of magnitude. Furthermore, a distribution in particle sizes causes a decrease in the stability ratio, but only slightly. Both effects are particularly pronounced at low salt concentrations. In the case of very high salt concentrations, W will be slightly increased. Boluk and van de Ven (1989) showed the importance of electroviscous forces which make W and the CDC dependent upon the size of the ions in solution.

Due to the lack of adequate models to describe surface heterogeneity, and not knowing how the transport equation can be solved when such factors are taken into account, it is extremely difficult to predict deposition rates on heterogeneous surfaces. However, because of the importance of the problem, Adamczyk *et al.* (1983b) performed some calculations on systems for which the collector surface is homogeneous, but the particles exhibit a spread in surface charge. They assumed that the potentials of the particles are scattered normally around the mean value with standard deviations equal to 20% and 30%. As these calculations had only an illustrative purpose, simple expressions were adopted for the van der Waals and electrical double-layer interactions. It was assumed that the total forces acting on the particles were composed of electrical double-layer forces, characterized by the parameters Dl and τ, dispersion forces, characterized by Ad and $\bar{\lambda}$, and gravitational forces, characterized by the gravity number Gr. This model was applied to the data of Hull and Kitchener (1969) for whose system Gr $\ll$ 1 and so had little influence on deposition. The Hamaker constant was assumed to be 10^{-20} J. The ζ potential of the substrate was treated as constant and equal to the experimentally measured value. A scatter in the particle potentials results in a scatter in their individual Dl values which, if higher than the critical value Dl_c (see Fig. 6.8), all other parameters being

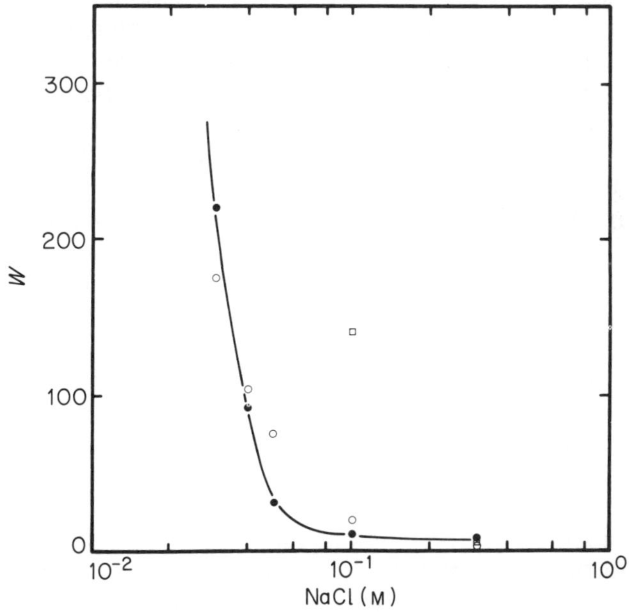

Fig. 6.20. Experimentally determined stability ratio, *W*, as a function of electrolyte concentration. ●, Experimental results (Hull and Kitchener, 1969); ○, calculated assuming a spread of 30% about the mean particle ζ potential; □, calculated assuming a spread of 20%.

constant, causes a strong decrease in the particle flux. Most of the deposition occurs for particles which have $Dl < Dl_c$.

Figure 6.20 shows the experimental results and the theoretically calculated values of the stability ratio, W, according to this model (but with neglect of electroviscous forces). As we can see, the results obtained by assuming a distribution of particle potentials are not too far from the experimental data, and it seems to make sense to incorporate similar corrections into a standard analysis for any experimental system. In doing so, the particle flux will depend on a larger number of experimental parameters, but it is by now clear that the number of parameters presently used is insufficient to describe fully the deposition phenomena, certainly when the electrical double-layer interactions are repulsive. In this case in particular, it is essential to know the energy of interaction at small particle–collector separations. Deposition experiments, together with further studies of the structure of solid interfaces, can provide essential information needed to formulate a more powerful theory of the interactions in the vicinity of a wall.

6.6.4 Distribution of particles on collector surfaces

It can be concluded from the many experiments described in the literature that colloidal stability has a large influence on the structure of the deposited phase. Collisions between deposited and flowing particles can cause growth of surface aggregates. On surfaces in fluids these aggregates are usually densely packed when intensive hydrodynamic forces are acting on them. In air, dendrites with a very complicated and fine shape can be observed.

The formation of surface aggregates and dendrites is an example of how deposited particles can further influence the deposition process by changing the geometry of the collector, as well as the hydrodynamic conditions and force fields near the collector surface. Except for numerical modelling of the dendrite formation in air (Payatakes and Gradon, 1980), little has been done to describe this important process quantitatively.

In the case of stable colloids, the efficiency of collisions between deposited and flowing particles is small. As long as the coating density is low, the influence of the deposited particles on the flux should be small, at least when the collector surface is homogeneous and particles already deposited do not block "active sites". A study of the distribution of particles on the surface can provide important information about the heterogeneity of the surface and the interactions between particles. On homogeneous surfaces and in the absence of any interactions (between particles) the distribution of particles should be Poissonian

$$P_N = \frac{s^N}{N!} \mathrm{e}^{-s} \tag{6.85a}$$

where P_N is the probability of finding N particles in a given area and s is the average number of particles per area. If particles interact, the Poisson distribution must be modified. We may assume that the particles are brought to the collector surface randomly, but the probability of deposition will

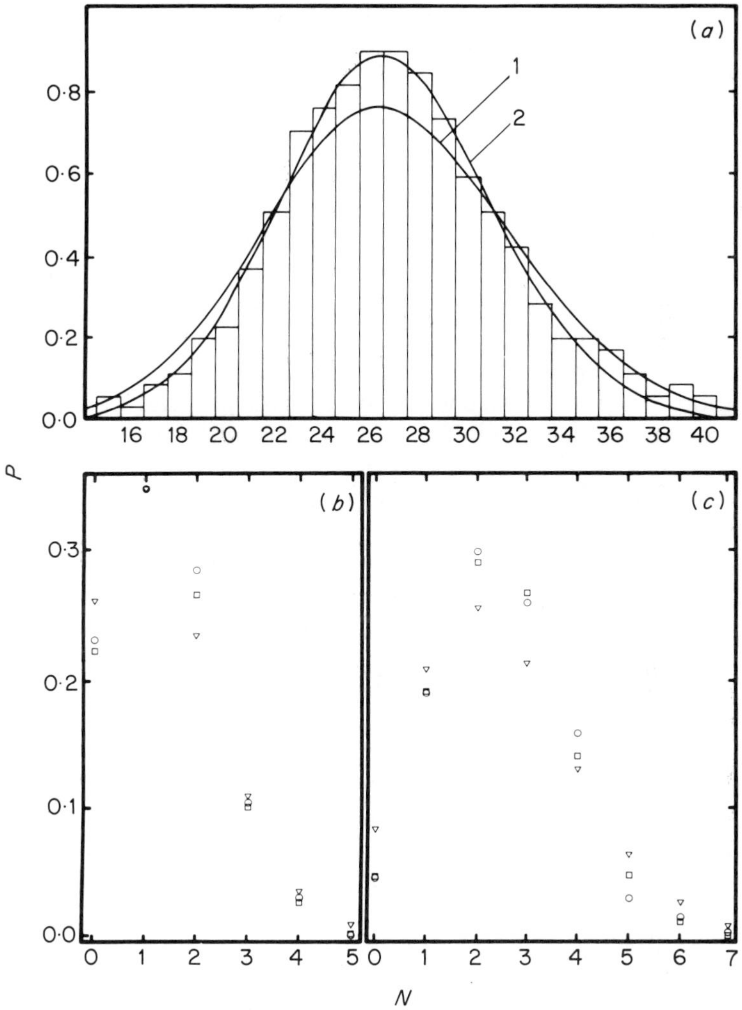

Fig. 6.21. Observed distribution of particles on collector surfaces.
(a) 0·86-μm $BaSO_4$ particles on a rotating glass disk (352 areas of $A = 3600$ μm^2 with an average of $s = 26·9$ particles per square). (1) The Poisson distribution (eq (6.85a)) and (2) the modified Poisson distribution (eq 6.85b)) (after Adamczyk and Pomianowski, 1980).
(b, c) 0·5-μm latex particles on glass in a stagnation-point flow collector; 385 areas of $A = 25$ μm^2 with $s = 1·35$ (*b*) and $s = 2·48$ (*c*). ○, Experimental observations; ▽, the Poisson distribution; □, the modified Poisson distribution (after Dabros and van de Ven, 1983c).

depend on the number of particles deposited previously in a given area. This probability can be expressed as $P_{dep} \propto (1 - N/N_m)$ or $P_{dep} \propto 1/(1 + \beta' N/s)$, where N is the actual number of particles per given area, N_m the maximum number, and β' a dimensionless empirical constant.

It can be shown (Adamczyk and Pomianowski, 1980) that for large values of N ($N > \sim 40$), the Poisson distribution with corrections for blocking effects turns into a Gaussian distribution

$$P_N = \sqrt{\frac{1+\beta}{2s\pi}} \exp\left[-\frac{(s-N)^2}{2s}(1+\beta)\right] \tag{6.85b}$$

As can be seen, the value of the variance is reduced by a factor $(1+\beta)$ due to blocking; β is a blocking coefficient, related to the masking coefficient γ by $\beta = s\gamma$. Examples of experimental data on the distribution of particles at the surface of a rotating disk and of a stagnation-point flow collector are shown in Fig. 6.21. Although different experimental methods were used, in both cases the deposition was on glass surfaces. Figure 6.21a represents the distribution function for deposition of 0·9 μm particles of $BaSO_4$. This histogram was obtained from 352 surface elements, each with an area of

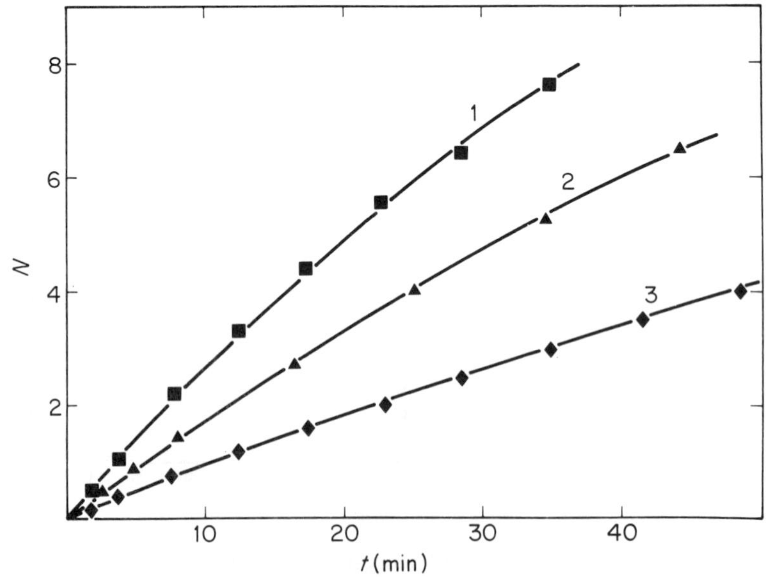

Fig. 6.22. Coating density, N, as a function of time, t, for various hydrodynamic conditions for 0·5-μm latex particles (in 10^{-2}M NaCl) in a stagnation-point flow collector. Curves 1 to 3 correspond to Reynolds numbers of 17·9, 12·1 and 3·4, respectively (after Dabros and van de Ven, 1983c).

3600 μm^2. N denotes the number of particles stuck to such an area and P is probability of finding areas containing N particles. The standard deviation of the distribution was $\sigma = 4 \cdot 75$ and the mean value $s = 26 \cdot 9$. Curve 1 shows the Poisson distribution and curve 2 was calculated taking blocking effects into account ($s = 27$ and $\sigma^2 = 27$). Analysis of the data shows that a particle is able to block an area which is much larger than its geometrical cross-section.

Similar results were obtained by direct microscopic observation for the deposition of 0·5-μm latex particles in the stagnation-point flow collector (Fig. 6.21b). In this case the analysis of the distribution was conducted on much smaller areas, A (385 areas, $A = 25$ μm^2 each). From the non-linear dependence of the coating density versus time curve (Fig. 6.22), it was found that for this system $\gamma \simeq 25\pi a^2$, i.e. one particle is able to block an area which is about 25 times larger than its cross-section (cf. eqs (6.67) and (6.68) with $k_0 = 0$). Using this value of γ in determining $N_m = \gamma / A$ and P_{dep} and correcting the Poisson distribution for blocking effects, a good agreement between experimental and theoretical distributions was obtained (Fig. 6.21b,c).

Further detailed studies on the distribution of particles on collector surfaces can provide valuable information on the nature of the surface as well as on interactions between deposited and flowing particles. This information is necessary for describing the kinetics of deposition for high coating densities.

6.6.5 Influence of colloid stability

Besides controlling the formation of surface aggregates, colloid stability is, in a direct way, responsible for the formation of various kinds of aggregates in the bulk of the dispersion. Since no exact theory for the deposition of aggregates of complex geometry exists, experimental data on such systems are especially important. Clint *et al.* (1973) studied the deposition of negatively charged 0·43-μm latex particles onto a negatively charged poly(styrene) surface, using the rotating-disk technique, for a variety of electrolyte concentrations. At high values of salt concentration, deposition of aggregates was observed. At each value of the electrolyte concentration the percentage of deposited singlets, doublets and triplets was measured by electron microscopy. By assuming that coagulation only influenced the diffusion coefficient, and that the aggregates were sufficiently spherical that one (mean) value of the diffusion coefficient was sufficient to describe the diffusion, and by adopting Smoluchowski's theory for coagulation to describe the concentration of i-fold aggregates, the observed deposition could be described reasonably well.

An analogy with the classical theory of coagulation suggests that the decrease in the value of the diffusion coefficient of aggregates is partially cancelled by the increase in the collision radius (cf. Section 6.4.5). In such a case, the kinetics of the coating formation can be described by the approximate formula eq (6.75a). It should be noted that in eq (6.75a) the product $J_0 T_s$ appearing in front of the logarithmic term is independent of the colloid concentration. Hence it may be expected that a plot of N versus $\ln(1+t/T_s)$ should give a straight line with a slope which is independent of the particle concentration. To test the applicability of this theory, the deposition of 1-μm $BaSO_4$ particles on a synthetic chalcocite was studied using the rotating-disk technique (Waligora and Dabros, 1978). In these experiments the values of the coagulation half-time, obtained from deposition data, was $T_s = 35 \cdot 5$ min ($n_1^0 = 2 \cdot 0 \times 10^{14}$ m^{-3}). In direct measurements (using a haemacytometer) it was found that $T_s = 33 \cdot 4$ min. Figure 6.23 shows the dependence of the coating density, N (expressed as the number of particles and aggregates per 100 μm^2) as a function of $\ln(1+t/T_s)$. The half-time of coagulation, T_s, for various concentrations was calculated by assuming that T_s was inversely proportional to the concentration (see eq (6.70)). Although the experimental points seem to be scattered around one single curve, a deviation from linearity is observed, especially at high values of the coating density. This deviation is primarily caused by the blocking effect;

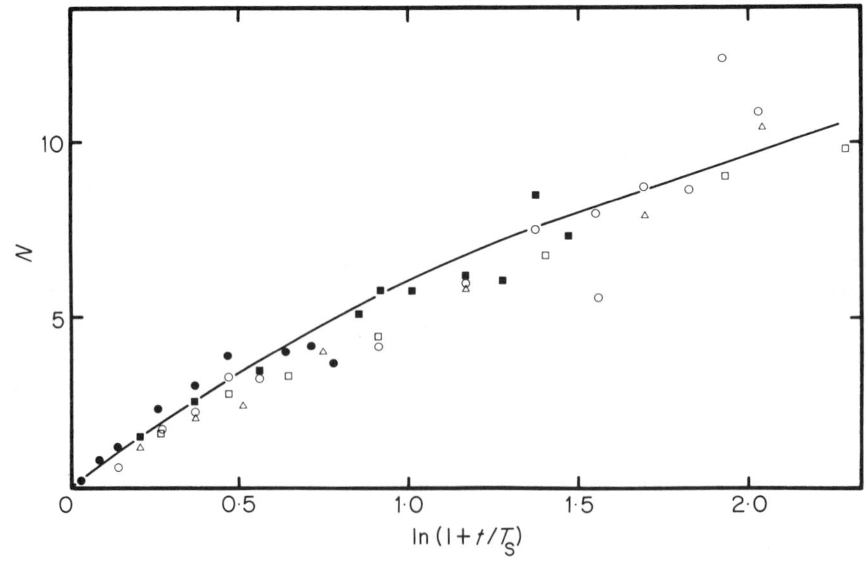

Fig. 6.23. The coating density, N, as a function of $\ln(1+t/T_s)$ for various concentrations of $1 \cdot 08$-μm $BaSO_4$ particles in 10^{-3}M $BaCl_2$ on a rotating disk with a chalcocite surface. The solid line is calculated from eq (6.75a), modified to incorporate blocking effects. ●, $n_1^0 = 2 \times 10^{14}$, ■, $n_1^0 = 5 \times 10^{14}$; ○, $n_1^0 = 1 \times 10^{15}$, □, $n_1^0 = 1 \cdot 5 \times 10^{15}$; △, $n_1^0 = 2 \times 10^{15}$ particles per m^3 (after Waligora and Dabros, 1978).

e.g. for $N \simeq 5$ particles per 100 μm^2 at least 5% of the collector surface is covered by particles. The solid line in Fig. 6.23 was calculated by adjusting J_1^0 in eq (6.75) to include corrections for blocking. As can be seen, a reasonably good agreement for a wide range of colloid concentrations is found.

Further study of the deposition of coagulating sols under well-controlled hydrodynamic conditions, especially with a separate analysis of the fluxes of various aggregates, could provide useful information on the mechanism of their deposition.

In this chapter we have outlined some of the problems associated with predicting deposition rates. Progress has been made in incorporating ageing and blocking effects into the theory; prediction of particle-detachment rates is still difficult because no adequate theory exists to predict the depths of the local energy minima. For heterogeneous systems, a modest beginning has been made by considering the spread about the mean value of the ζ potential of the particles, but usually a much better characterization of the surfaces will be required. Roughness can be approximately incorporated into the calculation of the dispersion forces, but little is known of how it affects double-layer interactions. As yet no theory exists to predict the magnitude of electroviscous forces. Dynamic effects are also poorly understood.

As a result of all these complicating factors, no exact agreement between experimental and theoretical data is found. When a deep energy minimum exists and no energy barrier is present, current theories predict deposition rates which differ typically by 10–20% from experimental observations. In the presence of energy barriers, the discrepancies between theory and experiment are usually much larger.

Some of the factors that complicate the prediction of deposition rates could, in principle, be incorporated in an approximate way into the immobilization reaction approach (Section 6.3.5), if sufficient information about the surfaces were available. Roughness and surface heterogeneity enter the colloidal- and external-force terms, while ageing and blocking, as we have seen, influence the reaction term, Q. Furthermore, deposited particles affect the velocity field near the collector and hence a time-dependent velocity field needs to be introduced into the transport equation. All hydrodynamic correction terms are as yet unknown functions of ion size, surface potential and double-layer thickness. Dynamic effects also need to be considered. However, even if surfaces were sufficiently characterized, and if the hydrodynamic (Navier–Stokes) equations could be solved exactly near surfaces partially coated by particles, and if all the effects mentioned could be incorporated in the convective-diffusion equation, solving the convective-diffusion equation could still present insurmountable difficulties. In such a case, dynamic-simulation and Monte Carlo methods might provide the only viable alternatives.

Chapter 7

Multiparticle Interactions in Colloidal Dispersions

7.1. General remarks

The properties of colloidal dispersions depend crucially on the volume fraction of the suspended particles. So far we have considered either very dilute dispersions in which all particle interactions are absent, or dilute dispersions where two-particle interactions play an important role, but where three- or more-body interactions could be neglected. Increasing the volume fraction further results in semi-dilute dispersions, in which multi-body interactions are significant. This concentration range is the most difficult to describe, because multibody hydrodynamic and colloidal interactions are extremely complex. Concentrated dispersions (discussed in Chapter 8) are easier to deal with in the sense that various approximate methods can be used to describe them.

In principle the expansion in volume fraction of any suspension property, S, can be continued to higher order:

$$S = S_0(1 + k_1 c + k_2 c^2 + k_3 c^3 + k_4 c^4 + \dots) \tag{7.1}$$

Judging by the difficulties involved in calculating k_2 for dilute dispersions, the determination of higher-order terms will be much more difficult and soon becomes prohibitive. A further complication is that although each coefficient, k_i, is determined by i-body interactions, the i-body distribution function is in part determined by higher-order interactions. In particular, the pair distribution of two-body interactions, at high Péclet numbers, also depends on three-body interactions. In simple shear, three-body interactions determine how many particles are on closed orbits, and in sedimentation they affect the mean sedimentation velocity indirectly up to order c.

To appreciate the effects that multibody interactions have on the pair-distribution function, let us consider the probability $P_N \mathrm{d}\boldsymbol{r}_1 \mathrm{d}\boldsymbol{r}_2 \dots \mathrm{d}\boldsymbol{r}_N / N!$ of finding N spheres at positions $\boldsymbol{r}_1$, $\boldsymbol{r}_2$, . . .$\boldsymbol{r}_N$ within a volume dV. The normalization constant

$$N! = \int P_N \mathrm{d}\boldsymbol{r}_1 \ldots \mathrm{d}\boldsymbol{r}_N \tag{7.2}$$

arises because all spheres are considered to be indistinguishable. Conservation of the number of spheres yields (cf. eq (2.85))

$$\frac{\partial P_N}{\partial t} = -\sum_{j=1}^{N} \nabla_i \cdot \boldsymbol{J}_i \tag{7.3}$$

where (cf. eq (5.31))

$$\boldsymbol{J}_i = \boldsymbol{D}_i \cdot \nabla P_N + \left(\boldsymbol{u}_i + \sum_{j=1}^{N} \boldsymbol{M}_{ij} \cdot \boldsymbol{F}_j\right) P_N \tag{7.4}$$

where $\boldsymbol{M}_{ij}$ is the mobility matrix, $\boldsymbol{F}_j$ the interparticle force, and $\boldsymbol{u}_i$ the velocity of particle i. When $\boldsymbol{u}_i = 0$ and steady-state conditions prevail, the solution of the above equation is the well-known Boltzmann distribution

$$P_N^0 = \mathrm{e}^{-V_{\mathrm{int}}/kT} \tag{7.5}$$

By defining a pair-distribution function as

$$P_2 = \frac{1}{(N-2)!} \int P_N \mathrm{d}\boldsymbol{r}_3 \ldots \mathrm{d}\boldsymbol{r}_N \tag{7.6}$$

and higher-order distribution functions in a similar manner, we can integrate eq (7.3) over all but s spheres, resulting in an expression for $\partial P_s/\partial t$ in terms of the divergence of the fluxes up to order s, plus terms containing integrals of fluxes of a higher order (Ohtsuki, 1981). It is possible, by writing down equations for all $\partial P_i/\partial t$, to obtain an hierarchy of equations in which $\partial P_i/\partial t$ is expressed in terms of P_i and P_{i+1}. This implies that it is impossible to determine P_2 without knowledge of P_3, which, in turn, cannot be determined without a knowledge of P_4, etc. To obtain a solution of this set of equations we must truncate them at some level and make some plausible assumptions about higher-order levels. Since the exact determination of P_3 is already exceedingly difficult, the set of equations is usually truncated at P_2, while for P_3 the superposition approximation is assumed (for the normalized functions P_i):

$$P_3(1,2,3) = P_2(1,2)\, P_2(1,3)\, P_2(2,3) \tag{7.7}$$

In this approximation P_3 is expressed as the product of pair-distribution functions. Making this approximation regarding P_3 and neglecting terms of the form $\int P_3 \mathrm{d}\boldsymbol{r}_3$, we can reduce eq (7.3) to $P_1 = n_0$ and (Russel and Gast, 1986)

$$\frac{\partial P_2}{\partial t} = \nabla \cdot (\bar{\boldsymbol{D}} \cdot \nabla P_2) - \nabla \cdot (\bar{u}_{\text{hydr}} P_2) + \nabla \cdot (\boldsymbol{D} \cdot \bar{F}_{\text{int}} P_2 / kT) \tag{7.8}$$

with

$$\bar{F} = -\nabla \bar{V}_{\text{int}} \tag{7.9}$$

which is identical to the usual convective-diffusion equation for the pair-distribution function (eq (5.31)), but in which $\boldsymbol{D}$, $\boldsymbol{u}_{\text{hydr}}$ and V_{int} are replaced by average functions, defined as

$$\bar{V}_{\text{int}}(\boldsymbol{r}) = \frac{1}{(N-2)!} \int V_{\text{int}}(\boldsymbol{r}_1, \ldots, \boldsymbol{r}_N)\, P(\boldsymbol{r}_3, \ldots, \boldsymbol{r}_N | \boldsymbol{r}_1, \boldsymbol{r}_2)\, \mathrm{d}\boldsymbol{r}_3 \ldots \mathrm{d}\boldsymbol{r}_N \tag{7.10}$$

where $\boldsymbol{r} = \boldsymbol{r}_2 - \boldsymbol{r}_1$ is the vector separation between sphere 1 and 2; P is the (conditional) probability of finding spheres at $\boldsymbol{r}_3, \ldots, \boldsymbol{r}_N$, given that there are two spheres at $\boldsymbol{r}_1$ and $\boldsymbol{r}_2$. Similar expressions can be written down for $\bar{\boldsymbol{u}}_{\text{hydr}}$ and $\bar{\boldsymbol{D}}$. The potential, $\bar{V}_{\text{int}}$, from which $\bar{F}_{\text{int}}$ is derived, is usually referred to as the potential of mean force (see Section 8.4). It can be seen from the definition of these terms (eq (7.10)) that $\bar{u}$, $\bar{\boldsymbol{D}}$ and $\bar{V}_{\text{int}}$ depend on the location of all the particles around the pair taken as reference. In the limit of low-volume fractions, these functions reduce to the usual two-body functions discussed in Chapter 5.

The advantage of the above approach is that it makes it possible to express suspension properties in terms of the pair-distribution function, even for concentrated systems. An inherent difficulty is that the diffusion tensor, the hydrodynamic particle velocity vector, and the energy of interaction depend explicitly on many-body effects. However, various methods exist that allow us to make reasonable approximations for these functions, some of which will be discussed in this chapter.

7.2 Singularity methods for calculating resistance tensors and particle velocities in multiparticle systems

Several methods exist for calculating resistance or diffusion tensors and particle velocities for many-body systems. Gluckman *et al.* (1971, 1972) developed a multipole-expansion technique in which the flow field is approximated by a truncated series of multi-lobular disturbances. Unknown coefficients in this series are determined from the no-slip boundary condition on the surface of each particle. An alternative method, developed by Dabros (1984), is a singularity method based on singular forces and sources situated inside each particle, the strength of which is again determined from the no-slip boundary condition on the particle surfaces. Since this method is very general and can be applied to any number of particles of any shape, we will discuss it in some detail.

Let us consider N particles in a flow field described by the creeping-flow and continuity equations:

$$\mu \nabla^2 v = \nabla p \tag{7.11}$$

$$\nabla \cdot v = 0 \tag{7.12}$$

Here, we compare this flow field with a hypothetical flow field in a fluid without particles, but in which singular point forces are acting and in which sources are present at positions within the volume corresponding to the removed particles. These forces and sources will be chosen such that, outside this volume, the flow field is identical to the flow field in the presence of particles.

In an unbounded fluid (a fluid with no bounding walls) the flow disturbance at a point $\boldsymbol{x}$ produced by a point force $\boldsymbol{k}(\boldsymbol{y})$ acting at a point $\boldsymbol{y}$ can be expressed as

$$v_i^*(\boldsymbol{x}) = t_{ij}(\boldsymbol{x},\boldsymbol{y})\, k_j(\boldsymbol{y}) \tag{7.13}$$

where the components t_{ij} of the Oseen tensor are given by

$$t_{ij} = \frac{1}{8\pi\mu}\left(\frac{\delta_{ij}}{r} + \frac{r_i r_j}{r^3}\right) \tag{7.14}$$

where $r_i = x_i - y_i$ and $r = |\boldsymbol{r}|$. Equation (7.13) is also valid in the presence of a bounding wall, but with modified components t_{ij} (Dabros, 1984). For a velocity due to a point source with a volume flux Q (volume of liquid per unit time)

$$v_i^* = \frac{Q}{4\pi}\, s_i = \tilde{Q}\, s_i \tag{7.15}$$

where for an unbounded fluid the source, $\boldsymbol{s}$, is given by

$$s_i = r_i / r^3 \tag{7.16}$$

Again, for a bounded fluid $\boldsymbol{s}$ needs modification.

Due to the linearity of the creeping-flow equation, the disturbance velocity $\boldsymbol{v}^*(\boldsymbol{x})$ at a point $\boldsymbol{x}$ for a system containing N particles, each with M force and source singularities, can be written as

$$v_i^*(x) = \sum_{mn} t_{ij}(\boldsymbol{x}, \boldsymbol{y}_{mn})\, k_j(\boldsymbol{y}_{mn}) + \sum_{mn} s_i(\boldsymbol{x}, \boldsymbol{y}_{mn})\, \tilde{Q}(\boldsymbol{y}_{mn}) \tag{7.17}$$

where $\boldsymbol{y}_{mn}$ denotes the position of the mth singularity in the nth particle, m ranging from 1 to M and n from 1 to N. The fluid velocity is given by

$$\boldsymbol{v} = \boldsymbol{v}_0 + \boldsymbol{v}^* \tag{7.18}$$

where $\boldsymbol{v}_0$ is the undisturbed external flow field. At each point on the surface of each particle, the velocity can be expressed as

$$\boldsymbol{u} = \boldsymbol{u}_0 + \boldsymbol{\omega}_0 \times \boldsymbol{r} \tag{7.19}$$

where $\boldsymbol{u}_0$ and $\boldsymbol{\omega}_0$ are the translational and angular velocities of the particle and $\boldsymbol{r}$ is the position of the surface point with respect to the origin of the particle. At each point on the surface it is required that

$$\boldsymbol{v} = \boldsymbol{u} \tag{7.20a}$$

By taking S points on the surface of each particle, eq (7.20a) represents a set of $3SN$ equations with $4MN$ unknowns. From eqs (7.17) to (7.19) it can be seen that eq (7.20a) can be formally written as

$$\boldsymbol{C} \cdot \boldsymbol{a} = \boldsymbol{b} \tag{7.20b}$$

Here the vector $\boldsymbol{a}$ stands for the $(4MN)$ unknown values of the intensity of the point forces and sources $k_i(\boldsymbol{y}_{mn})$ and $Q(\boldsymbol{y}_{mn})$, $\boldsymbol{C}$ is a tensor which depends solely on the geometry of the system, and $\boldsymbol{b}$ is a vector determined by the external flow field and the movement of the particle. By choosing a sufficient number, S, of surface points ($3NS > 4MN$), eq (7.20) represents an over-determined system of linear equations that can be solved using various standard numerical methods. The more point forces and sources taken, the more accurate the numerical solution becomes.

With this method it is possible to determine the hydrodynamic resistance tensor, $\boldsymbol{f}$, of a particle in any configuration of surrounding particles. Letting the external fluid velocity be a uniform velocity v_∞, then the force on the nth particle is

$$F^{(n)} = \sum_{m=1}^{M} k(y_{mn}) = f \cdot v_\infty \tag{7.21}$$

Similarly, the rotary friction tensors can be obtained from the torque exerted on a particle

$$L^{(n)} = \sum_{m=1}^{M} r \times k(y_{mn}) \tag{7.22}$$

Besides calculating resistance, or equivalently, diffusion tensors by this approach, it is also possible to calculate particle velocities directly. If $F_{\text{ext}}^{(n)}$ and $L_{\text{ext}}^{(n)}$ denote the external force and torque acting on the nth particle, then

$$F_{\text{ext}}^{(n)} = \sum_{m} k(y_{mn}) \text{ and } L_{\text{ext}}^{(n)} = \sum_{m} r \times k(y_{mn}) \tag{7.23}$$

These six equations can be regarded as constraints imposed on the singular forces and sources. We must now solve eq (7.20) subject to the constraints of eq (7.23). This can be achieved by several methods, discussed by Dabros (1984): the method of Lagrangian multipliers, the method of orthogonal transformations, and a direct least-squares method. It was found that the least-squares method was the easiest to implement. By solving eq (7.20) subject to eq (7.23), we can calculate the translational velocity, u_0, and the translational velocity, ω_0 (cf. eq (7.19)).

This singularity method has been applied to several systems, mainly to bounded systems for which the expressions for the components t_{ij} and s_i are more complex than those given by eqs (7.14) and (7.16). A relatively simple example is the rotation of a spheroid in a simple shear flow near a bounding wall. In principle this problem should have been discussed in the previous chapter, which dealt with particle–wall interactions, but at that stage the necessary theoretical background had not been discussed. Of course the method discussed here, valid for N-body interactions, also applies to two-body interactions. In Section 3.1.5 we saw that far from bounding walls the motion of a spheroid in simple shear is described by

$$\frac{\mathrm{d}\theta}{\mathrm{d}t} = \frac{1}{4} BG \sin 2\theta \sin 2\phi \tag{7.24a}$$

$$\frac{\mathrm{d}\phi}{\mathrm{d}t}=\frac{1}{2}G(1+B\cos 2\phi) \tag{7.24b}$$

When a spheroid is rotating near a wall, eqs (7.24) need modification. Dabros (1984) calculated the rotational velocity $\mathrm{d}\phi/\mathrm{d}t$ for a prolate spheroid of axis ratio 2 with an orbit constant $C=\infty$ (corresponding to $\theta=90°$), using 14 singular point forces (and no sources) distributed symmetrically near the centre of the spheroid and 66 surface points to define its shape. The results are shown in Fig. 7.1 for distances between the centre of the spheroid and the wall of 200 a and 2·1 a, a being the semi-major axis of the spheroid. It can be seen that the deviations from Jeffery's equations (solid line) are small, even for short separation distances.

As a second example we will discuss the result for a three-body interaction, namely two spheres in the neighbourhood of a wall (considered to be the third body). This problem is relevant to the deposition of colloidal particles onto a solid surface (discussed in Chapter 6), because it is found that the

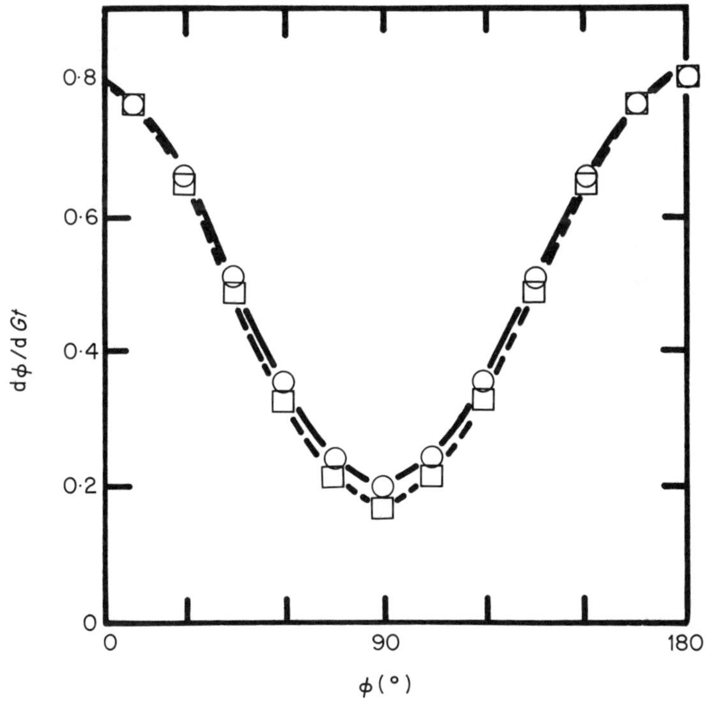

Fig. 7.1. Rotational motion of a spheroid of axis ratio 2 in simple shear near a solid wall. ○, Separation distance 200; □, separation distance 2·1. Solid line: eq (7.24b) (after Dabros, 1984).

interactions between freely mobile particles and particles deposited on a wall determine the coating density on a surface by determining the magnitude of the blocking parameter γ (cf. eqs (6.67b) and (6.68a)). Using 13 singular forces and sources distributed symmetrically about the centre of each sphere and 42 surface points per sphere, Małysa *et al.* (1986) calculated the trajectories of a sphere falling past a second sphere attached to a wall. For the stationary sphere, no constraints are imposed on eq (7.20), but for the freely mobile sphere the sum of the appropriate components of the singular forces must be balanced against the external force (cf. eq (7.21)). Since no external torque is imposed on the mobile sphere, the singular forces must fulfill the condition that the torque caused by them is zero (cf. eq (7.22)). Some theoretical results, together with experimental observations on a system of two 6·4 mm spheres, are presented in Fig. 7.2, which shows the sedimentation velocity, u, of the mobile sphere past the attached one, as a function of the vertical distance, y, from the centre of the fixed sphere for various initial distances, z_0, of the sphere from the wall. The solid lines are

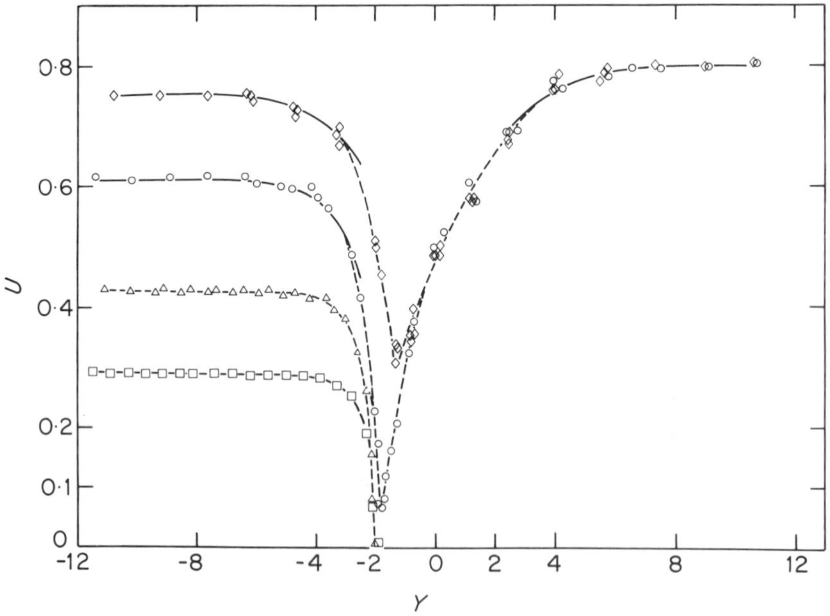

Fig. 7.2. Velocity, U (non-dimensionalized by the free-fall velocity), of a sphere sedimenting past a second sphere attached to a wall as a function of the vertical distance, Y, of the fixed sphere for various initial distances z_0 from the wall (non-dimensionalized by the sphere radius). ◇, $-z_0=2\cdot00$; ○, $-z_0=1\cdot35$; △, $-z_0=1\cdot05$; □, $z_0<1\cdot01$ (sphere pressed against the wall). (———) Calculated from the singularity method; (----) smooth experimental fits. For $z_0>2\cdot4$, the trajectories were found to be symmetric (not shown) (after Małysa *et al.*, 1986).

calculated from eq (7.20) with constraints (eq (7.23)). The figure shows some interesting discrepancies from theory. From the linearity of the creeping-flow equation, it follows that the trajectory of recession must be the mirror image of the trajectory of approach. However, it was observed that all trajectories, on which the mobile sphere was close to the wall, were asymmetric. Further from the wall symmetric trajectories were found, in agreement with theory. The asymmetry is attributed to the surface roughness of the spheres. According to the trajectory equations, the minimum distance of approach between the two spheres is extremely short when the mobile sphere is initially close to the wall, much shorter than the extension of the surface asperities on the rough spheres. It is the size of the surface roughness and not the hydrodynamic forces that determines the minimum distance of approach, and hence all the spheres initially close to the wall have the same trajectory of recession. This behaviour is similar to the behaviour of two rough spheres in shear flow (see Section 5.1.1).

The singularity method breaks down when the spheres come very close to each other, but, as mentioned before, the method can be made more accurate by including more singular forces and sources. These examples show that the singularity method is a useful technique for treating multibody interactions. Its main disadvantage is that the computer time required increases steeply with the number of particles.

A closely related method for dealing with multiparticle interactions has been given by Durlofsky *et al.* (1987) who expanded the disturbance flow in terms of point forces (but no sources), but located on the surface S_n of each sphere n:

$$v_i^* = \sum_{n=1}^{N} \int_{S_n} t_{ij}(\boldsymbol{x},\boldsymbol{y})\, k_j(\boldsymbol{y})\, \mathrm{d}S_y \tag{7.25}$$

This integral equation can be expanded in moments about the centre $\boldsymbol{x}^{(n)}$ of each particle. The zeroth moment of the force density corresponds to the total force exerted on the particle (eq (7.21)). The first moment can be split into a symmetric and an antisymmetric part, the antisymmetric part corresponds to the total torque (eq (7.22)), while the symmetric part is known as the stresslet, $\boldsymbol{Z}$, with the components

$$Z_{ij}^{(n)} = \frac{1}{2}\int_{S_n} [(y_i - x_i^{(n)})k_j + (y_j - x_j^{(n)})k_i - \frac{2}{3}\,\delta_{ij}(y_k - x_k^{(n)})k_k]\, \mathrm{d}S_y \tag{7.26}$$

This expansion in moments allows the disturbance flow to be expressed in terms of forces, torques and stresslets exerted by the fluid onto the particles:

$$v^* = \sum_n \left(1+\frac{1}{6}a^2\nabla^2\right) \boldsymbol{t}\cdot\boldsymbol{F}^{(n)} + \boldsymbol{C}\cdot\boldsymbol{L}^{(n)} + \left(1+\frac{1}{10}a^2\nabla^2\right)\boldsymbol{K}:\boldsymbol{Z}^{(n)} + \ldots \tag{7.27}$$

where the quadrupole contribution to the force $(1/6)a^2\nabla^2\boldsymbol{t}\cdot\boldsymbol{F}$ and a similar contribution to the stresslet arise from the finite size of the spheres. The tensor $\boldsymbol{C}$ is known as couplet or rotlet and has components given by

$$C_{ij}(\boldsymbol{r}) = \epsilon_{ijk} r_k / r^3 \tag{7.28}$$

and the components of $\boldsymbol{K}$ are

$$K_{ijk} = \frac{1}{2}(\nabla_k t_{ij} + \nabla_j t_{ik}) \tag{7.29}$$

The motion of the nth particle, located at $x^{(n)}$, in a flow given by eq (7.27) can be found from the Faxén formulae (Batchelor and Green, 1972):

$$\begin{aligned} \boldsymbol{u}^{(n)} &= v_0^{(0)} + \frac{\boldsymbol{F}^{(n)}}{6\pi\mu a} + \left(1+\frac{1}{6}a^2\nabla^2\right) v' \\ \omega^{(n)} &= \omega_0^{(n)} + \frac{\boldsymbol{L}^{(n)}}{8\pi\mu a^3} + \frac{1}{2}\nabla\times v' \\ \boldsymbol{S}^{(n)} &= \frac{3\boldsymbol{Z}^{(n)}}{20\pi\mu a^3} + \left(1+\frac{1}{10}a^2\nabla^2\right)\boldsymbol{S}' \end{aligned} \tag{7.30a}$$

where the prime indicates that we consider the disturbance field (eq (7.27)) caused by all the particles other than the nth particle itself; $\boldsymbol{u}$ and ω are the translational and rotational velocities of the particle and v_0 and ω_0 of the undisturbed flow at $\boldsymbol{x}^{(n)}$, while $\boldsymbol{S}$ is the rate of strain tensor of the undisturbed flow. These equations can be written in symbolic form:

$$\begin{pmatrix} U \\ \boldsymbol{S} \end{pmatrix} = -\boldsymbol{M}^* \begin{pmatrix} F \\ \boldsymbol{Z} \end{pmatrix} \tag{7.30b}$$

where U is the generalized velocity and F the generalized force vector (cf. eq (2.107b)). In this approximation the grand mobility tensor $\boldsymbol{M}^*$ has been truncated at the level of stresslets and thus is a far-field approximation of the true mobility matrix $\boldsymbol{M}$. On inversion of eq (7.30b) we obtain

$$\begin{pmatrix} F \\ Z \end{pmatrix} = -\tilde{\boldsymbol{R}} \begin{pmatrix} U \\ S \end{pmatrix} \tag{7.31}$$

Since the expansion (eq (7.27)) in combination with eq (7.30a) contains only leading terms of the disturbance flow around a sphere, no two- or more-body interactions are included in $\boldsymbol{M}^*$. To obtain a more accurate representation of $\boldsymbol{R}$, exact two-body terms (denoted by 2B) are included:

$$\tilde{\boldsymbol{R}} = (\boldsymbol{M}^*)^{-1} + \boldsymbol{R}_{2\mathrm{B}} - \boldsymbol{R}_{2\mathrm{B}}^{\infty} \tag{7.32}$$

where $\boldsymbol{R}_{2\mathrm{B}}^{\infty}$ contains the far-field contributions which were already included in $\boldsymbol{M}^*$. The inclusion of $\boldsymbol{R}_{2\mathrm{B}}$ results in the correct lubrication behaviour when two spheres come close together, because at short distances, this lubrication behaviour dominates all other contributions.

Writing the matrix $\tilde{\boldsymbol{R}}$ in terms of submatrices

$$\tilde{\boldsymbol{R}} = \begin{pmatrix} \boldsymbol{R}_{\mathrm{FU}} & \boldsymbol{R}_{\mathrm{FS}} \\ \boldsymbol{R}_{\mathrm{ZU}} & \boldsymbol{R}_{\mathrm{ZS}} \end{pmatrix} \tag{7.33}$$

where the subscripts refer to the relations between the various forces, velocities, stresslets and rate-of-strain tensors, we can write the motion of the particles with prescribed forces, torques and a linear shear flow as

$$U = -\boldsymbol{R}_{\mathrm{FU}}^{-1} \cdot (F + \boldsymbol{R}_{\mathrm{FS}} : \boldsymbol{S}) \tag{7.34}$$

In eq (7.34) many-body interactions are approximately included because the solution of eq (7.31), i.e. determining U and $\boldsymbol{Z}$ from F and $\boldsymbol{S}$, introduces many-body interactions. This inversion process couples all the particles together.

This method, applied to the sedimentation of a horizontal chain of spheres, produces results for the friction coefficient of each sphere that are very close to those obtained by the singularity method of Dabros (1984) and those obtained by Ganatos *et al.* (1978), using the method of Gluckman *et al.* (1971, 1972).

An example of the method of Durlofsky *et al.* is shown in Fig. 7.3, which shows the sedimentation of three spheres initially located on a line in a horizontal plane. It can be seen that the trajectories of a three-sphere system are quite complex; the spheres strongly interact hydrodynamically and even change their rank during the interaction. It was found that the trajectories at long times depend crucially on the initial location of the particles. Very small

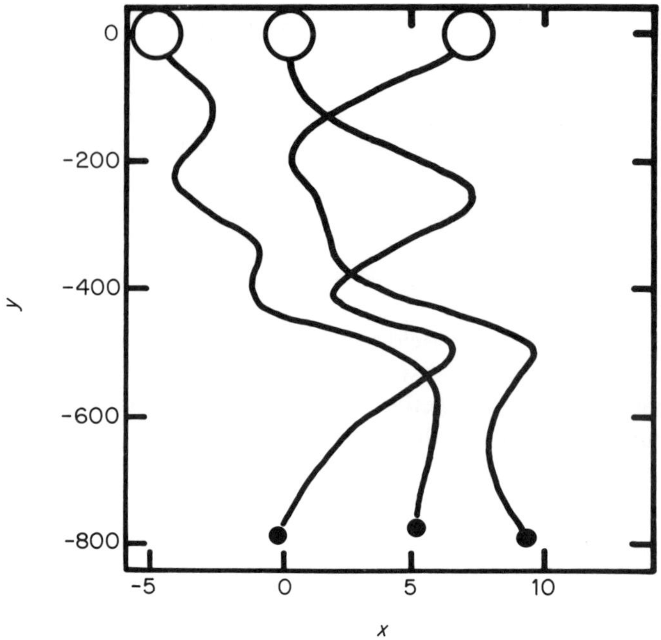

Fig. 7.3. Trajectories of three identical sedimenting hard spheres. Initially the spheres lie on a horizontal line $y=0$ at positions $x=-5$, 0 and 7. Trajectories are shown up to time $t=1200\ a/u$ (after Durlofsky *et al.* 1987).

changes in initial location can result in very different trajectories after the strong interaction.

Other methods for the sedimentation of three spheres are described by Kim (1987) and by Janse van Rensburg (1986).

7.3 Pairwise additivity

7.3.1 Pairwise additivity of velocities

A useful approximation which can be made when multi-particle interactions are present is to assume that the relative velocities between particles are additive. From eqs (7.8) and (7.10) we see that, in order to include multibody interactions, we must know the velocity of sphere 2 relative to sphere 1, averaged over the configuration of the $N-2$ remaining spheres. If pairwise additivity of relative velocities is assumed, $\boldsymbol{u}$ can be approximated by

$$\bar{\boldsymbol{u}}_{12} \simeq \sum_{j\neq 2} \boldsymbol{u}_{2j} \tag{7.35}$$

where u_{2j} is the relative trajectory of sphere j with respect to sphere 2, assuming all other spheres are absent. Thus u_{ij} can be calculated from the known hydrodynamic functions of two-sphere interactions. Schematically, the generalized velocity vector, U (cf. eq (7.34)), for a particle subjected to a linear shear flow can be written as

$$U = -\mu\, R^{-1} \cdot (\Phi : S) \tag{7.36}$$

where $\mu R^{-1} = R_{FU}^{-1}$ (having $6N \times 6N$ components) and $\mu\Phi = R_{FS}$ (having $6N \times 3 \times 3$ components). The tensor Φ is called the shear-resistance tensor. As all the components of R and Φ are known (Arp and Mason, 1977a; Adler, 1981a), u_{ij} can be obtained from eq (7.36). In calculating particle trajectories, no matrices need be inverted. The particle velocities for a given configuration of N particles can be calculated from eq (7.36) for each particle pair and subsequently the velocity of each particle relative to a reference particle can be calculated from eq (7.35). Thus if the position of each particle at time t is known their subsequent location at time $t + \Delta t$ can be calculated. This process can be continued numerically and particle trajectories can be generated.

When the spheres are far apart, eq (7.35) is quite accurate, but it fails when two or more particles come close to the test particle. The reason is that, in

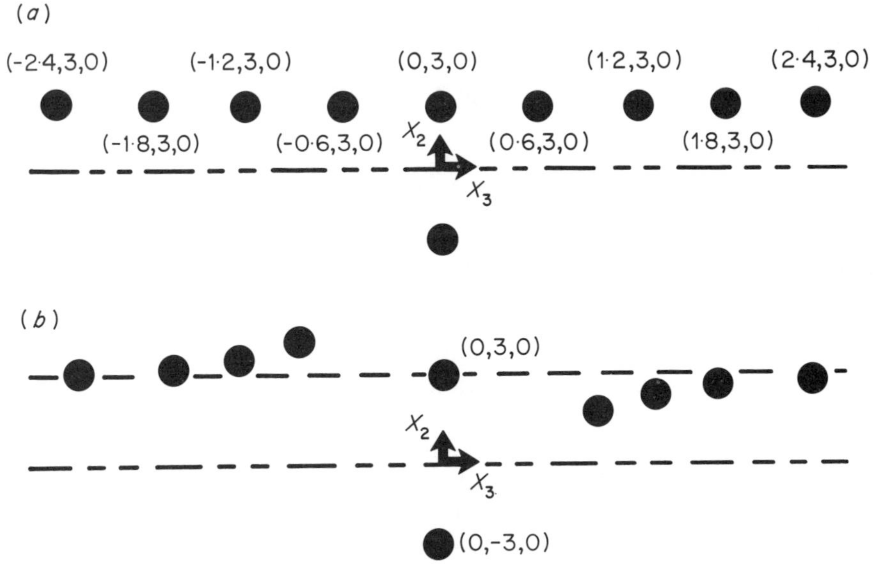

Fig. 7.4. Motion of 10 spheres in Poiseuille flow.
(a) Initial position.
(b) Position of spheres at a later time (lateral migration exaggerated) (after Schonberg *et al.*, 1984).

calculating the velocity of sphere 2 according to eq (7.35), nothing prevents it from overlapping with sphere 1.

As an example consider the motion of 10 spheres placed in a plane Poiseuille flow as shown in Fig. 7.4a. From the far-field solution of two spheres, it can be shown that the velocity of sphere i due to sphere j in the k direction is (Schonberg *et al.*, 1984)

$$(u_{ij})_k = -\frac{5}{2}\,\frac{1}{x^5}\,x_k x_\ell x_m\,(\tilde{S}_j)_{k\ell}\,\alpha\beta^2 \qquad (7.37)$$

where $\boldsymbol{x} = r_{ij}$, $r = \beta r$, $\tilde{r} = \alpha r$, $\alpha = a/d$ and $\beta = a/\ell$, with ℓ the characteristic interparticle distance and d the channel width. $\tilde{\boldsymbol{S}}_j$ is the rate-of-strain tensor of the undisturbed flow at sphere j in terms of $\tilde{r}$. It is assumed that $\alpha \ll \alpha\beta \ll 1$. Using eq (7.37), the position of each particle can be calculated as a function of time. The relative position of the 10 spheres at a subsequent time is shown in Fig. 7.4b. It may be seen that many-body interactions can induce the lateral migration of particles.

It is possible to include colloidal interaction forces in the approximation of pairwise additivity of velocities. When colloidal forces (and torques) are acting they balance the hydrodynamic forces (and torques):

$$\boldsymbol{F}_{\text{coll}} + \boldsymbol{F}_{\text{H}} = 0 \qquad (7.38)$$

with (cf. eq (7.36))

$$\boldsymbol{F}_{\text{H}} = -\mu(\boldsymbol{R}\cdot \boldsymbol{U} + \boldsymbol{\Phi}:\boldsymbol{S}) \qquad (7.39)$$

and hence

$$\boldsymbol{U} = -\mu\boldsymbol{R}^{-1}\cdot(\boldsymbol{F}_{\text{coll}} + \boldsymbol{\Phi}:\boldsymbol{S}) \qquad (7.40)$$

7.3.2 Pairwise additivity of forces

To avoid the physically unrealistic eventuality of overlap between two spheres resulting from the assumption of pairwise additivity of velocities, we can make the more realistic assumption of pairwise additivity of forces. This has the advantage that the proper lubrication limit is included in the analysis, preventing the overlap of spheres. Again the basic equations are eqs (7.39) and (7.36) or (7.40), depending on whether or not colloidal forces are present. If addivitity of forces is assumed, the force exerted on each particle can be calculated (using the two-sphere hydrodynamic functions), and the $6N\times 6N$ grand resistance matrix and the $6N\times 3\times 3$ shear-resistance matrix constructed. We can then invert eq (7.39) and obtain the velocities of all the

N particles. Thus, knowing the positions of N particles at time t, their position at time $t+\Delta t$ can be determined. We can construct $\boldsymbol{R}$ and $\boldsymbol{\Phi}$ anew and continue the process, and in this way generate particle trajectories. It is clear that this procedure takes much more computer time than does the simple approximation of additivity of velocities, since in each integration step a $6N\times 6N$ matrix must be inverted. However, this method has the great advantage that the proper lubrication behaviour is included at short separations and thus it is much more useful, especially for more concentrated dispersions. This method was used by Bossis and Brady (1984) to simulate the behaviour of a monolayer of spheres subjected to simple shear.

7.4 Many-sphere hydrodynamic interactions

In Section 7.2 it was shown that the disturbance flow created by N spheres can be expanded in terms of forces, torques and stresslets (cf. eq (7.27)). This is equivalent to saying that the surface forces, k_i, can be expanded in terms of forces, torques, etc. As shown by Mazur (1987), this expansion can be extended to any order of irreducible force multipoles. Higher-order terms in this expansion correspond to two-, three- and more-body interactions. By comparing the general flow field with the multipole expansion, the various terms corresponding to multibody interactions can be evaluated. The method can also take into account spherical boundaries by considering N spheres within a spherical container. The approach applies equally well to the container as to the particles and thus the interactions of the spheres with the container wall can also be found. The expansions become quite complicated and for more details the reader is referred to the paper by Mazur (1987).

In applying this method to the (short time) self-diffusion coefficient of Brownian hard spheres Mazur found:

$$D_s^0 = D_0(1 - 1\cdot 73c + 0\cdot 88c^2 + \ldots .) \tag{7.41}$$

where the term $-1\cdot 73c$ is due to two-sphere interactions and the term $0\cdot 88c^2$ due to three-sphere interactions. This result shows that hydrodynamic interactions are non-additive because calculating the term of order c^2 from the assumption of additivity results in $-0\cdot 93c^2$. Experiments on index-matched dispersions (van Veluwen *et al.*, 1987) show that eq (7.41) is a reasonable approximation. Assuming a value of $-1\cdot 83$ (cf. Section 4.2) for the first virial coefficient, van Veluwen *et al.* found that the second virial coefficient was $0\cdot 7 \pm 0\cdot 3$.

7.5 Viscosity of semi-dilute dispersions

7.5.1 Stable dispersions

Several approximate theories exist to predict the viscosity of stable dispersions of spheres. None of these theories take account of the exact hydrodynamic interactions among three or more particles. These theories are based either on a statistical mechanical approach (eq (7.8)), or on effective medium theories. Since these approaches apply equally well to concentrated and semi-dilute dispersions, we shall delay describing them till the next chapter. Instead we will look here at semi-dilute dispersions of slender rods because this is one of the few systems for which results are available for anisotropic particles.

At infinite dilution, the viscosity of a dispersion of slender rods (spheroids with $r_e \gg 1$), can be found from eqs (3.291) and (3.283). For slender rods the leading terms are those containing the constants C_1 and C_4, given by

$$C_1 = \frac{r_e^2}{4 \ln r_e} \; ; \; C_4 = \frac{3}{2} \frac{r_e^2}{\ln r_e} \tag{7.42}$$

Averaging eq (3.291) and using eq (2.143) yields, in the limit $\mathrm{Pe}_r \to 0$

$$[\mu] = \frac{4 r_e^2}{15 \ln r_e} \tag{7.43}$$

From eq (2.104) the rotary-diffusion constant ($a_2 = a_3$; $r_e \to \infty$) is found to be:

$$D_r = \frac{3kT}{8\pi\mu a_1^3} \ln r_e = \frac{kT}{2\mu V} \frac{\ln r_e}{r_e^2} \tag{7.44}$$

where V is the volume of the rod. By combining eqs (7.43) and (7.44) and expressing the volume fraction, c, in terms of the number concentration, n, of particles ($c = nV$), we can write for the viscosity of a dispersion of slender rods:

$$\mu = \mu_0 + \mu_0[\mu]\, c = \mu_0 + \alpha \frac{kT}{D_r} n \tag{7.45}$$

where α is a constant which is 2/15 for slender rods. Equation (7.45) is more generally true if α is allowed to be a function of r_e. For spheres ($r_e = 1$), $\alpha = 5/12$.

From the above equation it can be seen that the zero-shear viscosity (i.e. the viscosity in the limit Pe $\to 0$), is independent of the shear rate G, and depends only on D_r (for slender rods). This equation holds for very dilute dispersions in which particle interactions are absent. If interactions are included, then the rotary-diffusion constant, D_r, becomes a function of particle concentration. As a first approximation we can assume that eq (7.45) is still valid, but with a modified rotary-diffusion coefficient to allow for particle interactions:

$$\mu = \mu_0 + \alpha \frac{kT}{D_r^{\text{eff}}} n \tag{7.46}$$

This assumption was made by Doi and Edwards (1978b) to describe the viscosity of solutions of rigid linear polymers. These authors also provided an approximate expression for D_r^{eff}, based on purely geometrical arguments, which is valid in the concentration range

$$1/L^3 \ll n \ll 1/dL^2$$

or, in terms of the axis ratio L/d,

$$r_e^{-2} \ll c \ll r_e^{-1}$$

where L is the length and d the diameter of the slender rod. The argument goes as follows. Consider a reference rod in a suspension of rods and imagine a cylinder around it with a radius a_c equal to the mean distance between nearest neighbours. The rotation of the test rod is restricted by the presence of the surrounding rods. Because of this restriction, during the time τ_0 in which a reference rod remains in the reference cylinder, the solid angle through which the reference rod rotates is about $\Delta\Omega \simeq (a_c/L)^2$, and hence the effective rotary-diffusion constant is approximately given by (provided $a_c \gg d$),

$$D_r^{\text{eff}} \simeq (a_c/L)^2/\tau_0 \tag{7.47}$$

The time τ_0 can be estimated from

$$\tau_0 \simeq L^2/2D_t \tag{7.48}$$

where D_t is the translational-diffusion coefficient parallel to the rod's axis, for slender rods related to the rotary-diffusion constant by (cf. eqs (2.103) and (2.104)):

$$D_r = \frac{2D_t}{3L^2} \tag{7.49}$$

In a random orientation and distribution of rods, it can be shown from the geometry that the mean distance between nearest neighbours is

$$a_c \simeq \frac{2}{n\pi L^2} \tag{7.50}$$

In general, a_c is a fucntion of the orientation distribution. Notice that the condition $a_c \gg d$ is equivalent to $r_e c \ll 1$. Combining eqs (7.47) and (7.50) yields

$$D_r^{\text{eff}} = \frac{\beta D_r}{n^2 L^6} \tag{7.51}$$

where β is an unknown constant of order unity. Inserting this into eq (7.46) yields for the viscosity

$$\mu = \mu_0 + \tilde{\alpha}\,\frac{kT}{D_r}\,L^6 n^3 \tag{7.52}$$

where $\tilde{\alpha} = \alpha/\beta$. Usually the second term on the right-hand side of eq (7.52) is much larger than the first and the term μ_0 can be dropped. In this case, μ simply becomes proportional to the cube of the volume fraction

$$\mu \propto \mu_0 \,\frac{r_e^6}{\ln r_e}\, c^3 \tag{7.53}$$

which shows that it is possible to determine the axis ratio of slender rods by measuring the viscosity, even in semi-dilute dispersions.

Doi and Edwards (1978b) also calculated the viscosity of a suspension of slender rods as a function of shear rate (or rotary Péclet number, Pe_r), in which case the rotary-diffusion constant depends on the orientation distribution, p, because a_c depends on p. To solve for p, the term $D_r \nabla^2 p$ in the rotary convective-diffusion equation (cf. eqs (3.102) and (3.131)) must be replaced by $\boldsymbol{\nabla}\cdot(D_r \boldsymbol{\nabla} p)$, which makes the equation much more difficult to solve. Instead this term is approximated as $\bar{D}_r \nabla^2 p$ where $\bar{D}_r$ is defined by an equation analogous to eq (3.152). Since this expression for $\bar{D}_r$ is difficult to integrate, Doi and Edwards defined $\bar{D}_r$ as

$$\bar{D}_r = 1/\langle 1/\sqrt{D_r}\rangle^2$$

which is easier to evaluate.

For $\boldsymbol{\omega}$, Jeffery's equations are assumed to be valid. After solving for $\boldsymbol{p}$, the stress tensor, which can be expressed in terms of orientational averages (cf. Section 3.1.11.5(a)), is evaluated. The stress tensor used by Doi and Edwards is slightly different from that used here, as they used a molecular stress tensor derived for rod-like molecules. In the limit $\mathrm{Pe}_r \to 0$, the result (eq (7.45)) is recovered, but with $\alpha = 1/10$. This difference is immaterial because the constant $\tilde{\alpha}$ in eq (7.52) in unknown. For higher Péclet numbers the equations of Doi and Edwards predict shear thinning, similar to eq (3.291) (cf. Fig. 3.41). Although the approach taken by Doi and Edwards is generally correct, the assumption that in semi-dilute dispersions ω is given by Jeffery's equation is rather doubtful. Instead it seems better to define an effective shear rate, G^{eff}, to account for many-body interactions. An approximate expression for G^{eff} can be found by realizing that the rotational motion of a rod in simple shear can be thought of as a torque balance between an applied torque $L_1 \propto \mu G V$ and a resistance torque $L_2 \propto f_r \omega$, where $f_r = kT/D_r$ is the rotary friction coefficient. In dilute systems

$$\mu_0 G V \propto f_r \omega_0 \tag{7.54a}$$

and in semi-dilute systems

$$\mu_{\mathrm{eff}}\, G V \propto f_r^{\mathrm{eff}} \omega \tag{7.54b}$$

from which it follows that

$$\frac{\omega}{\omega_0} = \frac{\mu_{\mathrm{eff}}}{\mu_0}\frac{f_r}{f_r^{\mathrm{eff}}} = \frac{G^{\mathrm{eff}}}{G} \tag{7.55}$$

For low Péclet numbers, using eqs (7.53) and (7.51):

$$G^{\mathrm{eff}} = \tilde{\beta}\, G \frac{r_e^2}{\ln r_e} c \tag{7.56}$$

$\tilde{\beta}$ being an unknown constant of order unity. For higher values of Pe_r, μ_{eff} and G_{eff} become a function of Pe_r.

To summarize, in the concentration range $r_e^{-2} \ll c \ll r_e^{-1}$, it is possible to use the theory for the viscosity (and normal stress differences) for dilute dispersions of slender rods, provided that we use a modified rotary-diffusion constant or, equivalently, an effective equivalent axis ratio, r_e^{eff}, defined as

$$D_r^{\text{eff}} = \frac{kT}{2\mu V} \frac{\ln r_e^{\text{eff}}}{(r_e^{\text{eff}})^2} \tag{7.57}$$

and a modified rotary Péclet number

$$\text{Pe}_r = G^{\text{eff}}/D_r^{\text{eff}} \tag{7.58}$$

Besides rheological properties, it is also possible to predict optical (and rheo-optical) properties with this theory (Doi and Edwards, 1978a).

An example of experimental measurements of the viscosity of suspensions of slender rods is given in curve 4 of Fig. 7.5, which also contains particles of other shapes for comparison. The axis ratio of the rods indicated as $r_e = 375$, was estimated from an expression approximately equivalent to eq (7.43). It can be seen that the viscosity becomes proportional to c^3 at a volume fraction of about 10^{-3}. Estimating the equivalent axis ratio from eq (7.53) we find that $(\ln r_e)/r_e^6 \simeq 10^{-11}$ or $r_e \simeq 100$. This appears to be a low value and is

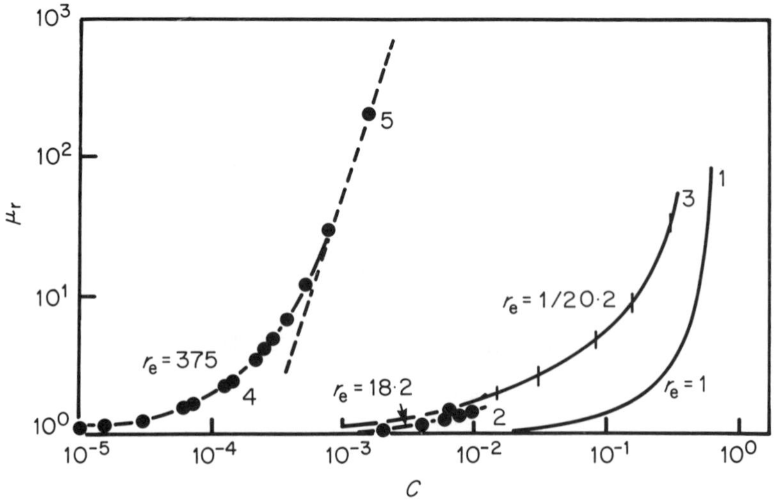

Fig. 7.5. Comparison of the relative viscosity, μ/μ_0, of suspensions consisting of various shaped particles. (1) Spheres, $r_e = 1$ (Thomas, 1965); (2) glass fibres in sucrose solution, $r_e = 18{\cdot}2$ (Attanasio *et al.*, 1972); (3) mica flakes in poly(ethylene), $r_e = 1/20{\cdot}2$ (Fisa and Utracki, 1984); (4) Xantan in salted water, $r_e = 375$ (Chauvateau, 1982); (5) theoretical prediction of Doi and Edwards theory (eq (7.53)) (after Utracki, 1984).

probably due to polydispersit diffusion constant, and to flexi twin strands of macromolecules certainly bend in a shear flow.

It is of interest to note from eq (7 slender rods depends almost entirely independent of its width, because the p minor axis a_2. Increasing a_2 to $a_2(1+\delta$ smaller, but terms of this magnitude we main reason is that it was assumed that When $r_e c \simeq 1$, all rods touch and the viscosi As a first attempt to include the effects of pa ...ne restriction $r_e c \ll 1$ to $r_e c < 1$. In this case a_c m ... $_c - 2a_2$ or $a_c - 2a_2(1+\delta)$ if the particle has an electrical dou ...lymer layer of thickness δ. This results in a modified rotary-dif ... constant

$$D_r^{\text{eff}} = D_{ro}^{\text{eff}}[1 - 2(1+\delta)\, a_2 L^2 n/\pi] = D_{ro}^{\text{eff}}\,[1 - \beta^*(1+\delta)\, r_e c] \tag{7.59}$$

where D_{ro}^{eff} is as given by eq (7.51) and β^* is a constant depending on the precise geometry of the slender rod (cylinder, spheroid, square rod, etc.). With this assumption we can write

$$\frac{\mu}{\mu_0} \propto \frac{r_e^6\, c^3}{[1 - \beta^*(1+\delta)\, r_e c] \ln r_e} \tag{7.60}$$

Notice that when $\beta^*(1+\delta) r_e c \to 1$, $\mu \to \infty$. Before this happens, such systems will phase-separate into liquid crystalline and disordered phases. Equation (7.60) predicts that the viscosity of a dispersion of stable electrically charged rods increases with decreasing electrolyte concentration.

7.5.2 Coagulated dispersions

When an unstable colloidal dispersion is subjected to shear, orthokinetic coagulation occurs. In general, the rheological properties of the system will change until a dynamic equilibrium is established between the formation and break-up of aggregates. This steady-state behaviour depends on the sizes and distribution of flocs present in the dispersion. It is observed that coagulated dispersions show non-Newtonian behaviour, shown schematically in Fig. 7.6, where the shear stress τ ($= P_{23}$, cf. eq (3.282)) is plotted against the shear rate G. It can be seen that such systems are shear thinning (similar to stable dispersions), the viscosity being given by the slope of the τ versus G plot. Behaviour of the type shown in Fig. 7.6 is called pseudoplastic. At shear

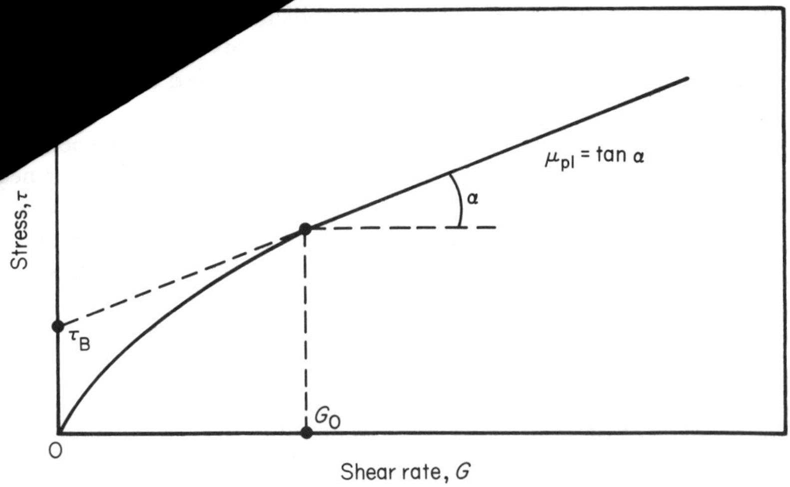

Fig. 7.6. Schematic flow diagram (rheogram) of a coagulated dispersion.

rates larger than a critical shear rate G_0, the stress increases linearly with the shear rate and can be described by the equation

$$\tau = \tau_B + \mu_{pl} G \tag{7.61}$$

where τ_B is called the Bingham yield value and μ_{pl} the plastic viscosity. In order to understand and to predict the behaviour of coagulated dispersions G_0, τ_B and μ_{pl} must be expressed in terms of the colloidal forces holding the particles together, the structure of the flocs and the strength of the flow.

The pseudoplastic behaviour can be explained by the elastic floc model (Firth and Hunter, 1976b; van de Ven and Hunter, 1979) in which it is assumed that a dynamic equilibrium is established between the formation and break-up of small compact flocs or flocculi at $G > G_0$ and of large open floc aggregates at $G < G_0$. This can be deduced from the floc density which remains nearly constant at $G > G_0$, but decreases with decreasing shear rate at $G < G_0$. All flocs are supposed to be elastic, i.e. they can stretch and contract when subjected to shear. At G_0 all the large flocs are just broken up. If we consider a doublet of two flocs, it will most likely break up at an orientation of $\theta = 90°$, $\phi = 45°$ at which the extension is maximum. At this orientation the internal stress σ is approximately

$$\sigma \simeq 5\mu_0 G \tag{7.62}$$

When this stress exceeds the bond strength of the doublet it will break up. The bond strength per unit area can be estimated as $n_\ell F_{max}$, where n_ℓ is the number of bonds per unit area and F_{max} the force needed to break a bond, given by

$$F_{max} = \left(\frac{dV_{int}(h)}{dh} \right)_{h_m} \simeq a\,Q \tag{7.63}$$

where h_m is the maximum distance a bond can be stretched until it breaks, a is the radius of the primary particle, and Q is approximately given by (cf. eq (1.135))

$$Q \simeq \frac{A}{12h_m^2} + B(h_m)\zeta^2 \tag{7.64}$$

where A is the Hamaker constant, ζ the zeta potential, and B a function depending on ionic strength. It follows that the critical shear rate at which a doublet breaks up can be found from

$$5\mu_0 G_0 \simeq n_\ell F_{max} \simeq n_\ell a Q \tag{7.65}$$

The plastic viscosity, $\mu_{p\ell}$, can be estimated by considering the floc, together with the liquid immobilized within it, to be a single spherical particle of radius a_f. The volume fraction of flocs is then $c_f = 4\pi a_f^3/3$. Several expressions exist in the literature for describing the volume-fraction dependence of suspension of spheres (cf. eq (5.55) or eq (8.4)), which we can write as $\mu(c)$. The plastic viscosity is then

$$\mu_{p\ell} = \mu(c_f) = \mu(C_{fp} c_p) \tag{7.66}$$

where $C_{fp} = c_f/c_p$ is the ratio of the volume fraction of flocs to the volume fraction of the primary particles (c_p). Equation (7.66) also holds approximately when the primary particles are non-spherical because aggregates of non-spherical particles are often approximately spherical. From measurements of $\mu_{p\ell}$ and c_p and the theoretical relation $\mu(c)$, we can determine C_{fp} experimentally. C_{fp} is a measure of the (inverse) floc compactness. For very compact flocs C_{fp} is close to one, while for open flocs $C_{fp} \gg 1$. The floc density or compactness is usually determined by the highest shear rate, G_{max}, to which the suspension was subjected. It thus depends strongly on the history of the suspension. At G_{max} a stress balance yields

$$5\mu_0 G_{max} \simeq n_c F_{max} \tag{7.67}$$

where n_c is the number of particle chains crossing a unit area in a given direction. Flocs with too few chains are not strong enough to survive at G_{max}. From geometrical arguments we can relate n_c to the compactness of the floc

$$n_c \simeq \frac{1}{4C_{fp}a^2} \tag{7.68}$$

and thus, combining eqs (7.67), (7.68) and (7.63)

$$C_{fp} \simeq \frac{Q}{20\mu_0 a G_{max}} \tag{7.69}$$

Finally it can be shown that the Bingham yield stress, τ_B, depends on the energy dissipation during the rupture of floc doublets and the rate with which doublets are formed. This energy dissipation depends on the energy needed to stretch the bonds, to rearrange the location of the particles in the floc during stretching and the energy dissipated due to internal fluid motion inside the floc when it changes shape and volume during a collision. It can be shown that the energy dissipation due to fluid motion inside a colliding floc is the dominant contribution to τ_B. The energy dissipated per floc can be estimated from

$$E_f \simeq \frac{3n_s}{4\pi\bar{\rho}a_f^3}\,\alpha\int_0^{a_f} F(r)\ell(r)\rho\pi r^2\mathrm{d}r \tag{7.70}$$

where n_s is the number of spheres per floc, ρ the number of spheres in a floc per unit volume, α the fraction of the liquid volume inside the floc that is mobile, estimated to be $\alpha \simeq \delta/a$, δ being the distance a bond is stretched; $F(r) = 6\pi\mu_0 a\lambda u(r)$ is the force on a single particle in the floc at a distance r from the floc centre, λ being a correction factor to Stokes' law allowing for screening by neighbouring particles, and $u(r)$ is the velocity of the particle during a collision, estimated as $u(r) \simeq \frac{1}{2}Gr$; and $\ell(r)$ is the distance a sphere in a floc travels during a collision, estimated as $\ell \simeq \frac{1}{2}\pi r$. Evaluating E_f using these assumptions, as well as assuming that λ and ρ are constant, results in

$$E_f \simeq 0{\cdot}9\pi^2\lambda\mu_0 G a^2\delta n_s \tag{7.71}$$

The number of spheres per floc can be found from C_{fp}:

$$n_s = a_f^3/C_{fp}a^3 \tag{7.72}$$

The doublet frequency equals (cf. eqs (5.21) and (5.22))

$$f = \frac{16}{3}\,\alpha_0 n_f^2 G a_f^3 = \frac{3\alpha_0 c_f^2 G}{\pi^2 a_f^3} \tag{7.73}$$

The energy dissipation associated with doublet rupture is thus

$$E_{DR} \simeq 2fE_f \tag{7.74}$$

and the Bingham yield value is $\tau_B = E_{DR}/G$, or

$$\tau_B \simeq 0 \cdot 54\, \lambda\alpha_0 C_{fp} G\mu_0 (a_f^2\delta/a^3) c_p^2 \tag{7.75}$$

Although α_0 and a_f are functions of G, it can be shown, as a consequence of the dynamic equilibrium between the formation and break-up of compact flocs at $G > G_0$, that $\alpha_0 G a_f^2$ is approximately constant (Hunter and Frayne, 1980).

A large body of experimental data is available (measurements of G_0, τ_B and $\mu_{p\ell}$), which can be satisfactorily explained using the elastic floc model (Firth and Hunter, 1976a), despite the crudeness of the underlying assumptions. The elastic floc model can also be used to predict the viscoelastic behaviour of coagulated sols (van de Ven and Hunter, 1979).

In the elastic floc model it is implicitly assumed that the density of a floc is a constant. Recent work on floc structure has shown that flocs often have a fractal geometry (Lindsay *et al.*, 1987; Meakin, 1987). In experiments performed to verify the elastic floc model, compact flocs were formed at high shear prior to the measurements of the rheological properties performed at lower shear rates. Under such conditions the fractal nature of flocs is probably not very pronounced. However, in general, this fractal nature must be considered in describing the rheology of coagulated dispersions.

7.6 Break up of aggregates in flow

Depending on the hydrodynamic conditions, particles dispersed in a suspending medium can either grow into aggregates or clusters, or particle clusters can break up into smaller clusters or single particles. An example of this was given in Section 5.1.4 where the stability of dilute dispersions of spheres subjected to simple shear was discussed. As indicated in Fig. 5.14, at shear rates above a critical shear rate, G_{crit}, all doublets of spheres will break up. The magnitude of G_{crit} is determined by the bond strength between the

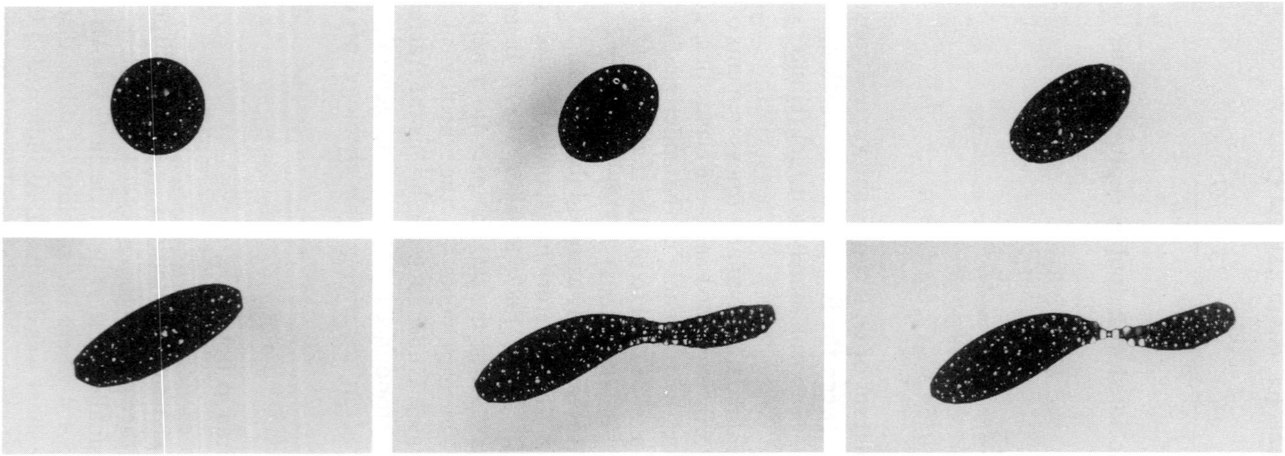

Fig. 7.7. Deformation and break-up of a poly(styrene)/water cluster in silicone oil subjected to simple shear. The initial radius of the cluster in the first frame is 0·78 mm. The shear rates in sequential frames are 0, 0·49, 0·90, 1·25, 1·68 and 1·68 s^{-1}, respectively. The volume fraction of spheres is ca. 15% (after Smith and van de Ven, 1985a).

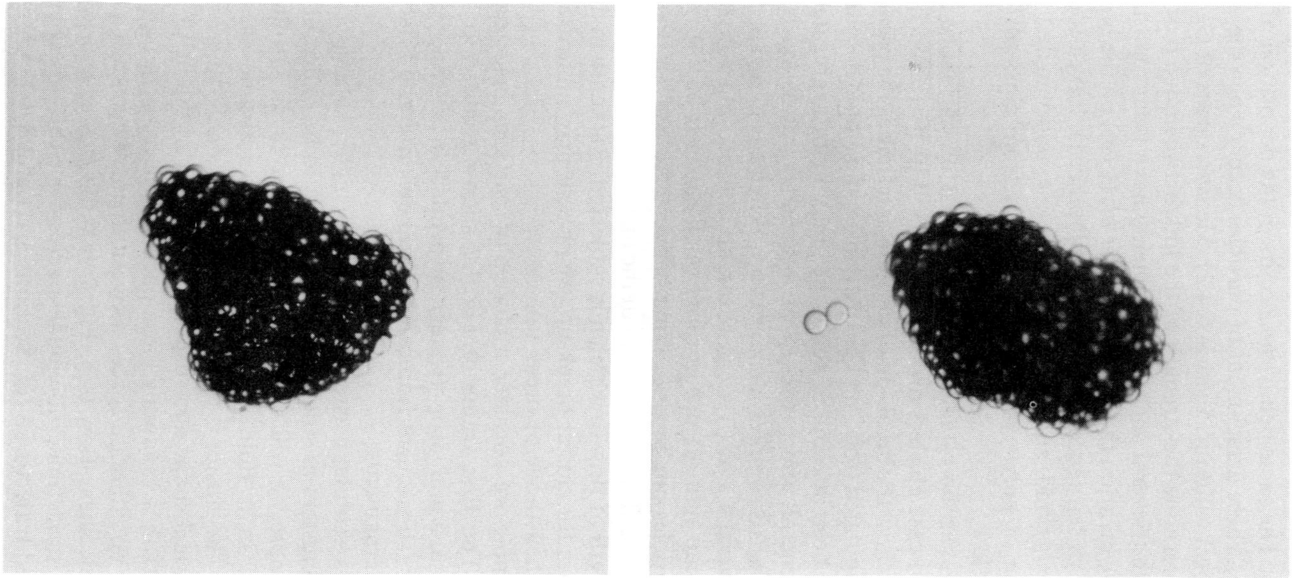

Fig. 7.8. Poly(styrene)/water cluster in silicone oil subjected to a simple shear. Maximum cluster dimension $\simeq 2\cdot 3$ mm; shear rate $4\cdot 41\ s^{-1}$; volume fraction of spheres ca. 50% (after Smith and van de Ven 1985a).

two spheres or, equivalently, by the depth of the primary minimum in which they are captured. The same behaviour occurs in the deposition of particles on a solid surface, where detachment of particles occurs when the wall shear rate exceeds a critical value (cf. Section 6.4.2). Measurements of G_{crit} is a convenient way to determine the strength of the link between two particles or a particle and a wall. Several such measurements have been reported in the literature. Here we mention the work by Tha *et al.* (1986) on the break-up in shear of doublets of hardened spherized red blood cells, linked together by an antigen–antibody bond, and the work of Varennes and Van de Ven (1987) on the detachment of poly(styrene) latex particles from glass surfaces in a stagnation-point flow.

The same ideas can be extended to interactions among more than two particles, in which case clusters are formed instead of doublets. In Section 7.5.2 it was shown how the break-up of clusters determines the pseudoplastic behaviour of coagulated sols, and in the elastic floc model some simple assumptions have been introduced about the break-up of aggregates of flocs or flocculi. In general, the break-up of aggregates consisting of a large number of particles is a complex phenomenon. When an aggregate breaks up it can do so in several ways: (i) it can break up into two more-or-less equal fragments; (ii) it can split off one or more small clusters; or (iii) it can split off individual particles one by one. What happens for a given floc depends on the structure of the floc, the strength of the links between the particles and the type and strength of flow to which it is subjected. Rather than discussing the various theories for cluster break-up, we present here some examples of model flocs and their mode of break up.

As a first example we consider a cluster of particles held together by capillary forces. The first frame of Fig. 7.7 shows a cluster of particles contained within a liquid drop (water) suspended in silicone oil. When subjected to shear (subsequent frames) the drop deforms and finally breaks up into two large fragments.

Figure 7.8 shows a similar cluster, but at a much higher volume fraction, in which case small clusters or single spheres are seen to be pulled off.

As a final example we will discuss the behaviour of an infinitely weak cluster suspended in a fluid subjected to either a simple shear or an extensional flow. The cluster is formed by moulding small beads with some liquid into a spherical shape and suspending them in a volume of the same liquid. This results in a randomly close-packed spherical configuration of hard spheres with no adhesion forces between them. When such a cluster is sheared in a simple shear flow (see. Fig. 7.9a), the cluster as a whole will rotate, but, at the same time, individual spheres leave the cluster and rotate around it on closed trajectories (cf. Sections 3.1.1 and 5.1.1), slowly filling the whole region of closed trajectories. Since the cluster diminishes in size,

Fig. 7.9. Photographs showing aggregates of small spherical beads. $G = 0{\cdot}44\ s^{-1}$ and $t = 7$ min. The X_1 axis is directed towards the viewer. The shear is in the X_2X_3 plane (after Kao and Mason, 1975).

(a) In simple shear.

(b) In extensional flow.

the single spheres will eventually get on open trajectories and separate away from the cluster. In extensional flow (Fig. 7.9b), no cluster rotation occurs and all the spheres that are pulled off the cluster are convected away immediately. From this example it can be seen that the break-up of flocs is more efficient in extensional flow than in simple shear. In general, the less vorticity there is in a flow, the more efficient break up is.

Chapter 8

The Flow of Concentrated Dispersions

8.1 Introduction

Concentrated dispersions are of interest for a variety of reasons. Besides being of practical importance, such as in paints, paper coatings, drilling muds, blood, etc., they can also serve as model systems which mimic many atomic or molecular systems. Since colloidal particles move at a much slower rate than do atoms or molecules, several phenomena, such as phase transitions, restructuring in external fields, the behaviour in shear, etc., are easier to observe in colloidal than in atomic systems.

When the volume fraction of a dispersion of hard spheres is increased, a critical concentration ($c_{crit} \simeq 0 \cdot 5$) is reached where the system spontaneously phase separates into ordered and disordered regions. If this did not occur the motion of a particle above c_{crit} would become severely restricted due to frequent collisions with other particles. By phase separating, the particles in the disordered phase increase their entropy, but those in the ordered region decrease their entropy, while still being allowed to move freely in a "cage" formed by neighbouring particles. Together with the corresponding changes in enthalpy, the net result is a decrease in the free energy of the system. The ordered region resembles a crystal, the structure of which depends on the volume fraction. Usually a bcc structure is formed first, while at higher volume fractions the structure becomes fcc.

For suspensions of electrically charged spherical particles the situation is similar. Particles surrounded by an electrical double layer can be thought of as hard spheres with an effective radius $a_{eff} = a + \delta$, where δ is some measure of the double-layer thickness (not necessarily equal to κ^{-1}). It follows that suspensions of charged spheres exhibit a phase transition at volume fractions lower than the critical-volume fraction of hard-sphere suspensions. The phase transition in dispersions of charged spheres was studied by Hachisu *et al.* (1973) and their results are shown in Fig. 8.1.

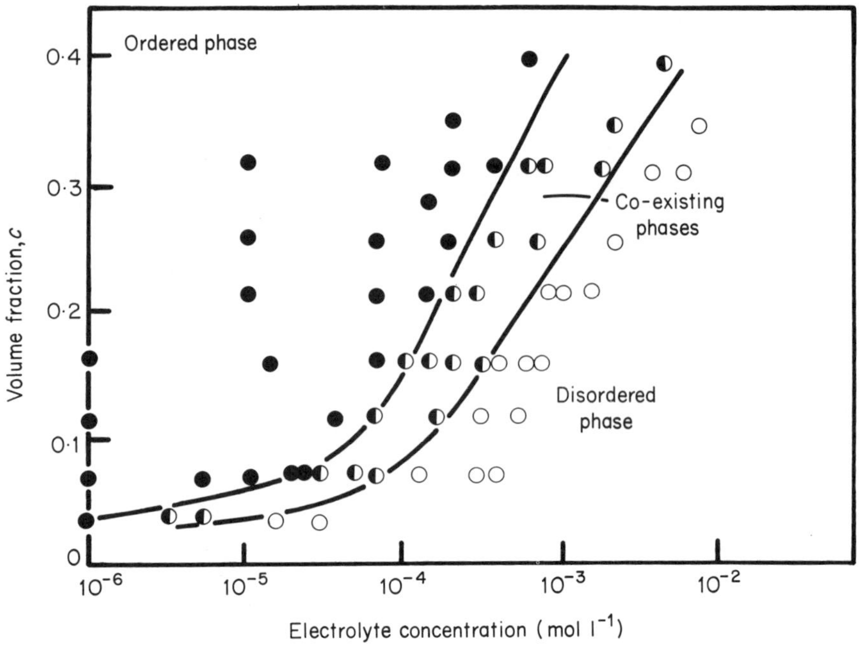

Fig. 8.1. Experimental phase diagram of a poly(styrene) latex dispersion. Radius of the spheres $a = 85$ nm (after Hachisu *et al.*, 1973).

It can be seen from this figure that even at volume fractions of a few percent dispersions can be ordered, provided there is a sufficiently strong long-range repulsion between the particles. Depending on the volume fraction, c, and the salt concentration, κa, the dispersion can be completely ordered (high c, low κa), completely disordered (low c, high κa), or it can consist of these two phases in thermodynamic equilibrium.

When an external field is applied to the dispersion, the structure can change and under certain conditions it is possible to cross one of the phase boundaries shown in Fig. 8.1. In some cases it is possible to order a disordered phase by applying a field (field-induced crystallization) or to destroy order (field-induced melting). It is also possible to restructure the ordered phase by an external field (by changing the crystal structure).

For suspensions of non-spherical particles we can expect the same behaviour as occurs with spheres. At some critical volume fraction an ordered structure appears which resembles the structure of liquid crystals. Depending on particle shape and volume fraction, several liquid crystalline structures are possible. For suspensions of spheres, the viscosity, μ, is a smooth function of the volume fraction, c, μ increasing with increasing c, and the order–disorder phase transition does not show up as a discontinuity in the $\mu(c)$ dependence, probably because of gradual changes when the region of co-existing phases is traversed. It is not clear that the same applies to suspensions of non-spherical particles. For polymer solutions which can be modelled as hard-rod suspensions, such as solutions of hydroxypropyl cellulose (HPC) in water, it is observed that after passing through the disorder liquid–crystal phase transition ($c_{crit} \simeq 0 \cdot 4$), the viscosity actually decreases with increasing volume fraction (Ernst *et al.*, 1988). It is not known whether similar effects can occur for suspensions of rods.

Since the literature on concentrated dispersions is extensive and could easily warrant a whole book in itself, this chapter is restricted to some basic approaches and models which can be applied to a variety of concentrated systems. Some useful experimental techniques, especially adapted for concentrated dispersions, are also discussed.

8.2 Effective-medium theory

In attempting to describe the behaviour of concentrated dispersions, a useful idea is to treat the dispersion as an effective medium to which effective rheological, optical and other properties can be ascribed. Here, as an example, we will consider the viscosity of concentrated dispersions.

If a small number of particles is added to a dispersion already containing a certain number of particles, as a first approximation we can assume that

$$\mu(c_1 + c_2) \simeq \mu(c_1)\,(1 + [\mu]c_2) \tag{8.1}$$

where c_2 is the volume fraction of the added particles ($c_2 \ll 1$) and c_1 that of the particles already present. However, because of the presence of particles that were there before the volume fraction c_2 was added, not all the space in the liquid is available for the newly added particles. The fraction of the volume available to the newly added particles is $1 - kc_1$ where k is a crowding factor. Hence, allowing for the presence of previously added particles, eq (8.1) should be modified to

$$\mu(c_1 + c_2) = \mu(c_1)\left(1 + \frac{[\mu]c_2}{1 - kc_1}\right) \tag{8.2}$$

Denoting the increase in μ due to an increase, Δc, in volume fraction c by $\Delta\mu$, the above equation can be written as

$$\mu + \Delta\mu = \mu \left(1 + \frac{[\mu]\Delta c}{1 - kc}\right) \tag{8.3a}$$

or

$$\frac{d\mu}{\mu} = [\mu] \frac{dc}{1 - kc} \tag{8.3b}$$

which upon integration yields

$$\mu = \mu_0 (1 - kc)^{-[\mu]/k} \tag{8.4}$$

This equation was first derived by Krieger and Dougherty (1959). The equation shows that $\mu \to \infty$ as $c \to 1/k$. It follows that we can interpret $1/k$ as the volume fraction, c_m, at which the viscosity reaches an infinite value. The magnitude of c_m depends on the packing of the particles in the suspension. For randomly packed spheres $c_m \simeq 0 \cdot 6$, while for hexagonally close-packed spheres $c_m \simeq 0 \cdot 74$. In general, c_m depends on the rate of shear to which the particles are subjected since shearing will affect the distribution of particles in the suspension.

Expanding eq (8.4) for small c results in

$$\frac{\mu}{\mu_0} = 1 + [\mu]c + \frac{[\mu]([\mu] + \mathrm{k})}{2} c^2 + \ldots \tag{8.5}$$

which shows that up to order c the exact result is recovered. The second virial coefficient $k_2 = \frac{1}{2}[\mu]([\mu] + k)$ varies, when $[\mu] = 2 \cdot 5$, from $4 \cdot 8$ to $5 \cdot 2$ for c_m $(= 1/k)$ in the range $0 \cdot 6$–$0 \cdot 74$. This is smaller than the actual value for hard spheres, as can be seen from Table 5.2, except for dispersions subjected to simple shear at an infinite Péclet number. Furthermore, the effective-medium theory does not distinguish between the various types of linear shear flow. The effective-medium theory could be made more accurate by allowing the crowding factor, k, to be a function of the volume fraction, c. As $c \to 0$, a crowding factor given by

$$k_0 = k(c = 0) = \frac{2k_2}{[\mu]} - [\mu] \tag{8.6}$$

would result in the correct limiting behaviour up to order c^2; as $c \to c_m$, k must approach $1/c_m$.

Since at least 100 (semi)empirical relations between μ and c can be found in the literature, we will take the liberty of suggesting one or two new relations, based on the effective-medium theory, but with the modification that k be a function of c. Since no information about this function is available, we simply let k vary linearly from the value k_0 at $c=0$ to $1/c_m$ at $c=c_m$:

$$k = k_0 - \left(\frac{k_0}{c_m} - \frac{1}{c_m^2}\right) c \tag{8.7}$$

Note that k is a weak function of c, for hard spheres varying from about 2 to 1·4 as c goes from 0 to c_m. In principle, all the many-body interactions are incorporated in this weak $k(c)$ dependence.

Inserting eq (8.7) into eq (8.3b) and integrating, results in

$$\mu = \mu_0 \left[\frac{1 - c/c_m}{1 - (k_0 c_m - 1)\, c/c_m}\right]^{-\frac{c_m[\mu]}{2 - k_0 c_m}} \tag{8.8}$$

Note also that, since k_0 is given by eq (8.6), eq (8.8), like eq (8.4), contains only one adjustable parameter (c_m). When $k_0 = 1/c_m$, eq (8.8) reduces to eq (8.4). Equation (8.8) yields the correct asymptotic limit

$$\mu = \mu_0(1 + [\mu]c + k_2 c^2) \tag{8.9}$$

and, when $k_0 c_m < 2$, approaches infinity as $c \to c_m$.

To see how reliable the predictions of eq (8.8) are at high volume fractions, we can take an alternative choice for $k(c)$, e.g.

$$k = \frac{k_0 c_m}{c_m + (k_0 c_m - 1)c} \tag{8.10}$$

Equation (8.10) can only be applied when $k_0 > 0$. Inserting this equation into eq (8.3b) and integrating results in

$$\mu = \mu_0(1 - c/c_m)^{-k_0 c_m^2[\mu]}\, e^{-(k_0 c_m - 1)[\mu]c} \tag{8.11}$$

This equation also reduces to eq (8.4) when $k_0 = 1/c_m$ and to eq (8.9) when $c \ll 1$. It also approaches infinity as $c \to c_m$.

Equations (8.4), (8.8) and (8.11) are compared with each other in Fig. 8.2, which shows $\mu_r = \mu/\mu_0$ versus the volume fraction, c, for $k_0 = 2$ and $c_m = 0.7$. It can be seen that at a volume fraction of 30%, eqs (8.8) and (8.11) differ by less than 1·5%, while they both predict a value which is about 10% higher than eq (8.4). At $c = 50\%$, eq (8.8) and (8.11) differ by about 10% and are

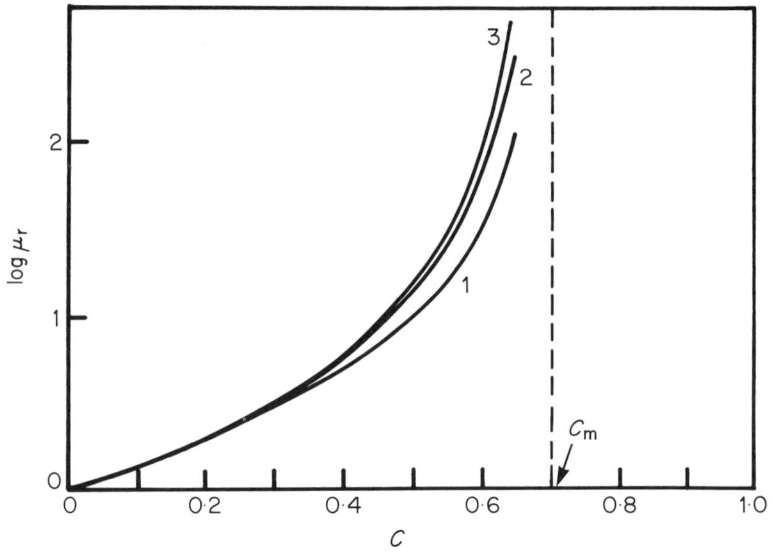

Fig. 8.2. Comparison between various predictions for the viscosity of concentrated dispersions of hard spheres: (1) eq (8.4); (2) eq (8.11); (3) eq (8.8). $k_0=2$, $c_m=0 \cdot 7$.

about 50% higher than predictions of eq (8.4). The advantages of eq (8.8) and (8.11) over others is that they are based on very plausible assumptions regarding concentrated dispersions. In contrast to most empirical relations, they show the correct behaviour up to order c^2.

Similar ideas can be applied to the normal stress differences $\tau_{ij}=P_{ii}-P_{jj}$ (cf. eq (3.287)). For spheres the first-order contribution is of order c^2 and thus when $c \ll 1$

$$\tau_{ij}=K_{ij}c^2 \tag{8.12}$$

where the constant K_{ij} can be found from the second virial stress tensor $\boldsymbol{B}$ (cf. eq (5.54)). Applying the effective-medium theory to this case results in

$$\mathrm{d}\tau_{ij}=\frac{2K_{ij}c\,\mathrm{d}c}{1-kc} \tag{8.13}$$

which upon integration yields

$$\tau_{ij}=-\frac{2K_{ij}}{k}\left[c+\frac{\ln(1-kc)}{k}\right] \tag{8.14}$$

For non-spherical particles the first-order contribution to τ_{ij} is of order c ($\tau_{ij}=M_{ij}c+\ldots$; cf. eq (3.287)) and the same argument yields

$$\tau_{ij}=-\frac{M_{ij}}{k}\ln(1-kc) \tag{8.15}$$

To force the term of order c^2 to agree with exact results (as yet unknown), we again can let k be a function of c, resulting in a modification of the above equation.

As stated above, c_m (or k) will, in general, depend on the rate of shear, and so will k_0 (cf. eq (8.6)). As we saw in Section 5.2.1, up to order c^2 suspensions of hard spheres are shear thinning in simple shear but shear thickening in a pure straining motion (extensional flow). Hence the shear dependence of k_0 depends on the type of flow to which the dispersion is subjected. For simple shear, eq (8.8) predicts shear thinning under all conditions, while for a pure straining motion eq (8.8) predicts shear thickening at low-volume fractions. At higher-volume fractions there are four possibilities, depending on the concentration and the shear dependence of k_0 and c_m: (i) shear thickening at all shear rates; (ii) shear thickening at low shear rates followed by shear thinning at higher rates; (iii) shear thinning at low shear rates followed by shear thickening at high rates; and (iv) shear thinning at all shear rates. Since the shear dependence of k_0 and c_m is unknown it is not known which possibility actually occurs, even for hard spheres.

From a dimensionless analysis it follows that for neutrally buoyant spheres in the absence of colloidal or other forces

$$\mu_r=f(\mathrm{Pe},\gamma,c) \tag{8.16}$$

where Pe is the translational Péclet number (Ga^2/D) and γ characterizes the type of flow. In the case of simple shear ($\gamma=0$), several semi-empirical relations of this type have been proposed. One often used is that given by Krieger (1972), which can be written as

$$\mu_r=\mu_r^{\infty}+\frac{(\mu_r^0-\mu_r^{\infty})}{1+0\cdot123\ \mathrm{Pe}} \tag{8.17}$$

where μ_r^{∞} is the high shear viscosity given by eq (8.4) with $c_m=1/k=0\cdot68$, and μ_r^0 the zero shear viscosity given by the same equation with $c_m=0\cdot57$.

An example of the shear-thinning behaviour of dispersions of hard spheres is given in Fig. 8.3, which shows that viscosities of various systems, when plotted in dimensionless form, all fall on the same curve, given approximately by eq (8.17). Figure 8.3 shows that the viscosity varies

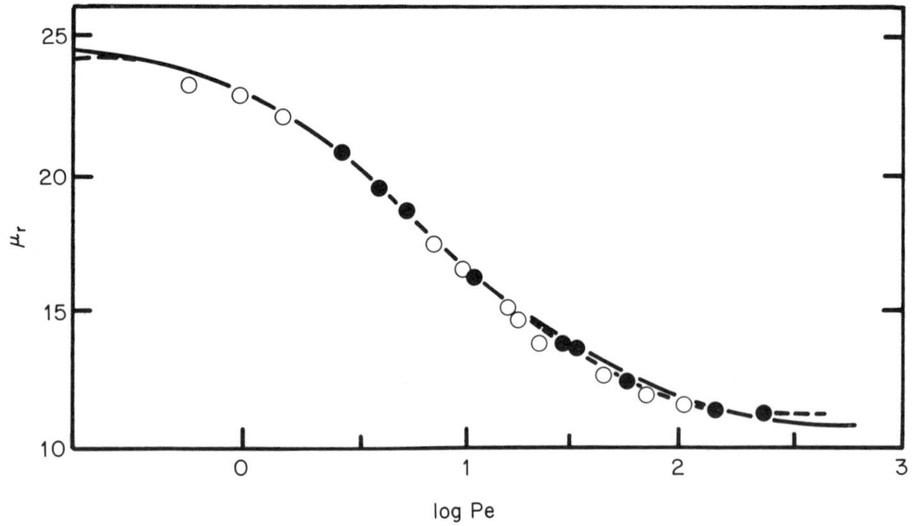

Fig. 8.3. Relative viscosities, μ_r, as a function of the Péclet number, Pe, for monodisperse suspensions of poly(styrene) spheres of various sizes in various media. The solid line is calculated from eq (8.17) with $\mu_r^0 = 24 \cdot 7$ and $\mu_r^\infty = 13 \cdot 9$. Volume fraction $c = 0 \cdot 5$. ○, Spheres in benzyl alcohol; ●, spheres in an aerosol; (----) spheres in water (after Krieger, 1972).

smoothly from its zero-shear value when Pe $\ll 1$ and Brownian motion dominates the distribution of particles (characterized by $c_m \simeq 0 \cdot 57$) to its high-shear value when Pe $\gg 1$ and shear dominates the particle distribution. The inflection point in the curve occurs when $\text{Pe} = 1/0 \cdot 123 \simeq 8$. Alternatively, we can use eq (8.8) to calculate the zero-shear and high-shear viscosities. When Pe $\ll 1$, $[\mu] = 2 \cdot 5$, $k_2 = 6 \cdot 2$ and thus $k_0 = 2 \cdot 46$. When Pe $\gg 1$, $[\mu] = 2 \cdot 5$, $k_2 \simeq 5 \cdot 2$ and $k_0 \simeq 1 \cdot 7$. Table 8.1 contains viscosities calculated from eq (8.8) together with comparisons with experimental data. For the zero-shear viscosity the value of k_0 (which is sensitive to $[\mu]$), has been adjusted to $2 \cdot 4$ to obtain somewhat better agreement with experiment. The conclusion we can draw from Table 8.1 is that c_m can always be chosen as the value for hexagonal close packing, seemingly because the concentration $c = c_m$ is always above the disorder/order phase transition and thus in the limit $c \to c_m$ sheared suspensions are always ordered when steady-state conditions prevail. In the modified effective-medium theory the shear thinning seems to be entirely due to the shear dependence of k_0 (or k_2, cf. eq (8.6)). This suggests that shear thinning is a result of the energy dissipation associated with the particles, hydrodynamic interactions, which at low shear rates is larger than at high ones.

TABLE 8.1
Comparison between experimental and theoretical values for the viscosities of suspensions of hard spheres

Volume fraction, c	μ_r^0 (exp.)[a]	μ_r^0 (theory)[b]	μ_r^∞ (exp.)[a]	μ_r^∞ (theory)[b]
0·50	24·7	23·7	10·8	10·3
0·45	10·7	11·7	7·2	6·8
0·40	6·3	6·9	5·0	4·8
0·35	4·3	4·5	3·7	3·6

[a] Data from Krieger (1972).
[b] From eq (8.8) with $k_0 = 2{\cdot}4$ and $c_m = 0{\cdot}74$.
[c] From eq (8.8) with $k_0 = 1{\cdot}7$ and $c_m = 0{\cdot}74$.

When eq (8.4) is used to predict μ_r^0 and μ_r^∞, rather low values of c_m must be used (0·57 and 0·68, respectively), because it underestimates the viscosity (cf. Fig. 8.2)). It is of interest to note that if c_m is indeed constant and independent of shear, then eq (8.8) predicts shear thickening for dispersions subjected to extensional flow, no matter what their volume fraction.

For dispersions of electrically charged spheres, a dimensional analysis shows that, besides the parameters appearing in eq (8.16), μ_r also depends on the parameters Dl, τ, Ad, λ^*, Φ_s and b_i (cf. Sections 6.3.3.1 and 2.2.1). In this case no general expressions for μ_r are available. It is observed experimentally that at low electrolyte concentrations (small τ), the zero-shear viscosity can attain very high values, often several orders of magnitude higher than the viscosity of suspensions of hard spheres at the same volume fractions (Goodwin, 1982). This behaviour can be explained qualitatively by eq (8.8). For dispersions of charged particles, the secondary electroviscous effect (cf. Section 5.2.2) can be very large at low salt concentrations, resulting in large values of k_0. For example, when $k_0 = 10$, eq (8.8) predicts that μ approaches infinity when $c \simeq 12\%$. Since k_0 is expected to decrease with increasing shear rate (it depends on the ratio of hydrodynamic to colloidal forces), eq (8.8) predicts that, when k_0 is large at low shear rates, a critical shear rate exists beyond which the dispersion will start to flow, i.e. such dispersions are expected to have a yield stress. In reality, dispersions of charged spheres are found to have an apparent yield stress with a well-defined (but very large) zero-shear viscosity. Nevertheless, eq (8.8) predicts the general steep increase in viscosity with increasing k_0 (corresponding to decreasing τ). Note that this prediction depends strongly on the form of $k(c)$ because eq (8.11) does not predict such a strong increase. To predict the value of the apparent yield stress, we must know both the relations $k(c)$ and $k_0(G)$;

$k_0(G)$ could, in principle, be calculated from the theory outlined in Section 5.2.2 by removing the restriction Pe $\ll 1$, but no theory exists to calculate $k(c)$.

At high shear rates, concentrated dispersions of charged spheres show shear thickening, after having passed through a shear-thinning regime (von Laun, 1988). The reason for this shear thickening is not well understood. One possible explanation is that the shear thinning at intermediate shear rates is more pronounced for soft than for hard spheres and, since the viscosity becomes more like that of a dispersion of hard spheres as $G \to \infty$, shear thickening occurs as a result. Another possibility is that for soft spheres $k_2(G)$ passes through a minimum at intermediate Péclet numbers.

Suspensions of spheres coated with polymer behave, at low-volume fractions, similar to hard spheres, but at high volume fractions ($c > \sim 0{\cdot}4$), the zero-shear viscosity is observed to be substantially larger than for hard spheres. The high shear viscosity follows closely the hard-sphere behaviour (Choi and Krieger, 1986). A possible reason is that at low shear rates and high-volume fractions the particles are prevented from rotating in the shear flow due to the interpenetration of polymer segments of neighbouring particles. At high shear the hydrodynamic forces are large enough for particles to rotate. It is known that when particle rotation is inhibited, the viscosity increases (cf. Section 3.3.2)

8.3 Cell models

Cell models are based on the concept that the properties of concentrated dispersions can be derived from the flow field around a reference particle. Dispersions with a periodic structure, such as ordered arrays of spheres or cylinders, have well-defined unit cells in which, because of symmetry, the flow is the same. An example of such an array of cylinders is shown in Fig. 8.4, from which it can be seen that knowledge of the flow field in the two-dimensional unit cell BCDEF is sufficient to determine the complete flow field in the array. For an array of spheres the unit cell is, of course, three-dimensional.

For a dispersion of spheres in which there is a random distribution of particles, we can construct a spherical unit cell of radius R around a reference particle of radius a, where the size of R is determined by the condition that the volume fraction in the unit cell equals the volume fraction in the dispersion, i.e. $(a/R)^3 = c$. Similar two-dimensional unit cells can be constructed for suspensions of aligned cylinders. The flow in such a unit cell can be found from the solution of the creeping-flow and continuity equations, together with the appropriate boundary conditions. For a dispersion that

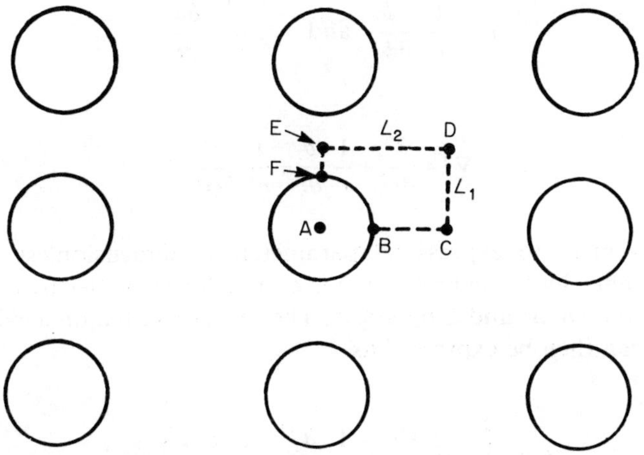

Fig. 8.4. Schematic representation of a portion of an infinite array of cylinders (extending horizontally and vertically to infinity). Because of symmetry, only the flow in the region BCEF need be considered.

exhibits local partial order, we can construct several shells around a reference particle instead of just one.

Many examples of cell models exist in the literature, starting with Happel (1959), who studied the flow through assemblages of spheres and cylinders as well as the viscosity of concentrated dispersions. Here, as an illustrative example, we will discuss an example of our own work dealing with the rotational friction coefficients of cylinders in a concentrated suspension of parallel cylinders (Masliyah and van de Ven, 1987).

To obtain the rotary-friction coefficient we can apply a known torque, L, to a reference cylinder and calculate the resulting angular velocity, ω, with which the cylinder rotates. The rotary (or spin) friction coefficient, for motion about the axis of revolution, is found from

$$L = f_s \omega \tag{8.18}$$

In order to calculate the torque on a reference cylinder, we need only consider the flow in region BCDEF of Fig. 8.4. The flow field in this unit cell is governed by the creeping-flow equation which, in terms of the stream function, ψ, can be written as

$$\nabla^4 \psi = 0 \tag{8.19}$$

where ψ is defined in cylindrical coordinates by

$$v_r = \frac{1}{r}\frac{\partial \psi}{\partial \phi} \quad \text{and} \quad v_\phi = -\frac{\partial \psi}{\partial r} \tag{8.20}$$

and where

$$\nabla^2 = \frac{\partial^2}{\partial r^2} + \frac{1}{r}\frac{\partial}{\partial r} + \frac{1}{r^2}\frac{\partial^2}{\partial \phi^2} \tag{8.21}$$

It is convenient to express all parameters in dimensionless units, by dividing distance by the cylinder radius a, angular velocities by a reference velocity ω^*, ψ by $a^2\omega^*$ and L by $\mu a_2 \omega^*$. The torque exerted on a cylinder per unit length can then be expressed as

$$L = \int_0^{2\pi} r^2 \left(\frac{\partial^2 \psi}{\partial r^2} - \frac{1}{r}\frac{\partial \psi}{\partial r} - \frac{1}{r^2}\frac{\partial^2 \psi}{\partial \phi^2} \right) \mathrm{d}\phi \tag{8.22}$$

which is derived from the general relation $\boldsymbol{L} = \int_A \boldsymbol{\Pi} \times \boldsymbol{r} \cdot \mathrm{d}\boldsymbol{A}$.

The problem now consists of determining a function $\psi(r, \phi)$, that satisfies eq (8.19) and produces a given torque, given by eq (8.22), subject to the following no-slip and symmetry boundary conditions:

on BF:

$$\psi = 0 \qquad \frac{\partial \psi}{\partial r} = -\omega$$

on BC and DE:

$$\frac{\partial \psi}{\partial y} = 0 \qquad \frac{\partial^3 \psi}{\partial y^3} = 0$$

on CD and EF:

$$\frac{\partial \psi}{\partial x} = 0 \qquad \frac{\partial^3 \psi}{\partial x^3} = 0$$

The general solution that satisfies the symmetry boundary conditions along BD and EF is:

$$\psi = g_0 + \sum_{n=1}^{\infty} g_n \cos 2n\phi \tag{8.23}$$

where

$$g_0 = Ar^2 \ln r + Br^2 + C \ln r + D \tag{8.24}$$

and

$$g_n = a_n r^{2n+2} + b_n r^{2n} + c_n r^{-2n+2} + d_n r^{-2n} \tag{8.25}$$

From the boundary conditions on BF and eq (8.22), it follows that

$$A = 0 \ ; \ B = L/8\pi - \omega/2 \ ; \ C = -L/4\pi \ ; \ D = -L/8\pi + \omega/2$$

$$c_n = -(2n+1)\, a_n - 2nb_n \ ; \ d_n = 2na_n - (2n-1)\, b_n$$

Hence, the general solution is

$$\psi = (r^2 - 1)(L/8\pi - \omega/2) - (L/4\pi)\ln r + \sum_{n=1}^{\infty} [a_n \{r^{2n+2} - (2n+1) r^{-2n+2} + 2nr^{-2n}\} + b_n \{r^{2n} - 2nr^{-2n+2} + (2n-1)\, r^{-2n}\}] \cos 2n\phi \tag{8.26}$$

In this equation, ω and the coefficients a_n and b_n are, as yet, unknown. They can be evaluated from the boundary conditions along DE and CD. This can be accomplished by choosing M points on DE and CD at which the boundary conditions must be satisfied. When the series in eq (8.26) is truncated at $n = N$, this leads to $2M$ equations with $2N+1$ unknowns. By choosing $2M > 2N+1$, these equations can be solved numerically using standard least-squares techniques for ω and the coefficients a_n and b_n. After having found ω, the rotational-friction coefficient, f_s, can be found from eq (8.18). Up to $c \simeq 20\%$, the results can be represented fairly accurately by

$$f_s = 4\pi\mu a^2 \ell (1 + k_1 c + k_2 c^2) \tag{8.27}$$

where a and ℓ are the radius and the length of a cylinder. The coefficients k_1 and k_2 depend on the way the cylinders are ordered, i.e. they depend on the ratio L_1/L_2, L_1 being the distance DC and L_2 the distance AC. The first virial coefficient, k_1, is very close to 1 over a wide range of L_1/L_2, while k_2 shows a minimum of about $1 \cdot 1$ at $L_1/L_2 = 1 \cdot 0$. Both k_1 and k_2 approach infinity as $L_1/L_2 \to 0$ or ∞. In principle, calculations can be performed for all volume fractions up to close packing.

This model can be extended to calculate the flow of a one-dimensional array of cylinders sandwiched between two plates (Masliyah and van de Ven, 1986), which is a model for a single sheared layer of cylinders in a suspension of ordered cylinders (aligned along the vorticity axis), which in turn resembles a single layer of spheres in a sheared ordered suspension of spheres.

The above example shows the general feature of cell models, namely the solution of the hydrodynamic equations in a unit cell, subject to the appropriate boundary and symmetry conditions. In principle it is possible to extend these calculations to charged particles by solving the coupled set of equations

$$\begin{aligned}
&\mu\nabla^2 v - \nabla p = \rho\nabla\psi \;\; ; \;\; \rho = \sum_i n_i z_i e \\
&\nabla\cdot v = 0 \\
&\nabla^2\psi = -\rho/\epsilon \\
&\nabla\cdot j_i = 0 \;\; ; \;\; j_i = -D_i\nabla n_i - \frac{z_i n_i e}{f_i}\nabla\psi + v n_i
\end{aligned} \tag{8.28}$$

together with the appropriate boundary conditions (here ψ is the electrical potential, not the stream function; for the definitions of other symbols see Section 2.2.2). However, no exact solutions for charged particles are yet available.

As an example of a cell model of a disordered dispersion, we take a dispersion of cylinders which are aligned, but whose axes are distributed randomly. As before, we are interested in the rotational-friction coefficient,

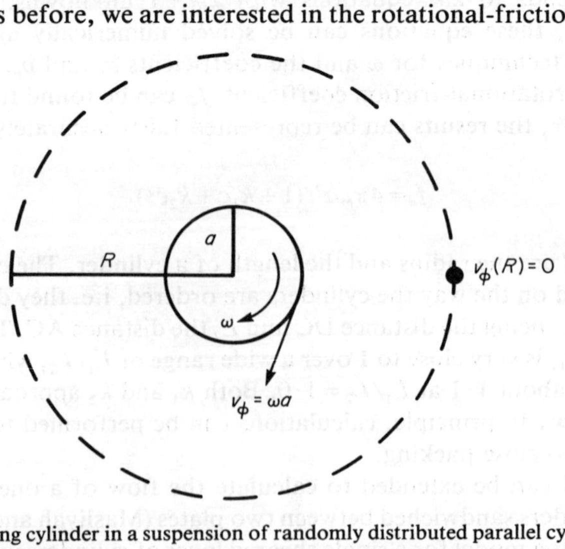

Fig. 8.5. Rotating cylinder in a suspension of randomly distributed parallel cylinders. The flow is calculated in a cell of radius R around a reference cylinder; R is determined by the volume fraction of particles.

f_s. We now consider a cylindrical cell around a reference cylinder as shown schematically in Fig. 8.5. The radius, R, is chosen such that $(a/R)^2 = c$. Suppose the cylinder is rotating with an angular velocity ω. The flow created by this rotating cylinder is described by the creeping-flow equation $\mu\nabla^2 v = \nabla p$, which for the present geometry ($v_r = 0$ and constant pressure) reduces to

$$\frac{\mathrm{d}}{\mathrm{d}r}\left[\frac{1}{r}\frac{\mathrm{d}}{\mathrm{d}r}(rv_\phi)\right] = 0 \tag{8.29}$$

This equation has the solution

$$v_\phi = Ar + \frac{B}{r} \tag{8.30}$$

where A and B are integration constants which can be found from the conditions $v_\phi(a) = \omega a$ and $v_\phi(R) = 0$. The boundary condition at $r = a$ is the no-slip boundary condition, while the condition at $r = R$ reflects the assumption that, on average, the disturbance flow around the cylinder does not extend beyond the mean distance between cylinders. Depending on the problem under consideration, several choices of the boundary condition(s) at $r = R$ can be made, resulting in different predictions. Since the cell model is an approximation of the real situation, no general rules for choosing the boundary conditions at the cell's surface exist and we must be guided by physical intuition. In the problem at hand, calculating the torque exerted on the cylinder from eq (8.22) and comparing the result to eq (8.18) yields, for the rotary-friction coefficient,

$$f_s = \frac{4\pi\mu a^2 \ell}{1 - c} \tag{8.31}$$

which for small-volume fractions is very similar to eq (8.27). It can be seen from eq (8.31) that the cell model does not apply to highly concentrated dispersions, because it would be expected that $f_s \to \infty$ as $c \to c_m$ (the concentration at close packing). In many problems improvements can be made by choosing alternative boundary conditions at $r = R$.

The examples given here are relatively simple, since the flow field is two-dimensional, but the same principles apply to three-dimensional cells. The examples show that such models can be applied both to ordered and disordered dispersions. As stated previously, multiple-shell cell models can be developed for dispersions exhibiting local order. Examples can be found in a review by Castillo *et al.* (1984).

As a last example we will consider the dynamic rigidity of concentrated dispersions. Besides exhibiting non-zero normal stress differences (cf. Section 8.2), concentrated dispersions also show viscoelastic behaviour, described by the complex viscosity μ^*, defined as

$$\mu^* = \mu' - i\mu'' = \mu' - iG'/\omega \tag{8.32}$$

where μ' is the dynamic viscosity, G' is the dynamic rigidity (cf. Section 3.4), and ω is the angular frequency of the oscillating strain, γ, to which the dispersion is subjected:

$$\gamma = \epsilon \sin\omega t \tag{8.33}$$

ϵ being the amplitude of the strain. The rate of shear is

$$\mathrm{G} = \dot{\gamma} = \epsilon\omega\cos\omega t \tag{8.34}$$

and the shear stress (Walters, 1975)

$$\tau = \epsilon\omega\left(\mu'\cos\omega t + \frac{G'}{\omega}\sin\omega t\right) = \tau' + \tau'' \tag{8.35}$$

The energy stored and recovered per unit time per unit volume equals

$$E = \tau''\dot{\gamma} \tag{8.36}$$

As a model of a concentrated dispersion, we imagine the dispersion to be made up of unit cells in which the central reference sphere has m nearest neighbours. We choose the reference sphere (at the centre of the unit cell) to be located at the origin of a Cartesian-coordinate system X_i and denote the position of one of its nearest neighbours by (R, θ, ϕ), R being the centre-to-centre distance between neighbouring particles. The second sphere is subjected to an oscillating shear given by

$$v_1 = v_2 = 0\,;\; v_3 = Gx_2 = \epsilon x_2\omega\cos\omega t \tag{8.37}$$

Neglecting hydrodynamic interactions, the second sphere will, during a time interval Δt, be displaced by an amount Δx_3, given by

$$\Delta x_3 = Gx_2\Delta t = G\Delta t R\sin\theta\cos\phi \tag{8.38}$$

As a time interval we choose a quarter oscillation:

$$\Delta t = \frac{T}{4} = \frac{\pi}{2\omega} \tag{8.39}$$

During this time interval the average shear rate equals (cf. eq (8.34))

$$\bar{G} = \bar{\dot{\gamma}} = 2\epsilon\omega/\pi \tag{8.40}$$

The displacement in the X_3 direction corresponds to an increase in the distance between the two particles of magnitude (cf. eqs (5.1a) and (5.3a))

$$\Delta R = \epsilon R \sin^2\theta \sin\phi \cos\phi = \Delta x_3 \sin\theta \sin\phi \tag{8.41}$$

The restoring force acting along the line of sphere centres is

$$F_R = \Delta R \frac{\partial^2 V}{\partial R^2} = \Delta x_3 \frac{\partial^2 V}{\partial R^2} \sin\theta \sin\phi \tag{8.42}$$

and along the X_3 direction:

$$F_3 = \Delta x_3 \frac{\partial^2 V}{\partial R^2} \sin^2\theta \sin^2\phi \tag{8.43}$$

where V is the energy of interaction between two spheres. The above equation is only valid in the limit $\omega \to \infty$, in which case particles are displaced by a very small distance from their equilibrium positions.

The energy stored per unit cell during a time interval Δt is

$$E_s = m \int_{X_3}^{X_3 + \Delta X_3} F_3 \mathrm{d}x_3 = \frac{1}{2} m \, \Delta R^2 \frac{\partial^2 V}{\partial R^2} \tag{8.44}$$

(m being the number of nearest neighbours) and hence the energy stored and recovered per unit volume per unit time is

$$E = \frac{\omega m n}{k\pi} \Delta R^2 \frac{\partial^2 V}{\partial R^2} = \frac{\omega m n}{4\,\pi k} \epsilon^2 R^2 \sin^4\theta \sin^2 2\phi \frac{\partial^2 V}{\partial R^2} \tag{8.45}$$

n being the number concentration of particles and k the number of particles per unit cell.

Comparing this to eq (8.36) and noting that, on average, $\tau'' = 2\epsilon G'/\pi$ and $\dot{\gamma} = 2\omega\epsilon/\pi$, yields for the dynamic rigidity:

$$G'_{\infty}=\frac{\pi mn}{16k}\,R^{2}\sin^{4}\theta\sin^{2}2\phi\,\frac{\partial^{2}V}{\partial R^{2}} \tag{8.46}$$

where the subscript "∞" indicates that the result is expected to apply to very high frequencies.

Because of the relations

$$\left(\frac{R}{2a}\right)^{3}\simeq\frac{c_{\mathrm{m}}}{c} \tag{8.47}$$

and $c=\dfrac{4}{3}\pi a^{3}n$, eq (8.46) can be written as

$$G'_{\infty}=\frac{\alpha}{R}\,\frac{\partial^{2}V}{\partial R^{2}} \tag{8.48}$$

where

$$\alpha\simeq\frac{3m}{8k}\,c_{m}\sin^{4}\theta\sin^{2}2\phi \tag{8.49a}$$

Since we can expect a local (fluid-like) order in concentrated dispersions, it may be assumed that the orientation of the unit cell is random, which results in an average value of α given by

$$\alpha=\frac{3m}{8k}\,c_{m}\langle\sin^{4}\theta\sin^{2}2\phi\rangle=\frac{mc_{m}}{20k} \tag{8.49b}$$

Hence for a body-centred cubic (bcc) structure ($m=8$, $k=2$, $c_{\mathrm{m}}=0\cdot 68$), $\alpha=0\cdot 14$ and for a face-centred cubic (fcc) or hexagonal structure ($m=12$, $k=4$, $c_{m}=0\cdot 74$) $\alpha=0\cdot 11$.

A comparison with experimental data is given in Fig. 8.6, which shows data on 69-nm poly(styrene) latex particles suspended in aqueous media of various salt concentrations. The solid lines are calculated from eqs (8.48) and (1.131) using $\alpha=0\cdot 833$ (this value was taken by Buscall *et al.* (1982) to be the value for a fcc structure; however, correcting their averaging procedure yields a value of $0\cdot 283$). The value of $0\cdot 833$ is higher than the value $\alpha=0\cdot 11$ given here (or the corrected value of $0\cdot 283$). The discrepancy is probably due to the stretching of bonds between next-nearest neighbours, which are neglected in the derivation and shortcomings of eq (1.131) when applied to a concentrated dispersion. When $\alpha=0\cdot 833$, the slopes of the lines correspond to a surface potential ψ_0 of 50 mV. Using $\alpha=0\cdot 11$ results in a value of ψ_0 which is somewhat more than twice as large. Keeping in mind the

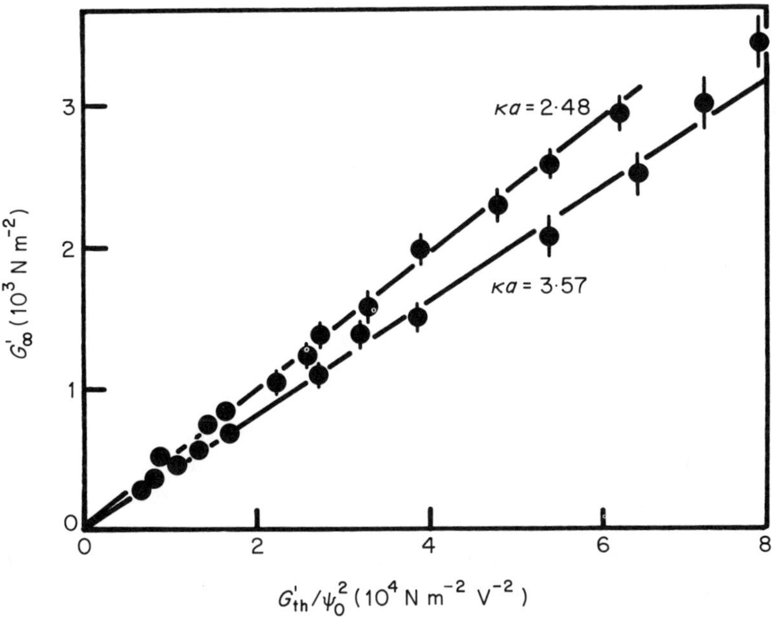

Fig. 8.6. Experimentally observed values of G'_∞ plotted against the theoretical values, G'_{th}, divided by the square of the surface potential, ψ_0, for various values of κa (after Buscall *et al.*, 1982).

limitations of the model, these are reasonable values, which shows that eq (8.48) predicts the correct order of magnitude for G'_∞.

8.4 Statistical mechanics of concentrated dispersions

In the introduction of Chapter 7 we touched upon the statistical-mechanical description of concentrated dispersions. We showed that the pair-distribution function is governed by the convective-diffusion equation

$$\frac{\partial P_2}{\partial t} = \nabla \cdot (\bar{\boldsymbol{D}} \cdot \nabla P_2) - \nabla \cdot (\bar{\boldsymbol{u}} P_2) - \nabla \cdot (\bar{\boldsymbol{D}} \cdot \bar{\boldsymbol{F}}_{int} P_2 / kT) \tag{8.50}$$

in which $\bar{\boldsymbol{D}}$, $\bar{\boldsymbol{u}}$ and $\bar{\boldsymbol{F}}_{int}$ are average quantities which take into account multibody interactions. In the absence of flow ($\bar{\boldsymbol{u}} = 0$), the steady-state solution of this equation is

$$P_2 = P_2^0 e^{-\bar{V}_{int}/kT} \tag{8.51}$$

where $\bar{V}_{int}$ is the potential of mean force. $\bar{V}_{int}$ is usually expressed in terms of the radial distribution function $g(r)$, which is related to P_2 by

$$P_2(r) = n^2 g(r) \tag{8.52}$$

n being the number concentration of particles in the bulk. From eqs (8.51) and (8.52) it follows that the potential of mean force can be defined as

$$\bar{V}_{int} = -kT \ln g_0(r) \tag{8.53}$$

where $g_0(r)$ denotes the radial distribution function in the absence of flow. The radial distribution function $g(r)$ is the all-important ingredient in the statistical description of concentrated dispersions; from it almost all dispersion properties can be derived. As its name suggests, the radial distribution function indicates how, on average, particles are distributed around an arbitrarily chosen reference sphere. In general, $g(r) = 0$ when $r \leqslant 2a$, a being the sphere radius, and $g(r) \to 1$ as $r \to \infty$, reflecting the fact that, on average, far from the reference sphere the concentration of particles is uniform. For concentrated systems $g(r)$ exhibits damped oscillations, the maxima of which occur near multiples of the nearest-neighbour distance, as shown schematically in Fig. 8.7. At low concentrations, $g_0(r)$ is simply a monotonically increasing function and $\bar{V}_{int}$ is equal to V_{int}, the interaction energy between two spheres (cf. Section 1.4.4). We are usually interested in stable dispersions in which the dominant contribution to V_{int} is the electrostatic repulsion, often approximated as

$$\begin{aligned} V_{int} &= Cr^{-1} e^{-\kappa r} \qquad r > 2a \\ V_{int} &= \infty \qquad r \leq 2a \end{aligned} \tag{8.54}$$

with $C = 4\pi\epsilon a^2 \psi_0 e^{2\kappa a}$. In the statistical-mechanical literature, such potentials are referred to as Yukawa potentials.

For concentrated dispersions, the potential of the mean force is often written as

$$\bar{V}_{int} = V_{int} + \Phi_{int} \tag{8.55}$$

where all the many-body interactions are incorporated in Φ_{int}. The magnitude of Φ_{int}, relative to V_{int}, is an indication of how concentrated a dispersion is.

A large body of literature exists on the ways in which $g_0(r)$ can be predicted

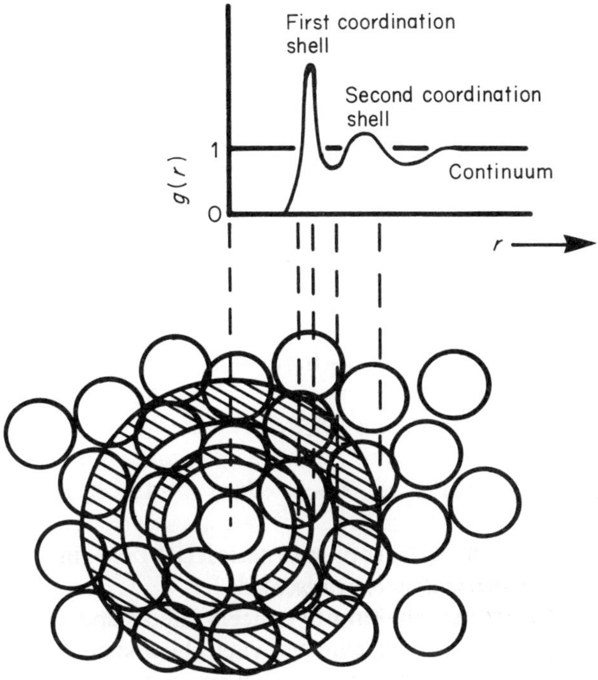

Fig. 8.7. Schematic representation of the radial distribution function $g(r)$.

or measured. It is outside the scope of this book to treat this topic in detail. Here we will outline some of the most important features. The most accurate methods for determining $g_0(r)$ stem from integral equation theories, perturbation theories, and Monte Carlo calculations. The Monte Carlo calculations are probably the most accurate but require vast amounts of computer time. The perturbation theories are based on evaluating Φ_{int} in eq (8.55). The best know perturbation schemes are referred to as BH (Barker and Henderson, 1967a,b) and WCA (Weeks, Chandler and Anderson, 1971). In the integral-equation theories, the total correlation function

$$h(r) = g(r) - 1 \tag{8.56}$$

is split into a direct correlation function, $c(r)$, and an indirect part which reflects the influence of the reference particle 1 on particle 3, which in turn affects particle 2. Averaging over all possible positions of particle 3 results in the Ornstein–Zernike equation:

$$h(r_{12}) = c(r_{12}) + n \int c(r_{13})\, h(r_{23})\, \mathrm{d}\mathbf{r}_3 \tag{8.57}$$

where r_{ij} is the distance between sphere i and sphere j. To evaluate $h(r)$, we need a relation between $c(r)$ and $h(r)$. Several such closing relations have been proposed (cf. Castillo *et al.*, 1984), e.g. the Percus–Yevick (PY) condition,

$$c(r) = g(r)[1 - \mathrm{e}^{-V_{\mathrm{int}}/kT}] \tag{8.58}$$

the hypernetted chain (HNC) condition,

$$c(r) = h(r) - \ln g(r) - V_{\mathrm{int}}/kT \tag{8.59}$$

the mean spherical aproximation (MSA)

$$c(r) = -V_{\mathrm{int}}/kT,\ r > 2a\ ;\ h(r) = -1,\ r \leqslant 2a \tag{8.60}$$

and the rescaled mean spherical approximation (RMSA), in which the actual diameter of the reference sphere is rescaled.

Since we are interested in the flow of concentrated dispersions, our interest is more in knowing how $g(r)$ is changed when the dispersion is subjected to flow, than to determine $g_0(r)$. In what follows we assume that $g_0(r)$ is known, either experimentally (cf. Section 8.5.2 below) or theoretically (from one of the theoretical approaches outlined above).

Writing $\bar{\boldsymbol{D}}$ as $\boldsymbol{d}D$ (D being the relative-diffusion coefficient, i.e. $D = 2D_0$), $\bar{\boldsymbol{u}}$ as $G\boldsymbol{u}^*$, and $\bar{\boldsymbol{F}}_{\mathrm{int}}/kT$ as $\boldsymbol{F}^*$ and non-dimensionalizing distances by the sphere radius a, eq (8.50) can be written, in steady-state conditions, as

$$\nabla\cdot(\boldsymbol{d}\cdot\nabla g) - \nabla\cdot(\boldsymbol{d}\cdot\boldsymbol{F}^* g) = \mathrm{Pe}\,\nabla\cdot(\boldsymbol{u}^* g) \tag{8.61}$$

When the Péclet number $\mathrm{Pe} = Ga^2/D$ is small, we can make an expansion in Pe:

$$\boldsymbol{d} = \boldsymbol{d}_0 + \mathrm{Pe}\,\boldsymbol{d}_1 \tag{8.62}$$

$$\boldsymbol{u}^* = \boldsymbol{u}_0^* + \mathrm{Pe}\,\boldsymbol{u}_1^* \tag{8.63}$$

$$\boldsymbol{F}^* = \boldsymbol{F}_0^* + \mathrm{Pe}\,\boldsymbol{F}_1^* \tag{8.64}$$

and

$$g = g_0 + \mathrm{Pe}\,g_1 \tag{8.65}$$

Equation (8.65) represents the radial distribution function when the dispersion is subjected to a low-intensity flow. Since g_0 is supposed to be known, we need to evaluate the function g_1. Substituting eqs (8.62) to (8.65) into eq (8.61) and collecting zero-order terms results in

$$\nabla\cdot(\boldsymbol{d}_0\cdot g_0) - \nabla\cdot(\boldsymbol{d}_0\cdot \boldsymbol{F}_0^* g_0) = 0 \tag{8.66}$$

the solution of which is eq (8.53). Collecting terms of order Pe and noting also that

$$\nabla\cdot(\boldsymbol{d}_1\cdot \nabla g_0) - \nabla\cdot(\boldsymbol{d}_1\cdot \boldsymbol{F}_0^* g_0) = 0$$

results in

$$\nabla\cdot(\boldsymbol{d}_0\cdot \nabla g_1) - \nabla\cdot(\boldsymbol{d}_0\cdot \boldsymbol{F}_0^* g_1 + \boldsymbol{d}_0\cdot \boldsymbol{F}_1^* g_0) = \nabla\cdot(\boldsymbol{u}_0^* g_0) \tag{8.67}$$

This is the fundamental equation governing g_1. Notice that up to order Pe the functions $\boldsymbol{d}_1$ and $\boldsymbol{u}_1^*$ do not enter the problem. The function $\boldsymbol{F}_1^*$ can be expressed in terms of the functions $\boldsymbol{F}_0^*$, g_0 and g_1 if we restrict ourselves to three-body interactions. Using the definition of $\bar{V}_{\text{int}}$ (cf. eq (7.10)), but truncating it at $N=3$, expressing the conditional probability, P, in terms of regular probabilities:

$$P(\boldsymbol{r}_3/\boldsymbol{r}_1, \boldsymbol{r}_2)\, \mathrm{d}\boldsymbol{r}_3 = \frac{P_3(\boldsymbol{r}_1, \boldsymbol{r}_2, \boldsymbol{r}_3)\mathrm{d}\boldsymbol{r}_1\mathrm{d}\boldsymbol{r}_2\mathrm{d}\boldsymbol{r}_3}{P_2(\boldsymbol{r}_1, \boldsymbol{r}_2)\mathrm{d}\boldsymbol{r}_1\mathrm{d}\boldsymbol{r}_2} \tag{8.68}$$

and making use of the superposition approximation (eq (7.71)) and the definition of g (eq (8.52)), leads to a relation between $\boldsymbol{F}_1^*$, g_0, g_1 and $\boldsymbol{F}_0^*$. Substituting this relation in eq (8.67) results in a rather complicated integro-differential equation for g_1, which, in principle, can be solved numerically.

One of the problems associated with eq (8.67) is that usually only g_0 and $\boldsymbol{F}_0^*$ are (approximately) known, while $\boldsymbol{d}_0$ and $\boldsymbol{u}_0^*$ are unknown, even for dilute systems, because the electroviscous contributions to $\boldsymbol{d}_0$ and $\boldsymbol{u}_0^*$ (cf. Sections 4.3 and 5.1.2.2) have not, as yet, been worked out.

As an example let us apply the statistical-mechanical theory to a dispersion subjected to an oscillating simple shear characterized by a shear rate $G\mathrm{e}^{i\omega t}$. The particle velocity is then given by

$$\boldsymbol{u}_0^* = (\mathrm{d}r/\mathrm{d}t,\, r\mathrm{d}\theta/\mathrm{d}t,\, r\sin\theta\mathrm{d}\phi/\mathrm{d}t)\, \mathrm{e}^{i\omega t} \tag{8.69}$$

with

$$\begin{aligned}
\frac{\mathrm{d}r}{\mathrm{d}t} &= \overline{A(r)}\, \sin^2\theta\sin2\phi \\
\frac{\mathrm{d}\theta}{\mathrm{d}t} &= \frac{1}{4}\, \overline{B(r)}\, \sin2\theta\sin2\phi \\
\frac{\mathrm{d}\phi}{\mathrm{d}t} &= \frac{1}{2}\, (1 + \overline{B(r)}\, \cos2\phi)
\end{aligned} \tag{8.70}$$

where the functions $\bar{A}$ and $\bar{B}$ are defined by equations analogous to eq (7.10). Denoting the radial and tangential components of $\boldsymbol{d}_0$ by $d_\parallel^0(r)$ and $d_\perp^0(r)$, respectively, and making the substitutions

$$g_1 = f(r)\,\sin^2\theta \sin 2\phi\; e^{i\omega t} \tag{8.71}$$

$$F_1^* = -\frac{1}{kT}\frac{\partial V_1}{\partial r}\sin^2\theta \sin 2\phi\; e^{i\omega t} \tag{8.72}$$

and reintroducing the time-dependent term $\partial g/\partial t$ because of the fluctuating nature of the flow, eq (8.67) results in

$$\frac{d}{dr}\left[r^2 d_\parallel^0\left(\frac{df}{dr}-\frac{f}{g_0}\frac{dg_0}{dr}\right)\right]-6d_\perp^0 f+\frac{d}{dr}\left(r^2 d_\parallel^0\frac{g_0}{kT}\frac{\partial V_1}{\partial r}\right)-K(r)=\frac{i\omega a^2 f r^2}{D} \tag{8.73}$$

where

$$K(r)=\frac{d}{dr}(\bar{A}g_0 r^2)-\frac{3}{2}\bar{B}g_0 r^2 \tag{8.74}$$

Equation (8.73) is subjected to the no-penetration boundary condition at $r=2$ and the condition $f\to 0$ as $r\to\infty$.

The steady-shear behaviour can be obtained by setting $\omega=0$. For an oscillating flow the above equations apply to all frequencies, ω, provided the shear rate, G, is small enough for the condition Pe $\ll 1$ to hold.

As mentioned, V_1 can be expressed in terms of g_0 and f, in which case eq (8.73) becomes an integro-differential equation. Alternatively, if a procedure is used to calculate $\bar{V}_{\text{int}}$ from $g(r,\theta,\phi)$, e.g. by using the assumption of pair additivity of colloidal forces, then eq (8.73) can be solved by iteration (as can the integro-differential equation). Solving eq (8.73) with $V_1=0$ yields a first approximation to f. From the relation $g=g_0+\text{Pe}\,g_1$, it is then possible to obtain an approximate value of V_1. With this approximation for V_1, eq (8.73) can be solved again, yielding an improved estimate of the function f. This procedure can be repeated until futher iterations no longer change f.

Ohtsuki (1981) calculated the complex viscosity, μ^*, of concentrated dispersions of spheres using the approach outlined here. He (implicitly) assumed that $d_\parallel^0=d_\perp^0=1$, while for the particle velocity, $\boldsymbol{u}_0$, he chose the undisturbed fluid flow field, corresponding to $\overline{A(r)}=\frac{1}{2}r$ and $\overline{B(r)}=1$. With this choice all hydrodynamic interactions are negelected, but the many-body colloidal interactions are properly taken into account. Hence it is expected that these approximations hold well for systems that are ordered at very low-volume fractions (highly charged particles at low salt concentrations). Ohtsuki replaced the term containing the perturbation potential, V_1, by an integral of the type

$$\int_2^\infty F(r,s)\, f(s)\mathrm{d}s$$

where F is a complicated kernel depending on the function g_0.

Russel and Gast (1986), after arriving at an equation equivalent to eq (8.67), made the assumptions that $d_{\parallel}^0$, $d_{\perp}^0$, $\bar{A}$ and $\bar{B}$ are all determined by the hydrodynamic interactions between two spheres (cf. Sections 4.12 and

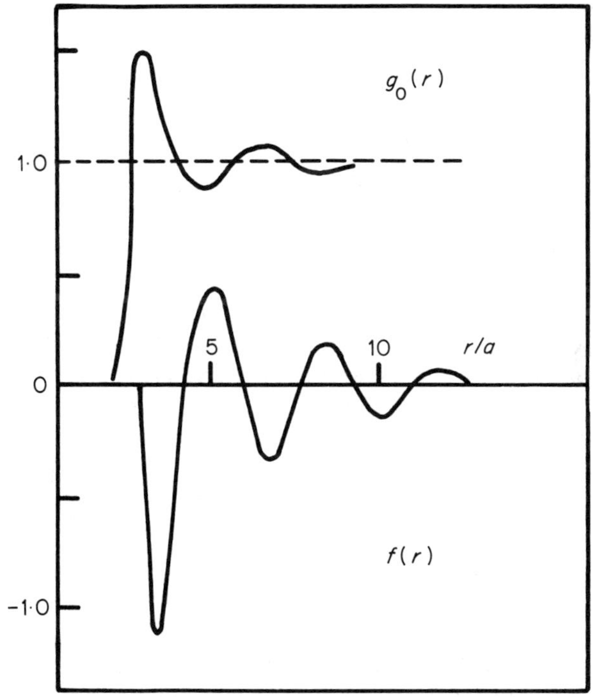

Fig. 8.8. Example of the function $f(r)$, describing the perturbation of the radial distribution function due to shear for a dispersion of spherical particles of 125-nm diameter with a surface charge density of 4·0 mC m^{-2} at a volume fraction of 0·117 and at a 1 − 1 electrolyte concentration of 1 mM. The corresponding radial distribution function $g_0(r)$ is also shown (top) (after Ohtsuki, 1981).

5.1.1) and that $V_1 = 0$. With this choice, all the many-body hydrodynamic interactions and the electroviscous coupling forces are neglected. The colloidal many-body interactions are taken to be those corresponding to an equilibrium structure and the variation of $\bar{V}_{int}$ with structure is neglected. However, this variation contributes a term of order unity to eq (8.73), and thus the colloidal many-body interactions are not taken into account correctly.

Approximate solutions of eq (8.73) show that $f(r)$, like $g_0(r)$, is a damped oscillating function of r. An example is shown in Fig. 8.8.

From the functions g_0 and f it is possible to calculate the dynamic viscosity

$$\mu^* = \mu' - i\mu'' = \mu_0(1 + 2{\cdot}5c + k_2 c^2) \tag{8.75}$$

where the complex coefficient k_2 is given by

$$k_2 = \frac{5}{2} + \frac{15}{2}\int_2^\infty J g_0 r^2 \mathrm{d}r + \frac{9}{40kT}\int_2^\infty (1-\tilde{A})\,\frac{\partial \bar{V}_{int}}{\partial r}\, f r^3 \mathrm{d}r - \frac{9}{40}\int_2^\infty W f r^2 \mathrm{d}r \tag{8.76}$$

Here J, $\tilde{A}$ and W are known functions of r, defined by Russel and Gast (1986). Ohtsuki (1981) took only the third term on the right-hand side of eq (8.76) into account, neglecting hydrodynamic interactions and the direct contribution of Brownian motion to the viscosity.

Equations (8.73) to (8.76), and the definitions given by eqs (8.65) and (8.71), together with the equilibrium theories for determining g_0, completely describe the behaviour of concentrated dispersions of spheres, subjected to a weak shear flow. For stronger flows, the full eq (8.61) must be solved, but as yet no solutions are available.

Results of the theories of both Ohtsuki (1981), and Russel and Gast (1986) predict qualitatively correct behaviour. The dynamic and steady shear viscosities show the expected shear thinning. The dynamic viscosity varies smoothly from a low shear viscosity, μ_0', to a high shear viscosity, μ_∞', when the dimensionless frequency $\omega a^2/D$ is varied from 0 to ∞. The frequency at which the transition from μ_0' to μ_∞' occurs is reasonably well predicted, as is the maximum in μ''. The theory of Ohtsuki predicts a strong increase in μ' as the electrolyte concentration is decreased, which is in qualitative agreement with observations. No exact agreement with experiment is found with either theory because of the many simplifying assumptions which they contain.

8.5. Experimental techniques

8.5.1 Diffraction methods

8.5.1.1 Bragg diffraction

Bragg diffraction is a convenient technique for studying the structure of ordered concentrated dispersions. The technique is based on the interference of light (or other electromagnetic radiation) which is reflected from the various layers of particles in the dispersion. Depending on the angle of incident light and the spacing between the layers, the interference is either destructive or constructive. It follows that for a given wavelength of light, the reflected light has a maximum intensity at a certain angle, referred to as the Bragg angle. From a measurement of the Bragg angle the spacing can be determined. Alternatively, the angle of incidence can be kept constant and the wavelength varied, in which case the intensity of the reflected light is a maximum at a certain wavelength, from which we can again calculate the spacing. For concentrated colloidal dispersions, the spacing between the layers is usually close to the wavelength of visible light so that common lasers can be used in Bragg-diffraction experiments. The fact that the spacing between particle layers and the wavelength of light is very close is also the reason why concentrated colloidal dispersions often show beautiful iridescent colours.

By analogy to X-ray diffraction, Krieger and O'Neill (1968) proposed the following relation for the interpretation of Bragg-diffraction data:

$$m\lambda = 2n_{\mathrm{d}} d \sin\theta \tag{8.77}$$

where m is a positive integer, λ the wavelength in air, n_{d} the mean refractive index of the dispersion, d the spacing between layers, and θ the Bragg angle. Taking into account the phase change when a beam of light passes through a layer of particles, Tomita and van de Ven (1984b) derived an equation that can be written as

$$m\lambda = 2dn_0 \left(1 + \frac{n_{\mathrm{d}} - n_0}{n_0 \sin^2\theta}\right) \sin\theta \tag{8.78}$$

where θ is the diffraction angle (at which the intensity of the reflected light is maximum), which differs from the Bragg angle θ_0, given by $m\lambda = 2n_0\, d\sin\theta_0$; and n_0 is the refractive index of the suspending medium. n_{d} is related to n_0 by

$$n_{\mathrm{d}} \simeq n_0 \left(1 + \frac{3}{2}\,\frac{m_{\mathrm{r}}^2 - 1}{m_{\mathrm{r}}^2 + 2}\,c\right) \tag{8.79}$$

where $m_r = n_p/n_0$, n_p being the refractive index of the suspended particles; and c is the volume fraction. For Bragg diffraction at right angles ($\theta = 90°$), eqs (8.77) and (8.78) become identical.

If the structure is known, eq (8.78) can be used to determine the refractive index, n_d, of the dispersion. If the structure and/or n_p is unknown, the refractive index, n_d, can be found by normally illuminating (with a laser) a dispersion covered by a flat transparent plate (see. Fig. 8.9). A bright halo appears around the point of illumination, the outer radius of which depends on the refractive index of the dispersion. If the thickness of the transparent plate (usually glass) is denoted by h and its refractive index by n_g, then from Snell's law and simple geometric arguments it can be shown that the refractive index, n_d, of the dispersion is given by (Tomita *et al.*, 1983)

$$n_{\mathrm{d}} = \frac{r_0 n_{\mathrm{g}}}{(r_0^2 + 4h^2)^{1/2}} \tag{8.80}$$

where r_0 is the outer radius of the halo. The refractive index, n_g, can be found from the inner radius of the halo. Equation (8.80) applies to both ordered and disordered dispersions.

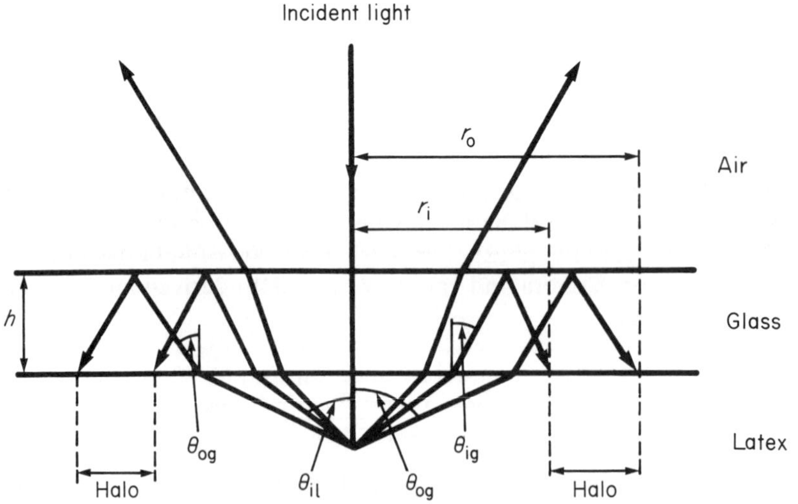

Fig. 8.9. Schematic representation showing the origin of the halo around the point of illumination at a latex/glass interface. The drawing is not to scale; the light is actually scattered by the latex very near to the latex/glass interface.

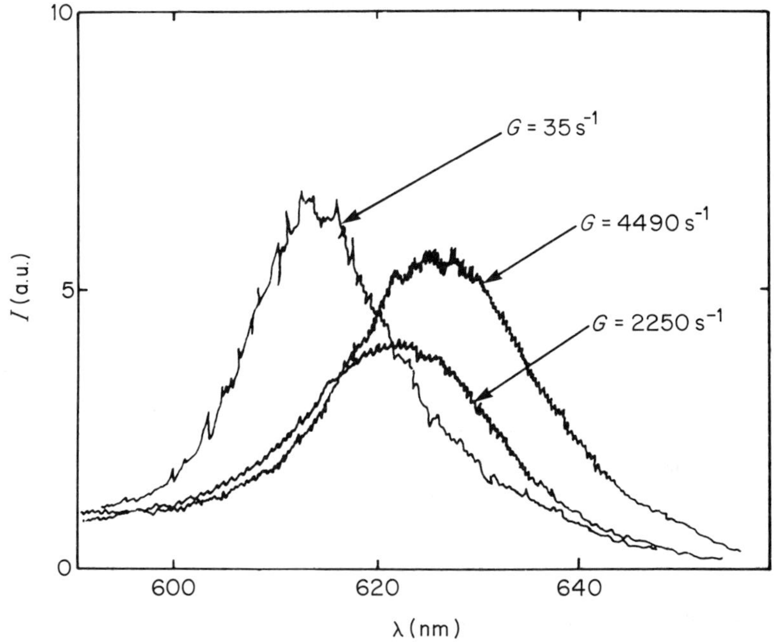

Fig. 8.10. Bragg-diffraction curves at various shear rates, observed using right-angle Bragg diffraction ($a = 64$ nm, $\phi = 0 \cdot 064$) (after Tomita and van de Ven, 1984a).

An example of the results of right-angle Bragg-diffraction experiments on ordered poly(styrene) latex dispersions subjected to simple shear is shown in Fig. 8.10. The figure shows that the wavelength at which the intensity is maximum depends on the rate of shear. The data indicate that near the wall the spacing between layers increases with increasing shear rate. From such experiments it is impossible to determine how far from the wall the structure extends into the dispersion.

It is of interest to note that Bragg diffraction can also cause the occurrence of a dark ring inside the bright halo, seen when illuminating the dispersion normally (cf. Fig. 8.9). Scattered light satisfying the Bragg condition is diffracted away from the halo resulting in a dark band, the radius of which is determined by the spacing, d, between the layers.

8.5.1.2. Two-dimensional and three-dimensional diffraction

When a laser beam passes through a latex dispersion (contained in a transparent cell), some of the light scatters backwards at the cell/dispersion interface, while the remaining light is transmitted through the cell or is scattered forwards. It can be shown that the back-scattered light contains information predominantly about the first layer of particles, immediately adjacent to the cell wall. The interference of this back-scattered light leads to a two-dimensional diffraction pattern. Since an ordered layer at the wall can be assumed to consist of parallel linear arrays of particles (in various directions), such a diffraction pattern is similar to that produced by a grating of fine parallel grooves. The forward-scattered light leads to a three-dimensional diffraction pattern, containing information about the three-dimensional structure in the cell.

The two-dimensional diffraction pattern can be observed either on the cell/dispersion interface or on a screen above (see Fig. 8.11). The pattern usually consists of a number of spots from which the distance between the layers of particles can be determined, using the two-dimensional grating equation (Tomita *et al.*, 1983)

$$\lambda = D \sin\theta_{s\ell} \tag{8.81}$$

where D is the spacing between rows of particles and the angle $\theta_{s\ell}$ is determined by

$$\sin\theta_{s\ell} = \frac{n_g}{n_d} \sin\left(\tan^{-1}\frac{r_s}{2h}\right) \tag{8.82a}$$

when the spots appear on the glass (Fig. 8.11b) or by

$$h' \tan\left[\sin^{-1}(n_d \sin\theta_{s\ell})\right] + h \tan\theta_{s\ell} = r_s \tag{8.82b}$$

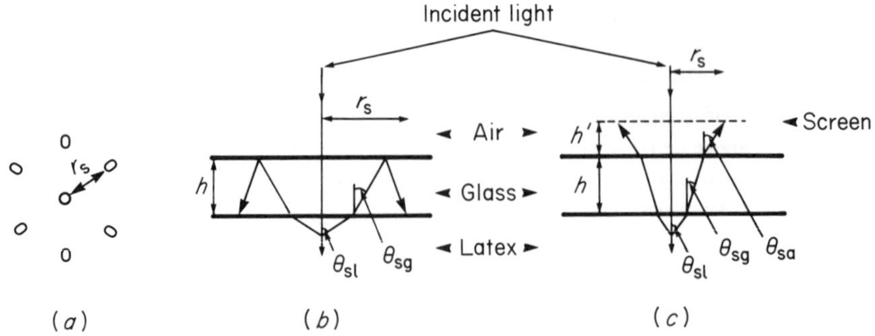

Fig. 8.11. Schematic representation showing the hexagonal pattern of diffraction spots.
(a) Top view.
(b) The appearance of spots on the latex/glass interface.
(c) The appearance of the spots on a screen in front of the glass/air interface.

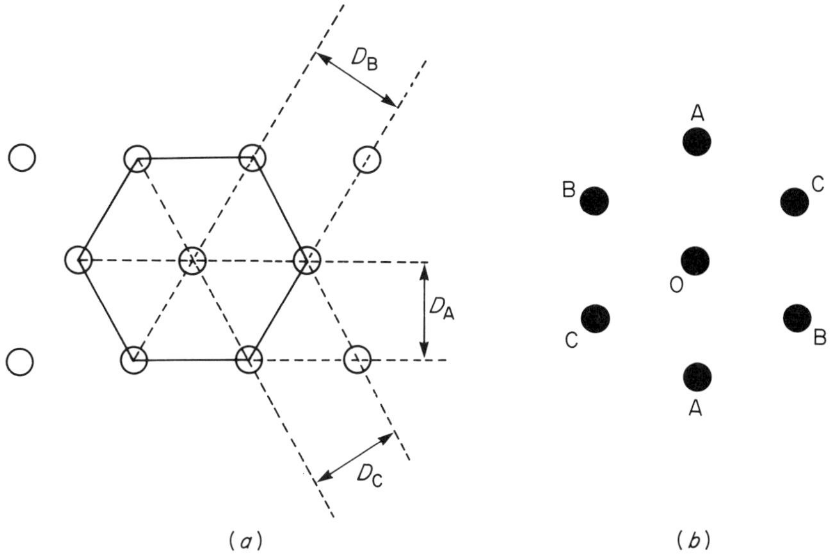

Fig. 8.12. Schematic representation of *(a)* two-dimensional hexagonally packed latex particles and *(b)* corresponding hexagonal pattern consisting of diffraction spots.

when they appear on the screen (Fig. 8.11c). The parameters appearing in eqs (8.82a) and (8.82b) are defined in Fig. 8.11.

The relationship between the location of the diffraction spots and the structure of the surface layer for a hexagonal or fcc structure is shown in Fig. 8.12. It is observed experimentally that when a concentrated poly(stryene) latex dispersion is subjected to shear, the hexagonal pattern deforms, resulting in spacings D_A and $D_B = D_C$ (cf. Fig. 8.12). Figure 8.13 shows experimental observations of D_A and D_B for various rates of shear. It can be seen that at low and high shear rates $D_A = D_B$, while at intermediate shear rates $D_A > D_B$. This implies that the hexagonal structure at the wall is compressed in the flow direction and expanded in the direction normal to the flow. From the decrease in D_A and D_B at high shear we can conclude that at high shear rates the particles are more densely packed at the wall. Assuming that the dispersion consists of stacked hexagonally close packed layers, it follows, since the total number of particles must remain constant, that the spacing, *d*, between the layers must increase when each layer increases its number of particles. The increase in *d*, calculated with this assumption, is included in Fig. 8.13. This increase is consistent with the shift in the Bragg-diffraction peaks shown in Fig. 8.10.

The three-dimensional diffraction pattern can be observed by placing a

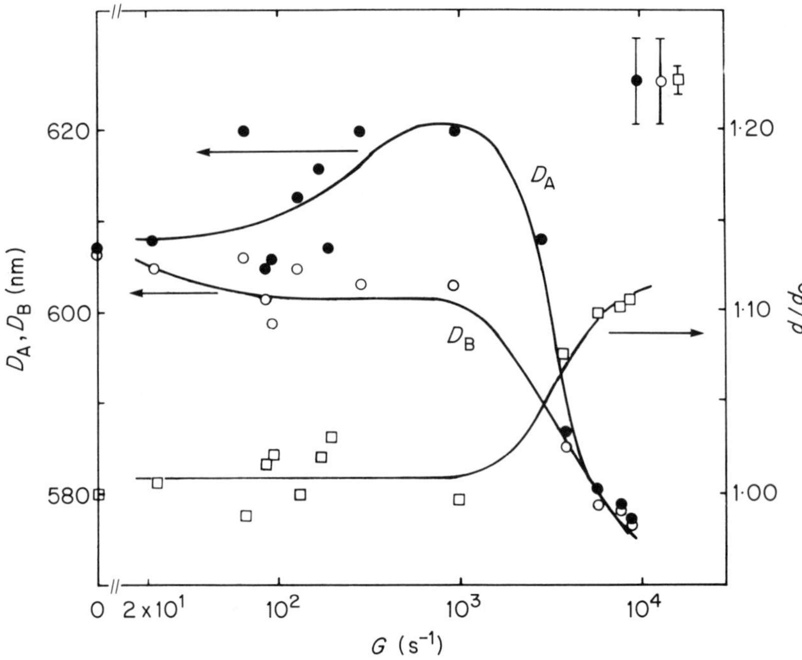

Fig. 8.13. Values of D_A and D_B, calculated from the positions of the six diffraction spots, as a function of the rate of shear. $a = 249$ nm, $\phi = 0{\cdot}261$. Also shown is the spacing, d, between layers parallel to the latex/glass interface, calculated from D_A and D_B (after Tomita and van de Ven, 1984a).

translucent screen behind the cell (Clark and Hurd, 1979). In case of random stacks of hexagonally close-packed layers parallel to the bounding walls, the forward scattering (of monochromatic light) produces a hexagonal pattern of spots, similar to the two-dimensional diffraction.

This type of order is typical for ordered suspensions. Under certain conditions (low volume fractions, low salt concentrations) such a structure can be destroyed by shearing the dispersion (shear-induced melting) and the hexagonal pattern degenerates in a Debye–Sherrer ring. Such a ring is caused by an overlap of diffraction spots originating from many arbitrarily oriented regions in the dispersions which show local order. After switching off the shear, the hexagonal pattern reappears. If the dispersion is allowed to stand for several days, the crystal structure becomes body-centred cubic and, besides diffraction spots, Kossel lines can be observed. These lines are caused by the interference of light rays that satisfy the Bragg angle for a given set of crystal planes, similar to the dark ring in the halo observed in two-dimensional diffraction.

8.5.2 Neutron, X-ray and light scattering

Depending on the system to be investigated, it is possible to obtain the radial distribution function, $g(r)$, by measuring the intensity of the radiation scattered by the particles from neutron beams, X-rays or laser beams as a function of scattering angle. For all these types of radiation, the theories are essentially equivalent. The intensity of the radiation observed at a scattering angle θ_s is given by

$$I_s(\theta_s) = KNP(\theta_s)S(\theta_s) \tag{8.83}$$

where K is an experimental constant, N the number of scatterers in the scattering volume, $P(\theta_s)$ the form factor, and $S(\theta_s)$ the structure factor. For dilute systems, $S(\theta_s) = 1$ and eq (8.83) reduces to eq (3.316). The form factor $P(\theta_s)$ and the constant K can be determined by performing experiments at very low concentrations of particles. When K and P are known, the structure factor S can be obtained by measuring $I_s(\theta_s)$ at higher volume fractions. To avoid the occurrence of multiple scattering (in which case eq (8.83) no longer applies), the refractive index (or its equivalent) of the particles must be closely matched to that of the medium. In terms of the scattering vector $\boldsymbol{q}$ (cf. eq (2.215)), the structure factor can be expressed as

$$S(q) = N^{-1} \sum_{k=1}^{N} \sum_{\ell=1}^{N} \langle \exp i\boldsymbol{q} \cdot (\boldsymbol{r}_k - \boldsymbol{r}_\ell) \rangle = 1 + N^{-1} \sum_{k \neq \ell}^{N} \sum^{N} \exp i\boldsymbol{q} \cdot (\boldsymbol{r}_k - \boldsymbol{r}_\ell) \tag{8.84}$$

The equation shows that $S(q)$ measures the correlations between the positions of the suspended particles. In the absence of such correlations $S(q) = 1$. For an isotropic distribution of particles, eq (8.84) can be expressed in terms of $g(r)$, which is a measure of the degree of correlation between neighbouring particles:

$$S(q) = 1 + n \int_0^\infty [g(r) - 1] \frac{\sin qr}{qr} 4\pi r^2 \mathrm{d}r \tag{8.85}$$

n being the number concentration of particles. The radial distribution, $g(r)$, can be found by taking the Fourier transform of eq (8.85):

$$g(r) = 1 + \frac{1}{2\pi^2 rn} \int_0^\infty [S(q) - 1]\, q \sin qr \mathrm{d}q \tag{8.86}$$

Thus it is possible to obtain $g(r)$ be measuring $S(q)$. Equations (8.85) and (8.86) are only valid for isotropic distributions of particles

For a non-isotropic distribution, e.g. for a distribution of particles in a suspension subjected to simple shear, eqs (8.85) and (8.86) need modification. From the Fourier transform of the non-isotropic distribution

$$g(r) = g_0(r) + \mathrm{Pe}f(r)\,\sin^2\theta\sin 2\phi \tag{8.87}$$

and a Taylor expansion of the structure factor, we obtain

$$S(q) = S_0(q) - \frac{2\mathrm{Pe}\,f(r)}{r\mathrm{d}g_0/\mathrm{d}r}\,\frac{q_2 q_3}{q}\,\frac{\mathrm{d}S_0}{\mathrm{d}q} \tag{8.88}$$

where $S_0(q)$ is the isotropic structure factor when $\mathrm{Pe}=0$, and q_2 and q_3 are the components of $\boldsymbol{q}$ in the X_2 and X_3 directions, respectively. Notice that in this case the structure factor depends on the direction of the incident beam relative to the direction of the flow. From eq (8.88) it follows that it is possible to obtain the shear perturbation function $f(r)$ by measuring the static structure factor $S_0\,(q)$, from which we can obtain $\mathrm{d}g_0/\mathrm{d}r$ and $\mathrm{d}S_0/\mathrm{d}q$, and the structure factor $S(q)$ under shear.

Besides measuring the absolute intensity of the scattered radiation, it is also possible to measure fluctuations in intensity, from which, for example, it is possible to determine the various diffusion coefficients as a function of volume fraction (see Sections 4.2 and 4.0). Again, either the systems must be closely index-matched or an optical probe can be inserted into the dispersion to measure the heterodyne spectrum of the back-scattered light ($\theta_s = 180°$) (Horn *et al.*, 1984).

References

Ackerson, B. J. (1978). *J. Chem. Phys.* **69**, 684.

Ackerson, B. J. and Clark, N. A. (1981). *J. Phys.* **42**, 929.

Adamczyk, Z. and Pomianowski, A. (1980). *Power Technol.* **27**, 125.

Adamczyk, Z. and van de Ven, T. G. M. (1981a). *J. Colloid Interface Sci.* **80**, 340.

Adamczyk, Z. and van de Ven, T. G. M. (1981b). *J. Colloid Interface Sci.* **84**, 497.

Adamczyk, Z. and van de Ven, T. G. M. (1982). *Chem. Eng. Sci.* **37**, 869.

Adamczyk, Z. and van de Ven, T. G. M. (1983). *Int J. Multiphase Flow* **9**, 203.

Adamczyk, Z. and van de Ven, T. G. M. (1984). *J. Colloid Interface Sci.* **97**, 68.

Adamczyk, Z., Dabros, T. and van de Ven, T. G. M. (1982a). *Chem. Eng. Sci.* **37**, 1513.

Adamczyk, Z., Dabros, T. and van de Ven, T. G. M. (1982b). *PhysicoChem. Hydrodynam.* **3**, 215.

Adamczyk, Z., Adamczyk, M. and van de Ven, T. G. M. (1983a). *J. Colloid Interface Sci.* **96**, 204.

Adamczyk, Z., Dabros, T., Czarnecki, J. and van de Ven, T. G. M. (1983b). *Adv. Colloid Interface Sci.* **19**, 183.

Adamczyk, Z., Dabros, T., Czarnecki, J. and van de Ven, T. G. M. (1984). *J. Colloid Interface Sci.* **97**, 91.

Adler, P. M. (1981a). *J. Colloid Interface Sci.* **84**, 461.

Adler, P. M. (1981b). *J. Colloid Interface Sci.* **84**, 489.

Adler, P. M., Takamura, K., Goldsmith, H. L. and Mason, S. G. (1981). *J. Colloid Interface Sci.* **83**, 502.

Albery, W. J., Kneebone, G. R. and Foulds, A. W. (1985). *J. Colloid Interface Sci.* **108**, 193.

Allison, S A., Chang, E. L. and Schurr, J. M. (1979). *Chem. Phys.* **38**, 29.

Altenberger, A. R. (1979). *J. Chem. Phys.* **70**, 1994.

Altenberger, A. R., and Deutch, J. M. (1973). *J. Chem. Phys.* **59**, 894.

Anderson. J. L. and Reed, C. C. (1976). *J. Chem. Phys.* **64**, 3240.

Arp, P. A. and Mason, S. G. (1977a). *J. Colloid Interface Sci.* **61**, 21.

Arp, P. A. and Mason, S. G. (1977b). *J. Colloid Interface Sci.* **61**, 44.

Arp, P. A. and Mason, S. G. (1977c). *Colloid Polym. Sci.* **255**, 566.

Arp, P. A., Foister, R. T. and Mason, S. G. (1980). *Adv. Colloid Interface Sci.* **12**, 295.

Attanasio, A., Bernini, U., Gallopo, P. and Sergé, G. (1972). *Trcns. Soc. Rheol.* **16**, 147.

Baloch, K. M. and van de Ven, T. G. M. (1989). *J. Colloid Interface Sci.* **129**, 91.

Barker, J. A. and Henderson, D. (1967a) *J. Phys. Chem.* **47**, 2856.

Barker, J. A. and Henderson, D. (1967b) *J. Phys. Chem.* **47**, 4714.

Bart, E. (1968). *Chem. Eng. Sci.* **23**, 103.

Barthes-Biesel, D. and Acrivos, A. (1973) *J. Fluid Mech.* **61**, 1.

Bartok, W. and Mason, S. G. (1959). *J. Colloid Sci.* **14**, 13.
Batchelor, G. K. (1972). *J. Fluid Mech.* **52**, 245.
Batchelor, G. K. (1976). *J. Fluid Mech.* **74**, 1.
Batchelor, G. K. (1977). *J. Fluid Mech.* **83**, 97.
Batchelor, G. K. (1982). *J. Fluid Mech.* **119**, 379.
Batchelor, G. K. and Green, J. T. (1972). *J. Fluid Mech.* **56**, 375.
Batchelor, G. K. and Janse van Rensburg, R. W. (1986). *J. Fluid Mech.* **166**, 379.
Batchelor, G. K. and Wen, C. S. (1982). *J. Fluid Mech.* **124**, 495.
Bell, D. N., Teirlinck, H. C. and Goldsmith, H. L. (1984). *Microvasc. Res.* **27**, 297.
Bell, G. M., Levine, S. and McCartney, L. N. (1970). *J. Colloid Interface Sci.* **33**, 335.
Belouschek, P., Maier, S., Suppa, M., Sattler, U. and Lorenz, D. (1988). *Mater. Sci. Forum* **25/26**, 145.
Benoit, H. (1951). *Ann. Phys. Ser. 12* **6**, 561.
Bentley, B. J. and Leal, L. G. (1986). *J. Fluid Mech.* **167**, 219.
Berne, B. J. and Pecora, R. (1976). "Dynamic Light Scattering", Wiley-Interscience, New York.
Boluk, Y. and van de Ven, T. G. M. (1989). *PhysicoChem. Hydrodynam.* (in press).
Booth, F. (1950). *Proc. R. Soc. London, Ser. A* **203**, 533.
Bossis, G. and Brady, J. F. (1984). *J. Chem. Phys.* **80**, 5141.
Boussinesq, J. (1913). *Compt. Rend. Acad. Sci. (Paris)* **156**, 124.
Bowen, B. D. and Epstein, N. (1979). *J. Colloid Interface Sci.* **72**, 81.
Boyce, W. E. and Diprima, R. C. (1977). "Elemenetary Differential Equations and Boundary Value Problems", 3rd ed. p. 121. Wiley, New York.
Brenner, H. (1961). *Chem. Eng. Sci.* **16**, 242.
Brenner, H. (1964). *Chem. Eng. Sci.* **1**, 599.
Brenner, H. (1967). *J. Colloid Interface Sci.* **23**, 407.
Brenner, H. (1974). *Int. J. Multiphase Flow* **1**, 195.
Brenner, H. (1980). *PhysicoChem. Hydrodynam.* **1**, 91.
Brenner, H. (1982). *PhysicoChem. Hydrodynam.* **3**, 139.
Brown, R. (1828). *Philos. Mag.* **4**, 161.
Brown, J. C., Pusey, P. N. and Dietz, R. (1975). *J. Phys. Chem.* **62**, 1136.
Brugmann, D. A. G. (1935). *Ann. Phys.* **24**, 636.
Burgess, J. M. (1941). *Proc. K. Ned. Akad. Wet* **44**, 1045.
Buscall, R. Goodwin, J. W., Hawkins, M. W. and Ottewill, R. H. (1982). *J. Chem. Soc. Faraday Trans. I* **78**, 2889.
Carter, J. M. and Phillies, G. D. J. (1985). *J. Phys. Chem.* **89**, 5118.
Castillo, C. A., Rajagopalan, R. and Hirtzel, C. S. (1984). *Rev. Chem. Eng.* **2**, 175.
Cerda, C. M. and van de Ven, T. G. M. (1983). *J. Colloid Interface Sci.* **93**, 54.
Cerda, C. M., Foister, R. T. and Mason, S. G. (1981). *J. Phys. Chem.* **85**, 1075.
Chaffey, C. E. and Mason, S. G. (1964). *J. Colloid Sci.* **19**, 525.
Chaffey, C. E. and Porter, R. S. (1984). *J. Rheol.* **28**, 249.
Chandrasekhar, S. (1943). *Rev. Mod. Phys.* **15**, 1.
Chauvateau, C. E. (1982) *J. Rheol.* **26**, 111.
Choi, G. N. and Krieger, I. M. (1986). *J. Colloid Interface Sci.* **113**, 101.
Chow, R. S. and Takamura, K. (1988a). *J. Colloid Interface Sci.* **125**, 212.
Chow, R. S. and Takamura, K. (1988b). *J. Colloid Interface Sci.* **125**, 226.
Christenson, H. K. and Horn, R. G. (1983). *Chem. Phys. Lett.* **98**, 45.
Clark, N. A. and Hurd, A. J. (1979). *Nature* **281**, 657.
Clint, G. E., Clint, J. H., Corkill, J. M. and Walker, T. (1973). *J. Colloid Interface Sci.* **44**, 1973.
Cohen, C. and Leal, L. G. (1978). *J. Chem. Phys.* **68**, 5348.
Cohen Stuart, M. A., Waajen, F. H. W. H., Cosgsrove, T., Vincent, B. and Crowley, T. L. (1984). *Macromolecules* **17**, 1825.

Cox, R. G. (1969). *J. Fluid Mech.* **37**, 601.
Cox, R. G. and Brenner, H. (1971). *Chem. Eng. Sci.* **26**, 65.
Cox, R. G., Zia, I. Y. Z. and Mason, S. G. (1968). *J. Colloid Interface Sci.* **27**, 7.
Crank, J. (1956). "The Mathematics of Diffusion". Clarendon Press, Oxford.
Curtis, A. S. G. and Hocking, L. M. (1970). *Trans. Faraday Soc.* **66**, 1381.
Czarnecki, J. and Dabros, T. (1980). *J. Colloid Interface Sci.* **78**, 25.
Dabros, T. (1977). Ph.D. Thesis, Jagiellonian University, Krakow, Poland.
Dabros, T. (1984). *J. Fluid Mech.* **156**, 1.
Dabros, T. and Adamczyk, Z. (1979). *Chem. Eng. Sci.* **34**, 104.
Dabros, T. and Czarnecki J. (1980). *Fizykochem. Prob. Mineral.* **12**, 47.
Dabros, T. and van de Ven, T. G. M. (1982). *J. Colloid Interface Sci.* **89**, 232.
Dabros, T. and van de Ven, T. G. M. (1983a). *J. Colloid Interface Sci.* **92**, 403.
Dabros, T. and van de Ven, T. G. M. (1983b). *J. Colloid Interface Sci.* **93**, 576.
Dabros, T. and van de Ven, T. G. M. (1983c). *Colloid Polymer Sci.* **261**, 694.
Dabros, T. and van de Ven, T. G. M. (1987). *PhysicoChem. Hydrodynam.* **8**, 161.
Dahneke, B. (1975). *J. Colloid Interface Sci.* **50**, 194.
Davis, C. N. (1973). "Air Filtration". Academic Press, London.
Davis, H. M. (1964). *Quart. J. Mech. Appl. Math.* **17**, 499.
de Gennes, P. G. (1987). *Adv. Colloid Interface Sci.* **27**, 189.
Derjaguin, B. V. (1934). *Kolloid. Z.* 69, 155.
Derjaguin, B. V. (1940). *Trans. Faraday Soc.* **36**, 203.
Derjaguin, B. V. and Landau, L. (1941). *Acta Physicochim. USSR* **14**, 633.
De Witt, J. and van de Ven, T. G. M. (1989) (to appear).
Dickenson, E. (1983). *Ann. Rep., R. Soc. Chem., London, Ser. C*, p. 3.
Dill, L. H. and Brenner, H. (1982a). *J. Colloid Interface Sci.* **85**, 101.
Dill, L. H. and Brenner, H. (1982b). *PhysicoChem. Hydrodynam.* **3**, 267.
Doi, M. and Edwards, S. F. (1978a). *J. Chem. Soc. Faraday Trans. II*, **74**, 560.
Doi, M. and Edwards, S. F. (1978b). *J. Chem. Soc. Faraday Trans. II*, **74**, 918.
Durlofsky, L., Brady, J. F. and Bossis, G. (1987). *J. Fluid Mech.* **180**, 21.
Dzyaloshinskii, I. E., Lifshitz, E. M. and Pitaevski, L. P. (1959). *Adv. Phys.* **10**, 165.
Einstein, A. (1905). *Ann. Phys.* **17**, 549.
Einstein, A. (1906). *Ann. Phys.* **19**, 289.
Einstein, A. (1911). *Ann. Phys.* **34**, 591.
Elrick, D. E. (1962). *Austr. J. Phys.* **15**, 283.
Ennis, G. J., Okagawa, A. and Mason, S. G. (1978). *Can. J. Chem.* **56**, 2824.
Ernst, B., Navard, P. and Handin, J. M. (1988). *J. Polym. Sci., B* **26**, 211.
Evans, E. (1988). ACS Symposium, Toronto "Macromolecules" (to appear).
Exner, F. (1900). *Ann. Phys.* **2**, 843.
Feke, D. L. and Schowalter, W. R. (1983). *J. Fluid Mech.* **133**, 17.
Feke, D. L. and Schowalter, W. R. (1985). *J. Colloid Interface Sci.* **106**, 203.
Felderhof, B. U. (1977). *Physica A* **89**, 373.
Felderhof, B. U. (1978). *J. Phys., Ser. A.* **11**, 929.
Firth, B. A. and Hunter, R. J. (1976a). *J. Colloid Interface Sci.* **57**, 248.
Firth, B. A. and Hunter, R. J. (1976b). *J. Colloid Interface Sci.* **57**, 266.
Fisa, F. and Utracki, L. A. (1984). *Polym. Composites* **5**, 36.
Fitts, D. O. (1962). "Nonequilibrium Thermodynamics". McGraw-Hill, New York.
Fitzpatrick, J. A. and Spielmann, L. A. (1973). *J. Colloid Interface Sci.* **43**, 350.
Fleer, G. J. and Scheutjens, J. M. H. M. (1982) *Adv. Colloid Interface Sci.* **16**, 341.
Flory, P. (1969). "Statistical Mechanics of Chain Molecules". Interscience, New York.
Foister, R. T. and van de Ven, T. G. M. (1980). *J. Fluid Mech.* **96**, 105.

Frattini, P. L. and Fuller, G. G. (1984). *J. Colloid Interface Sci.* **100**, 506.

Frens, G. and Overbeek, J. Th. G. (1972). *J. Colloid Interface Sci.* **38**, 376.

Ganatos, P., Pfeffer, R. and Weinbaum, S. (1978). *J. Fluid Mech.* **84**, 79.

Gierszewski, P. J. and Chaffey, C. E. (1978). *Can. J. Phys.* **56**, 6.

Gluckman, M. J., Pfeffer, R. and Weinbaum, S. (1971). *Fluid Mech.* **50**, 705.

Gluckman, M. J., Weinbaum, S. and Pfeffer, R. (1972). *Fluid Mech.* **55**, 677.

Goldman, A. J., Cox, R. G. and Brenner, H. (1967). *Chem. Eng. Sci.* **22**, 637.

Goldsmith, H. L. and Marlow, J. (1972). *Proc. R. Soc. London, Ser. B*, **181**, 351.

Goodwin, J. W. (1982). *In* "Colloidal Dispersions", Chap. 8 (J. W. Goodwin *ed*). Royal Society of Chemistry, London.

Goren, S. L. and O'Neill, M. E. (1971). *Chem. Eng. Sci.* **26**, 325.

Grace, H. P. (1971). Eng. Found Res. Conf. Mixing, 3rd, Andover, N. H.

Haber, S. and Brenner, H. (1984). *J. Colloid Interface Sci.* **97**, 496.

Hachisu, S., Kobayashi, Y. and Kose, A. (1973). *J. Colloid Interface Sci.* **42**, 342.

Hadamard, J. S. (1911). *Compt. Rend. Acad. Sci. (Paris)* **152**, 1735.

Hadamard, J. S. (1912). *Compt. Rend. Acad. Sci. (Paris)* **154**, 109.

Hannon *et al.* (1982).

Happel, J. (1959). *Assoc. Ind. Chem. Eng. J.* **5**, 174.

Happel, J. and Brenner, H. (1983). "Low Reynolds Number Hydrodynamics". Martinus Nijhoff, Boston.

Harris, J. B., Nawaz, M. and Pittman, J. F. T. (1979). *J. Fluid Mech.* **95**, 415.

Harris, S. (1976). *J. Phys. Ser. A.* **9**, 1985.

Henry, D. C. (1931). *Proc. R. Soc. London, Ser. A* **133**, 106.

Hess, W. and Klein, R. (1976). *Phys. Status Solids: A* **85**, 509.

Hiemenz, P. C. (1977, 2nd edn 1986). "Principles of Colloid and Surface Chemistry". M. Dekker, New York.

Hill, T. L. (1960). "An Introduction to Statistical Mechanics", Chap. 19. Addison-Wesley, Reading, Mass.

Hinch, E. J. and Acrivos, A. (1980). *J. Fluid Mech.*, **98**, 305.

Hinch, E. J. and Leal, L. G. (1973). *J. Fluid Mech.* **57**, 753.

Hinch, E. J. and Leal, L. G. (1979). *J. Fluid Mech.* **92**, 591.

Hochrainer, D., Hidy, G. M. and Zebel, G. (1969). *J. Colloid Interface Sci.* **30**, 553.

Hogg, R., Healy, T. W. and Fuerstenau, D. W. (1966). *Trans. Faraday Soc.* **62**, 1638.

Honig, E. O., Roebersen, G. J. and Wiersema, P. H. (1971). *J. Colloid Interface Sci.* **36**, 97.

Horn, R. G. and Israelachvili, J. N. (1981). *J. Chem. Phys.* **75**, 1400.

Horn, D., Auweter, H., Ditter, W. and Eisenhauer, J. (1984). Tenth International Conference in Organic Coatings Science and Technology, Athens, Greece.

Huggins, M. L. (1942a). *J. Phys. Chem.* **46**, 151.

Huggins, M. L. (1942b). *J. Am. Chem. Soc.* **64**, 1712.

Hull, M. and Kitchener, J. A. (1969). *Trans. Faraday Soc.* **65**, 3093.

Hunter, R. J., 1987, "Foundations of Colloid Science", Clarendon Press, Oxford.

Hunter, R. J. and Frayne, J. (1980). *J. Colloid Interface Sci.* **76**, 107.

Israelachvili, J. N. (1985). "Intermolecular and Surface Forces". Academic Press, New York.

Ivanov, Y., van de Ven, T. G. M. and Mason, S. G. (1982). *J. Rheol.* **26**, 213.

Janse van Rensburg, R. W. (1986). Ph.D. Thesis, University of Cambridge, Cambridge, England.

Janse van Rensburg, R. W. (1989). *J. Fluid Mech.* (in press).

Jayasuriya, D. S. and van de Ven, T. G. M. (1985). *J. Phys. Chem.* **89**, 3368.

Jayasuriya, D. S. and van de Ven, T. G. M. (1986a). *J. Chem. Soc., Faraday Trans. I* **82**, 457.
Jayasuriya, D. S. and van de Ven, T. G. M. (1986b). *J. Chem. Soc., Faraday Trans. I* **82**, 473.
Jayasuriya, D. S. and van de Ven, T. G. M. (1987). *J. Colloid Interface Sci.* **115**, 17.
Jeffery, G. B. (1922). *Proc. R. Soc. London, Ser. A* **102**, 161.
Jeffrey, D. J. and Onishi, Y. (1984). *J. Fluid Mech.* **139**, 261.
Johnson, S. J., Frattini, P. L. and Fuller, G. G. (1985). *J. Colloid Interface Sci.* **104**, 440.
Jones, R. C. (1945). *Phys. Rev.* **68**, 93.
Kao, S. V. and Mason, S. G. (1975). *Nature* **253**, 619.
Kao, S. V., Cox, R. G. and Mason, S. G. (1977). *Chem. Eng. Sci.* **32**, 1505.
Karnis, A., Goldsmith, H. L. and Mason, S. G. (1966). *Can. J. Chem. Eng.* **44**, 181.
Kim, S. (1986). *Int. J. Multiphase Flow* **12**, 469.
Kim, S. (1987). *Phys. Fluids* **30**, 2309.
Kim, J. S. and Rajagopalan, R. (1982). *Colloids Surfaces* **4**, 17.
Kops-Werkhoven, M. M. and Fijnaut, H. M. (1981). *J. Chem. Phys.* **74**, 1618.
Krasney-Ergen, W. (1936). *Kolloid Z.* **74**, 172.
Krieger, I. M. (1972). *Adv. Colloid Interface Sci.* **3**, 111.
Krieger, I. M. and Dougherty, T. J. (1959). *Trans. Soc. Rheol.* **3**, 137.
Krieger, I. M. and O'Neill, F. M. (1968). *J. Am. Chem. Soc.* **90**, 3114.
Krupp, H. (1967). *Adv. Colloid Interface Sci.* **1**, 111.
Kuo, R. J. and Matijevic, E. (1980). *J. Colloid Interface Sci.* **78**, 407.
Lan, K. H., Ostrowsky, N. and Sornette, D. (1986). *Phys. Rev. Lett.* **57**, 17.
Langmuir, I. (1938). *J. Chem. Phys.* **6**, 893.
Laun, H. M. (1988). *Prog. Trends Rheology II*, 287.
Leal, L.G. and Hinch, E. J. (1971). *J. Fluid Mech.* **46**, 685.
Lever, D. A. (1979). *J. Fluid Mech.* **92**, 421.
Levich, V. G. (1962). "Physicochemical Hydrodynamics". Prentice-Hall, Englewood Cliffs, New Jersey.
Lifshitz, E. M. (1956). *Sov. Phys. JETP* **2**, 73.
Lindsay, H. M., Lin, M. Y., Weitz, D. A., Sheng, P., Chen, Z., Klein, R. and Meakin, P. (1987). *Faraday Discuss. Chem. Soc.* **83**, 153.
Lyklema, J. and van Leeuwen, H. P. (1982). *Adv. Colloid Interface Sci.* **16**, 127.
MacInnes, D. A. (1939). "The Principles of Electrochemistry". Dover, New York.
Małysa, K., Dabros, T. and van de Ven, T. G. M. (1986). *J. Fluid Mech.* **162**, 157.
Mandelbrot, B. (1975). "Les Objets Fractals". Flammarion, Paris.
Marshall, J. K. and Kitchener, J. A. (1966). *J. Colloid Interface Sci.* **2**, 342.
Mashelkar, R. A. and Venkatsubramanian, C. V. (1975). *Appl. Sci. Res.* **30**, 321.
Masliyah, J. H. and van de Ven, T. G. M. (1986). *Int. J. Multiphase Flow* **12**, 791.
Masliyah, J. H. and van de Ven, T. G. M. (1987). *J. Chem. Soc., Faraday Trans I* **87**, 547.
Masliyah, J. H., Neale, G., Malysa, K. and van de Ven, T. G. M. (1987). *Chem. Eng. Sci.* **42**, 245.
Mason, S. G. (1973). *Chem. Can.* **25**, 29.
Mazur, P. (1987). *Faraday Discuss. Chem. Soc.* **83**, 33.
McMillan, W. G., and Mayer, E. J., 1945, J. Chem. Phys., **13**, 276.
Meakin, P. (1987). *Faraday Discuss. Chem. Soc.* **83**, 113.
Melik, D. H. and Fogler, H. S. (1984a). *J. Colloid Interface Sci.* **101**, 72.
Melik, D. H. and Fogler, H. S. (1984b). *J. Colloid Interface Sci.* **101**, 84.

Muthukumar, M. and Freed, K. F. (1983). *J. Chem. Phys.* **78**, 497 or 511.
Napper, D. H. (1983). "Polymeric Stabilization of Colloidal Dispersions". Academic Press London.
Natanson, G. L. (1957). *Dokl. Akad. Nauk. SSSR* **112**, 696.
Nir, A. and Acrivos, A. (1973). *J. Fluid Mech.* **59**. 209.
O'Brien, R. W. (1975). Ph.D. Thesis, University of Cambridge, Cambridge, England.
O'Brien, R. W. (1981). *J. Colloid Interface Sci.* **81**, 234.
O'Brien, R. W. and White, L. R. (1978). *J. Chem. Soc., Faraday Trans. II* **74**, 1607.
Ohshima, H., Healy, T. W. and White, L. R. (1983). *J. Chem. Soc. Faraday Trans. II* **79**, 1613.
Ohshima, H., Healy, T. W., White, L. R. and O'Brien, R. W. (1984). *J. Chem. Soc., Faraday Trans II* **80**, 1299.
Ohtsuki, T. (1981). *Physia A* **108**, 441.
Okagawa, A. and Mason, S. G. (1973). *Science* **181**, 159.
Okagawa, A. and Mason, S. G. (1974). *J. Colloid Interface Sci.* **47**, 568.
Okagawa, A. and Mason, S. G. (1977). *Can. J. Chem.* **55**, 4243.
Okagawa, A., Cox, R. G. and Mason, S. G. (1973). *J. Colloid Interface Sci.* **45**, 303.
Okagawa, A., Cox, R. G. and Mason, S. G. (1974). *J. Colloid Interface Sci.* **47**, 536.
Okagawa, A., Cox, R. G. and Mason, S. G. (1978). *J. Chem. Soc., Faraday Trans. I* **74**, 1242.
Parsegian, V. A. (1975). In "Physical Chemistry: Enriching Topics from Colloid and Surface Science". (van Olphen, H. and Mysels, K. J., Eds) Theorex, La Jolla, California, p. 27.
Pashley, R. M. and Israelachvili, J. N. (1981a). *J. Colloid Interface Sci.* **80**, 153.
Pashley, R. M. and Israelachvili, J. N. (1981b). *J. Colloid Interface Sci.* **83**, 531.
Paterson, A. L. (1939). *Phys. Rev.* **56**, 972.
Payatakes, A. C. (1977). *Assoc. Ind. Chem. Eng. J.* **23**, 192.
Payatakes, A. C. and Chi-Tien (1976). *J. Aerosol Sci.* **7**, 85.
Payatakes, A. C. and Gradon, L. (1980). *Chem. Eng. Sci.* **35**, 1083.
Payatakes, A. C., Chi-Tien and Turian, R. M. (1974). *Assoc. Ind. Chem. Eng. J.* **20**, 889.
Perrin, F. (1934). *J. Phys. Radium* **5**, 497.
Perrin, J. (1913). "Les Atomes". Librarie Félix Alcan, Paris.
Peschel, G., Belouschek, P. Müller , M. M., Müller, M. R. and Konig, R., (1982). *Colloid Polym. Sci.* **260**, 444.
Phillies, G. D. J. (1982). *J. Chem. Phys.* **77**, 2623.
Powell, R. L. (1983). *J. Colloid Interface Sci.* **95**, 148.
Powell, R. L. and Mason S. G. (1982). *Assoc. Ind. Chem. Eng. J.* **28**, 286.
Prieve, D. C. and Alexander, B. M. (1986). *Science*, **231**, 1269.
Prieve, D. C. and Lin, M. M. J. (1980). *J. Colloid Interface Sci.* **76**, 32.
Prieve, D. C. and Lin, M. M. J. (1982). *J. Colloid Interface Sci.* **86**, 17.
Prieve, D. C. and Ruckenstein, E. (1974). *Assoc. Ind. Chem. Eng. J.* **20**, 1178.
Proudman, I. and Pearson, J. R. A. (1962). *J. Fluid Mech.* **12**, 591.
Provencher, S. W. (1979). *Macromolec. Chem.* **180**, 201.
Pyun, C. W. and Fixman, M. (1964). *J. Chem. Phys.* **41**, 937.
Rallison, J. M. (1984). *Ann. Rev. Fluid Mech.* **16**, 45.
Reed, L. D. and Morrison, F. A. (1976). *J. Colloid Interface Sci.* **54**, 117:
Reerink, H. and Overbeek, J. Th. G. (1954). *Discuss. Faraday Soc.* **18**, 74.
Ruckenstein, E. (1986). ACS Conference, New York, USA.
Ruckenstein, E. and Prieve, D. C. (1973). *J. Chem. Soc. Faraday Trans. II* **69**, 1522.
Rumscheidt, F. D. and Mason, S. G. (1962). *J. Colloid Sci.* **17**, 260.

Russel, W. B. (1978). *J. Fluid Mech.* **85**, 209.
Russel, W. B. (1980). *J. Rheol.* **24**, 287.
Russel, W. B. and Gast, A. (1986). *J. Chem. Phys.* **84**, 1815.
Russel, W. B. and Glendinning, A. B. (1981). *J. Chem. Phys.* **74**, 948.
Rybczynski, W. (1911). *Bull. Acad. Sci. Cracovie, A* p. 40.
Salem, A. J. and Fuller, G. G. (1985). *J. Colloid Interface Sci.* **108**, 149.
Sasaki, S. (1985). *Colloid Polymer Sci.* **263**, 935.
Schenkel, J. M. and Kitchener, J. A. (1960). *Trans. Faraday Soc.* **56**, 161.
Scheutjens, J. M. H. M. and Fleer, G. J. (1982). *Adv. Colloid Interface Sci.* **16**, 361.
Schonberg, J. A., Drew, D. A. and Belfort, G. (1984). 58th Colloid and Surface Science Symposium, Pittsburgh, U. S. A.
Schumacher, G. A. and van de Ven, T. G. M. (1987). *Faraday Discuss. Chem. Soc.* **83**, 75.
Sheraga, H. A. (1955). *J. Chem. Phys.* **23**, 1526.
Smith, P. G. and van de Ven, T. G. M. (1985a). *Colloids Surfaces* **15**, 191.
Smith, P. G. and van de Ven, T. G. M. (1985b). *Colloids Surfaces* **15**, 211.
Smoluchowski, M. (1903). *Bull. Int. Acad. Sci. Cracovie* **1**, 403.
Smoluchowski, M. (1906). *Ann Phys.* **21**, 756.
Smoluchowski, M. (1916). *Koll. Zeitschr.* **18**, 190.
Smoluchowski, M. (1917). *Z. Phys. Chem.* **92**, 129.
Sozou, C. (1972). *Proc. R. Soc. London, Ser. A* **331**, 263.
Sozou, C. (1973). *Proc. R. Soc. London, Ser. A* **334**, 343.
Spielman, L. A. and Cukor, P. M. (1973). *J. Colloid Interface Sci.* **43**, 51.
Spielman, L. A. and Fitzpatrick, J. A. (1973). *J. Colloid Interface Sci.* **42**, 607.
Stein, H. N., Logtenberg, E. H., van Diemen, A. J. G. and Peters, P. J. (1986). *Colloids and Surfaces* **10**, 223.
Suda, H. and Imai, N. (1985). *J. Colloid Interface Sci.* **104**, 204.
Takamura, K., Goldsmilth, H. L. and Mason, S. G. (1979). *J. Colloid Interface Sci.* **72**, 385.
Takamura, K., Goldsmith, K. L. and Mason, S. G. (1981a). *J. Colloid Interface Sci.* **82**, 175.
Takamura, K., Goldsmith, H. L. and Mason, S. G. (1981b). *J. Colloid Interface Sci.* **82**, 190.
Takamura, K., Adler, P. M., Goldsmith, H. L. and Mason, S. G. (1981c). *J. Colloid Interface Sci.* **83**, 516.
Taylor, G. I. (1934). *Proc. R. Soc. Ser. A* **146**, 501.
Taylor, G. I. (1953). *Proc. R. Soc. Ser. A* **219**, 186.
Thomas, D. G. (1965). *J. Colloid Sci.* **20**, 267.
Tha, S. P., Shuster, J. and Goldsmith, H. L. (1986). *Biophys. J.* **50**, 1117.
Tomita, M. and van de Ven, T. G. M. (1984a). *J. Colloid Interface Sci.* **99**, 374.
Tomita, M. and van de Ven, T. G. M. (1984b). *J. Colloid Interface Sci.* **100**, 112.
Tomita, M., Takano, K. and van de Ven, T. G. M. (1983). *J. Colloid Interface Sci.* **92**, 367.
Torza, S., Cox, R. G. and Mason, S. G. (1971a). *Phil. Trans. R. Soc. London, Ser. A* **269**(1198), 295.
Torza, S., Henry, C. P., and Cox, R. G. and Mason, S. G. (1971b). *J. Colloid Interface Sci.* **35**, 529.
Torza, S., Cox, R. G. and Mason, S. G. (1972). *J. Colloid Interface Sci.* **38**, 395.
Trevelyan, B. J. and Mason, S. G. (1951). *J. Colloid Interface Sci.* **6**, 354.
Uhlenbeck, G. E. and Ornstein, L. S. (1930). *Phys. Rev.* **36**, 823.

Usui, S. (1972). *Progress in Surface and Membrane Sci.* (Danielli, J. F., Ed.) Academic Press, **5**, 223.
Usui, S. (1973). *J. Colloid Interface Sci.* **44**, 107.
Utracki, L. A. (1984). *Rubber Chem. Technol.* **57**, 267.
Uzgiris, E. (1981). *Adv. Colloid Interface Sci.* **14**, 75.
Vadas, E. B., Goldsmith, H. L. and Mason, S. G. (1973). *J. Colloid Interface Sci.* **43**, 630.
Vadas, E. B., Cox, R. G., Goldsmith, H. L. and Mason, S. G. (1976). *J. Colloid Interface Sci.* **57**, 308.
van den Broeck, C. Lostak, F. and Lekkerkerker, H. N. W. (1981). *J. Chem. Phys.* **74**, 2006.
van de Ven, T. G. M. (1975). Ph.D. Thesis, McGill University, Montreal, Canada.
van de Ven, T. G. M. (1977). *J. Colloid Interface Sci.* **62**, 352.
van de Ven, T. G. M. (1981). *J. Colloid Interface Sci.* **81**, 290.
van de Ven, T. G. M. (1982). *Adv. Colloidal Interface Sci.* **17**, 105.
van de Ven, T. G. M. (1984). *J. Chem. Soc., Faraday Trans. I* **80**, 2677.
van de Ven, T. G. M. (1988a). *J. Colloid Interface Sci.* **124**, 138.
van de Ven, T. G. M. (1988b). *PhysicoChem. Hydrodynam.* **10**, 97.
van de Ven, T. G. M. and Firth, B. A. (1978). *Rheol. Acta* **17**, 606.
van de Ven, T. G. M. and Hunter, R. J. (1977). *Rheol. Acta* **16**, 534.
van de Ven, T. G. M. and Hunter, R. J. (1979). *J. Colloid Interface Sci.* **68**, 135.
van de Ven, T. G. M. and Mason, S. G. (1976a). *J. Colloid Interface Sci.* **57**, 505.
van de Ven, T. G. M. and Mason, S. G. (1976b). *J. Colloid Interface Sci.* **57**, 517.
van de Ven, T. G. M. and Mason, S. G. (1977a). *Colloid Polym. Sci.* **225**, 468.
van de Ven, T. G. M. and Mason, S. G. (1977b). *Colloid Polym. Sci.* **255**, 794.
van de Ven, T. G. M., Takamura, K. and Mason, S. G. (1981a). *J. Colloid Interface Sci.* **82**, 373.
van de Ven, T. G. M., Zuzovsky, M. and Mason, S. G. (1981b). *J. Chem. Soc. Faraday Trans I* **77**, 1661.
van Veluwen, A., Lekkerkerker, H. N. W., de Kruif, C. G. and Vrij, A. (1987). **87**, 4873.
Varennes, S. and van de Ven, T. G. M. (1987). *PhysicoChem. Hydrodynam.* **9**, 537.
Verwey, E. J. and Overbeek, J. Th. G. (1948). "Theory of the Stability of Lyophobic Colloids". Elsevier, Amsterdam.
Visser, J. (1972). *Adv. Colloid Interface Sci.* **3**, 331.
Voet, A. (1947). *J. Phys. Colloid Chem.* **51**, 1037.
Wacholder, E. and Sather, N. F. (1974). *J. Fluid Mech.* **65**, 417.
Waligora, B. and Dabros, T. (1978). *Wiss. Z. Friderich-Schiller Univ., Jena, Math-Nat. R.*, Chap. 5/6.
Walters, K. (1975). "Rheometry". Chapman and Hall, London.
Ware, B. R. (1974). *Adv. Colloid Interface Sci.* **4**, 1.
Weeks, J. D., Chandler, D. and Andersen, H. C. (1971). *J. Chem. Phys.* **54**, 5237.
Workman, H. J. and Hollingsworth, C. A. (1969). *J. Colloid Interface Sci.* **29**, 664.
Yoshida, N. (1983). *Chem. Phys. Lett.* **101**, 555.
Zebel, G. (1965). *J. Colloid Sci.* **20**, 522.
Zeichner, G. R. and Schowalter, W. R. (1977). *Assoc. Ind. Chem. Eng. J.* **23**, 243.
Zeichner, G. R. and Schowalter, W. R. (1979). *J. Colloid Interface Sci.* **71**, 237.
Zuck, E. and van de Ven, T. G. M. (1989) (to appear).
Zuck, E., Cox, R. G. and van de Ven, T. G. M. (1989) (to appear).
Zukoski, C. F. and Saville, D. A. (1987). *J. Colloid Interface Sci.* **115**, 422.
Zuzovsky, M., Priel, Z. and Mason, S. G. (1980a). *J. Colloid Interface Sci.* **75**, 230.
Zuzovsky, M., Priel, Z. and Mason, S. G. (1980b). *J. Rheol.* **24**, 705.

Subject Index